THE HIGH LATITUDE HELIOSPHERE

THE HIGH LATITUDE HELIOSPHERE

Proceedings of the 28th ESLAB Symposium, 19–21 April 1994, Friedrichshafen, Germany

Edited by

R. G. Marsden

Solar System Division,
Space Science Department of ESA, ESTEC,
Noordwijk, The Netherlands

Reprinted from *Space Science Reviews*, Vol. 72, Nos. 1–2, 1995

SPRINGER-SCIENCE+BUSINESS MEDIA, B.V.

A C.I.P. Catalogue record for this book is available from the Library of Congress

ISBN 978-94-010-4075-4 ISBN 978-94-011-0167-7 (eBook)
DOI 10.1007/978-94-011-0167-7

Printed on acid-free paper

All Rights Reserved
© 1995 Springer Science+Business Media Dordrecht
Originally published by Kluwer Academic Publishers in 1995
Softcover reprint of the hardcover 1st edition 1995

No part of the material protected by this copyright notice may be reproduced or utilized in any form or by any means, electronic or mechanical, including photocopying, recording or by any information storage and retrieval system, without written permission from the copyright owner.

TABLE OF CONTENTS

Session 3: ENERGETIC PARTICLES IN THE HELIOSPHERE

Session 4: COSMIC RAYS IN THE HELIOSPHERE

Foreword

The launch in October 1990 of the joint ESA-NASA Ulysses mission marked the start of a new era in the study of the heliosphere. For the first time, in-situ observations are being made covering the full range of heliographic latitudes. Following the successful gravity-assist manoeuvre at Jupiter in February 1992, Ulysses left the ecliptic plane in a southerly direction and headed back toward the Sun, passing over the southern solar pole in mid-1994.

To mark these unique events, the 28th ESLAB Symposium, held in Friedrichshafen, Germany, on 19-21 April 1994, was devoted to "The High Latitude Heliosphere". Following on from the highly successful 19th ESLAB Symposium "The Sun and the Heliosphere in Three Dimensions" (Les Diablerets, 1985), the purpose of the meeting was to review out-of-ecliptic results from the Ulysses mission obtained to date, and to provide a focus for the first polar pass. Relevant results from other space missions, as well as ground-based and theoretical studies, were also included.

Attended by 130 scientists, the main themes of the Symposium were The Sun and Corona, Large-Scale Heliospheric Structure, Energetic Particles in the Heliosphere, Cosmic Rays in the Heliosphere, and Interstellar Gas and Cosmic Dust. The scientific programme consisted of a number of Topical Review papers introducing various aspects of these themes, supplemented by a large number of contributed papers (72 in total) presented either orally or as posters. Undoubtedly, the excellent poster sessions formed one of the highlights of the meeting.

Of the contributed papers presented at the Symposium, all but eight are included in these proceedings. The material presented by R. Schwenn in his Topical Review paper 'The 3-Dimensional Structure of the Solar Wind During the Solar Activity Cycle' has been published in R. Schwenn and E. Marsch (eds.), *Physics of the Inner Heliosphere 1*, Springer-Verlag (1990). The review paper by G. Morfill entitled 'Cosmic Dust in the Heliosphere' will be published elsewhere. Every paper submitted for publication in these proceedings has been refereed.

It is my pleasure to thank the members of the Scientific Programme Committee for their help in putting together the scientific programme. The assistance of the referees in the preparation of this volume is gratefully acknowledged. Thanks are due to colleagues from the Solar System Division, in particular Karl-Heinz Trattner, Gerhard Schwehm, Trevor Sanderson, Bernard Fleck and Peter Wenzel, for help at various stages. Special thanks are due to the Symposium Secretaries, Clare Bingham and Marion Radford, for their invaluable assistance in handling the local arrangements.

Space Science Reviews **72**: xi–xii.
© 1994 *Kluwer Academic Publishers.*

The success of the meeting was due in no small part to the excellent facilities at the Graf-Zeppelin-Haus and the services provided by Mrs. I. Rehm and her colleagues. Thanks are also due to Mrs. B. Glatzel of the Tourist Information Office, Friedrichshafen. The financial contribution of Dornier is gratefully acknowledged.

The 28th ESLAB Symposium was organised by the Space Science Department of ESA.

Noordwijk, June 1994 R.G. Marsden

Group Photograph, 28th ESLAB Symposium

1. C.P. Sonett
2. R. Woo
3. J. Solomon
4. M.C.E. Huber
5. J. Geiss
6. S. Habbal
7. A. Barnes
8. C. Maclennan
9. R. Gold
10. L.J. Lanzerotti
11. M. Radford
12. R. Marsden
13. C. Bingham
14. F. Neubauer
15. N. Murphy
16. J. Giacalone
17. A. Galvin
18. I. Mann
19. -
20. B.E. Goldstein
21. L. Fisk
22. N. Lin
23. U. Mall
24. K.-P. Wenzel
25. W. Coles
26. J.R. Jokipii
27. M. Pick
28. B. Sanahuja
29. T. Sanderson
30. P. Beech
31. D. Winterhalter
32. F. de Castro
33. P. Kellogg
34. P. Lantos
35. M. Burton
36. C. Paizis
37. C. Rastoin
38. R. Ducros
39. J. Gosling
40. P. Ferrando
41. R. Schwenn
42. H. Kunow
43. A. Raviart
44. S. Bame
45. S. Hoang
46. G. Erdös
47. A. Löffler
48. D. Hovestadt
49. G. Gloeckler
50. G. Simnett
51. R. Bodmer
52. Y.-M. Wang
53. E. Massey
54. J. Brandt
55. M. Potgieter
56. R. B. McKibben
57. W.V. Jones
58. R. Müller-Mellin
59. J.D. Anglin
60. M. Lee
61. M. Reuss
62. D.E. Page
63. E. Lorenz
64. J. Phillips
65. J.T. Hoeksema
66. D. Ponyavin
67. M. Reiner
68. S. Hawkins III
69. E. Smith
70. A. Balogh
71. P. Riley
72. M. Velli
73. J.-L. Bougeret
74. B. Tsurutani
75. J. Simpson
76. J. Karl
77. V. Bothmer
78. -
79. J. Cooper
80. C. Tranquille
81. F. McDonald
82. E. Keppler
83. B. Fleck
84. R. v. Steiger
85. E. Roelof
86. G. Schwehm
87. E. Marsch
88. J.L. Culhane
89. E. Grün
90. S. Krimigis
91. M. Bird
92. G. Siscoe
93. M. Witte
94. N. Crooker
95. K. Szegö
96. M. Schulz
97. T. Summanen
98. M. Fränz
99. P. Canu
100. J. Woch
101. R. Forsyth
102. N. Krupp
103. R. Wimmer
104. K. Staines
105. P. Zlobec
106. W. Meeks
107. B. Heber
108. P. Gazis
109. J. Kohl
110. R. MacDowall
111. M. Bzowski
112. D. Meyer
113. P. Király
114. B. Klecker
115. K. Trattner
116. I. Veselovsky
117. M. Kane
118. H. Fahr
119. M. Banaszkiewics
120. D. Rucinski
121. D. MComas
122. M. Yamauchi
123. A. Usmanov

Participants in the 28th ESLAB Symposium, held in Friedrichshafen, Germany, 19–21 April, 1994.

OPENING REMARKS

MARTIN C.E. HUBER
Space Science Department of ESA, ESTEC, Noordwijk, The Netherlands

Ladies and Gentlemen, Dear Colleagues, Good Morning!

In the name of the European Space Agency and its Space Science Department, I should like to welcome you to the 28th ESLAB Symposium on 'The High Latitude Heliosphere'.

It is a special pleasure to welcome this impressive gathering of space scientists: many pioneers of space research are present, yet I also see a good number of young faces.

We are happy that we could assemble participants from three continents - from Europe, Africa and America. We further note with pleasure that twelve European countries - among them Hungary, Poland and Russia - are represented here and that we have among us over fifty researchers from the United States.

To those of you who attend an ESLAB Symposium for the first time; the name ESLAB may be unknown. In the early days, when European space research was organised through the European Space Research Organisation, ESRO, there was a separate laboratory called ESLAB, the European Space Research Laboratory. ESLAB later became ESA's Space Science Department (SSD), now located at the European Space Research and Technology Centre, ESTEC. The old name survives in the ESLAB Symposia, which are organised annually for the benefit of the wide community, but also in view of the 'education permanente' of SSD staff.

SSD's prime function within the Directorate of Scientific Programmes of the European Space Agency is to provide the Project Scientists. They work closely with the Project Manager (who belongs to the other Department of our Directorate, the Scientific Projects Department), and are the advocates of science in the continuous process of optimising a mission between scientific, technical and financial considerations. SSD's Project Scientists thus provide the scientific in-house expertise and guidance - a task that can be achieved only thanks to the Department's extensive interfaces with the wide scientific community, i.e. the scientists working in Universities and Research Institutes in our 14 member countries - and beyond, as beautifully demonstrated with the Ulysses mission.

After launch, SSD staff assume the task of Project Manager as well. The Department, however, also provides a scientific environment in which the Project (and Study) Scientsts can perform their own, original research. This, in our view, is mandatory, if our Project Scientists are to stay in touch with the reality of the scientific enterprise - the arduous process of step-by-step progress towards achieving new knowledge. Indeed, SSD has not only an infrastructure for data analysis and computational modeling,

Space Science Reviews **72**: 1–2.
© 1995 *Kluwer Academic Publishers.*

we also maintain the technical infrastructure for builidng flight hardware. A product of our laboratories flies as part of the COSPIN experiment on Ulysses.

With this mention of the mission that is central to the topic of this Symposium, the high-latitude heliosphere, we now turn to the present ESLAB Symposium. It is, indeed, timely to take stock of the observations made by the Ulysses probe in its journey out of the ecliptic in good time before its first passage over a solar pole this Autumn.

I am especially pleased to see that a natural interdisciplinary collaboration is developing between the heliospheric and solar communities. This morning's session bears witness to the merit of a broader approach towards a scientific topic: a full understanding of heliospheric processes requires a good knowledge (and also understanding) of the physical and chemical conditions and processes at, and near, its inner boundary - our central star.

At the outset of this ESLAB Symposium we will have the pleasure of listening to a keynote address by Professor Fisk. I cannot resist mentioning the remarkable fact that Len Fisk, who is a Co-Investigator on the Ulysses Mission, has maintained this status throughout, also during his highly successful tenure as NASA's Associate Administrator of Space Science and Applications, and as the Administration's Chief Scientist. A scientist by profession, he obviously has a special relation to this mission, and in his keynote address he will remind us of the fundamental scientific aims to be achieved with the Ulysses Mission.

Let me now wish you a stimulating and rewarding discussion during the Symposium and three enjoyable spring days here in Friedrichshafen.

ADDRESS BY ESA'S DIRECTOR OF SCIENTIFIC PROGRAMMES

R.M. BONNET
Director of Scientific Programmes, ESA

This is an extremely exciting time for those involved in heliospheric research. Ulysses is now at 60 degrees south of the solar equator, rapidly approaching the first-ever passage above the Sun's polar regions. To date, the mission has proceeded extremely well. The spacecraft has performed as intended with the exception of a few minor problems that are under control. All experiments are performing without fault and have returned essentially continuous measurements since launch. Thanks are due to the ESA spacecraft operations team at JPL in this regard.

Even though, strictly speaking, the key phase of the mission has yet to begin, Ulysses has already produced many impressive scientific results:

- during the journey out to Jupiter, Ulysses collected a record of in-ecliptic phenomena between 1.0 and 5.4 AU of unprecedented detail and completeness. These observations include important new results on the evolution of the solar wind in this region of the heliosphere.
- although not the topic of this Symposium, the encounter with Jupiter produced a host of new magnetospheric observations during the traversal of the previously unexplored dusk sector.
- Ulysses' unique orbit has also enabled exciting new measurements of interstellar material to be made, the results of which will be reported at this meeting.
- latitude gradients in the solar wind have already been observed, with the disappearance of the magnetic sector structure beyond 30 degrees south and the appearance of a persistent fast solar wind.

The great interest in the mission is clearly reflected in the large number of participants at this meeting and the large number of abstracts submitted, and the success of the mission to date is particularly rewarding for the members of the science teams, many of whom have been with the project from its earliest days. Looking ahead, an important mission milestone will occur in September of this year when the spacecraft reaches its maximum solar latitude in the southern hemisphere. To celebrate this event, ESA is hosting a Workshop at Estec at which the science teams will compare their latest results.

Looking yet further ahead, it is clear that Ulysses has a unique role to play in the context of the ISTP programme. In particular, coordinated observations with SOHO hold the promise of producing many exciting results. To this end, ESA has already made a conditional commitment to continue operating Ulysses for a second solar orbit after its nominal prime mission ends in September 1995. This offers the exciting pros-

Space Science Reviews **72**: 3–4.
© 1995 *Kluwer Academic Publishers.*

pect of obtaining high-latitude observations at the time of the next solar maximum.

I am sure that no one will dispute that, by the time its unique exploratory journey has come to an end, Ulysses will have changed the way we view the heliosphere. I wish you all an enjoyable and stimulating three days at this ESLAB Symposium.

EXPLORING THE HELIOSPHERE IN THREE DIMENSIONS A KEYNOTE PRESENTATION

L. A. FISK
Department of Atmospheric, Oceanic and Space Sciences,
University of Michigan, Ann Arbor, MI 48109 U.S.A.

Abstract. The 28th ESLAB Symposium marks the beginning of the sprint of the Ulysses mission to the very high heliographic latitudes and the pole-to-pole passage. The more than twenty-year quest to understand the Sun and the heliosphere in three dimensions is about to be realized. It is perhaps worthwhile, as we are poised to begin this journey, to ask how history is likely to judge this mission, or equivalently: what questions need to be answered so that the judgment will be kind?

1. Introduction

This Symposium marks the beginning of the sprint to high latitudes for the Ulysses spacecraft. We will assess where we are, having just completed the journey through the mid-heliographic latitudes, and position ourselves for the sprint to the very high heliographic latitudes, after which we will execute the pole-to-pole passage, and, hopefully, thereafter have a continued mission to study the heliosphere during solar maximum conditions.

It is, perhaps, worthwhile to remember how long it has taken us to get here. It was twenty years ago this year when a small group assembled in ESTEC to conceive the modern out-of-the-ecliptic mission. There had been considerable thought and encouragement given to out-of-the-ecliptic missions earlier, dating back even into the 1960's. However, the modern mission, the present-day Ulysses, had its origins in that ESTEC meeting.

Ian Axford (1994) has written a marvelous article for the 75th anniversary of the American Geophysical Union, reminiscing on his career, in which he discusses the events surrounding the conception and advocacy of the Out-of-the-Ecliptic mission, which became known as the International Solar Polar Mission, and now Ulysses. Those of us who attended that ESTEC meeting also have our recollections, with Ian standing at the blackboard, drawing the two-spacecraft concept, with one going over the north pole of the Sun and the other the south pole, and noting how they looked like two railroad trains racing in opposite directions at each other.

One of the trains, of course, was derailed along the way. However, the other has fulfilled our hopes over twenty years and longer, of finally being able to answer the simple, but compelling question: What does the heliosphere look like in three dimensions? In many ways it is such a basic question. The heliosphere is obviously a three-dimensional structure. We have every reason to expect that conditions vary markedly with latitude. We know that prior to Ulysses our sampling of the heliosphere near the equatorial plane of the Sun, although extensive, is non-representative of

Space Science Reviews **72**: 5–14.
© 1995 *Kluwer Academic Publishers.*

heliospheric conditions, and, moreover, that conditions at other latitudes, beyond our current observations, will influence what happens near the equatorial plane. It is, perhaps, a wonder that it has taken us so long to fill this obvious gap in our knowledge and understanding of what has been a central subject in space research, the Sun and the heliosphere.

Rather than dwell on this past, however, it is probably more worthwhile to consider Ulysses and the exploration of the three-dimensional heliosphere from the perspective of the future. As we begin the sprint to high latitudes and the pole-to-pole survey, we should ask what we expect to learn, or more appropriately, we should consider that we are reading the history of this mission twenty years from now, and ask what is likely to be said about our impact. These exercises are useful if for no other reason than to call our attention to the larger issues. We all tend to work on problems that most interest us, even if they are very focused and of a detail interesting only to a few. We will be judged, however, on whether we addressed the larger issues. Was this a seminal mission that laid to rest uncertainties that had plagued the field for many years? Did we make new discoveries that opened up whole new areas of research? Did we have an impact on fields beyond our own?

Consider a possible list of the larger issues, or equivalently, the several themes which unify the several investigations that Ulysses will pursue—and which, if successfully completed, history will judge this mission kindly. Essentially there are four: The first is to determine the overall morphology of the solar wind, its embedded magnetic field, and its accompanying waves and turbulence. The second is to determine the behavior of galactic cosmic rays. The third is to trace the behavior of the interstellar gas which enters the heliosphere as neutral particles and which departs as the anomalous cosmic ray component. The fourth is to understand the source of the solar wind and energetic particles in the solar corona and lower solar atmosphere.

The first theme—the overall morphology of the solar wind—is a matter of making *in situ* observations. The fourth theme—inferring coronal processes—is a matter of making remote observations. The second theme—understanding cosmic ray behavior—is a matter of inferring where the particles have been in physical space. The third theme—the interstellar ions—is a matter of inferring where the particles have been in energy space. Consider each of these problems separately.

2. The Overall Morphology of the Solar Wind

This theme embodies the exploration aspect of Ulysses—go where we have never been before and provide a complete description of the heliosphere at all latitudes for at least this portion of the solar cycle. The observations should fall in three categories:

First, we will confirm what we expect to find, and, as is evident in the papers of this Symposium, have found to date with Ulysses. We have every expectation that the solar wind is faster over the solar poles originating in the large coronal hole present there. We expect that there will be fewer effects from stream-stream interactions. We expect that the polarity of the magnetic field will be uniform. We expect that the magnetic field, on average, will be more radial. Indeed, if there are marked deviations near the solar poles from these expectations, that will be significant.

The second category of observations is the one that will settle controversies. There is some expectation (Jokipii and Kota, 1989) that although the magnetic field is on average radial over the solar poles, its weakness, its presence in a strong radial outflow, and with strong rotational effects over the poles, there could be significant non-radial terms present. These deviations of the magnetic field from its average configuration, if they occur, could have a strong influence on galactic cosmic ray propagation, but it will remain for Ulysses to determine whether there is evidence that the fluctuations exist.

Finally, we have the category of observations that is characteristic of all exploratory missions—the observations that will be a surprise. We have a picture of MHD fluctuations in the solar wind, which results from our observations near the equatorial plane—the fluctuations appear to be more wave-like and of solar origin in the inner heliosphere and more characteristic of locally generated turbulence in the azimuthal magnetic fields of the outer heliosphere (see, e.g. Roberts and Goldstein, 1991). What will Ulysses observe admittedly at essentially one radial distance, in the radial field over the solar poles? We have an understanding near the equatorial plane of the influence of stream-stream interactions on the dynamics and internal structure of the solar wind. What will Ulysses see in the polar regions where stream-stream interactions are absent?

By simply going to high latitudes and documenting what is observed, this aspect of Ulysses is bound to be a success. The data obtained should make it possible to construct an overall model for the three-dimensional heliosphere during the current, solar minimum portion of the solar cycle.

3. The Behavior of Galactic Cosmic Rays

The second unifying theme—understanding at long last the behavior of galactic cosmic rays in the heliosphere—also has a reasonable chance of success. In many ways, this problem has the same twenty year history as does Ulysses. In the late 1960's and early 1970's it was recognized that adiabatic deceleration of the cosmic rays in the expanding solar wind would be important (see, e.g. Parker, 1965; Gleeson and Axford, 1967). Particles, on average, would loose several hundred MeV/nucleon in energy, which is not a major effect for high energy particles but which effectively precludes particles with energies of less than 100-200 MeV/nucleon in the interstellar medium from penetrating to the inner heliosphere near the equatorial plane (see, e.g. Goldstein *et al.*, 1970). In the mid-1970's it was recognized that latitude effects would be important. The intrinsic difference between low heliographic latitudes, where the magnetic field is primarily azimuthal, and the polar regions, where it is primarily radial, should give easier access over the solar poles. Thus, large latitude gradients should be introduced, which with cross-field transport, makes latitude effects significant at all latitudes (see, e.g. Fisk, 1976). Finally, in the late 1970's and throughout the 1980's, when the overall organization of the magnetic structure of the heliosphere was understood, with one polarity in one hemisphere and the opposite polarity in the other, and with a current sheet between whose orientation varies with the solar cycle, the role of gradient and curvature drifts, the ultimate transport mechanism in latitude, was recognized (see, e.g. Jokipii, 1986).

Throughout all this evolution in understanding, the basic physics, or equivalently, the governing equation for cosmic ray behavior has not changed:

$$-\nabla \cdot \mathbf{V}\left(\frac{p}{3}\right)\left(\frac{\partial F}{\partial p}\right) = \nabla \cdot (\mathbf{K} \cdot \nabla F) - \mathbf{V} \cdot \nabla F - \mathbf{v_D} \cdot \nabla F \quad (1)$$

This is the equation for the steady-state behavior of the cosmic ray distribution function, F, in terms of its momentum p. It is a balance between several effects: (i) spatial diffusion (the first term on the right), which is usually described by a phenomenologically determined diffusion tensor, $\mathbf{K}$, which governs the tendency of the cosmic rays to relax to isotropy in the frame of the solar wind; (ii) convection with the solar wind speed, $\mathbf{V}$ (the second term on the right); (iii) adiabatic deceleration in the expanding solar wind (the term on the left); and (iv) a term (the third on the right) which describes the drift of the particles with an effective drift speed, $\mathbf{v_D}$, in the mean magnetic field of the heliosphere. An explicit time dependence can be added to this equation through a time derivative term and with, for example, a diffusion tensor that varies with time. Care also needs to be exerted in dealing with the drift term near the equatorial current sheet.

This equation is believed to be an accurate description of the cosmic ray behavior. The issue is not its likely accuracy, but rather what parameters to choose in applying the equation, or equivalently: How do the cosmic rays observed at a given location in the heliosphere reach this point? The equation is second order due to the diffusion term; however, for relatively high energy particles where the spatial gradients are not large, only the first order terms may be significant. The resulting approximate equation, then, describes the average trajectories of the cosmic rays, or the streamlines, in physical and momentum space, from the interstellar medium to where they are observed. In models where drifts in the mean magnetic field are important, and in which the magnetic field over the solar poles is essentially radial, the trajectory of the cosmic rays during the portion of the solar cycle seen by the upcoming polar pass of Ulysses should be inward over the solar poles and downward in latitude onto the equatorial plane (see, e.g. Jokipii and Kopriva, 1979). However, if the magnetic field over the poles is less radial and more disordered, with a resulting increase in diffusion and a reduction in the mean drift, then the average trajectory of the particles could be altered significantly (Jokipii and Kota, 1989).

For several decades we have speculated on how cosmic rays behave in the solar wind and what causes their solar cycle variations. We have models with warped current sheets, which should be important when the magnetic cycle of the Sun is such that the drift of the cosmic rays is inward along the current sheet (see, e.g. Smith and Thomas, 1986). We have models in both halves of the magnetic cycle of the Sun in which large stream-stream interaction regions are important for cosmic ray behavior (see, e.g. Burlaga *et al.*, 1987). The details and the differences of these various models can be resolved only when we know the flow patterns of the cosmic rays—where they have been and what they have experienced—and this, through observations of the solar wind and its magnetic field, and the response of the cosmic rays, is information that Ulysses in principle can provide.

There may be an extra bonus also over the solar poles. If the access is easy, then adiabatic deceleration, which results in 100-200 MeV/nucleon energy loss for particles seen near Earth, may be reduced. The inward trajectory of cosmic rays over the solar poles may be sufficiently fast so that there is little time to lose energy, and low energy interstellar cosmic rays, which are unobservable elsewhere in the heliosphere, may be present. Similarly, the absence of solar energetic particles over the solar poles, should that be the case, may reveal heretofore unobserved portions of the galactic cosmic ray spectrum.

4. The Behavior of Interstellar Gas in the Heliosphere

The third unifying theme is to trace the behavior of interstellar gas in the heliosphere. Tracing the behavior of galactic cosmic rays was a matter of filling in the gap in our knowledge of their spatial behavior. Tracing the behavior of the interstellar gas is a matter of finding out where these particles have been in energy space.

Ulysses will make direct measurements of neutral interstellar helium as it is swept into the heliosphere by the motion of the Sun relative to the local interstellar medium (Witte *et al.*, 1992). However, there is much more to the behavior of the interstellar gas than just its neutral component. There is the behavior of pick-up ions—interstellar gas as it becomes ionized by photo-ionization from the Sun and charge-exchange with the solar wind (Gloeckler and Geiss, 1994)—and there is the anomalous component. This problem also spans a time period identical to the development of Ulysses. In 1973 and 1974, the anomalous component—cosmic rays with energies in the range 10-100 MeV/nucleon with the anomalous composition of mainly helium, nitrogen and oxygen—was observed (Garcia-Munoz *et al.*, 1973; Hovestadt *et al.*, 1973; McDonald *et al.*, 1974). Shortly afterwards the anomalous component was identified by Fisk *et al.* (1974) as likely to originate as interstellar neutral gas that is ionized and accelerated in the solar wind. It has remained, however, for two decades to find compelling evidence that this origin is correct. The principal prediction of Fisk *et al.* was that the particles would be singly-charged, a prediction now confirmed by observation (see, e.g. Adams *et al.*, 1991). It remained for Ulysses en route to Jupiter to observe the recently ionized neutrals, principally the hydrogen and oxygen, shortly after they are picked up by the solar wind—the so-called pick-up ions. (Gloeckler and Geiss, 1994).

There is still, however, a major issue with this problem. Somewhere in the solar wind, but not of certain location, a very effective acceleration mechanism is occurring. When the interstellar ions are picked up by the solar wind, they acquire energies of approximately 1 keV/nucleon. When they are observed as the anomalous component, they have energies of order tens of MeV/nucleon, or a four order of magnitude increase.

The prevailing explanation is that the interstellar particles are accelerated at the termination shock of the solar wind, as was suggested by Pesses *et al.* (1981). The acceleration has to occur in the outer heliosphere, since the intensity seen by Pioneer and Voyager increases with heliocentric distance (see, e.g. Cummings *et al.*, 1994). Presumably, there is a shock present since the supersonic solar wind has to become subsonic to merge with the interstellar medium. It surrounds the heliosphere giving a good shock surface for acceleration, and shocks are known to be good accelerators. However, we should perhaps always be somewhat cautious about prevailing explanations for which there is no direct observational support and only circumstantial evidence. Indeed, there may be some potential concerns with the termination shock being the sole accelerator of the anomalous component.

In a recent paper, Cummings *et al.* (1994) estimated the pressure in the anomalous component at the termination shock by extrapolating Voyager data out to the likely shock location and by assuming that the shock is strong, i.e. a jump in the flow velocity by a factor of 4. They found that the pressure was only about 2% of the ram pressure of the solar wind. Had they done their calculations for a weaker shock, e.g., a velocity jump by 2-3, they would have found that the required pressure was dramatically higher, perhaps 40-50% of the ram pressure.

On the other hand, observations of Gloeckler *et al.* (1994) from Ulysses indicate that there may be considerable pressure in the pick-up particles, particularly if, as seems to be the case, there is additional acceleration of the pick-up ions in, e.g., Corotating Interaction Regions. An extrapolation of the observed pick-up pressure to the location of the termination shock yields a possible pressure in the pick-up ions of 10-15% of the ram pressure.

Herein lies the concern about the termination shock as the sole accelerator of the anomalous component. If the pick-up ions are mobile in the solar wind and do not contribute significantly to the solar wind internal pressure, as is the case for galactic cosmic rays, they will not affect the solar wind Mach number. The termination shock will then be a strong shock, with a velocity jump of order 4. However, if the pick-up ions are mobile in the solar wind, they should also be readily injected into the shock acceleration mechanism. The resulting pressure then in the accelerated anomalous particles will exceed the 10-15% of the ram pressure in the pick-up ions, and be much larger than the 2% inferred by Cummings *et al.* (1994).

Conversely, if the pick-up ions are not able to move easily in the solar wind, as is probably the case, they will determine the solar wind Mach number. At a pressure of 10-15% of the ram pressure, they render the termination shock a relatively weak shock and their heating at the shock is responsible for pressure balance. For example, if there is considerable pressure in the anomalous component, there should be a strong fore shock region, which slows the upstream solar wind and adiabatically compresses the embedded pick-up ions. It can readily be shown that the pressure allowed in the anomalous component, so that the adiabatically compressed pick-up ions, the anomalous component, and the solar wind ram pressure are all in pressure equilibrium, is much less than the 40-50% required to explain the observations of Cummings *et al.* (1994) in the weak shock case.

These conclusions depend on the pressure in the pick-up ions being significant at the termination shock (10-15% of the ram pressure). Should this be the case, we should be concerned about the role the termination shock plays in the acceleration of the anomalous component.

The role that observations from Ulysses will have in determining the acceleration mechanism for the anomalous component will depend in part on what the mechanism is. If the acceleration occurs at the termination shock, Ulysses has the opportunity to probe conditions at the shock at different latitudes. The shock strength could vary with latitude, as well as the ability of particles to be injected into the shock acceleration mechanism since the direction of the magnetic field relative to the shock normal will vary with latitude. These differences may be revealed in the behavior of the anomalous component with latitude. Conversely, if the termination shock is not the only accelerator, and this is determined either by direct observations from Pioneer and Voyager or by inference because the magnitude of the pressure in pick-up ions and its role in the strength of the termination shock is understood, then the latitude observations of Ulysses could have the important role of determining what the principal acceleration mechanism actually is.

5. Coronal Processes

The fourth and final unifying theme is to use observations from Ulysses to access remotely coronal processes; in particular, the heating and acceleration of the solar wind and the acceleration and propagation of energetic particles.

The basic concepts of how the solar wind is accelerated have, of course, been known since the late 1950's (Parker 1958). However, "the devil is in the details", some of which are quite fundamental: How is the corona heated? How are high speed streams with the correct speed, density, and temperature generated? How are such basic properties as electron heat conduction included correctly, or what is the role of open and closed magnetic field lines in the transition region and lower corona? How do we model this dynamic, time-varying process, or, equivalently, how important are time-varying effects for the models which are essentially steady-state?

Ulysses, of course, gives us three advantages in being able to contribute to understanding the acceleration of the solar wind. The first is perspective. Ulysses gets to look down the throat of the solar wind by observing the solar wind as it comes directly from a large polar coronal hole, relatively undisturbed by non-radial flow through the corona or by interactions between high and low speed streams. Second, Ulysses carries very capable plasma instruments, including one that can measure the elemental and charge composition of the solar wind (Gloeckler *et al.*, 1992). Since the charge states of the solar wind are frozen-in in the corona and the elemental composition is altered by coronal processes and in the lower solar atmosphere, these observations can place serious constraints on acceptable models for coronal heating and solar wind acceleration. Finally, there should be an opportunity for detailed coordinated measurements between Ulysses measuring the product of the solar wind acceleration and Spartan 201 measuring the coronal conditions that give rise to this wind. It will be a rare opportunity to look at the solar wind acceleration process from both ends.

Studies of solar energetic particles will also be significant. The issue here should be primarily coronal propagation. The sources of solar cosmic rays are primarily at low and mid-latitudes. By observing at high latitudes, particularly over a large coronal hole, Ulysses should provide a unique perspective on the mechanisms by which energetic particles propagate in the corona.

6. Conclusions

Determine the overall morphology of the heliosphere—the large-scale variations of the solar wind and its magnetic field; the behavior of its turbulence. Trace where the galactic cosmic rays have been. Reveal the influence of interstellar ions and determine where and how the acceleration of the anomalous component is occurring. Contribute to our understanding of the acceleration of the solar wind and the coronal propagation of solar energetic particles. It is a pretty good list. Certainly not a complete one, for there are other significant investigations on Ulysses, e.g., the measuring of the properties of dust in the heliosphere and the studying of solar x-ray emission. However, a good list nonetheless, which if executed successfully will allow history to judge kindly the Ulysses mission and our twenty-year odyssey to explore the heliosphere in three dimensions.

References

Adams Jr., J. H. *et al.* : 1991, *Astrophys. J. (Letters)* **375**, L45.
Axford, W. I.: 1994, *J. Geophys. Res.*, in press.
Burlaga, L. F., Ness, N. F., and McDonald, F. B.: 1987, *J. Geophys. Res.* **92**, 13647.
Cummings, A. C., Stone, E. C., and Webber, W. R.: 1994, *J. Geophys. Res.*, in press.
Fisk, L. A.: 1976, *J. Geophys. Res.* **81**, 4646.
Fisk, L. A., Kozlovsky, B., and Ramaty, R.: 1974, *Astrophys. J. (Letters)* **190** L35.
Garcia-Munoz, M., Mason, G. M., and Simpson, J. A.: 1973, *Astrophys. J. (Letters)* **182**, L81.
Gleeson, L. J. and Axford, W. I.: 1967, *Astrophys. J. (Letters)* **149**, L115.
Gloeckler, G. and Geiss, J.: 1994, *Cosmic Winds and the Heliosphere*, ed. by J. R. Jokipii and C. Sonett, University of Arizona Press, Tucson, in press.
Gloeckler, G. *et al.* : 1992, *Astron. and Astrophys (Suppl)* **92**, 267.
Gloeckler, G. *et al.* : 1994, *J. Geophys. Res.*, in press.

Goldstein, M. L., Fisk, L. A. and Ramaty, R.: 1970, *Phys. Rev. Letters* **25**, 832.
Hovestadt, D. *et al.* : 1973, *Phys. Rev. Letters* **31**, 650.
Jokipii, J. R.: 1986, *The Sun and the Heliosphere in Three Dimensions*, Reidel Pub., Dordrecht, pp 375-388.
Jokipii, J. R. and Kopriva, D. A.: 1979, *Astrophys. J.* **234**, 384.
Jokipii, J. R. and Kota, J.: 1989, *J. Geophys. Res. Letters* **16**,1.
McDonald, F. B. *et al.* : 1974, *Astrophys. J. (Letters)* **249**, L71.
Parker, E. N.: 1958, *Astrophys. J.* **128**, 664.
Parker, E. N.: 1965, *Planet. Space Sci.* **13**, 9.
Pesses, M. E., Jokipii, J. R., and Eichler, D.: 1981, *Astrophys. J. (Letters)* **246**, L85.
Roberts, D. A. and Goldstein, M. L.: 1991, *U. S. Nat. Report. to Inter. Union of Geodesy and Geophysics 1987-1990*, pp 932-943.
Smith, E. J. and Thomas, B. T.: 1986, *J. Geophys. Res.* **91**, 2933.
Witte, M. *et al.*: 1992, *Astron and Astrophys (Suppl.)* **92**, 333.

SESSION 1
THE SUN AND CORONA

YOHKOH RESULTS IN THE CONTEXT OF THE HIGH-LATITUDE HELIOSPHERE

J.L. CULHANE

Mullard Space Science Laboratory
Department of Space and Climate Physics
University College London

Abstract: Designed primarily to study solar activity, Yohkoh includes an X-ray telescope that obtains full-sun coronal images which show a range of features. Coronal X-ray emission - exclusive of flares, is notable for its variability even in the largest structures. A mass ejection event is related to magnetic field reconnection. Such events exhibit both accelerated and decelerated behaviour. Coronal hole temperatures are estimated from the filter ratio method. A plasma component at around 2.10^6 K is identified. X-ray emission is detected from the South polar coronal hole. A preliminary comparison of Spartan coronagraph images with Yohkoh data suggests that polar plumes or rays are not connected to bright points.

1. Introduction

Yohkoh, launched into a near-earth orbit (~520km, 31.3°) on 30 August 1991, was built by Japan's Institute of Space and Astronautical Science (ISAS). It should be operational until the year 2000. Other Japanese institutes (National Astronomical Observatory, Tokyo and Rikkyo Universities) along with groups in the USA (Lockheed and NRL) and in the UK (MSSL and RAL) provided flight instruments. While mainly for studies of solar activity and flares, the mission includes a US-Japanese X-ray telescope which takes full-sun images. In studying the heliosphere, it is important to combine data from the solar corona with both remote and in-situ observations of the interplanetary medium. Ulysses offers an unparalleled opportunity for such a programme. However the SPARTAN flights are also obtaining highly relevant data while the ESA/NASA SOHO mission, following its launch in July 1995, will make a major contribution to the field.

The sun influences the interplanetary medium through the solar wind - originating principally in coronal holes, the ejection of matter in discrete episodes (CMEs) and the acceleration of particles. Yohkoh can observe the first two phenomena directly and can infer properties of the third through the detection of hard X-rays and gamma-rays. Following release of data to the wide community in October, 1993, the possibility exists for significant involvement of Ulysses investigators in the Yohkoh programme. This process has begun and will be extended to SOHO after its launch in July, 1995.

Space Science Reviews **72**: 17–28.
© 1995 *Kluwer Academic Publishers.*

Following a description of Yohkoh instruments, observations relevant to the heliosphere will be described. The X-ray appearance of the corona will be discussed with emphasis on variability and dynamic properties of large coronal structures. Results from studies of CMEs will be presented and an account given of coronal hole observations. During the SPARTAN flight, data were obtained simultaneously by Yohkoh and the Mark III coronameter on Mauna Loa in Hawaii. This work will be described briefly.

2. The Yohkoh Mission - Instruments and Observations

Yohkoh mainly studies solar activity with emphasis on flares at high photon energies. Instruments intended for flare work include the Hard X-ray Telescope (HXT, Kosugi et al., 1991); a Fourier transform telescope with 5 arc sec angular resolution in the 13 - 93 keV range. Plasma parameters e.g. temperature and velocity, come from the Bragg Crystal Spectrometer (BCS, Culhane et al., 1991); curved crystals taking spectra of highly ionized species (S XV, Ca XIX, Fe XXV, Fe XXVI) with 3s time resolution. The Wide Band Spectrometer (WBS, Yoshimori et al., 1991) provides non-dispersive spectra in the range 2 keV to 10 MeV. The Soft X-ray Telescope (SXT, Tsuneta et al., 1991), is also highly relevant for coronal studies. Its parameters are listed in Table I.

Table I Features of the Yohkoh Soft X-ray Telescope

Feature	Parameter Value
Field of View/Spatial Resolution:	42'x42' /2.5 arc sec.
Time Resolution:	2s in 2'x2' (Flares/Activity)
	260s in 42'x21' (Coronal Mode)
X-ray Detector:	CCD (1024x1024, 18.3μm)
Energy/Wavelength Range:	0.25 - 4.0 keV/3 - 45Å
Dynamic Range:	$\sim 10^9$ (CCD full well/exposure control)

The ability of SXT to obtain coronal images with a dynamic range of $\sim 10^9$ is of immense value for studies of active regions, quiet coronal structures and coronal holes. The greatly reduced surface roughness of the grazing incidence mirrors (Tsuneta et al., 1991) compared to the earlier Apollo SO-54 telescope (Vaiana et al., 1977) leads to much less scattering from wings to core of the Point Spread Function (PSF). This is clear in Figure 1a where the PSFs are compared. However for studies of faint coronal features surrounded by regions of high activity, the scattering contribution to the on-axis signal by radiation from the whole sun must be estimated. Before the launch of Yohkoh, the PSF was measured up to 90 arc sec off axis at 1.48 keV in a long beam facility where the existence of core ($I \sim r^{-4}$) and halo ($I \sim r^{-2}$) components was established (Martens et al. 1994). After launch, the PSF characterization was extended to more than 30 arc min off axis using radiation scattered from several solar flares (Hara et al., 1994).

The extended PSF is shown in Figure 1b. Use of the PSF is particularly important when analysing data from coronal hole observations.

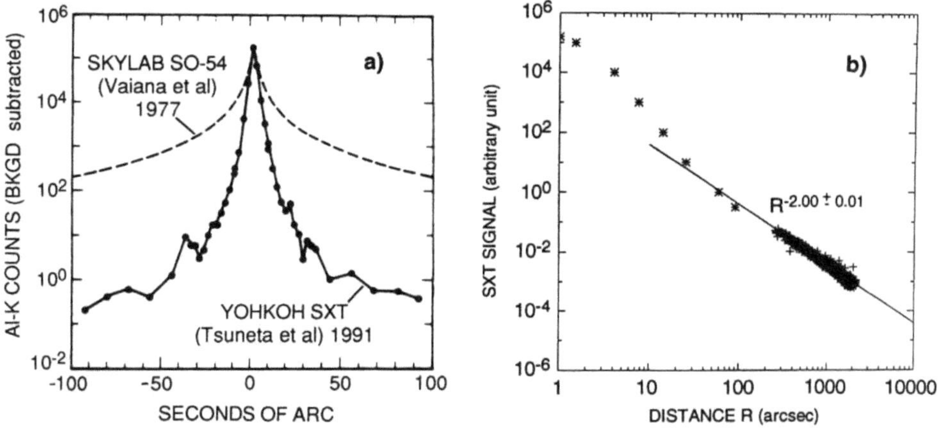

Figure 1. a) The Yohkoh SXT (Tsuneta et al., 1991) and Skylab SO-54 (Vaiana et al., 1977) X-ray telescope point spread functions (PSFs) measured with 1.48 keV X-rays. b) SXT PSF extended to large off-axis angles using flare data (Hara et al., 1994).

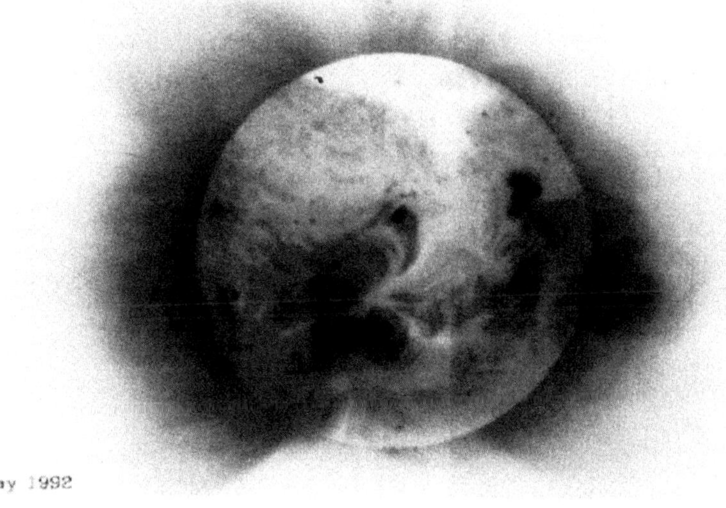

Figure 2. Composite image of the Sun for 8 May, 1992 from SXT for 3Å to 45Å. Data used from three different spacecraft pointings. Partial frame short exposure images are inserted in regions of intense solar emission to enhance the overall dynamic range.

a) X-ray Emission from the General Corona

The image in Figure 2, a mosaic from three different pointings, was obtained on 8 May, 1992. Over-exposed sections are replaced by data obtained with shorter exposures. This enhances dynamic range for viewing of active regions and faint coronal holes. Intense active region emission is seen on the disc and the West limb. However in other areas, active region loops and loops connecting centres of activity can be distinguished. Interconnecting loops and more dynamic features (jets; Shibata et al., 1992) often exhibit kinked morphology unlike the simple loops that were characteristic of the Skylab data. Bright points (Strong et al., 1992) are seen both in and out of coronal holes. In Figure 2 North and South polar coronal holes are well developed. Polar rays can be seen near the North polar hole which is otherwise clearly delineated. Large scale structures are evident on the South-East limb where the unresolved emission is described as a "general" corona. The South polar hole is "veiled" probably due to general emission from along the line of sight that passes through the hole but above the limb. Since the related structures are not resolved, general coronal emission is hard to explain. It may originate from large-scale closed structures related to flux tubes originating in the network boundary regions.

b) Coronal Variability

The image in Figure 2 appears static but use of several full sun images per orbit has shown that even the quiet corona and the largest coronal structures are exceptionally variable. Figure 3 (a, b) shows solar images taken about a year apart which reveal the declining phase of the cycle. The intensity plot (Figure 3c), shows the variability of coronal X-rays during 1.5 years. Flare related data have been eliminated. While there are shorter term variations due to active regions emerging and rotating onto and off the disc, longer term episodes lasting 6-8 weeks can be distinguished. Periods longer than a solar rotation may indicate a more global magnetic control of the general coronal emission.

A quantitative view of coronal variability is given by Hara (1993) where pixel brightness distributions indicate the spatial distribution of variability. At times of lower solar activity - 1993, June 16 as indicated in Figure 4, it is possible to show that while the active region distribution of $dN(>I)/dI$ versus I is of power law form, the quiet coronal distribution is exponential.

c) Large Scale Structures and Mass Ejections

Extended time coverage allows the study of large coronal structures. The formation of a helmet streamer on the South-West limb on 24 January, 1992 and its later evolution have been described by Hiei et al., 1994. Yohkoh data together with Mauna Loa prominence monitor and Mark III coronameter results were used in the analysis. The streamer formed on 24 January and replaced a closed magnetic structure which was positioned over a darker cavity and a large prominence. The Yohkoh observations in

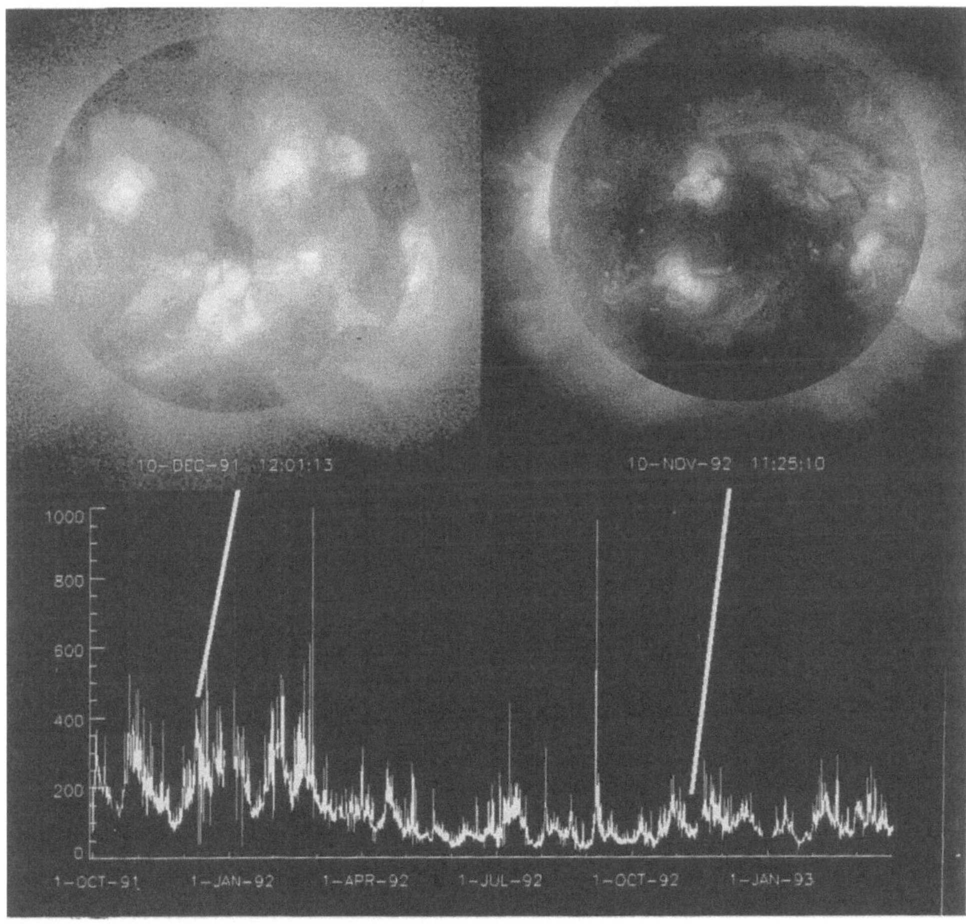

Figure 3. Yohkoh images. a) 10 December, 1991. b) 10 November 1992. Cycle related activity drop is clear. c) X-ray intensity - no flares; 1 October, 1991 - 1 April, 1993. Appearance and disappearance of active regions, the cycle-related decline and variability episodes of 6 - 8 weeks duration are seen. (Strong, K.T., 1994. Private Communication)

particular show that between 08:07 U.T. and 09:20 U.T., the X-ray emitting structure reformed. Later observations from Mauna Loa showed that the prominence had vanished while the coronameter indicated an identical structure to that seen in X-rays.

The disappearance of the prominence and the overlying coronal structure both of which had existed on 22 and 23 January, suggests a prominence eruption and a related CME.

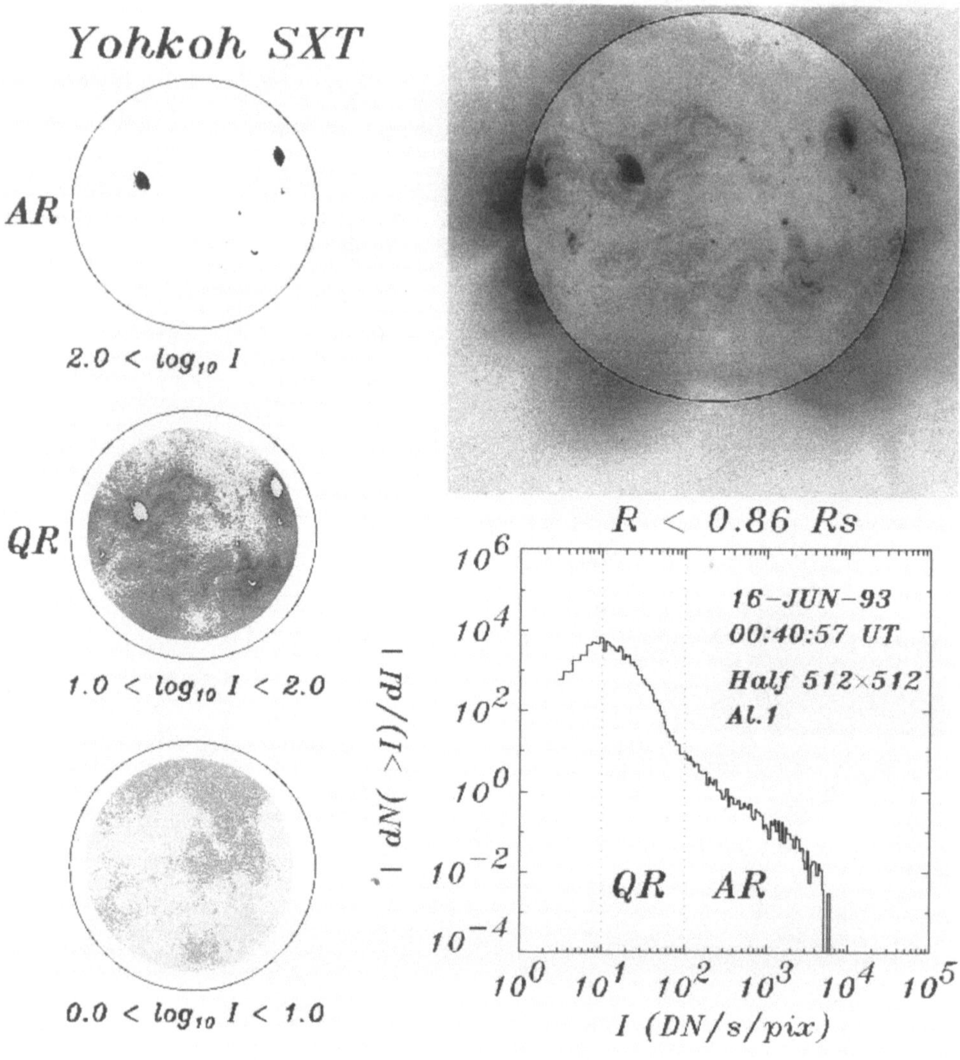

Figure 4. Yohkoh data from 16 June, 1993 (Hara, 1993); dN(>I)/dI vs I is for R < 0.86 R_0. Full image is at the top right. Images for the three ranges of intensity labelled in the histogram are on the left. Solid circles indicate the solar limb positions.

The newly formed helmet moved outwards at a speed of ~ 10 km s^{-1}. While the coronameter data show a rather uniform electron density, the X-ray image is brightest at its edges suggesting that the temperature is high around the outside of the streamer. These features characterize a magnetic field reconnection proceeding outwards along a vertical current sheet in a region of previously closed magnetic field blown open by the

CME (Hirayama, 1976; Kopp and Pneuman, 1976). The development of the X-ray brightness of the structure shows a gradual rise and a slow decay lasting for more than a day. Such a time profile would previously have been recognized as a Long Duration X-ray Event - a class of flare often characteristic of mass ejections and prominence eruptions. Thus this sequence of X-ray and ground based observations significantly clarifies the role of magnetic field reconnection in mass ejection phenomena.

Yohkoh has observed other eruptive events usually without simultaneous optical coverage. Klimchuk et al. (1994) have identified 29 events whose widths around the limb, apparent radial velocities and occurrence rates suggest that many are CMEs. For these events, the mean width and radial velocity are 32° and 48 km s^{-1}. The distributions are similar to those of CMEs observed by the SMM. (Burkepile and St. Cyr, 1993). While the mean width and velocity values are larger for the SMM events (46° and 349 km s^{-1}), the white light observations are made higher in the corona while a CME will accelerate with height. Klimchuk et al. studied four of these events in greater detail. They found that in two cases the height-time trajectory of the eruption indicates acceleration while in the remaining cases deceleration is observed. Both types of behaviour are shown in Figure 5 (a,b). For the accelerating case the average velocity is 71 km s^{-1} while at a height of 1 R_0 above the limb, the extrapolated velocity is 485 km s^{-1}. For the

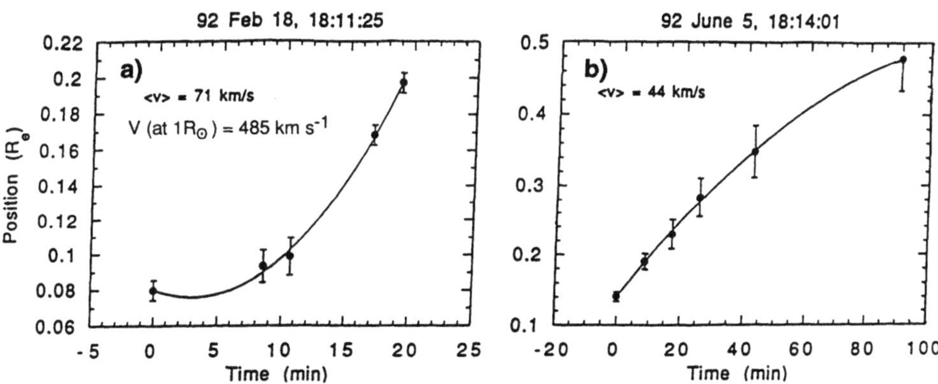

Figure 5. Height against time trajectories for accelerating (a) and decelerating (b) soft X-ray eruptions on 18 February and 5 June, 1992 (Klimchuk et al., 1993)

decelerating event the average and extrapolated velocities are 44 km s^{-1} and 104 km s^{-1}. For the accelerating events, an initially closed magnetic structure opens up fully as the event progresses. In the decelerating cases the eruption is only partial and the related magnetic field structures remain closed. The partial events may be examples of the active region expansion phenomenon first seen in Yohkoh data by Uchida et al., 1992.

d) Coronal Holes

Hara et al. (1994) observed several coronal holes on the disc with data obtained from four regions on 6 March, 21 March, 3 April and 2 September, 1992. The filter ratio method is used (Hara et al. 1992) but with higher precision than previously and with a correction for X-ray scattering from surrounding regions (Figure 1b). The filter pass bands are given by Tsuneta et al. 1991. The emissivity - $P(\lambda,T)$, of an isothermal plasma has been calculated for X-ray line and continuum emission by Mewe et al.(1985, 1986). Using these $P(\lambda,T)$ calculations with the coronal abundances of Meyer (1985), the intensity measured by SXT in each of the pass bands is of the form

$$I = CF(T) \int n_e^2 \, ds \tag{1}$$

where C is a constant, n_e is the electron density, s is length along the line of sight and if $\eta(\lambda)$ is the effective area of the SXT as a function of wavelength, then

$$F(T) = \int_{\lambda_1}^{\lambda_2} P(\lambda,T)\, \eta(\lambda) d\lambda \tag{2}$$

If I_a and I_b are observed from a coronal hole which contains isothermal plasma, then

$$I_b/I_a = F_b(T)/F_a(T) = f(T) \tag{3}$$

and temperatures for the hole material may be obtained from the intensity ratio. Given the uncertainties in the X-ray scattering correction, Hara et al. (1994) find T for coronal holes in the range 1.8×10^6 K to 2.4×10^6 K with emission measure values in the range $10^{25.5}$ cm^{-5} to $10^{26.2}$ cm^{-5}. While these temperature values are similar to those in quiet coronal regions, the emission measures are ten times less. Hence the coronal hole densities are ~ three times less than the surrounding corona. Since plasma at 10^6 K with previously reported emission measures (e.g. Raymond and Doyle, 1981) would have made only a 5% contribution to the SXT intensity in filter (a), these observations indicate additional hotter material and a flat differential emission measure distribution.

Since May, 1992, the South polar hole has continued to develop. Starting in February, 1994, the Yohkoh SXT has observed this feature with a series of southerly off-pointings by the spacecraft. A long exposure image (t = 483s; Acton et al., In Preparation), is shown in Figure 6. The data have been corrected for background and stray light but not for X-ray scattering. The limb is clearly evident as are two bright points on the edge of the disc. Further long exposure data sets have now been obtained and the X-ray intensity distribution in the hole will be analysed in detail.

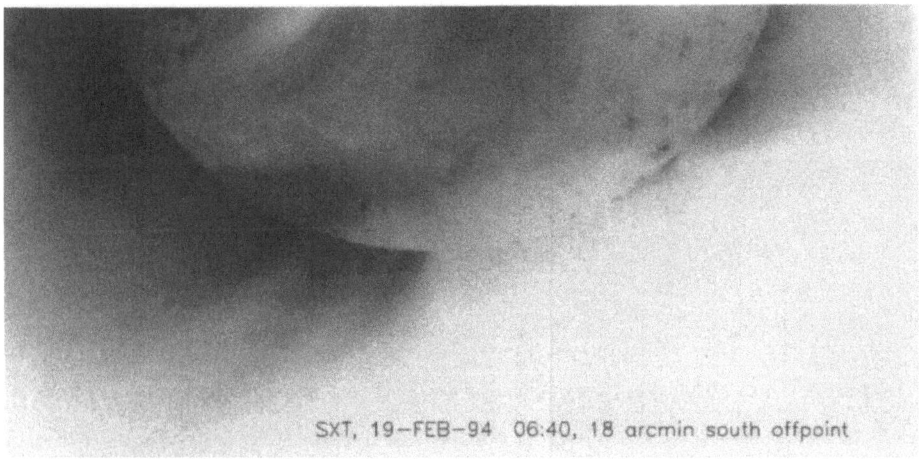

Figure 6. The South polar hole on 19 February, 1994; filter (b) and exposure 483s.

3. Yohkoh and the Spartan Mission

Yohkoh was operated in campaign mode during the Spartan mission. The retrievable spacecraft was launched on the shuttle Discovery on 7 April, 1993 and operated for 47 hours. It carried a White Light Coronagraph (Fisher and Guhathakurta, 1994) and an Ultraviolet Coronal Spectrometer (Kohl et al., 1994; see also Kohl, J., These Proceedings). Data were also available from the HAO Mark III Imaging Coronameter on Mauna Loa and from the NSO/GSFC Disk Spectromagnetograph. Images obtained on 11 April, 1993, are superimposed in Figure 8a. The dotted circles indicate that these instruments can observe the corona to a distance of 6 R_0. The streamer on the South-East limb is non-radial and inclined towards the equator at an angle increasing with height. Electron densities inferred from the measurements of polarized brightness (pB) vary from 3×10^7 cm^{-3} at 1.5 R_0 to about 6×10^4 cm^{-3} at 5 R_0 and are typical of those previously observed in such structures with variations of ~ 25% possibly due to uncertainties in filter calibration. All three images in Figure 7a indicate well developed North and South polar holes. A more detailed view of the North polar hole is given in Figure 7b which shows a number of faint structures - polar plumes or rays. These are about 300,000km wide and can be traced back to the Yohkoh image where the majority do not coincide with X-ray bright points. Although the density contrast between the structures and the coronal hole material is not yet established, the two show differing

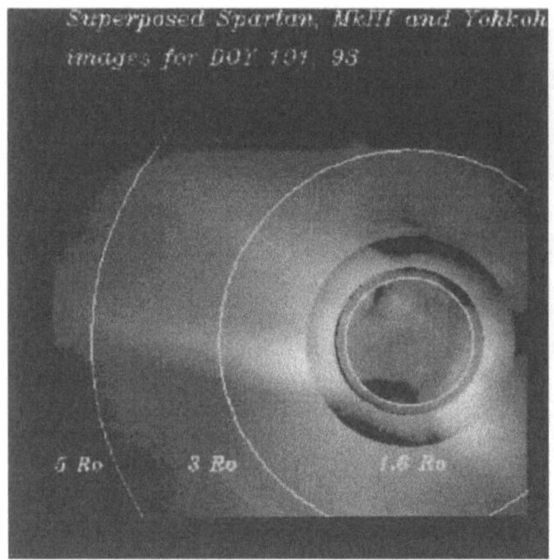

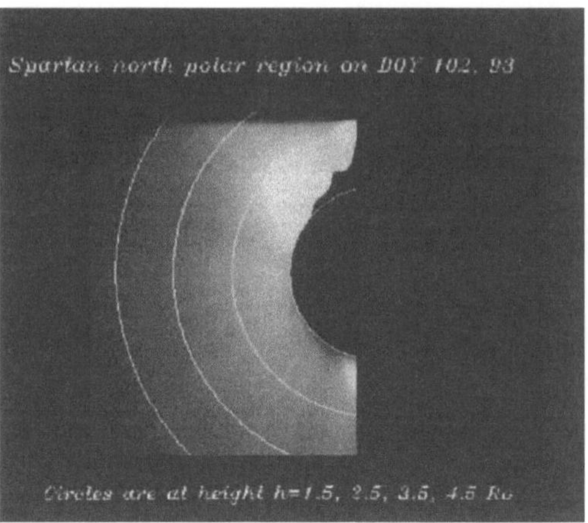

Figure 7. a) Images from SXT, the Mark III and Spartan white light coronameters. North top, East left. b) North polar hole - plume-like structures are seen. North is left.

radial density gradients. Fisher and Guhathakurta conclude that the plumes may be foreground or background structures while the darker regions represent the genuine coronal hole material. The latter show density values about a factor two lower than was measured during the Skylab mission. Further Spartan flights are planned to coincide with polar observations by Ulysses and Yohkoh should operate until reentry in 2000. The launch of SOHO in July, 1995 and its operation in orbit around L1 will provide an array of instruments - White Light and UV Coronagraphs, UV and EUV spectrometers, EUV Imagers, high energy particle detectors, mass spectrometers and a Hydrogen Ly_{α} photometer with 4π field of view. The improved capability of SOHO will transform our understanding of the Sun with a major increase in our knowledge of the Heliosphere.

4. Conclusions

We have described the Yohkoh Soft X-ray Telescope. Scattered X-rays from bright coronal regions near the fainter features must be allowed for in measurements of intensity. We have presented the telescope PSF established in orbit from observations of flares at angles up to 2000 arc sec off axis. The SXT, due to better time resolution and reduced surface roughness compared to Skylab era telescopes, permits study of a broader range of phenomena. One of the main conclusions to emerge has been recognition of the high degree of variability of both general corona and large scale structures. This work is being pursued quantitatively to help identify the coronal heating mechanism. Particularly high degree of variability of both general corona and large scale structures. This work is being pursued quantitatively to help identify the coronal heating mechanism. Particularly important examples of variability concern the evolution of large scale structures e.g. streamers, which seem related to well understood magnetic reconnection geometries. The division of Coronal Mass Ejection events into those genuine CMEs which are accelerated and leave the atmosphere and those which lack sufficient energy to escape represents a useful clarification. Emphasis on coronal hole studies is growing as the solar cycle declines Observation of several holes on the disc has revealed material at temperatures of ~ $2x10^6$ K. Recent observations of the well developed South polar coronal hole have detected the limb crossing and further long exposure data are being obtained for a more detailed study. White light and UV spectroscopic data from Spartan are being compared with Yohkoh images in an effort to find the origin of the polar plumes or rays and to clarify the evolution of large scale coronal structures. Further Spartan flights and the SOHO launch in July, 1995 will together with more Ulysses and Yohkoh data provide the basis for a major advance in Solar and Heliospheric science.

Acknowledgements

The major role of the Japanese Institute for Space and Astronautical Science (ISAS) is acknowledged with gratitude. I have benefitted from discussions with Loren Acton, Hirohisa Hara, Hugh Hudson, Jim Klimchuk and Keith Strong. I thank Madalika Guhathakurta for the data displayed in Figure 8. In addition to ISAS who were supported by Japan's Ministry of Education and Culture, the support of NASA and NRL, the UK SERC and the Lockheed Independent Research Programme is gratefully acknowledged. I also thank the British Council for support of visits to Japan and the UK.

References

Burkepile, J.T. and St.Cyr, O.C., 1993, *NCAR Tech. Note 369*, Boulder.

Culhane, J.L. et al., 1991, *Solar Phys.*, **136**, 89.

Fisher, R.R. and Guhathakurta, M., 1994, *Submitted to Space Sci. Rev.*

Hara, H., Tsuneta, S., Lemen, J.R., Acton, L.W., McTiernan, J., 1992, *PASJ*, **44**, L135.

Hara, H., 1993, *Proc. Kofu Symposium*, In Press.

Hara, H., Tsuneta, S., Acton, L.W., Bruner, M.E., Lemen, J.R., Ogawara, Y., 1994, *PASJ.*, In Press.

Hirayama, T., 1976, *Solar Phys.*, **34**, 323.

Klimchuk, J.A., Acton, L.W., Harvey, K.L., Hudson, H.S., Kluge, K.L., Sime, D.G., Strong, K.T., 1994, in "X-ray Solar Physics from Yohkoh", eds. Y. Uchida, T. Watanabe, K. Shibata, H. Hudson, Universal International Press Inc., p 181.

Watanabe, Takashi, 1993, *Proc. Kofu Symposium*, In Press.

Kohl, J., Gardner, L.D., Strachan, L., Hassler, D.M., 1994, *Submitted to Space Sci. Rev.*

Kopp, R.A., Pneuman, G.W., 1976, *Solar Phys.*, **50**, 85.

Kosugi, T., et al., 1991, *Solar Phys.,* **136**, 17.

Martens, P.C., Acton, L.W., Lemen, J,R., 1994, *Submitted to Solar Phys.*

Mewe, R., Gronenschild, E.H.B.M., van den Oord, G.H.J., 1985, *Astron. Astrophys. Suppl.*, **62**, 197.

Mewe, R., Lemen, J.R., van den Oord, G.H.J., 1986, *Astron. Astrophys. Suppl.*, **65**, 511.

Meyer, J.P., 1985, *Astrophys. J. Suppl.*, **57**, 173.

Raymond, J.C., Doyle, J.G., 1981, *Astrophys. J.*, **247**, 686.

Shibata, K., et al., 1992, *PASJ*, **44**, L173.

Strong, K., Harvey, K., Hirayama, T., Nitta, N., Shimizu, T., Tsuneta, S., *PASJ*, **44**, L161.

Tsuneta, S., et al., 1991, Solar Phys., **136**, 37.

Uchida, Y., McAllister, A., Strong, K.T., Ogawara, Y., Shimizu, T., Matsumoto, R., Hudson, H.S., 1992, *PASJ*, **44**, L155.

Vaiana, G.S., Van Speybroek, L., Zombeck, M.V., Krieger, A.S., Silk, J.K., Timothy, A., 1977, *Space Sci. Instr.,* **3**, 19.

Yoshimori, M., et al., 1991, *Solar Phys.*, **136**, 69.

SPARTAN 201 CORONAL SPECTROSCOPY DURING THE POLAR PASSES OF ULYSSES

J.L. KOHL, L.D. GARDNER and L. STRACHAN
Harvard-Smithsonian Center for Astrophysics
Cambridge, Massachusetts 02138 U.S.A.

and

R. FISHER and M. GUHATHAKURTA
NASA/Goddard Space Flight Center
Greenbelt, Maryland 20771 U.S.A.

Abstract. Spartan 201 is a shuttle deployed spacecraft that is scheduled to perform ultraviolet spectroscopy and white light polarimetry of the extended solar corona during two 40 hour missions to occur in September 1994 and August 1995. The spectroscopy is done with an ultraviolet coronal spectrometer which measures the intensity and spectral line profile of HI Lyα up to heliocentric heights of 3.5 solar radii. It also measures the intensities of the OVI doublet at 1032 and 1037 Å and of Fe XII at 1242 Å. The HI Lyα line profile measurements are used to determine the random velocity distribution of coronal protons along the line-of-sight. The absolute HI Lyα intensities can be used together with electron densities from the white light coronagraph to estimate electron temperatures from hydrogen ionization balance calculations, and bulk outflow velocities from models of Doppler dimmed resonant scattering. Intensities of minor ion lines are used to determine coronal abundances and outflow velocities of O^{5+}. Ultraviolet spectroscopy of extended coronal regions from the 11 April 1993 mission of Spartan 201 are discussed.

Key words: Solar wind, UV spectroscopy

1. Introduction

The Spartan Ultraviolet Coronal Spectrometer was developed for spectroscopic studies of the expansion of the solar corona into the solar wind. The observational goal is to obtain a more complete empirical description of the extended corona than has been derived in the past from broad band polarimetry, which determines electron density and the associated geometry and temporal behavior of observed coronal structures. Ultraviolet spectroscopy provides information about the thermal and nonthermal velocities of the protons including their bulk outflow velocity and provides information about the electron temperature (Withbroe, Kohl, Weiser, and Munro 1982). Ultraviolet spectroscopy also provides information about the velocity distribution, bulk outflow velocity and density of minor ions in the expanding coronal plasma (Kohl and Withbroe 1982; Noci, Kohl, and Withbroe 1987). It is best to combine the ultraviolet spectroscopy with visible polarimetry and *in situ* measurements of solar wind parameters in order to obtain as complete an empirical description of the expanding corona as possible. In the case of Spartan, there is a companion White Light Coronagraph (Fisher and

Space Science Reviews **72**: 29–38.
© 1995 *Kluwer Academic Publishers.*

Guhathakurta, 1994) which provides polarized radiance observations that are co-registered with the spectroscopic observations. Spartan 201 flights are planned for the times of the Ulysses polar passes in September 1994 and August 1995 when polar solar wind parameters will be measured by Ulysses together with the Spartan coronagraphic observations. The combined data sets will provide a unique opportunity to establish empirical constraints on theoretical models of solar wind flows from polar coronal holes. Later in 1995, ultraviolet spectroscopy will be carried out with the SOHO Ultraviolet Coronagraph Spectrometer (UVCS). That instrument will have higher spectral, spatial and time resolution than Spartan, and will observe a much broader spectral range (Kohl and Noci 1992).

2. Characteristics of the Spartan Ultraviolet Coronal Spectrometer

The Spartan Ultraviolet Coronal Spectrometer consists of an externally and internally occulted telescope system, a dual spectrometer, two uv detectors and electronics. There are two channels, one for observing the spectral line profile and intensity of the HI Lyα line and the intensity of Fe XII at 1242 Å; and the other for measuring line intensities of the O VI doublet at 1032 and 1037 Å.

The Lyα channel has a spherical telescope mirror with a focal length of 47.5 cm, which is coated with Al + MgF_2, and an Ebert Fastie spectrometer with a discrete multi-anode detector. There are separate entrance slits for measuring the line profile and the integrated intensity. The size of the former slit corresponds to a spatial resolution element of 0.5 by 2.5 arcminutes and the latter to 4.0 by 5.0 arcminutes. The detector has two linear arrays, a 42 pixel array with effective spectral resolution elements of 0.25 Å, which are used for profile measurements, and a 48 pixel array with 2.0 Å elements for measuring integrated intensities. The detector has a CsI photocathode, a MgF_2 window and a 2 l s^{-1} ion pump.

The telescope mirror for the O VI channel is similar to that for Lyα except it has an iridium coating. The spectrometer has a spherical diffraction grating on a Roland circle mount and a detector system which includes a rectangular aperture that acts as the spectrometer exit slit, and two channel electron multipliers (CEM's) which detect the light from each of the two O VI lines. The CEM's are housed in a vacuum canister with a vacuum door. The CEM's are coated with KBr. The entrance slit for O VI corresponds to a spatial resolution element of 2.5 by 5.0 arcmin. For observations of the solar disk, a mechanism inserts a pinhole of about 0.01mm diameter in front of the Lyα profile slit and another in front of the O VI slit.

The field-of-view is illustrated in Figure 1. At any instant of time, the instrument observes three spatial elements corresponding to the three en-

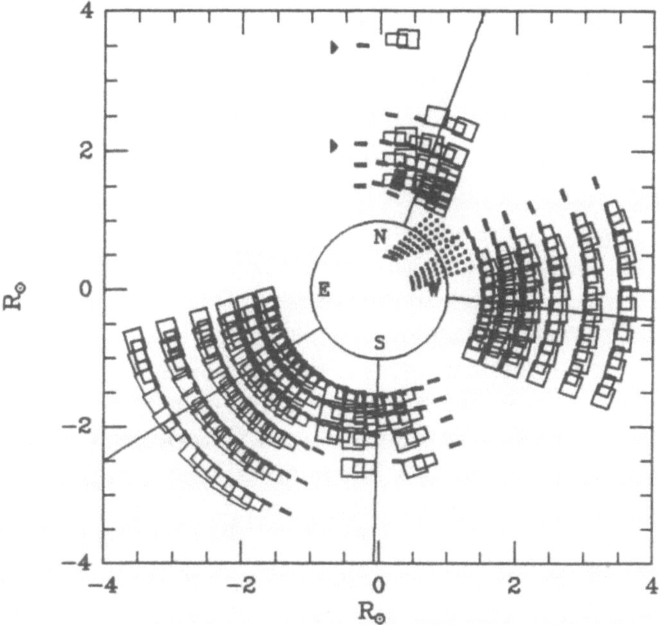

Fig. 1. Spatial elements observed by the Ultraviolet Coronal Spectrometer during the Spartan 201-1 mission are illustrated. At any instant, the instrument observes three spatial elements (e.g., the group of three rectangles at the extreme top of the figure). The smallest and largest rectangles are, respectively, for observations of the line profile and integrated line intensity of HI Lyα, and the intermediately sized one is for the intensity of OVI lines. Telescope motions are used to vary the heliocentric height and the entire spacecraft is rotated to vary the position angle.

trance slits. The telescope mirrors can be rotated to scan the slits parallel to a radial line that passes through the field-of-view. Heliocentric heights between about 1.5 and 3.5 $R_\odot$ can be selected. The spacecraft can be offset pointed to allow observations down to 1.39 $R_\odot$, and onto the solar disk. The spacecraft is rotated to place the field-of-view at the desired position angle.

The uv instrument, along with the White Light Coronagraph, is housed in an evacuated cylinder with an aperture door that is opened after the Spartan is released from the shuttle. Spartan is battery powered, has no command or data links, operates autonomously executing a preprogrammed set of commands, and stores the data on magnetic tape. Four coronal targets corresponding to four primary position angles, can be selected just prior to deployment. The first is determined by the Spartan orientation at release, and the others by attitude adjusts that are keyed into the spacecraft controller about 12 hours before deployment.

3. Correlated Spartan 201 and Ulysses Observations

The combination of EUV spectroscopy in the solar wind acceleration region with Spartan and *in situ* measurements with Ulysses is expected to enhance the understanding of coronal dynamics and the generation of the polar solar wind. For the first time, coordinated measurements will be carried out in two important domains of the expanding solar atmosphere, thus providing much more severe constraints on solar wind theories and numerical expansion models than was hitherto possible. Spartan 201 will give information on the most significant physical parameters in the solar wind acceleration region such as densities and temperatures for electrons and different ion species. Spartan is expected to observe some key ions gaining momentum as they form the solar wind. The *in situ* instruments will accurately and in detail establish the asymptotic properties of the polar solar wind, properties that to a large extent are determined by processes taking place in the coronal regions that are covered by Spartan 201. Solar wind properties of particular importance for comparative studies are the bulk speed and individual speeds of ion species, the charge state distributions for several different elements, and variations in elemental abundances.

It was shown many years ago (*cf.* Geiss *et al.* 1970) that Coulomb collisions contribute significantly to the acceleration of heavy ions into the solar wind. In an environment where Coulomb collisions are dominant, heavy ions (mass A, charge Z) tend to be slower than protons by $\Delta v \sim (2A\text{-}Z\text{-}1)/Z^2$. Evaluation of this expression for the coronal ions measured by Spartan shows that hydrogen (H^+, and thus also H^0) should have the highest velocity, and O^{5+} should be slower than most other ions. In contrast to this, the heavy ions in the solar wind (at 0.3 to 1 AU) sometimes have excess velocities over the proton velocity of up to the local Alfven speed (Neugebauer 1981, Schmidt *et al.* 1980), a fact that is attributed to acceleration by waves (*cf.* Marsh *et al.* 1982). So far it is not known how exactly these waves interact with the plasma (Isenberg and Hollweg 1983) and at what distance from the Sun this interaction begins to dominate, except that this has to be inside 0.3 AU (the Helios perihelion). Measuring the speeds of ions with such different collisional behavior as H and O^{5+} should therefore reveal at which radial range the transition from a collision dominated to a wave-dominated solar wind takes place.

Measurements of the solar wind composition by Ulysses will complement the EUV spectroscopy in the polar corona. The charge state distribution of the elements in the solar wind result from a freezing-in process occurring within the first few solar radii (Hundhausen *et al.* 1968, Bame *et al.* 1974, Burgi and Geiss 1986). They carry the signature of the coronal electron ionization balance at the freezing-in site. Comparison of those quantities to the same parameters derived by spectroscopic techniques is expected to

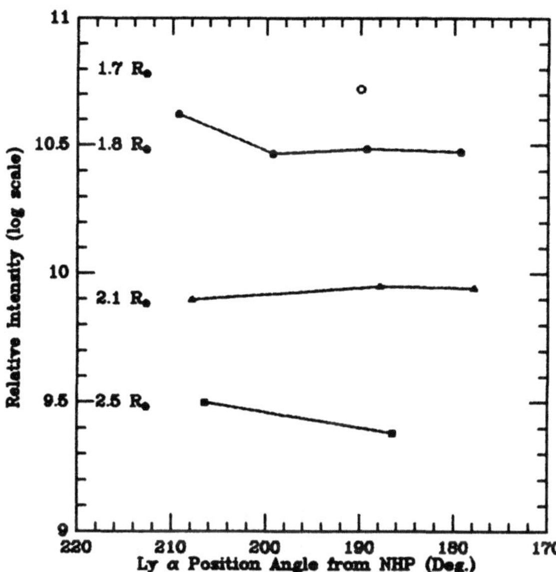

Fig. 2. Line-integrated HI Lyα intensity measurements obtained in the south coronal hole at heliocentric heights of 1.7, 1.8, 2.1, and 2.5 $R_{\odot}$. Position angles are measured counter-clockwise from Heliographic North.

provide new insights into the physical processes responsible for establishing the properties of the polar solar wind. Measurements of the charge state composition of many ions of C, N, O, Ne, Mg, Si and Fe in the solar wind, which are affected in different ways by the electron environment at various coronal altitudes, will therefore place many additional constraints on a coronal model compared to a model obtained from either coronal or solar wind spectroscopy alone.

Correlated Spartan and Ulysses observations during the Spartan missions will be used to map solar wind streams from their source regions through interplanetary space. This information will be used to identify source regions of high speed solar wind streams, compare mass fluxes measured near the Sun and at about 2 AU, and generate theoretical solar wind models with empirical constraints near the Sun and at 2 AU.

4. Preliminary results from the first flight of Spartan 201

Spartan 201-1 was deployed from the space shuttle on 11 April 1993 and spent about 40 hours observing the extended solar corona. The four primary targets for Spartan Mission 201-1 were the north polar region, the south polar region, a helmet streamer at position angle 135° and an active region above the west limb. Center to limb scans (from −8.0 arcmin to the limb) of the HI Lyα profile and the OVI intensity were made at position angles

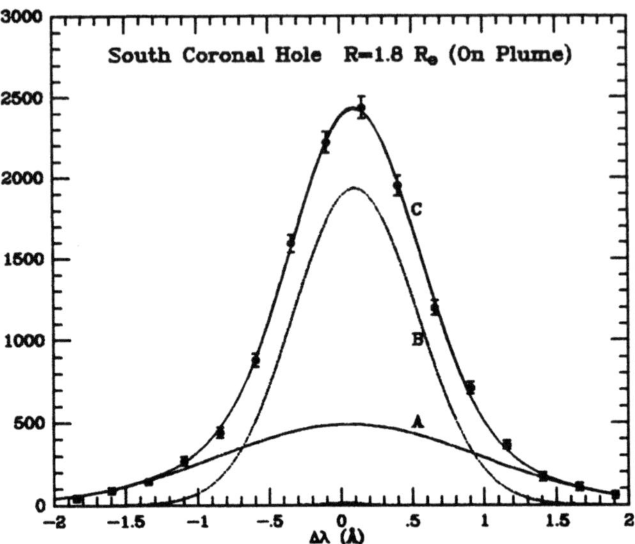

Fig. 3. The observed HI Lyα line at a position angle of 210° and a heliocentric height of 1.8 R$_\odot$. The solid points are the raw data (in counts/0.25 Å), which have 1 σ statistical uncertainties. The solid curve is a fit to the data by two Gaussian curves (A & B) which make up the coronal profile C (long dashes).

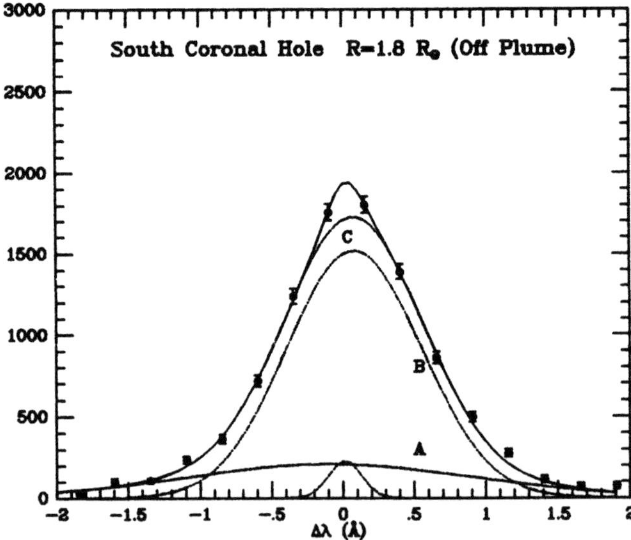

Fig. 4. The observed HI Lyα line at a position angle of 200° and a heliocentric height of 1.8 R$_\odot$. The solid points are the raw data (in counts/0.25 Å), which have 1 σ statistical uncertainties. The solid curve is a fit to the data by two Gaussian curves (A & B) which make up the coronal profile C (long dashes) and a Gaussian for the geocoronal contribution (short dashes).

between 281° and 352°. There also were measurements of the HI Lyα geocorona at five orientations between 95° and 160° from the Sun-center direction. Background measurements were made while the instrument was in the Earth's umbra and facing the Earth. In order to distinguish between a light signal and background, data were accumulated with two very different areas of the telescope mirrors uncovered.

Figure 1 indicates all of the solar spatial elements observed during the mission. Concentrating on the smallest rectangles, which indicate the HI Lyα profile measurements, it can be seen that the four targets were sampled extensively, but not completely. The most complete coverages were for the helmet streamer at 135° and for the south polar region.

Observations by the Spartan White Light Coronagraph indicate that the north and south polar regions are coronal holes with substructures consisting of polar plumes or rays that extend outward to heliocentric heights in excess of 5 $R_\odot$ (Guhathakurta, *et al.* 1994).

Figure 2 is a plot of measured HI Lyα intensities versus position angle in the south polar coronal hole for heliocentric heights between 1.7 and 2.5 $R_\odot$. Notice that the point at position angle 210° and height 1.8 $R_\odot$ appears to be significantly brighter than the other observations at the same height. Comparison to images from the White Light Coronagraph indicates that the observation at 210° is on a polar plume and the other observations at that height appear to be between plumes.

The observed HI Lyα profile at 1.8 $R_\odot$ and 210° is shown in Figure 3 and the HI Lyα profile at 1.8 $R_\odot$ and 200° is shown in Figure 4. The central portion of each profile must be corrected for the effects of the geocoronal HI Lyα line, although the correction may be negligible as for the profile in Figure 3.

In general, the solar coronal profiles have complex shapes with enhanced line wings relative to a Gaussian curve. There are several possible explanations for the enhanced line wings of the latter profile. One possibility suggested by the work of Scudder (1992) is that the protons have a non-Maxwellian velocity distribution. Another possibility is that the observed profile is formed in two distinct regions. The latter explanation can be pursued by fitting the observed profile to two Gaussian components (plus a narrow component due to the geocorona, Withbroe, *et al.* 1985). The result is a narrow component corresponding to a kinetic temperature of 1.6×10^6 K and a weaker broad component corresponding to a kinetic temperature of 7.6×10^6 K. Here the kinetic temperature, T_k, includes the effects of both thermal and nonthermal motions and is related to the 1/e width, $\Delta\lambda = \frac{\lambda_o}{c}\left(\frac{2kT_k}{m_p}\right)^{1/2}$, where the symbols k, c, m_p have their usual meanings and λ_o is the Ly α line center in the hydrogen rest frame (Withbroe, *et al.* 1985). Continuing with this line of reasoning, the narrow component might be taken as the coronal hole contribution and the broad component might be assigned to a high temperature region in the foreground/background or interspersed within the coronal hole. However, it is unlikely that a 7.6 million degree region would produce the broad Ly α component. A region with such a high proton kinetic temperature at 1.8 $R_\odot$ would be expected to have a similarly high electron temperature. However, the ionization fraction of neutral hydrogen in such a region would be very small and the density to account for the observed HI Lyα intensity would necessarily be much higher than typically observed in polar regions. Another explanation for the broad component is that the line wings are due to bulk outflow velocities in the broad coronal hole. The idea is that the line-of-sight component of the outflow velocity in the region of the coronal hole nearest and furthest from the observer tends to broaden the observed line profile. This is an unlikely explanation for the present observation because the Doppler half-width of 1.44 Å, corresponds to a velocity of about 150 km s^{-1} which is an unexpected and extremely high value for the component along the line-of-sight.

Although detailed modeling is required to reach a more definitive conclusion, it appears that the most likely explanation for the observed profile is a non-Maxwellian velocity distribution in the polar coronal hole. The wavelength scale in Figure 4 can be converted to a scale for line-of-sight velocities using the usual relationship among wavelength shift, velocity, and the speed of light. The resulting curve (plot not shown) is a good approximation to the velocity distribution of neutral hydrogen over the line-of-sight. (The correction for the angular dependence of resonant scattering is believed to be small.) Using this approach, the derived rms velocity along the line-of-sight is 160 km s^{-1}, which is strongly influenced by the line wings. An alternate parameterization is that of Scudder (1992), which provides parameters describing a central line core with enhanced wings.

Figure 5 is the difference between the profile observed on the plume (*cf.* Figure 3) and the one observed in the gap between plumes (*cf.* Figure 4). This profile is approximately attributable to the plume, where it was assumed that the coronal hole contribution to the profile in Figure 3 is identical to the profile in Figure 4. This assumes that the geometry of the coronal hole along both lines-of-sight are the same, and that the plume occupies a negligible fraction of the coronal hole along the line-of-sight. The difference profile has a complex shape with enhanced wings over a Gaussian profile. The two component fit shown in Figure 5 corresponds to kinetic temperatures of 8.9×10^5 and 5.3×10^6 K. An explanation in terms of two components has problems similar to those considered for the profile in Figure 4. Again, the implied electron temperature would result in a very small neutral hydrogen fraction that would require a very high density to explain. It appears that the plume also has a non-Maxwellian velocity distribution. The conversion to a velocity distribution gives a rms velocity of 137 km s^{-1}.

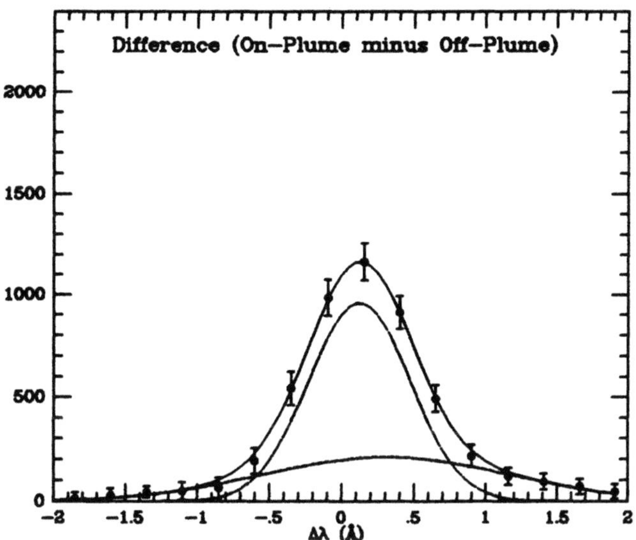

Fig. 5. The HI Lyα profile derived from the difference between the coronal profiles in Figs. 3 and 4. The solid points are the differences and have 1 σ statistical uncertainties. The solid curve is a two-component fit to the data with two Gaussian curves. The narrow component (long dashes) has a 1/e width of 0.49 Å and the broad component (short dashes) has a 1/e width of 1.20 Å.

This is smaller than the coronal hole value. It is also evident that the core of the distribution is narrower. Hence, we are left with the impression that the observed polar plume is cooler than the adjacent coronal hole or has smaller non-thermal velocities, and also likewise has an rms velocity that is much larger than that found in a Maxwellian distribution with a similar core distribution.

5. Plans For Spartan 201 Observations During The Ulysses South Polar Pass

The primary scientific goal of the Ultraviolet Coronal Spectrometer for the south polar pass is to characterize solar wind source regions near the south solar pole. At the time of the Spartan deployment from the space shuttle during Mission STS-64, Ulysses will be at approximately −80° heliographic latitude and about 2 AU from the Sun. Four coronal targets will be selected just prior to deployment from the Shuttle. The observing sequences for each target will be similar to those shown in Figure 1. The current plan is to observe the expected south polar coronal hole as was done on the first flight and the adjacent coronal structures which are expected to be coronal streamers. The fourth target will be either the north polar region or a second observation of the south polar region after a time lapse of about 24 hours.

Instruments on Ulysses will make measurements of the solar magnetic field in the vicinity of the spacecraft, the dynamic behavior of solar wind particles and the elemental and ionic-charge composition, temperatures and mean speeds of the most important solar wind ions from H through Fe. In addition to Spartan 201 and Ulysses, coordinated observations with other space-borne and ground-based instruments are anticipated and encouraged. For example, the Yohkoh satellite is expected to provide x-ray observations, the Mauna Loa Solar Observatory Mark III Coronagraph will provide polarization brightness observations of the visible corona up to 2 $R_\odot$, the Mees Solar Observatory is expected to provide magnetograms and the National Solar Observatory can provide coronagraphic observations in Fe X and Fe XIV radiations. It is also expected that radio observations will be carried out with the VLA radio telescope and with the EISCAT radio telescope network.

Acknowledgements

This work was supported by NASA under Grant NAG5-613 to the Smithsonian Astrophysical Observatory.

References

Bame, S. J., Asbridge, J. R., Feldman, W. C., and Kearney, P. D.: 1974, *Solar Physics* **35**, 137
Burgi, A. and Geiss, J.: 1986, *Solar Physics* **103**, 347
Fisher, R. and Guhathakurta, M.: 1994, *Space Science Reviews* , in press
Geiss, J., Hirt, P., and Leutwyler, H.: 1970, *Solar Physics* **12**, 458
Guhathakurta, M., Fisher, R., Strong, K., and Acton, L.: 1994, *EOS, Transactions, AGU, Suppl.* **75, no. 16**, 276
Hundhausen, A. J., Gilbert, T. E., and Bame, S. J.: 1968, *Astrophysical Journal* **152**, L3
Isenberg, P. A., and Hollweg, J. V.: 1983, *J. Geophys. Res.* **88**, 3923
Kohl, J. L., and Noci, G.: 1992, 'UVCS – The Ultraviolet Coronal Spectrometer for SOHO' in Coronal Streamers, Coronal Loops, and Coronal and Solar Wind Composition, ed(s)., *C. Mattoc*, ESA/ESTEC: Noordwijk, 23
Marsh, E., Goertz, C. K., and Richter, K.: 1982, *J. Geophys. Res.* **87**, 5030
Neugebauer, M.: 1981, *Fundamentals of Cosmic Physics* **7**, 131
Noci, G., Kohl, J. L., and Withbroe, G. L.: 1987, *Astrophysical Journal* **315**, 706
Scudder, J. D.: 1992, *Astrophysical Journal* **398**, 319
Schmidt, W. K. E., Rosenbauer, H., Shelley, E. G., Geiss, J.: 1980, *J. Geophys. Res. Lett.* **7**, 697
Withbroe, G. L., Kohl, J. L., Weiser, H., and Munro, R. H.: 1982, *Space Science Reviews* **33**, 17
Withbroe, G. L., Kohl, J. L., Weiser, H., and Munro, R. H.: 1985, *Astrophysical Journal* **297**, 324

ON THE DERIVATION OF EMPIRICAL LIMITS ON THE HELIUM ABUNDANCE IN CORONAL HOLES BELOW 1.5 R_S

SHADIA RIFAI HABBAL and RUTH ESSER
Harvard-Smithsonian Center for Astrophysics

Abstract. We present a simple technique describing how limits on the helium abundance, α, the ratio of helium to proton number density, can be inferred from measurements of the electron density, temperature and their gradients below 1.5 R_s. As an illustration, we apply this technique to emission line intensities in the extreme ultraviolet, measured in polar coronal holes. The example indicates that α can be significantly large in the inner corona. This technique could be applicable to the more extensive data to be obtained from coordinated ground and space-based observations during the Ulysses south polar passage and the Spartan flight, and subsequently during the SOHO mission. Limits on the helium abundance in the solar wind can thus be derived from its source region and compared to interplanetary values.

Key words: – Helium Abundance – Solar Wind

1. Introduction

The helium abundance, defined as the ratio of the helium to proton densities, as observed at 1 AU, is relatively constant in high speed solar wind streams, averaging around 4 to 5%. On the other hand, it is much more variable in low speed winds (see, e.g., Neugebauer and Snyder 1966). Studies of three fluid models of the solar wind show that observations both in the corona and at 1AU do not preclude the possibility that very large helium abundances, for example > 100%, could be present at the lower corona (see, e.g. Joselyn and Holzer 1978; Bürgi and Geiss 1986; and references therein). Furthermore, recent studies have shown that the presence of helium in the solar wind might reduce the sensitivity of the solar wind mass flux to the temperature at the coronal base (Bürgi 1992, Leer, Holzer and Shoub 1992).

So far, however, attempts to derive constraints on the helium abundance in the inner corona from observations have not been successful (see, e.g., Parkinson and Gabriel 1986). We present in this paper a simple technique illustrating how such limits can be derived from knowledge of the electron temperature, density and their gradients in the lower corona.

2. Theoretical Concept

For a multispecies solar wind plasma composed of electrons, e, protons, p, and fully ionized helium, (or alpha particles) α, the force balance equation in the very inner corona (below 1.5 R_s) where the flow speeds are negligibly

Space Science Reviews **72**: 39–44.
© 1995 *Kluwer Academic Publishers.*

small, is given by

$$0 = -\frac{dp}{dr} - \rho \frac{GM_s}{r^2} , \qquad (1)$$

where G is the solar constant, M_s the solar mass, and r the heliocentric distance. The mass density, ρ, is given by

$$\rho \approx m_p(n_p + 4n_\alpha) , \qquad (2)$$

The total pressure, p, is given by

$$p \approx (n_e + n_p + n_\alpha)\, k_B\, T , \qquad (3)$$

with the assumption that $T_e = T_p = T_\alpha = T$ is valid very close to the coronal base where the coupling between the species is expected to be strong. Charge neutrality dictates that

$$n_e = n_p + 2n_\alpha . \qquad (4)$$

Using the definition of the helium abundance, α, as the ratio of the helium ion to proton densities, i.e.

$$\alpha = \frac{n_\alpha}{n_p} , \qquad (5)$$

and eliminating n_p from the expressions for ρ and p, Eq. (1) can be rewritten as:

$$\frac{1}{n_\alpha}\frac{dn_\alpha}{dr} = (\frac{1+4\alpha}{\alpha})\frac{m_p G M_s}{k_B r^2}\frac{1}{T} + (\frac{2+3\alpha}{\alpha})\frac{1}{T}\frac{dT}{dr} + \frac{2(1+2\alpha)}{\alpha}\frac{1}{n_e}\frac{dn_e}{dr} . \qquad (6)$$

Eq. (6) can be used to derive limits on the helium abundance in the inner corona using empirical inferences of electron density and temperature and their gradients there, together with the condition that $dn_\alpha/dr < 0$. The procedure is as follows: For a given heliocentric distance, r, we substitute the variables and their gradients in Eq. (6) by their empirical values, then search for values of α for which the right hand side is negative. We note here that the temperature gradient in the inner corona plays an important role. If $dT/dr < 0$, Eq. (6) yields a *lower* limit to the helium abundance. For $dT/dr > 0$, an *upper* limit to the helium abundance is inferred.

3. Example: Inferences from EUV Data

There are two types of data sets which are expected to be acquired with the SPARTAN 201 (Kohl et al 1994) and SOHO missions (e.g., Domingo 1989), and from which inferences of electron density and temperature gradients in the inner corona will be derived. These consist of white light polarization brightness measurements and emission line intensities. As an illustration, we apply the procedure outlined above to the more complex case of extreme ultraviolet (EUV) line intensities.

We show first how the density gradient in Eq. (6) can be substituted by the gradient of the intensity of emission of an optically thin line. The intensity of an emission line is given by

$$I_\lambda = \int_L f(\lambda, T)\, n_e\, n_p\, ds \ , \tag{7}$$

where the emissivity, $f(\lambda, T)$, depends on the atomic data, the elemental abundance, the ionization balance, and the electron temperature (see, e.g. Raymond 1988, and references therein). We assume that the temperature and density are constant along the pathlength L which contributes to the emission along the line of sight. Expressing n_p in terms of n_e and α, the above equation can be rewritten as

$$I'_\lambda = \frac{I_\lambda}{f(\lambda, T)} = \frac{{n_e}^2}{(1+2\alpha)L}. \tag{8}$$

The literature commonly refers to I'_λ as emission measure. While I_λ is the measured quantity, $f(\lambda, T)$ can be computed theoretically. It then follows that Eq. (8) can be replaced by (for details, see Habbal and Esser 1994)

$$\frac{1}{n_\alpha}\frac{dn_\alpha}{dr} = -\frac{(1+\alpha)}{(1+3\alpha)}\left[\frac{(1+4\alpha)}{\alpha}\frac{m_p G M_s}{k_B r^2}\frac{1}{T} + \frac{(2+3\alpha)}{\alpha}\frac{1}{T}\frac{dT}{dr} + \frac{(1+2\alpha)}{\alpha(1+\alpha)}\left(\frac{1}{I'_\lambda}\frac{dI'_\lambda}{dr} - \frac{1}{L}\frac{dL}{dr}\right)\right]. \tag{9}$$

In what follows we assume that L is constant. Eq. (9) then replaces Eq. (6), and is solved as outlined in section 2, by substituting I'_λ, dI'_λ/dr, T and dT/dr by their empirically inferred values.

The data selected for the application of this technique were taken from the Harvard EUV experiment on *Skylab* (see Reeves et al. 1977). Measurements were made out to 1.08 R_s off the limb in a south polar coronal hole where two different radial directions, referred to as panels I and II in Figure 1, were selected for this study. Three emission lines were used: Mg X (625 Å), Ne VII (465 Å) and O VI (1032 Å), which have a maximum ionization fraction temperature of 1.12×10^6, 5×10^5 and 2.8×10^5 K respectively. The emission measure for each line, I'_λ, was derived from its measured intensity, I_λ, and its corresponding f_λ was computed using a code provided by J. Raymond. The average, I', of the emission measures from all three lines, is shown in panels Ia and IIa of Figure 1. The temperature profiles, shown in panels Ib and IIb of Figure 1, were inferred from the ratios of the line intensities. (See Habbal, Esser and Arndt, 1993, for details about the technique and

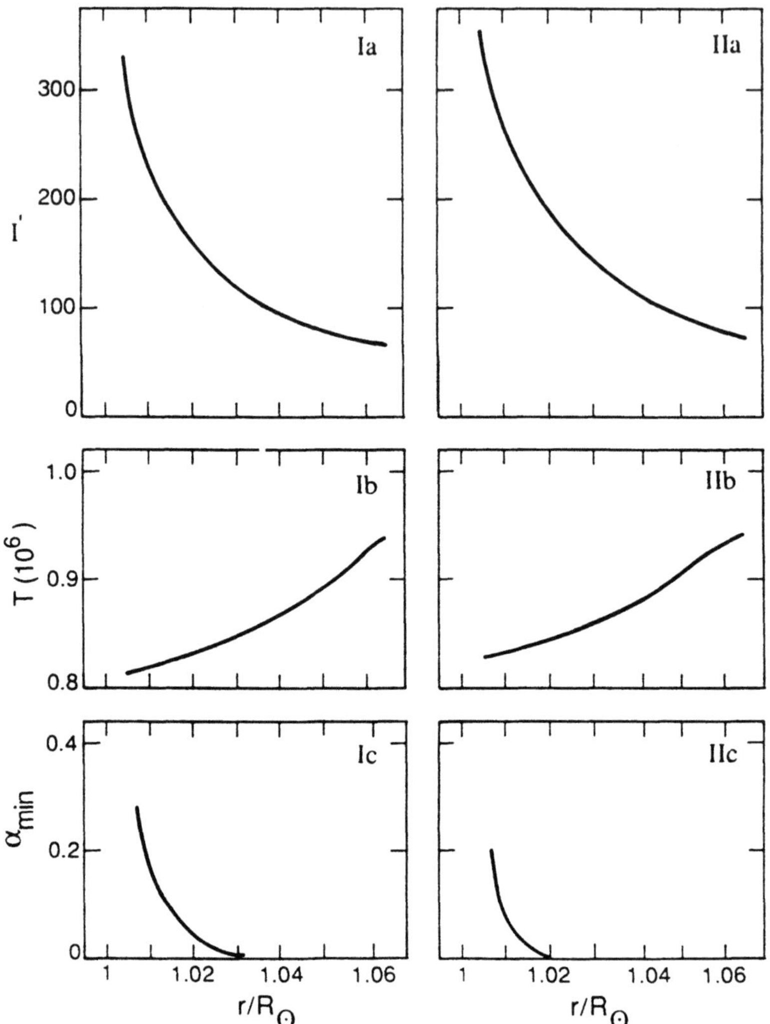

Fig. 1. Labels I and II refer to measurements and inferences along two different radial directions in a polar coronal hole. (a) Average emission measure, I', for the range of empirical values derived from emission line intensities, (b) average temperature profiles inferred from emission line ratios, (c) α_{min} inferred from the data in (a) and (b) using the technique described in the text.

uncertainties in this inference). Substituting for the values of I' and T, thus inferred, in Eq. (9), we derive a lower limit for α, α_{min}, as expected, since $dT/dr > 0$ (see panels Ib and IIb). The values of α_{min} for these two different radial directions, are shown in panels Ic and IIc. They indicate that α can be significantly large at the coronal base, i.e. at least 0.20, but drops very sharply to interplanetary values within a heliocentric distance of 1.05 R_s.

4. Discussion and Conclusion

The technique outlined in this paper shows that limits on the helium abundance, α, in the lower corona within coronal holes, where the flow speed is negligibly small, can be determined from the momentum balance equation using simultaneous empirical inferences of the coronal temperature and density as a function of heliocentric distance. Application of this technique to EUV data currently available indicates that the helium abundance, α, is at least 0.20 at the coronal base. This lower limit drops very rapidly to interplanetary values within a heliocentric distance of 1.1 R_s. In the example given, α_{min} carries at least the same uncertainty entering the empirical derivation of the profiles for I' and T, which are typically 25%.

In the derivation provided we neglected flow speeds and external body forces, as expected, for example, from MHD waves. We also made the simplifying assumption that electrons, protons and α's have the same temperature. For the distances considered in the example given, model computations indicate that such assumptions are appropriate. We also neglected the effects of thermal diffusion which might be important in the subsonic region (cf. Geiss et al 1970; Jokipii 1966). In the absence of any direct measurements, however, we cannot asses their importance except from numerical models. Ideally, self-consistent results can be obtained if model computations are carried out using the empirical derivations described above.

The other simplifying assumption used in our approach is the constancy of the density and temperature along the pathlength contributing to the emission along the line of sight. Such an assumption implies an approximately homogeneous atmosphere. To include the effects of local inhomogeneities along the line of sight would require extensive modeling. The relative importance of local inhomegeneities will be easier to evaluate when tomographic observations become available from the different instruments onboard the upcoming SOHO mission, scheduled for launch in 1995. More accurate temperature inferences are also expected from this mission, since they will be derived from ratios of emission lines produced by the same atomic species (such as with SUMER). We note, however, that the assumptions entering our computations are the same as those used to date in all inferences of plasma parameters. These assumptions will also be used in the inference of helium abundance from the measurements of helium lines if these become possible, for example, with the SOHO mission. With the very large data base to be acquired with the different instruments onboard SOHO, we expect that the technique presented here will yield reliable constraints on the helium abundance in the lower corona.

5. Acknowledgements

This work was supported by NASA grant NAGW-249, and Air Force grant AFOSR-91-0244. The authors extend their thanks to the referee for her/his useful comments.

6. References

Bürgi, A.: 1992, *J. Geophys. Res.* **97**, 3137.

Bürgi, A., and Geiss, J.: 1986, *Sol. Phys.* **103**, 347.

Domingo, V. (ed.): 1989, *The SOHO Mission - Scientific and Technical Aspects of the Instruments*, ESA SP-1104.

Geiss, J., Hirt, P. and Leutwyler, H.: 1970, *Sol. Phys.* **12**, 458.

Habbal, S. R., and Esser, R.: 1994, *Ap. J.* **421**, L59.

Habbal, S. R., Esser, R. and Arndt, M. B.: 1993, *Ap. J.* **413**, 435.

Joselyn, J. A. and Holzer, T. E.: 1978, *J. Geophys. Res.* **83**, 1019.

Jokipii, J. R.: 1966, 'Effects of diffusion on the composition of the solar corona and the solar wind' in R. J. Mackin and M. Neugebauer, ed(s)., *The Solar Wind*, Pergamon Press, New York, 215.

Kohl, J. L., Gardner, L. D, Strachan, L. and Hassler, D. M.: 1994, 'Ultraviolet spectroscopy of the extended solar corona during the SPARTAN 201 mission' in G. Poletto and G.-C. Noci, ed(s)., *Proceedings of the Second SOHO Workshop*, ESA Publication.

Lallement, R., Holzer, T. E., and Munro, R. H.: 1986, *J. Geophys. Res.* **91**, 6751.

Leer, E., Holzer, T.E., and Shoub, E. C.: 1992, *J. Geophys. Res.* **97**, 8183.

Neugebauer, M. and Snyder, C. W.: 1966, *J. Geophys. Res.* **71**, 4469.

Parkinson, J. H. and Gabriel, A. H.: 1986, *Adv. Space Res.* **6 (8)**, 243.

Raymond, J. C.: 1988, 'Radiation from Hot, Thin Plasmas' in R. Pallavicini, ed(s)., *Hot, Thin Plasmas in Astrophysics*, Kluwer Academic Publishers, 3.

Reeves, E. M., Huber, M. C. E., Timothy, J. G., and Withbroe, G. L.: 1977b, *Appl. Optics* **16**, 849.

LARGE SCALE STRUCTURE OF THE SOLAR CORONA IN THE DECLINING PHASE OF THE SOLAR CYCLE

P. LANTOS
Observatoire de Meudon, Meudon, France

and

C.E. ALISSANDRAKIS
National University of Athens, Athens, Greece

Abstract. Maps of the corona, obtained at meter wavelengths with the Nançay Radioheliograph (France), are used to study, on the disk, the radio counterpart of the coronal plasma sheet observed in K-corona on the limb. We study here the evolution of the coronal plasma sheet from the maximum of the activity cycle in 1980 to the minimum in 1986 and identify some of its large scale structures.

1. Introduction

The relationship between large scale structures of the corona, namely coronal holes and the coronal plasma sheet (the so-called streamer belt) with high speed and low speed solar winds is now well established. Nevertheless, the specific source of the slow solar wind remains unknown. Structures of the corona are observed, on the limb, in the optical range (K-corona and coronal lines) and, on the disk in X-rays, at radio wavelengths or with the Helium 1083 nm line. No measurements of the coronal magnetic field in quiet regions are presently available, but extrapolations of the photospheric field give a global structure in agreement with the observed density structure. The global coronal evolution during the sunspot cycle (Hundhausen et al, 1981) could thus be explained by the distribution of the large scale magnetic field (Hoeksema, 1984). The purpose of this paper is to consider the evolution of the corona during the activity cycle using information from radio mapping at meter wavelengths.

2. Advantages and limitations of the radio method

Disk observations at radio wavelengths have better resolution in longitude than limb observations of the K-corona. This facilitates the detailled comparison with underlaying solar chromospheric structures. Disk observations also open the possibility to follow the structure evolution from day to day. Nevertheless, since the aperture synthesis method needs a fairly constant brightness distribution over 6 to 8 hours, periods with high noise storm activity are missed. This is frequent during high solar activity.
The appearance of the corona at meter wavelengths, observed with the

Space Science Reviews **72**: 45–48.
© 1995 *Kluwer Academic Publishers.*

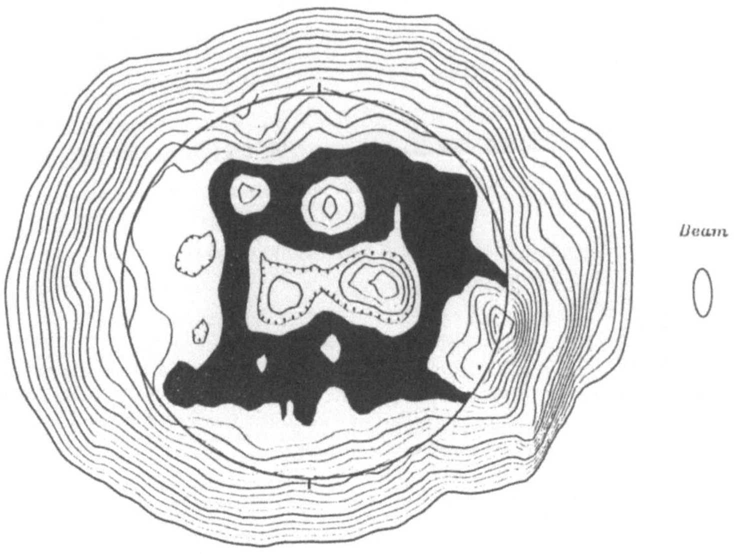

Fig. 1. Map of the Sun obtained on July 2, 1980 at 169 MHz. The coronal plateau is shown in black. Brightness depressions are indicated with hatched contours.

Nançay Radioheliograph, is roughly similar to that given by X-ray pictures (Lantos et al., 1994) but, in addition to differences due to the lower resolution of the radio instrument, structures are less precisely defined than in X-rays: scattering and refraction and even reflection (on high density structures) may affect radio propagation. Moreover, the radio brightness results from transfer through a wider range of electron temperatures than the X-rays, thus the contrast is lower. On the other hand, it appears than the coronal plasma sheet is easily detected at meter wavelengths as a bright "coronal plateau", presumably because of the integration over a wide range of temperatures and altitudes.

3. Evolution of the coronal plateau with the activity cycle

On the daily maps the coronal plateau appears as a region with constant brightness temperature surrounding almost all localized emission sources. Figure 1 shows a map obtained during the solar maximum on July 2, 1980 at 169 MHz. The brightness temperature (T_b) contours start at 10^5 K; multiples of 2. 10^5 K are shown with thick lines and the beam is given on the right side of the figure. At the center of the disk, two equatorial brightness depressions are shown with hatched isophotes. The deeper one, on the right (west), corresponds to a coronal hole detected in the Helium 1083 nm line. Two pairs of emission sources are located near the central meridian, north and south of the coronal hole. A fifth source, with T_b=1.2 10^6 K is located on the south-west limb; it is non-thermal and has been identified as a faint noise storm continuum (Alissandrakis et al., 1985). The coronal plateau, which surrounds emission sources and avoids coronal holes, is shown in black. Its shape, as expected during solar maximum, is irregular.

Very different is the shape of the coronal plateau during the solar minimum. Figure 2a shows the disk synoptic chart at 408 MHz for Carrington rotation 1776 (May 31 to June 25, 1986); the black region shows the border of the coronal plateau (T_b=4. to 4.5 10^5 K). Figure 2b shows the K-corona plasma sheet observed on the limb at Mauna Loa Solar Observatory (Sime et al., 1987). During this period of very low activity (monthly sunspot index of 1.1), the radio coronal plateau, as well as the K-corona plasma sheet straddle the solar equator. Note that the brightest radio source, located at a longitude of ~80°, corresponds to a density enhancement in the K-corona. The comparison of the two brightest radio sources in the north hemisphere with optical observations shows that they are located above old faculae in active regions without sunspots.

Figure 2c shows the disc synoptic chart at 169 MHz for Carrington rotation 1750 (from June 21 to July 16, 1984) during the declining phase of the solar cycle and figure 2d shows the corresponding K-corona chart (Fisher et al., 1985). On the radio chart, the black region indicating the border of the coronal plateau corresponds to T_b from 6.5 to 7. 10^5 K. A large coronal hole visible in the helium line is located in the northern hemisphere between longitudes 25° and 80°. It explains the depression features on both charts. The two brightest radio sources, located close to the equator at Carrington longitudes of 60° and 120°, are related to active regions with sunspots and are of non-thermal origin. The two radio sources located at 40° longitude have been identified (Lantos and Alissandrakis, 1992) as large scale loop systems having one footpoint on facular plages. They are at about the same location as the two dense K-corona regions. Thus the radio study provides identification of dense regions of the coronal plasma sheet. Note finally that because of the better resolution in longitude, radio charts indicate that the border of the plasma sheet in the northern hemisphere at longitude 100° is sharper than expected from K-corona observations and is aligned almost along a meridian. Both points are important when solar wind data are compared with coronal density structures.

4. Conclusions

The radio study of the quiet corona at meter wavelengths has proved its capability to detect the solar plasma sheet located at the base of the interplanetary neutral sheet. Radio disk observations facilitate the comparison

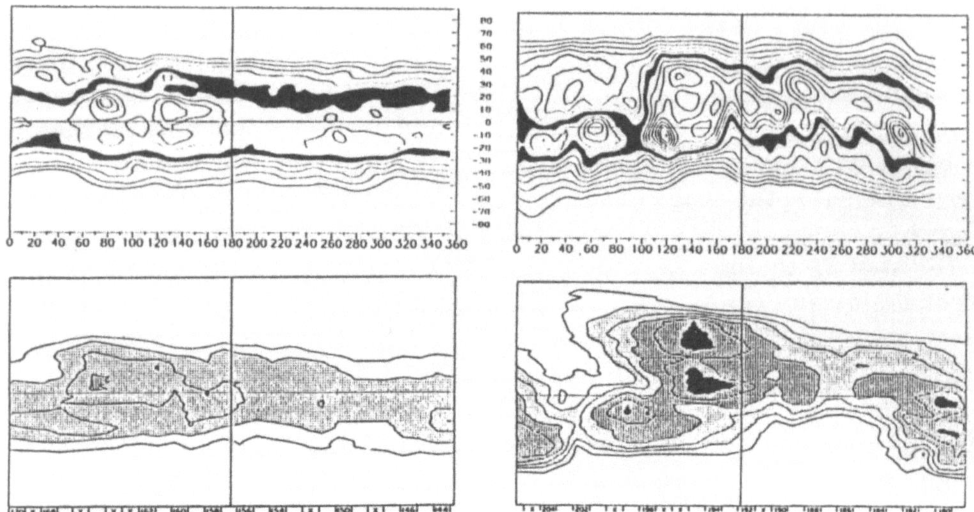

Fig. 2. Synoptic charts obtained at meter wavelengths (top) and in the K-corona (bottom). Charts 2a (at 408 MHz) and 2b are for Carrington rotation 1776 in 1986 (cycle minimum). Charts 2c (at 164 MHz) and 2d are for Carrington rotation 1750 in 1984 (declining phase of the cycle).

with optical chromospheric data and provide identifications of large scale dense structures. They also give a more realistic geometry of the coronal plasma sheet. In the future, observations at different radio wavelengths will provide, in complement to K-corona, EUV and X-ray observations, diagnostics of density and temperature in the regions which will help the study of the origin of the solar wind.

References

Alissandrakis C.E., Lantos P. Nicolaidis E.: 1985,*Solar Phys.* **97**, 267.

Fisher R., Garcia C., Lundin E., Seagraves P., Sime D.G. and Rock K.: 1985, *NCAR Technical Note TN-246*, High Altitude Observatory:Boulder, USA

Hundhausen A.J., Hansen R.T, Hansen S.F.: 1981,*J. Geophys. Res* **86**, 2079.

Hoeksema J.T.: 1984, *Thesis*, Stanford University: Palo Alto, USA

Lantos P., Alissandrakis C.E.: 1993,*Adv. Space Res.***14** , 4(97).

Lantos P., Alissandrakis C.E. and the Yohkoh Team: 1994, to be presented in COSPAR.

Sime D.G., Garcia C., Yasukawa E., Lundin E. and Rock K.: 1987, *NCAR Technical Note TN-280*, High Altitude Observatory: Boulder, USA

ORIGIN OF THE SOLAR WIND FROM COMPOSITION DATA

J. GEISS
Physikalisches Institut, University of Bern, Sidlerstrasse 5, CH–3012 Bern, Switzerland

G. GLOECKLER
Department of Physics and Astronomy, University of Maryland, College Park, MD 20742, USA

and

R. VON STEIGER
Physikalisches Institut, University of Bern, Sidlerstrasse 5, CH–3012 Bern, Switzerland

Abstract. The ESA/NASA spacecraft Ulysses is making, for the first time, direct measurements in the solar wind originating from virtually all places where the corona expands. Since the initial two polar passes of Ulysses occur during relatively quiet solar conditions, we discuss here the three main regimes of quasi-stationary solar wind flow: the high speed streams (HSSTs) coming out of the polar coronal holes, the slow solar wind surrounding the HSSTs, and the streamers which occur at B-field reversals. Comparisons between H-α maps and data taken by Ulysses demonstrate that as a result of super-radial expansion, the HSSTs occupy a much larger solid angle than that derived from radial projections of coronal holes. Data obtained with SWICS-Ulysses confirm that the strength of the FIP effect is much reduced in the HSSTs. The systematics in the variations of elemental abundances becomes particularly clear, if these are plotted against the time of ionisation (at the solar surface) rather than against the first ionisation potential (FIP). We have used a superposed-epoch method to investigate the changes in solar wind speed and composition measured during the 9-month period in 1992/93 when Ulysses regularly passed into and out of the southern HSST. We find that the patterns in the variations of the Mg/O and O^{7+}/O^{6+} ratios are virtually identical and that their transition from high to low values is very steep. Since the Mg/O ratio is controlled by the FIP effect and the O^{7+}/O^{6+} ratio reflects the coronal temperature, this finding points to a connection between chromospheric and coronal conditions.

1. Introduction

The source of the solar wind (SW) is the outer convective zone (OCZ) of the sun which is a large and well-mixed reservoir with a composition that remains stable over millions of years. Significant changes in composition are produced in the transport of solar matter from the OCZ to the corona, and in the process of lifting the coronal plasma out of the Sun's gravitational field. The question of the "origin of the solar wind" would be solved, if the ways and means of supplying matter, momentum and energy to corona and SW were quantitatively specified. Since Parker (1960, 1963) showed that under a broad range of physical conditions the solar corona expands into a supersonic outflow, the solar wind and – to a lesser extend – its composition have been experimentally investigated in considerable depth, leading to an understanding of many aspects of the dynamics of coronal expansion. However, we do not yet have a self consistent theory of material transport, coronal heating and solar wind acceleration, which reproduces the major observations of the flow of matter, momentum and energy, explaining at the same time the observed ion composition.

Space Science Reviews **72**: 49–60.
© 1995 *Kluwer Academic Publishers.*

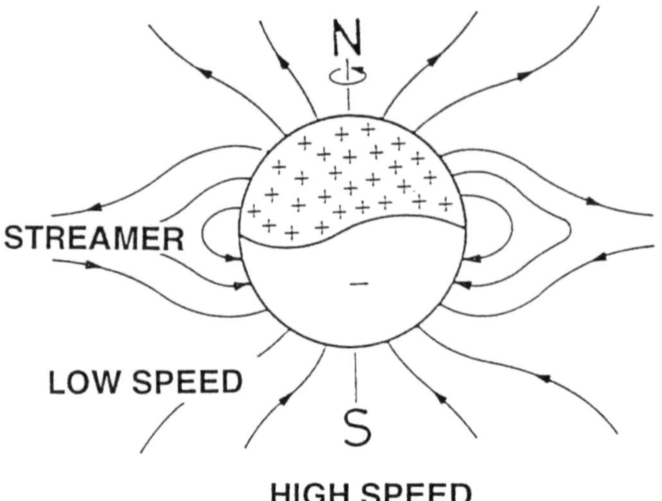

Fig. 1. Sketch of steady state solar wind expansion with dipole-type geometry.

Transient phenomena in the solar wind are often the rule rather than the exception. Is it then that the picture of a quiet SW is inappropriate and steady state theories are of questionable use even as an approximation? Observations show us that this is not the case: The solar wind never disappeared during the three decades it has been observed, and the flux variations were limited in range. In particular, in the 1970s the coronal holes were shown to emit high speed streams (HSSTs) (cf. Krieger et al., 1973) that were remarkably steady in their bulk properties and helium abundance (Bame et al., 1977). Since we approach now the solar minimum and since Ulysses' trajectory gives us the unique opportunity to study the solar wind at the high latitudes including the large HSSTs, we shall focus in this paper on the basic features of steady solar wind flow: The polar HSSTs, the slow SW surrounding it, and the streamers which occur at B-field reversals. More general and detailed accounts of the results of solar wind composition research are found in Neugebauer (1981), Schwenn and Marsch (1990), Geiss (1985), Bochsler and Geiss (1989), Gloeckler and Geiss (1989), and von Steiger and Geiss (1994).

A solar dipole field stretched out by the supersonic outflow (Fig. 1) is the simplest geometric configuration in which the three basic features of a steady SW flow can be described (cf. Hundhausen, 1972; Geiss and Bochsler, 1986). More complex geometries are often required for a realistic picture of the expansion (cf. Wang et al., 1993). However, for our discussion, the simple steady-state geometry we shown in Fig. 1 is a useful guide.

Models of coronal structure cannot be brought into agreement with the observations, unless the finite abundance of helium is taken into account (Noci and Porri, 1983). Thus a three-component plasma consisting of protons, alphas and electrons has to be used in corona expansion models, with the other ions treated as minor components (Bürgi and Geiss, 1986). Even before it enters the corona,

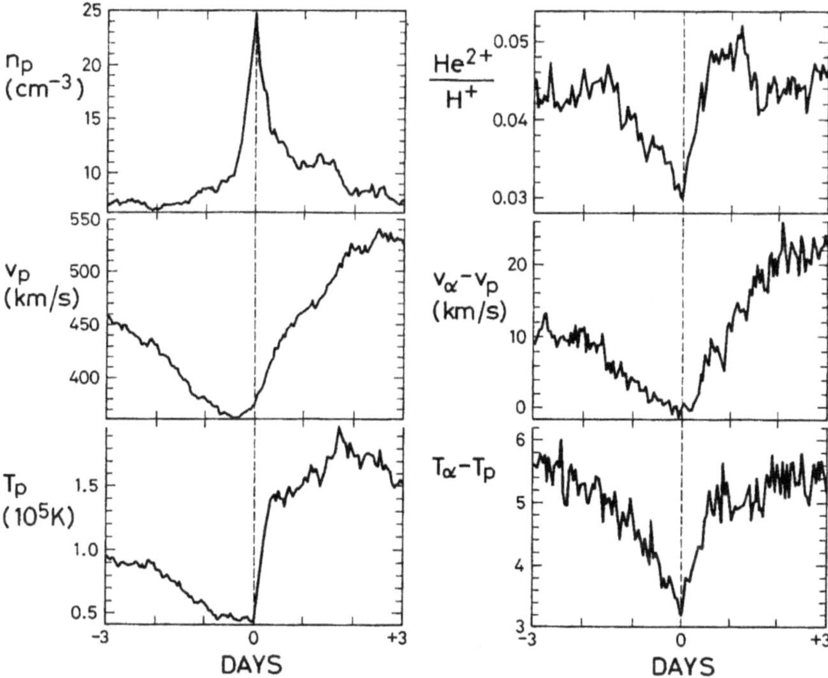

Fig. 2. Superposed epoch plots for 74 sector boundaries, showing a significant drop in the He/H ratio at the sector boundary, along with the characteristic changes of other solar wind parameters (from Borrini et al., 1981).

the solar gas is fractionated: Transport through the chromosphere into the corona causes an overabundance of elements with low first ionisation potential, the so called FIP-effect. Sofar, fractionation in the chromosphere has been considered independently from coronal heating and expansion. In this paper we present evidence for a causal connection.

2. The Slow Solar Wind in Streamers

There is a characteristic signature of the solar wind flow in the streamers, as has been demonstrated by Borrini et al. (1981) using the superposed epoch method (Fig. 2). Speed, kinetic temperature and He/H ratio have a minimum at the center of the current sheet which is identified by the B-field reversal. Schmid et al. (1988) found that also the iron abundance decreases near the B-field reversals. Although iron is much heavier than helium, the decrease for iron was not more pronounced than it was for helium. The reduced abundance of the heavier ions is – at least in part – due to the special B-field geometry: Near the sun, the expansion in streamers is probably supra-radial as Fig. 1 indicates, and consequently the SW speed there is exceptionally low allowing for gravitational de-mixing of helium and heavier ions from hydrogen. Bürgi (1992) has developed a model for the expansion of a three-fluid (p-α-e) plasma and shows that in streamer-type geometries strong depletions of helium are indeed to be expected. It would be worthwhile to resume the studies pioneered by Borrini et al. (1981) and include some more solar wind parameters. If, for instance, the O^{6+} and O^{7+} abundances could be measured as a function of the distance from the field reversal, modelling of the streamer would be strongly constrained. The O^{7+}/O^{6+} would narrow the choice for the coronal temperature, and the O/He/H abundance ratio would give direct information on the expansion kinematics, as may be demonstrated by the following simple estimates.

The degree of demixing is governed by a competition between the speed of settling and the expansion speed (or a turbulent mixing parameter, as in the Earth's atmosphere). In an isothermal proton-electron plasma V_s, the speed of settling of a heavier ion (A, Z) under the influence of the solar gravity, the electric field and non-resonant momentum transfer from waves is given approximately as (cf. Geiss et al., 1970)

$$V_s = \text{const.}\ \frac{2A - Z - 1}{Z^2} \tag{1}$$

For the main ions of He and O, we obtain $V_s(\mathrm{He}^{2+})/V_s(\mathrm{O}^{6+}) = 1.8$. Thus if the settling speeds barely overcome the expansion speed, oxygen would be less depleted than helium.

The opposite is expected if expansion speed and/or turbulent mixing are so slow that a real gravitational stratification takes place. This can be seen from comparing the scale heights of the heavy ions (again for an isothermal proton-electron plasma)

$$H = \frac{kT}{m^\star g^\star} \tag{2}$$

where $g^\star$ is the local gravitational acceleration reduced by the acceleration due to non-resonant wave interactions and $m^\star = (A - Z/2)m_0$ is the effective mass. Equation (2) gives $H(\mathrm{O}^{6+})/H(\mathrm{He}^{2+}) = 0.23$. The difference in the numerical factors in Eqs. (1) and (2) shows that the two different kinematic situations, which have been called "dynamic fractionation" and "static stratification" (Geiss, 1982), could be distinguished by H/He/O measurements at the B-field reversals. Acceleration and expansion in the streamers may well be more complex than indicated in Fig. 1. However, in any case, as shown by our simple estimates, measuring oxygen ions and perhaps other species would not simply provide redundancy, but would add additional, independent constraints on streamer models.

The dynamic fractionation and the static stratification mechanisms both predict local overabundances at some altitude in the corona. In fact, SW expansion models generally lead to low speeds and local enrichment of heavier species in the subsonic regime, quite independent of geometry and model assumptions (cf. Geiss et al., 1970; Bürgi and Geiss, 1986). The existence of such enrichments in the low corona is confirmed by the high abundances of helium and heavier ions in the "driver gas" that follows flare-generated interplanetary shocks (Hirshberg et al., 1970; Hundhausen, 1972; Galvin et al., 1987, 1993; Ipavich et al., 1986, 1992).

3. Corona Expansion at High Latitude

Extensions of the southern polar coronal hole regularly reached the ecliptic plane during the solar minimum in the 1970s, and thus important characteristics of the SW in HSSTs were established at that time. It was shown, for instance (Bame et al., 1977), that the He/H ratio in the HSSTs, though less less variable than in the slow SW, was still a factor of two below the solar ratio of 0.1. This important finding led to the suggestion that the systematic helium deficit in the SW is to a large part produced in the chromosphere (Geiss 1982). Measurements in flare particle populations by Reames (1994) and others have indirectly shown that the He/O ratio in the low-latitude corona is indeed lower than the solar ratio, which places a further constraint on mechanisms that produce the FIP effect (Sect. 4.).

In 1992 Ulysses, with the Solar Wind Plasma Experiment "SWOOPS" (Bame et al., 1992) and the Solar Wind Ion Spectrometer "SWICS" (Gloeckler et al., 1992) on board, began to pass regularly into and out of the HSST that originates in the Sun's southern coronal hole. In Fig. 3 we present SWICS data for the 13-month period of transition from the slow SW to the HSST, beginning with short excursions into the coronal hole stream and ending with permanent residency in the HSST. An anticorrelation between $T(O^{7+}/O^{6+})$ and solar wind speed is found, confirming that the high speed stream comes from a low-temperature region in the corona, i. e. the southern coronal hole. In Fig. 4 we have drawn the Ulysses trajectory into a sketch that is based on the H-α chart of Carrington rotation 1859, distinguishing times with V_α above or below 600 km/s.

Whereas Fig. 3 shows, by the relatively close correlation between solar wind velocity and $T(O^{7+}/O^{6+})$, that the source region of the HSST is coincident with the coronal hole, Fig. 4 demonstrates – as an example – that the solid angle filled by the HSST in interplanetary space is much larger than the solid angle occupied by the coronal hole right above the solar surface. This has been inferred some time ago from in-ecliptic HSST observations (cf. Kopp and Holzer 1976, Bame 1977, Geiss and Bochsler 1986), but a comparison of the Ulysses data with coronal maps should allow a rather direct estimate of the degree of super-radial expansion which is important for the modelling of high velocity streams. Bürgi and Geiss (1986) solved the momentum equation for a three-fluid plasma with heavier ions as minor components, allowing for super-radial expansion. Using their calculations of solar wind speed and freeze-in temperatures, they confirmed earlier conclusions (e. g. Hollweg 1978) that significant (non-resonant) momentum transfer from waves is needed to bring the low temperature and density conditions in coronal holes into agreement with the high velocities observed in

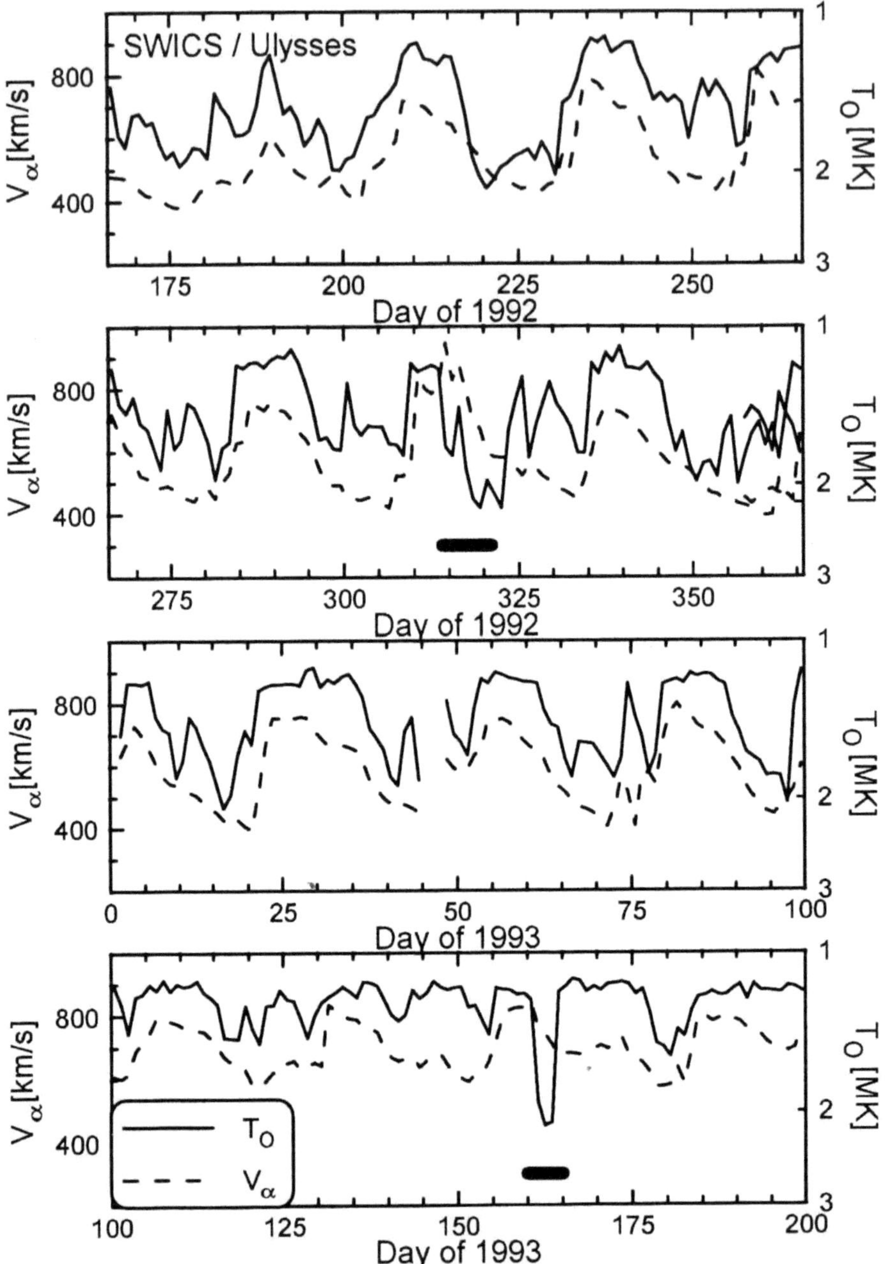

Fig. 3. Recurring transitions of Ulysses between the region with slow SW and the HSST originating in the southern polar corona hole. The SW speed is represented here by the α-speed. T_O, the O^{7+}/O^{6+} freezing-in temperature is drawn inversely for better visualizing the anticorrelation between V_α and T_O. Two periods are marked to indicate CMEs.

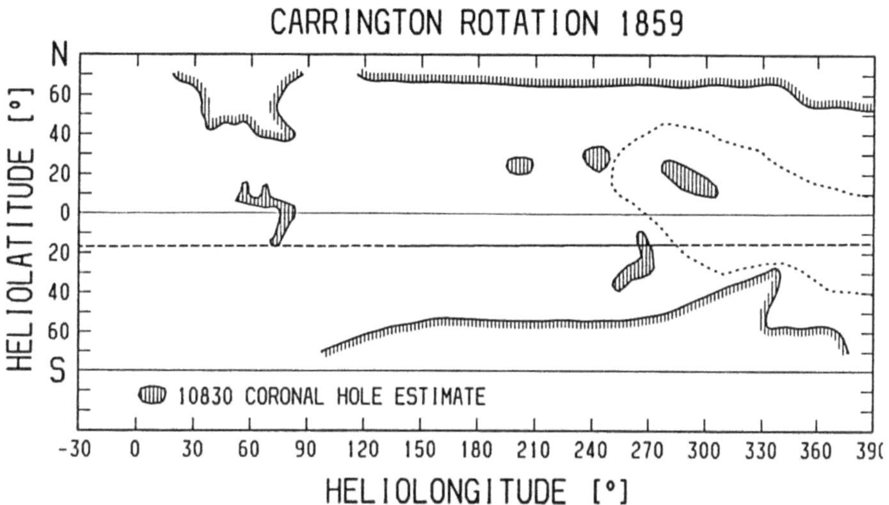

Fig. 4. Ulysses' trajectory (16° to 18° southern latitude) is drawn into the (simplified) H-α map for Carrington Rotation 1859 (NOAA 1992), distinguishing times with α-speed above (solid line) and below (dashed line) 600 km/s. The spacecraft event time was corrected for the difference between the Earth's and Ulysses' solar longitude, as well as for the SW travelling time. The extensions of the large polar coronal holes and some smaller ones are given. The solar activity is increased in the area marked by the dotted line as compared to other areas shown on this map.

the HSSTs.

There is an indication in Fig. 4 of an influence of active regions on the expansion of the coronal hole plasma into interplanetary space: A low SW velocity and high $T(O^{7+}/O^{6+})$ are observed by Ulysses relatively close to the edge of the coronal hole when there is significant activity at mid-latitude (250° to 390° longitude in Fig. 4). On the other hand, a high SW velocity and low $T(O^{7+}/O^{6+})$ are observed at latitudes far away from the edge of the coronal hole when there is no significant activity in the neighbourhood (130° to 240° longitude in Fig. 4). Thorough comparisons of this kind ought to be useful to understand the effects of lateral pressure gradients on the three-dimensional expansion of the corona.

4. Element Fractionation in the Chromosphere: Variation in the FIP Effect

Remote observations and measurements of the SW and of flare particles give a consistent picture of an overabundance of elements with low first ionisation potential (FIP), cf. Meyer (1992): Although there are some significant variations, a basic pattern of this FIP effect has been established: Elements with FIP roughly below 10 eV are overabundant by a factor of about 4 relative to the elements with FIP between $\sim$ 11 eV and 24 eV (cf. Garrard and Stone, 1993; Reames, 1994; Bochsler and Geiss, 1989). On the other hand, helium is underabundant (by a factor of $\sim$ 2) relative to these elements (cf. Geiss, 1982; Meyer, 1992;

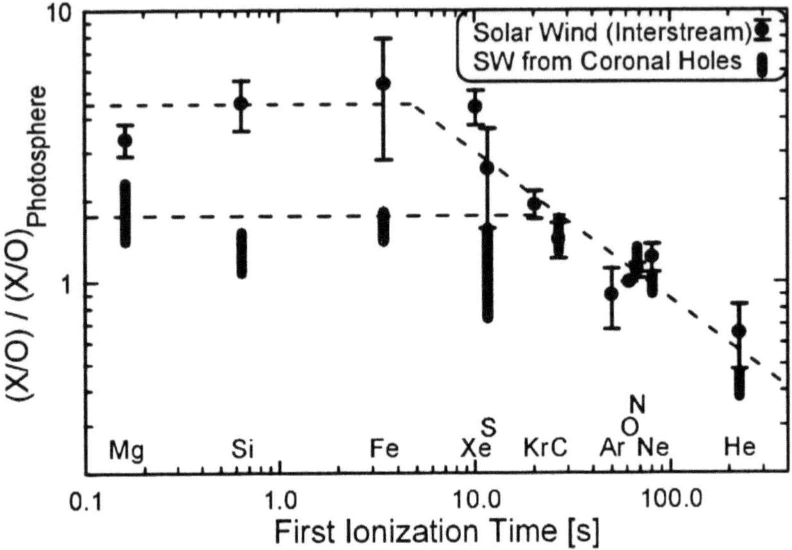

Fig. 5. Elemental abundances in the solar wind as a function of their "first ionisation time" (cf. text). Average patterns in the slow SW and in the HSST are indicated by dashed lines.

Reames, 1994). It was first discovered by investigations of the solar wind plasma in the magnetosheath with the CHEM instrument on AMPTE/CCE that the FIP effect is strongly reduced in fast solar wind streams (Gloeckler et al., 1989; von Steiger et al., 1992; see also the ICI/ISEE-3 results given by Ogilvie et al., 1992). This has been fully confirmed by the SWICS-Ulysses measurements (Galvin et al., 1992; Shafer et al., 1993; von Steiger and Geiss, 1993). It is now evident that the systematic differences between solar abundance and slow SW abundance are largely due to the FIP effect. Temporal or local SW abundance variations are caused in the corona (well documented for the He/H ratio, cf. Neugebauer, 1981), but systematic fractionations produced in the corona are less pronounced than those caused by the FIP effect which operates in the chromosphere. Since the FIP effect is much reduced in the HSSTs, the abundances there are relatively close to the solar abundances (cf. Gloeckler et al., 1989; von Steiger and Geiss, 1994), except probably for helium. Figure 5 shows the difference in the strength of the FIP effect between slow SW and HSST. The FIP effect does not produce a simple, monotonic relation of element abundances versus first ionisation potential which is most explicitly evident from the solar wind abundances of Kr and Xe derived from the gases trapped in lunar soils (Wieler et al., 1993). Geiss and Bochsler (1985) and von Steiger and Geiss (1989) have calculated ionisation times of various elements for zero optical depth at the solar surface. If we plot the elemental abundances as a function of these "first ionisation times (FIT)", we obtain rather well defined, nearly monotonous relationships clearly showing

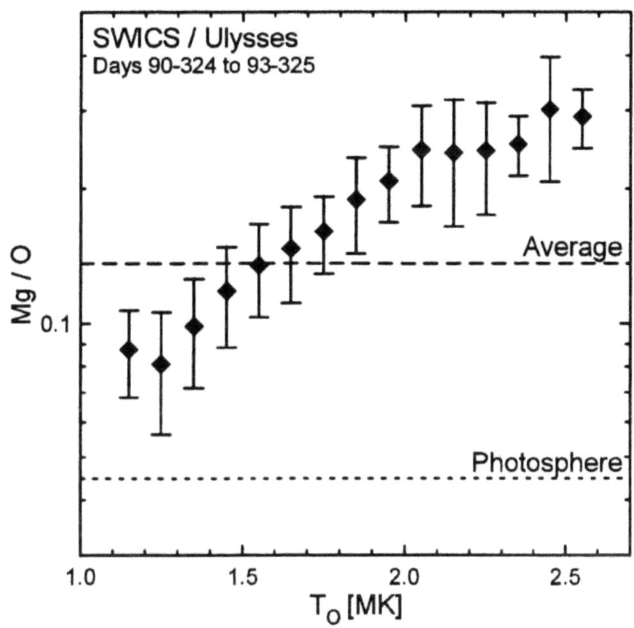

Fig. 6. The Mg/O ratio as a function of T_O, the freezing-in temperature $T(O^{7+}/O^{6+})$. Given are standard deviations (not standard errors, which would be much smaller).

the difference between slow SW and HSST (Fig. 5). These observations constitute important constraints for mechanisms that are designed to explain the FIP effect (cf. the discussions by Meyer, 1992; von Steiger and Geiss, 1994; Geiss, Gloeckler and von Steiger, 1994).

Oxygen and magnesium are a pair of a high- and a low-FIP element that can be particularly well studied with SWICS-Ulysses. The Mg/O ratio is found to decrease with increasing SW speed which reflects the weakening of the FIP effect in the HSST. However, the correlation is much weaker than the correlation between the Mg/O ratio and the freezing-in temperature $T(O^{7+}/O^{6+})$ shown in Fig. 6. This points to a relationship between coronal conditions and structure or processes in the chromosphere, which we shall further explore in the next section.

5. Relation between Chromospheric and Coronal Processes

We have used a superposed epoch method for studying systematics in the variations of solar wind properties across the border between the slow solar wind and the HSST during the period day 191.1992 to day 098.1993) when Ulysses spent similar times in the slow solar wind and in the HSST. One solar rotation (days 308–334.1992) was omitted, because a CME occurred. Using V_α = 600 km/s crossings as markers we empirically determined the average period of a full cycle (t = 25.8 days) and its increase with time (0.08 days per full cycle). As Ulysses flew to higher latitude and solar activity decreased, the partitioning of t

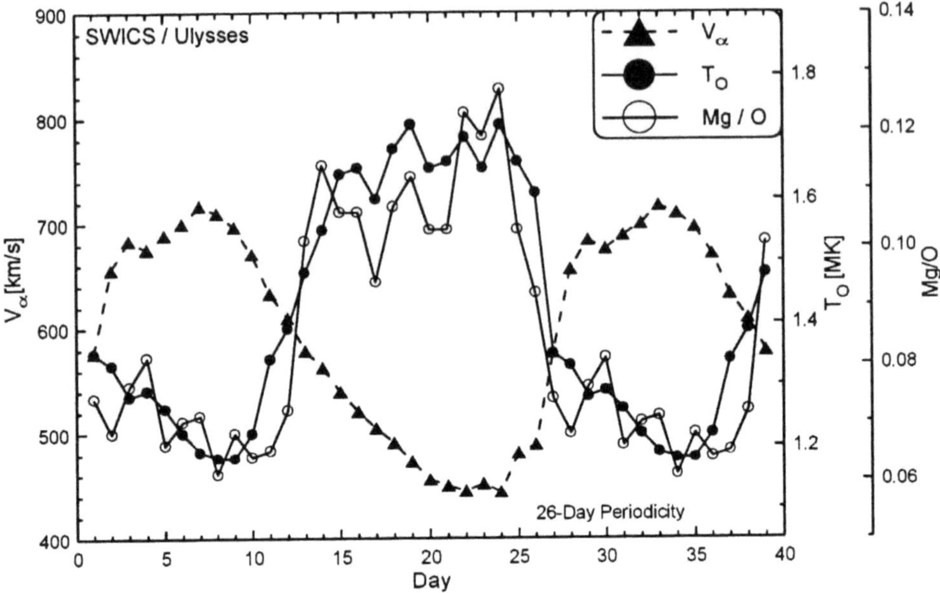

Fig. 7. Superposed epoch plot showing the systematic variations during the (effective) solar rotation period. The procedure used is outlined in the text. $T(O^{7+}/O^{6+})$ and Mg/O vary very steeply and trace each other almost perfectly, indicating a close relationship between corona temperature and chromospheric structure or processes.

between slow SW and HSST changed with time. Although this change is very irregular (cf. Fig. 3), we used a simple linear increase of 0.34 days per full cycle for the fraction of time in the HSST (determined again on the basis of the V_α = 600 km/s crossings). The data taken were our standard daily averages. The systematic changes in period and in fraction of time in and outside the HSST were taken into account by omitting or adding days with a random number method. Only 6% of the days were affected by this adjustment procedure. We have deliberately chosen this crude but straightforward procedure, which uses the minimum number of parameters, in order to avoid an inadvertent introduction of biases. A more refined procedure might improve, but certainly not destroy the systematics we find.

In Fig. 7 we show the result for three parameters: Plotted are, as a function of epoch-time, the solar wind speed (represented by the α-speed), coronal temperature (represented by $T(O^{7+}/O^{6+})$) and the strength of the FIP effect (represented by the Mg/O ratio). The resulting diagramme shows several remarkable features:

1. $T(O^{7+}/O^{6+})$ and the Mg/O ratio follow each other very closely, their relation to V_α is not as tight. This is of course consistent with the correlation shown in Fig. 6.
2. The figure indicates that even the fine variations of $T(O^{7+}/O^{6+})$ and Mg/O are correlated in the slow SW, but not so much in th HSST. As Ulysses

proceeds further into the southern coronal hole we should be able to further study variations in HSST abundances (or their absence).

3. The transition between low and high is very steep for $T(O^{7+}/O^{6+})$ and Mg/O. Statistical analyses show that it is almost as steep as a step function would be, when subjected to our procedure. Thus chromosphere and corona have a common, relatively sharp boundary, separating the low- from the high-FIP region in the chromosphere and the low- from the high-temperature region in the corona. This points to a causal relationship of the kind that conditions inside or even below the chromosphere determine the supply of energy into the corona, and underlines that discussions of the origin of the solar wind will have to include chromospheric processes.
4. The SW speed as a function of epoch-time shown in Fig. 7 with the characteristically slow decline of the speed in the trailing edge is similar to the pattern found by Bame et al. (1977) in the ecliptic during the solar minimum in the mid 1970s. When the trailing edge of a high speed stream is mapped back to the corona, it becomes clear, as shown by Nolte et al. (1977), that the large range of decreasing velocities comes from a relatively restricted range in longitude. This implies that the boundary between low and high $T(O^{7+}/O^{6+})$ and Mg/O at the eastern rim of the coronal hole is even sharper than is indicated in Fig. 7.
5. The HSST and the adjacent slow SW are embedded in interplanetary field-line configurations which do not seem to be grossly different. What causes then the reduction in kinetic energy by a factor of 4? More abundant closed magnetic field lines in the lower corona, weaker wave intensities or the reduced parallel component of the gravitational force might be factors worth exploring. Inclusion of other solar wind parameters into the superposed epoch analysis of Fig. 7 might help to understand why solar wind acceleration becomes less effective just outside the coronal holes.

Acknowledgements

The authors thank H. Balsiger, A. Bürgi, A. B. Galvin, and R. Wimmer for discussions. This work was supported by the Swiss National Science Foundation and by the US National Aeronautics and Space Administration (NASA/JPL contract 955460).

References

Bame, S.J., Asbridge J.R., Feldman W.C., and Gosling J.T.: 1977, *J. Geophys. Res.* **82**, 1487

Bame, S.J., et al.: 1992, *Astron. Astrophys. Suppl. Ser.* **92**, 237

Bochsler, P. and Geiss, J.: 1989, in J.H. Waite Jr., ed(s)., *Solar System Plasma Physics*, Geophysical Monograph 54, 133

Borrini, G., Gosling, J.T., Bame, S.J., Feldman, W.C., and Wilcox, J.M.: 1981, *J. Geophys. Res.* **86**, 4565

Bürgi, A.: 1992, in E. Marsch and R. Schwenn, ed(s)., *Solar Wind Seven*, Pergamon Press: Oxford, 333
Bürgi, A. and Geiss, J.: 1986, *Solar Physics* **103**, 143
Galvin, A.B., et al.: 1987, *J. Geophys. Res.* **92**, 12069
Galvin, A.B., Ipavich, F.M.,Gloeckler, G., von Steiger, R., and Wilken, B.: 1992, in E. Marsch and R. Schwenn, ed(s)., *Solar Wind Seven*, Pergamon Press: Oxford, 337
Galvin, A.B., Gloeckler, G., Ipavich, F.M., Shafer, C.M., Geiss, J., and Ogilvie, K.W.: 1993, *Adv. Space Res.* **13**, (6)75
Garrard, T.L., and Stone, E.C.: 1993, in *Proc. 23. Int. Cosmic Ray Conf.*
Geiss, J.: 1982, *Space Sci. Rev.* **33**, 201
Geiss, J.: 1985, in *Future Missions in Solar, Heliospheric and Space Plasma Physics*, ESA SP-235, 37
Geiss, J., Hirt, P., and Leutwyler, H.: 1970, *Solar Physics* **12**, 458
Geiss, J., and Bochsler, P.: 1985, in *Rapports isotopiques dans le système solaire*, Cepadues-editions: Toulouse, 213
Geiss, J., Gloeckler, G., and von Steiger, R.: 1994, in *Proc. Royal Society Symp. on the Solar System*, London, in press
Geiss, J., and Bochsler, P.: 1986, in R.G. Marsden, ed(s)., *The Sun and the Heliosphere in Three Dimensions*, Reidel: Dordrecht, 173
Gloeckler, G., Ipavich, F.M., Hamilton, D.C., Wilken, B. and Kremser, G.: 1989, *EOS Trans. AGU* **70**, 424
Gloeckler, G., and Geiss, J.: 1989, in C.J. Waddington, ed(s)., *Proc. Cosmic Abundances of Matter Symp.*, AIP Conf. Proc. **183**, 71
Gloeckler, G., et al.: 1992, *Astron. Astrophys. Suppl. Ser.* **92**, 267
Hirshberg, J., Asbridge, J.R., and Robbins, D.E.: 1972, *J. Geophys. Res.* **77**, 3583
Hollweg, J.V.: 1974, *J. Geophys. Res.* **79**, 1539
Hollweg, J.V.: 1978, *Rev. Geophys. Space Phys.* **16**, 689
Hundhausen, A.J.: 1972, *Solar Wind and Coronal Expansion*, Springer: New York
Ipavich, F.M., et al.: 1986, *J. Geophys. Res.* **91**, 4133
Ipavich, F.M., Galvin, A.B., Geiss, J., Ogilvie, K.W., and Gliem, F.: 1992, in E. Marsch and R. Schwenn, ed(s)., *Solar Wind Seven*, Pergamon Press: Oxford, 369
Kopp, R.A., and Holzer, T.E.: 1976, *Solar Phys* **49**, 43
Krieger, A.S., Timothy, A.F. and Roelof, E.C.: 1973, *Solar Phys* **29**, 505
Meyer, J.P.: 1992, in N. Prantzos, E. Vangioni-Flam and M. Cassè, ed(s)., *Origin and Evolution of the Elements*, Cambridge University Press, 26
NOAA: 1992, *Prelim. Report and Forecast of Solar Geophys. Data*, NOAA: Boulder
Noci, G., and Porri, A.: 1983, in *IAGA 18th Gen. Ass., Hamburg*, paper 4L.04
Nolte, J.T., Krieger, A.S., Roelof, E.C., and Gold, R.E.: 1977, *Solar Phys.* **51**, 459
Neugebauer, M.: 1981, *Fundamentals of Cosmic Physics* **7**, 131
Ogilvie K.W., Coplan M.A. and Geiss J.: 1992, in E. Marsch and R. Schwenn, ed(s)., *Solar Wind Seven*, Pergamon Press: Oxford, 379
Parker, E.N.: 1960, *Astrophys. J.* **132**, 821
Parker, E.N:: 1963, *Interplanetary Dynamical Processes*, Interscience: New York
Reames, D.V.: 1994, *Adv. Space Res.* **14**, (4)177
Shafer, C.M., Gloeckler, G., Galvin, A.B., Ipavich, F.M., Geiss, J., von Steiger, R., and Ogilvie, K.W.: 1993, *Adv. Space Res.* **13**, (6)79
Schmid, J., Bochsler, P., and Geiss, J.: 1988, *Astrophys. J.* **122**, 181
Schwenn, R., and Marsch, E. (eds.): 1990, *Physics of the Inner Heliosphere*, Springer: Heidelberg
von Steiger, R., Christon, S.P., Gloeckler, G., and Ipavich, F.M.: 1992, *Astrophys. J.* **389**, 791
von Steiger, R., and Geiss, J.: 1989, *Astron. and Astrophys.* **225**, 222
von Steiger, R., and Geiss, J.: 1993, *Adv. Space Res.* **13**, (6)63
von Steiger, R., and Geiss, J.: 1994, in J.R. Jokipii, C.P. Sonnett, and M.S. Giampapa, ed(s)., *Cosmic Winds and the Heliosphere*, University of Arizona Press: Tucson, in press
Wang, A.H., Wu, S.T., Suess, S.T., and Poletto, G.: 1993, *Solar Phys.* **147**, 55
Wieler, R., Baur, H., and Signer, P.: 1993, *Proc. Lunar and Planet. Sci.* **24**, 1519

SOLAR WIND HELIUM ISOTOPIC COMPOSITION FROM SWICS/ULYSSES

R. BODMER, P. BOCHSLER, J. GEISS and R. VON STEIGER
Physikalisches Institut, University of Bern, Switzerland

and

G. GLOECKLER
Department of Physics, University of Maryland, College Park, Md, USA

Abstract. This is the first study of the isotopic composition of solar wind helium with the SWICS time-of flight mass spectrometer. Although the design of SWICS is not optimized to measure ^{3}He abundances precisely, ^{4}He/^{3}He flux ratios can be deduced from the data. The long term ratio is 2290 ± 200, which agrees with the results obtained with the ICI magnetic mass spectrometer on ISEE-3 and with the Apollo SWC foil experiments.

The ULYSSES spacecraft follows a trajectory which is ideal for the study of different solar wind types. During one year, from mid-1992 to mid-1993, it was in a range of heliographic latitudes where a recurrent fast stream from the southern polar coronal hole was observed every solar rotation. Solar wind bulk velocities ranged from 350 km/s to 950 km/s which would, in principle allow us to identify velocity-correlated compositional variations. Our investigation of solar wind helium, however, shows an isotopic ratio which does not depend on the solar wind speed.

1. Introduction

Measurement of the ^{4}He/^{3}He ratio in the solar wind gives the best estimate for this ratio in the solar convective zone. Measurements of ^{3}He can be used

1. to derive the primordial D abundance, which is a sensitive measure of the baryonic density during the big-bang nucleosynthesis (Geiss and Reeves, 1972)
2. to set boundaries to nonstandard solar models with mixing processes in the solar interior (Bochsler *et al.*, 1990)
3. to test models of ionization, diffusion and acceleration in the solar atmosphere, because ^{3}He is a true test particle

SWICS/ULYSSES is able to measure $^3He^{2+}$ at high solar latitudes and for the whole range of solar wind velocities which was not the case for the measurements of ISEE-3/ICI (velocities were restricted to ≤ 620 km/s) (Coplan *et al.*, 1984).

If we want to deduce solar abundances from solar wind measurements, we have to know the relevant fractionation effects in the ionization and acceleration of the particles. The strongest fractionation effect for elements occurs in the chromosphere and depends on the first ionization potential (FIP). One possible mechanism was proposed by von Steiger and Geiss (1989), a different solution will be published by Marsch, von Steiger and Bochsler (1994).

Space Science Reviews **72**: 61–64.
© 1995 *Kluwer Academic Publishers.*

The important point to note is that neither of these models predicts a substantial fractionation between the two He isotopes.

2. Evaluation Method for $^3He^{2+}$

The Solar Wind Ion Composition Spectrometer (SWICS) consists of an electrostatic analyzer, a post acceleration, a time-of-flight telescope and a solid state detector (e.g. Gloeckler *et al.*, 1992). It measures energy per charge, time of flight and the total energy of an ion, which allows us to calculate the mass, charge and velocity.

Because of the large distance of other peaks from the $^3He^{2+}$ peak in the mass-mass/charge plane, the separation of $^3He^{2+}$ PHA words with energy information ('triples') from other species can be done easily. The background at the position of $^3He^{2+}$ is negligibly small.

The main problem with ^3He is its small count rate. The small abundance in the solar wind, the geometric factor and the efficiency of the instrument, the duty cycle and the stepping of voltages over a large range result in a count rate between 0 and 100 per day. This forces us to use an efficient evaluation method and reduces the ability of our instrument to detect short-term variations.

To derive the ^3He abundance we

1. determine flux and velocity distribution of $^4He^{2+}$, which can be done with a temporal resolution of 13 minutes.
2. assume the same velocity distribution for $^3He^{2+}$, which is justified, because most solar wind ions have approximately the same bulk velocity and the same kinetic temperature per mass.
3. assume the ratio of the helium isotopes to be constant for the accumulation period of one day.
4. find the best estimate for this ratio using a maximum-likelihood method.
5. calculate the resulting $^3He^{2+}$ flux using this ratio and the averaged flux of $^4He^{2+}$.

3. $^4He/^3He$ Results

Figure 1 shows daily averages of solar wind velocity, helium fluxes and the isotopic ratio. The relative errors of the single measurements are between 10% and 100%. Nevertheless the short term variations of the fluxes of the isotopes are correlated. The dependence of the isotopic ratio on solar wind speed has been investigated by splitting the data into velocity bins of 50 km/s width and analyzing the ratios for these groups separately (figure 2). To reduce the error of the resulting ratios we do not add up the estimates for the fluxes, but we assume the constancy of the ratio inside a velocity group

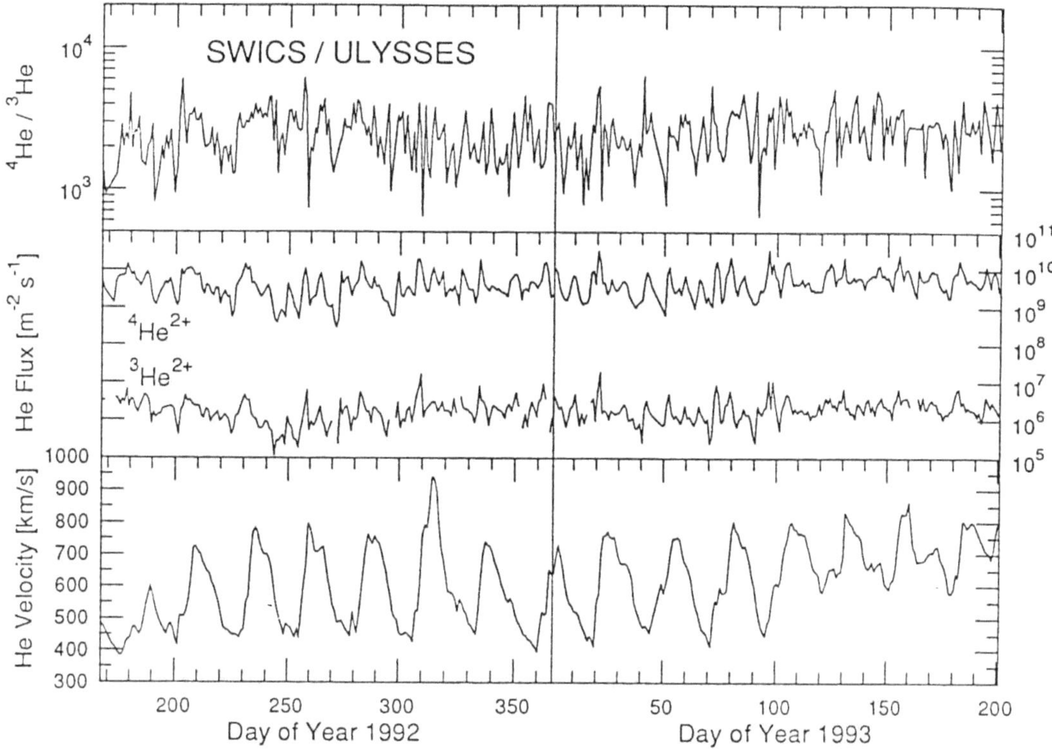

Fig. 1. Daily averages of solar wind velocity, helium fluxes and the isotopic ratio are plotted for a period from mid-1992 to mid-1993, where a recurrent fast stream from the southern coronal hole was seen every solar rotation.

and determine the best value using the same maximum-likelihood principle. The resulting ratios show no significant dependence on velocity.

If we analyze the whole data set under the assumption of a constant ratio, we get 2290 ± 200 as a best estimate of the ^{4}He/^{3}He ratio.

In ISEE-3/ICI data the ratio is apparently 3000 at 300 km/s and drops to 2100 for 400 km/s and then stays constant until at least 550 km/s (Bochsler, 1984). It can not completely be excluded, that an instrumental effect is the cause of this variation. In our data this dependence is not visible, but it would only affect one or two velocity bins. Days with an average solar wind velocity below 350 km/s did not occur often during the first three years of the ULYSSES mission, so it is difficult to reject a velocity dependence which affects only these very slow streams. The constancy of the ratio can now be extended to average velocities up to 900 km.

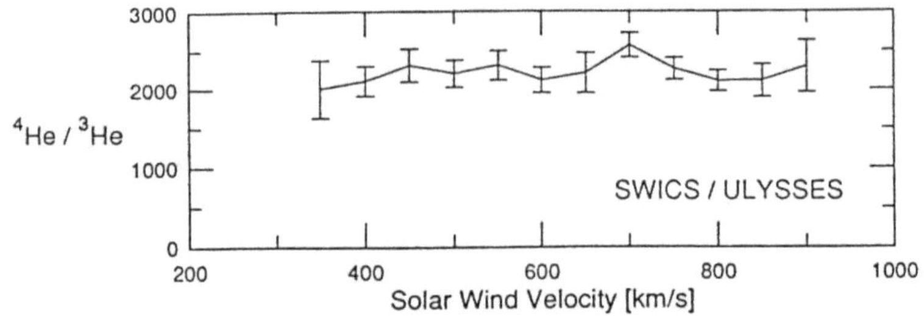

Fig. 2. Velocity dependence of the helium isotopic ratio.

4. Conclusions

The SWICS measurements allow us to extend the range of solar wind velocities for which a helium isotopic ratio can be deduced. The results show no clear correlation of the ratio with the solar wind velocity.

The statistics of ^{3}He measurements are poor, which makes it difficult to analyze short-term variations. The long-term isotopic ratio however gives a value, which is in very good agreement with results from other instrumental techniques:

- SWC Foil Collection Experiment on Apollo (Geiss *et al.*, 1972) $(^4\mathrm{He}/^3\mathrm{He}) = 2350 \pm 120$
- ICI Magnetic Mass Spectrometer on ISEE-3 (Coplan *et al.*, 1984) $(^4\mathrm{He}/^3\mathrm{He}) = 2050 \pm 200$
- ULYSSES SWICS, data from the first three years (1991-1993) $(^4\mathrm{He}/^3\mathrm{He}) = 2290 \pm 200$

References

Bochsler, P.: 1984, 'Helium and oxygen in the solar wind: dynamic properties and abundances of elements and helium isotopes as observed with the ISEE-3 plasma composition experiment', Habilitationsschrift, University of Bern.

Bochsler, P., Geiss, J., and Maeder, A.: 1990, 'The abundance of ^{3}He in the solar wind – a constraint for models of solar evolution', *Solar Phys.* **128**, 203.

Coplan, M. A., Ogilvie, K. W., Bochsler, P., and Geiss, J.: 1984, 'Interpretation of ^{3}He abundance variations in the solar wind', *Solar Phys.* **93**, 415.

Geiss, J. and Reeves, H.: 1972, 'Cosmic and solar system abundances of Deuterium and Helium-3', *Astron. Astrophys.* **18**, 126.

Geiss, J., Bühler, F., Cerutti, H., Eberhardt, P., and Filleux, C.: 1972, 'Solar-wind composition experiment', In *Apollo 16 Preliminary Science Report, NASA SP-315*, chapter 14, NASA.

Gloeckler, G., Geiss, J., and others: 1992, 'The solar wind ion composition spectrometer', *Astron. Astrophys. Suppl. Ser.* **92**, 267.

Marsch, E., von Steiger, R., and Bochsler, P.: 1994, 'Element fractionation by diffusion in the solar chromosphere'. submitted to Astron. Astrophys.

von Steiger, R. and Geiss, J.: 1989, 'Supply of fractionated gases to the corona', *Astron. Astrophys.* **225**, 222.

SOLAR WIND CHARGE STATES MEASURED BY ULYSSES/SWICS IN THE SOUTH POLAR HOLE

A.B. GALVIN, F.M. IPAVICH, C.M.S. COHEN, G. GLOECKLER
Dept. of Physics, University of Maryland, College Park, MD 20742, USA

R. VON STEIGER
Physikalisches Institut, Universität Bern, Sidlerstr. 5, CH-3012 Bern, Switz.

Abstract. The Ulysses mission now has an extensive data base covering several passes of the south polar coronal hole as the spacecraft proceeds to higher latitudes. Using composition measurements from the SWICS experiment on the Ulysses spacecraft, we have obtained charge state distributions, and hence inferred coronal ionization temperatures, for several solar wind species. In particular, we present an overview of Oxygen ionization temperature measurements, based on the O^{7+}/O^{6+} ratio, for the period January 1993 until April 1994 (~23°S to ~61°S heliographic latitude), and detailed Oxygen, Silicon and Iron charge state distributions of the south polar hole during a two month period of nearly continuous hole coverage, Dec 1993 - Jan 1994 (~45°S to 52°S heliographic latitude).

1. Introduction

The Solar Wind Ion Composition Spectrometer (SWICS) on Ulysses measures the mass, mass per charge, and incident energy of solar wind ions (Gloeckler *et al.*, 1992). Different solar wind elements, and different ionic charge states within a given element, are expected to "freeze-in" their relative charge state abundances at different coronal altitudes (e.g., Hundhausen *et al.*, 1968). By comparing Oxygen, Silicon, and Iron charge state distributions, inferences can be made regarding the coronal temperature and temperature gradient in the polar hole. Published measurements of heavy ion charge states in coronal holes prior to the Ulysses mission are limited. Galvin *et al.* (1984) made indirect charge state measurements of coronal hole CNO, based on the ionization states of accelerated diffuse ion events. Ipavich *et al.* (1986) reported the first direct measurements of solar wind Fe and CNO charge state distributions in a coronal-hole associated solar wind flow. In neither of these studies could individual charge states be resolved. Gloeckler *et al.* (1989), using a time-of-flight instrument (CCE/CHEM), measured magnetosheath distribution functions of several ions (including O, Si, and Fe) from which the relative charge state abundances for coronal hole solar wind were inferred. Ipavich *et al.* (1992) and Galvin *et al.* (1992) reported the composition of individual (i.e., resolved) ion species for O, Si, and Fe during the early phase of the Ulysses mission. These results included likely candidates for coronal-hole associated solar wind flows (based on speed, kinetic temperature and ionization state), but no attempt was made to establish specific solar source regions. All of these measurements were made in or near the ecliptic plane.

Space Science Reviews **72**: 65–70.
© 1995 *Kluwer Academic Publishers.*

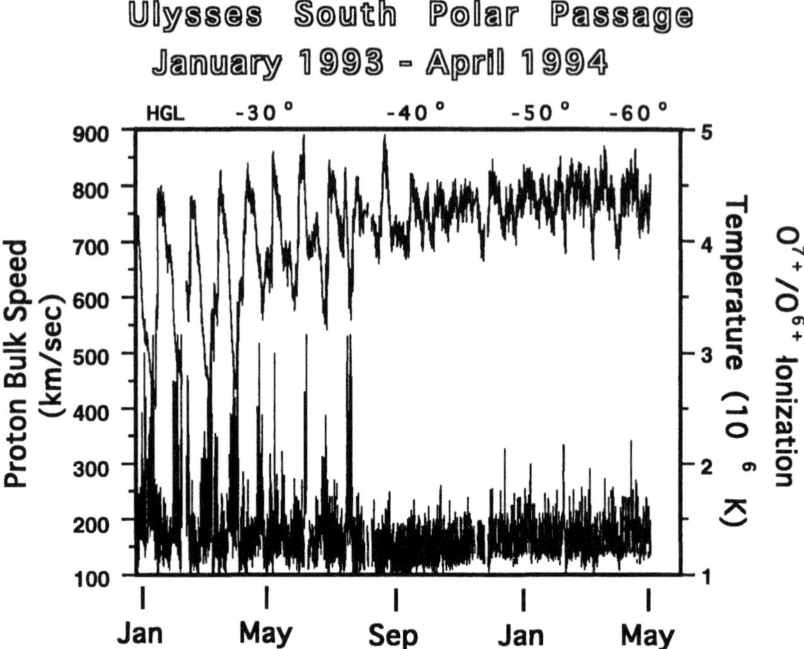

Fig. 1. Overview of selected solar wind parameters measured by SWICS from January 1993 to mid-April 1994. The upper trace is the proton bulk speed; the lower trace shows the ionization temperature derived from the charge state ratio O^{7+}/O^{6+}. The heliographic latitude (HGL) of Ulysses is given at the top.

In this paper we report the first measurements of charge states in the south polar hole for heliographic latitudes between ~ 20°S and 60°S for Oxygen, and from ~ 45°S to 52°S for Silicon and Iron. These constitute the first direct measurements of solar wind Silicon charge states from an identified coronal hole source.

2. Observations

The out-of-the ecliptic phase of the Ulysses mission began with the fly-by of Jupiter in February 1992. Beginning in July 1992, Bame *et al.* (1993) reported the first appearance of a long-term recurrent high speed stream, the source of which they identified as an equatorward extension of the south polar hole. They tracked the heliographic latitudinal evolution of this high speed stream through 14 solar rotations, from July 1992 (~13°S) through June 1993 (~34°S), which was the extent of the available data base at that time. A decrease in the bulk speed's peak-to-valley ratio was observed after April 1993 (Bame *et al.*, 1993; also see Fig 1), which was associated with the disappearance of the heliospheric current sheet and associated streamer belt at ~30°S (Smith *et al.*, 1993, Bame *et al.*, 1993).

In Fig. 1 we present the solar wind proton bulk speed and Oxygen ionization temperatures measured by Ulysses/SWICS from January 1993 through mid-April 1994. Data in Fig. 1 are given at the full temporal resolution of the experiment (13-min duty cycle). The Oxygen ionization temperatures are derived from the measured O^{7+}/O^{6+} ratio, assuming the ionization equilibria tables of Arnaud and Rothenflug (1985). The heliographic latitude is indicated at the top: Ulysses traveled from ~23°S to ~61°S during this period, as the spacecraft's heliocentric distance decreased from 5.1 to 3.2 AU. As seen in the figure, the high-speed stream returned for several additional solar rotations beyond the Bame *et al.* reporting period. The distinction between peaks and valleys in the bulk speed became increasingly obscure after September 1993, possibly implying that Ulysses was completely immersed in the polar hole. The high speeds of the recurrent stream (Fig. 1) correlate with higher proton kinetic temperatures T_{kin} (not shown), and anti-correlate with lower Oxygen ionization temperatures, $T_c(O)$. High speed solar wind flows associated with Coronal Mass Ejections are also present, and are often indicated by an increase in the $T_c(O)$ signature: for example, a well-defined CME passage on doys ~200 to ~206 1993, superimposed on the coronal hole stream, is indicated in Fig. 1 by a rapid increase followed by a steady decay in the bulk speed profile, decreased T_{kin} (not shown), and increased $T_c(O)$. The highest speed observed during the 1992 through early 1994 recurrent high speed stream time interval occurred on doy 314 1992. As noted by Bame *et al.* (1993) this was associated with a CME event. Otherwise, the highest speeds have been under 900 km/s, and more recently have been ≤ 850 km/s, indicating that the latitudinal velocity gradient observed by Bame *et al.* (1993) at mid-latitudes (i.e., below ~34°S) has decreased.

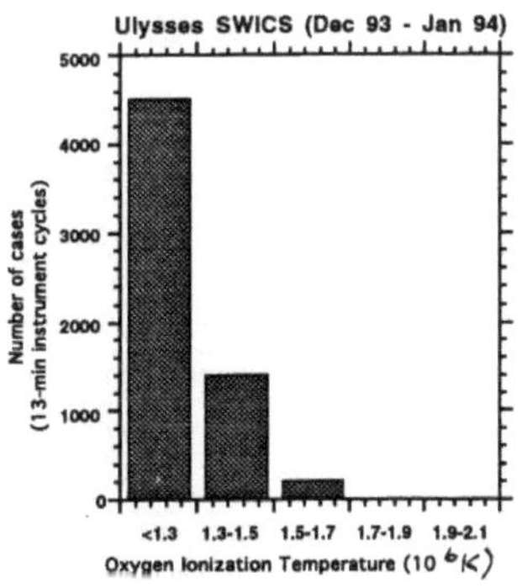

Fig. 2. Frequency distribution of the O^{7+}/O^{6+} ionization temperature during the Dec 93 - Jan 94 polar hole.

The time interval 1 December (doy 335) 1993 through 31 January (doy 031) 1994 was chosen for detailed charge state analysis. This interval appears to be free of any excursions (CME or otherwise) from the polar hole. The solar wind proton speed averaged ~770 km/s, varying from ~ 690 km/s to 850 km/s. The Ulysses spacecraft's heliocentric distance ranged from 4.0 to 3.7 AU, and the heliographic latitude covered 45.6°S to 51.5°S. This period maps back (assuming the constant velocity approximation) to portions of Carrington Rotations 1876, 1877, and 1878 on the sun, during which a segment of the south polar hole extended equatorward to 15°S heliographic latitude (Hα synoptic charts, *Preliminary Solar Geophysical Data Reports*).

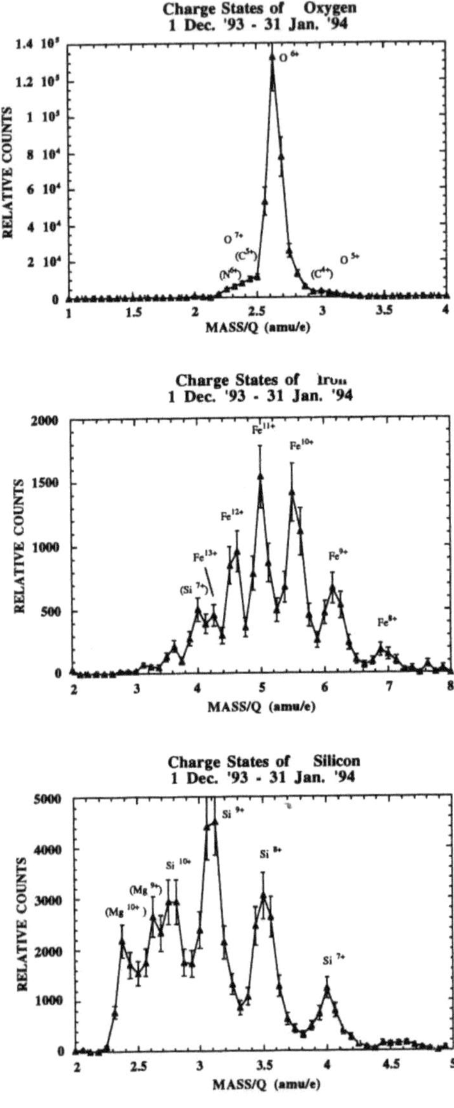

Fig. 3. Charge state distributions for solar wind Oxygen, Iron, and Silicon accumulated during the south polar hole period (Dec 1993 - Jan 1994).

Fig. 2 shows the frequency distribution of the O^{7+}/O^{6+} ionization temperatures derived for each 13-min instrument cycle over the December 1993 through January 1994 time interval. Over 6000 cases were accumulated under the selection criteria that there be no data gaps within the cycle period, and that there were non-zero count accumulations for each charge state over the 13-min interval. Under these restrictions (which are the same as those used for the $T_C(O)$ values shown in Fig. 1), 73% of the selected cases are represented by $T_C(O) \leq 1.3 \times 10^6$K. There can be a bias in the case selection toward higher temperatures since the incident flux of O^{7+} must be high enough to accumulate counts within the 13-min cycle. This selection effect is removed with the use of longer accumulation times. When the O^{7+}/O^{6+} data are accumulated over the entire 2-month period, the composite derived $T_C(O)$ is ~1.1×10^6K.

Fig. 3 displays the charge state distributions of O, Fe, and Si ions accumulated over the two month south polar hole interval. The distributions cover representative mass ranges of the respective species, however "spillover" from adjacent ions of similar mass can occur. Some of the possible " contaminants " are given parenthetically. (These spillover contributions are corrected for in the final determination of charge state abundances.) Oxygen is almost exclusively in charge state +6. Iron charge states Fe^{8+} through Fe^{12+} are all well-resolved; Fe^{13+} appears as a shoulder to some spillover from Si^{7+} and S^{8+}. The dominant Fe charge states are +10 and +11. In the Silicon ion spectrum, Si^{7+} through Si^{9+} are well-resolved; Si^{10+} and Si^{11+} have some spillover effects from Magnesium. The dominant Si charge states are +9 and +8.

3. Summary

We report the first measurements of solar wind Oxygen, Silicon and Iron charge states measured at high heliographic latitudes in the south polar hole. These also constitute the first published measurements of solar wind Silicon charge states in a firmly identified coronal hole flow. Iron measurements in the polar hole indicate a peak which is slightly shifted (by ≤ one charge state) from previous (in-ecliptic) results. Ipavich *et al.* (1986) found a peak near Fe^{10+} for one near-equatorial coronal hole and a peak at Fe^{9+} for a second near-equatorial hole. The previous SWICS results (Ipavich *et al.*, 1992) during the in-ecliptic phase of the Ulysses mission also suggested Fe charge states peaking at +9 and +10. Table 1 gives the coronal ionization temperatures derived for adjacent charge states of O, Si, and Fe assuming the ionization equilibria tables of Arnaud and Rothenflug (1985) and Arnaud and Raymond (1990). The estimated uncertainty in the temperatures is ~ 10 - 15%.

TABLE 1. Polar Hole Observations assuming Arnaud and Raymond (1992) & Arnaud and Rothenflug (1985) Ionization Equilibria:

		Si^{7+}		Fe^{9+}	
O^{6+}		Si^{8+}	$1.1x10^6$K	Fe^{10+}	$1.3x10^6$K
O^{7+}	$1.1x10^6$K	Si^{9+}	$1.3x10^6$K	Fe^{11+}	$1.2x10^6$K
		Si^{10+}	$1.5x10^6$K	Fe^{12+}	$1.3x10^6$K

Due to the lower electron densities in coronal hole regions, existing models (e.g., Feldman, 1977; Bürgi and Geiss, 1986) predict that solar wind heavy ion charge states in the coronal hole should "freeze-in" at altitudes below that of the coronal temperature maximum (predicted to occur at ~2-3 solar radii). The maximum temperature is predicted to be lower than that of the normal corona (e.g., Munro and Withbroe, 1972; Kopp and Orrall, 1976; Mariska, 1978). The ionization temperatures reported here, ranging from ~ 1.1 to 1.5 x 10^6K, are consistent with model expectations of lower (compared to the normal quiet sun) coronal temperatures predicted for coronal holes. The combined measurements of O, Si, and Fe imply that the electron temperature gradient over the range of freezing-in altitudes is quite small in the polar hole.

The Iron charge states given here are well represented by a single equilibria coronal temperature of ~1.3 x 10^6K. There is somewhat less consistency in the Silicon ionization temperatures under the assumption of equilibrium charge states.

The Oxygen results are consistent with the wave-driven coronal hole model of Bürgi and Geiss (1986). The Silicon results are not. Their model, under the assumed boundary conditions, predicts 67% of the Silicon in charge state +8, and 23% in charge state +7. This is not seen in the present data.

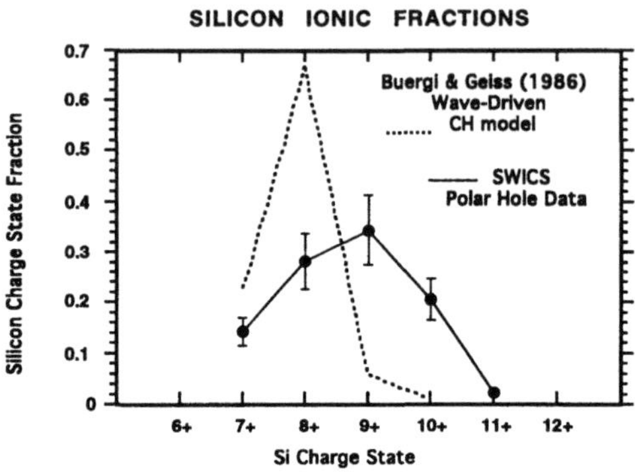

Fig. 4. Ionic charge fractions of Silicon measured during the south polar hole period (Dec 93 - Jan 94) compared to the wave-driven coronal hole model prediction of Bürgi and Geiss (1986).

Acknowledgments

This work was supported by NASA/JPL contract 955460 and by the Swiss National Science Foundation.

References

Arnaud, M. and Rothenflug, R.: 1985, *Astron. Astrophys. Suppl. Ser.*, **60,** 425.

Arnaud, M. and Raymond, J.: 1990, *Astrophys. J.*, **398**, 394.

Bame, S.J., Goldstein, B.E., Gosling, J.T., Harvey, J.W., McComas, D.J., Neugebauer, M., and Phillips, J.L.: 1993, *Geophys. Res. Lett.*, **20**, 2323.

Galvin, A.B., Ipavich, F.M., Gloeckler,G., Hovestadt, D., Klecker, B., and Scholer, M.: 1984, *J. Geophysical Res.*, **89**, 2655.

Galvin, A.B., Ipavich, F.M., Gloeckler, G., von Steiger, R., and Wilken, B.: 1992 in E. Marsch and R. Schwenn (eds.), *Solar Wind Seven*, Pergamon Press, pp. 337-340.

Gloeckler, G., Ipavich, F.M., Hamilton, D.C., Wilken, B., and Kremser, G.: 1989, *EOS*, **70,** SH41A-33, 424.

Gloeckler, G., *et al.*,: 1992, *Astron. Astrophys. Suppl. Ser.*, **92,** 267.

Hundhausen, A.J., Gilbert, H.E., and Bame, S.J.: 1968, *J. Geophysical Res.*, **73**, 5485.

Ipavich, F.M., Galvin, A.B., Gloeckler, G., Hovestadt, D., Bame, S.J., Klecker, B., Scholer, M., Fisk, L.A., and Fan, C.Y.: 1986, *J. Geophys. Res.*, **91**, 4133.

Ipavich, F.M., Galvin, A.B., Geiss, J., Ogilvie, K.W., and Gliem, F.: 1992 in E. Marsch and R. Schwenn (eds.), *Solar Wind Seven*, Pergamon Press, pp. 369-373.

Kopp, R.A., and Orrall, F.Q.: 1976, *Astron. Astrophys.*, **53**, 363.

Munro, R.H., and Withbroe, G.L.: 1972, *Astrophys. J.*, **176**, 511.

Mariska, J.T.: 1978, *Astrophys. J.*, **225,** 252.

Smith, E.J., *et al.*: 1993, *Geophys. Res. Lett.*, **20**, 2327.

KINETIC PROPERTIES OF HEAVY IONS IN THE SOLAR WIND FROM SWICS/ULYSSES

R. VON STEIGER and J. GEISS
Physikalisches Institut, Universität Bern, Sidlerstr. 5, CH–3012 Bern, Switzerland

and

G. GLOECKLER and A. B. GALVIN
Department of Physics and Astronomy, University of Maryland, College Park, MD 20742, USA

Abstract. The kinetic properties of heavy ions in the solar wind are known to behave in a well organized way under most solar wind flow conditions: Their speeds are all equal and faster than that of hydrogen by about the local Alfvén speed, and their kinetic temperatures are proportional to their mass. The simplicity of these properties points to a straightforward physical interpretation; wave-particle interactions with Alfvén waves are the probable cause. With the SWICS sensor on board Ulysses, it is now possible to investigate the kinetic properties of many more ion species than before. Furthermore, the transition of Ulysses into the fast stream emanating from the south polar coronal hole since 1992 allows us to study these properties both in the slow, interstream solar wind, as well as in an unambiguously identified fast stream. We present data from SWICS/Ulysses on the dominant ions of He, C, O, Ne, and Mg. As a result we find that, both in the slow wind and in fast streams, the isotachic property is obeyed even better than it could be determined by the ICI instrument on ISEE-3. The mass proportionality of $T_{\rm kin}$ is also shown to hold for these ions, including the newly identified C and Mg.

Key words: Solar wind – heavy ions – Ulysses

1. Introduction

The solar wind bulk velocity of protons at 1 AU varies between $v_{\rm p} = 300 - 800\,{\rm km/s}$, and their kinetic temperature, $T_{\rm p}$, between $\sim 10^4\,{\rm K}$ to a few times $10^5\,{\rm K}$. While the velocity stays constant at increasing heliocentric distances, the temperature drops by a power-law in r.

Solar wind helium and heavy ions are known not to be in thermal equilibrium with the protons. Their bulk velocities v_i are equal and vary between $v_{\rm p}$ at low speeds, and $v_{\rm p}+v_{\rm A}$ at high speeds, where $v_{\rm A}$ is the local Alfvén velocity (Marsch *et al.*, 1982). Their kinetic temperatures, T_i, tend to be higher than $T_{\rm p}$ by a factor of the ion's mass, m_i, such that two ion temperatures obey

$$T_i/T_j \simeq m_i/m_j \tag{1}$$

(Schmidt *et al.*, 1980; Ogilvie *et al.*, 1980). H does not normally fit in this rule very well, with a most probable $T_\alpha/T_{\rm p}$ ratio of 2–4 (Neugebauer, 1981), but we will show below that it is well obeyed by He and the heavy ions.

The simplicity of these rules, and the fact that they are observed under nearly all solar wind conditions, point to a straightforward physical interpretation. Wave-particle interactions with Alfvén waves are a candidate cause for both effects,

Space Science Reviews **72**: 71–76.
© 1995 *Kluwer Academic Publishers.*

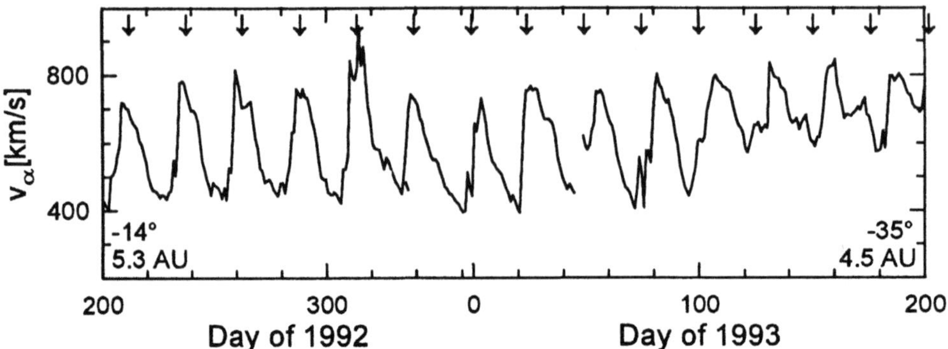

Fig. 1. Overview of the survey period. The solar wind velocity varied between 400 and 800 km/s, with one high-speed stream per solar rotation. The heliolatitude increased from 14° to 35° south, while the heliocentric distance decreased from 5.3 AU to 4.5 AU. Arrows indicate transits of 0° Carrington longitude as seen from Ulysses.

accelerating the heavy ions to speeds in excess of v_p and at the same time heating them to $T_j \propto m_i$. However, a quantitative theory explaining both observations has not been worked out as yet (but see Isenberg and Hollweg, 1983).

In this paper, we report on new data obtained with the SWICS instrument on Ulysses (Gloeckler *et al.*, 1992). These data are particularly suited for a fresh look at the heavy ion velocities and temperatures for two reasons: (1) The mass resolution of SWICS allows to unambiguously identify many more ion species, including those hidden by more abundant species in mass per charge spectra, such as C^{6+} and Mg^{10+}. (2) The post-acceleration (by $\sim$ 23 kV) of ions after the energy per charge selector guarantees that all ions have comparable energies — and thus similar detection efficiencies — in the sensor, irrespective of the original solar wind velocity. This facilitates comparisons of high-speed to low-speed solar wind, with little risk of systematic biases.

In this work, we concentrate on comparing bulk velocities and kinetic temperatures of He and heavy ions among themselves, deferring the comparison of these parameters to those of the main constituent, protons, to a later paper (but see Goldstein *et al.*, 1994).

2. Data Selection and Analysis

For our survey, we have chosen a period of one year duration, from day 200, 1992, to day 200, 1993 (see Fig. 1). During this period, Ulysses traversed the range between 14° and 35° southern heliographic latitude. From Fig. 1, it is evident that the spacecraft encountered a high-speed stream each solar rotation (periodicity 25.7 days, indicated by arrows), emanating from the low-latitude portion of the south polar coronal hole (Bame *et al.*, 1993). Our survey period is thus ideal for comparing the low-speed to the high-speed solar wind.

Even though SWICS completes one energy per charge (E/q) cycle every 13 minutes, it is not possible to use this time resolution for most heavy ions due to low statistics. We have thus summed over 32 E/q spectra, resulting in a time resolution of $\sim$ 8 hours, of the ions He^{2+}, C^{6+}, C^{5+}, C^{4+}, O^{8+}, O^{7+}, O^{6+}, Ne^{8+}, and Mg^{10+}. Individual ion species were picked from the mass-mass per charge matrix using the ellipse method (von Steiger, 1992), resulting in a distribution function for each species. Each ion spectrum was then fitted to a Maxwellian, from which the two parameters, v_0 and T, were obtained. Only the central portion of the distribution function, consisting of the E/q step with the maximum count rate plus all adjacent steps with count rates $\geq$ 20% thereof, was used in the fits, in order not to bias the parameters by the possible presence of suprathermal tails. Still, this procedure assumes that the solar wind conditions do not substantially vary over the accumulation period. If they do, the velocity obtained may be regarded as the average over the period, but the temperature is systematically increased by the velocity shift. Note, however, that a velocity shift affects all heavy ion temperatures by the same amount, not disturbing their correlation. Furthermore, since the solar wind velocity is nearly bi-stable, with quick transitions between slow and fast wind, only a small number of spectra are affected by this effect, and their parameters have large error bars, resulting in a low statistical weight in our fitting procedure (see below). We have therefore not cleaned up the data set and include all spectra in the results reported here.

Since the three-dimensional distribution functions are not necessarily isotropic, we need to further characterize the temperature value obtained by our analysis. Since the solar wind is supersonic, the E/q spectra essentially measure T parallel to the bulk velocity. The Parker spiral angle is nearly 90° at the heliocentric distance of our survey period, so this amounts to nearly the perpendicular temperature, $T_\perp$. Assuming that the possible anisotropies are similar and have the same direction for all ion species considered, the temperatures reported here are representative for the full distribution functions.

The above procedure resulted in $>$ 1000 data points for v_i and T_i of nine ion species from He^{2+} to Mg^{10+}. We then correlated the parameters between all ion pairs, using the Mahalanobis method, which is symmetric in the dependent and the independent variable, and which takes into account the errors in both directions (cf. appendix of von Steiger *et al.*, 1992). One such correlogram is shown in Fig. 2. The velocities were fitted with a linear function, while for the temperatures we made two power-law fits, one with both parameters free, and the other with the exponent fixed to one, giving the average temperature ratio.

3. Results

It is evident from Fig. 2 that (1) the velocities of the individual ion species are equal to a very high degree of accuracy, and (2) the kinetic temperatures are proportional, with those of the heavier species being enhanced by a roughly mass

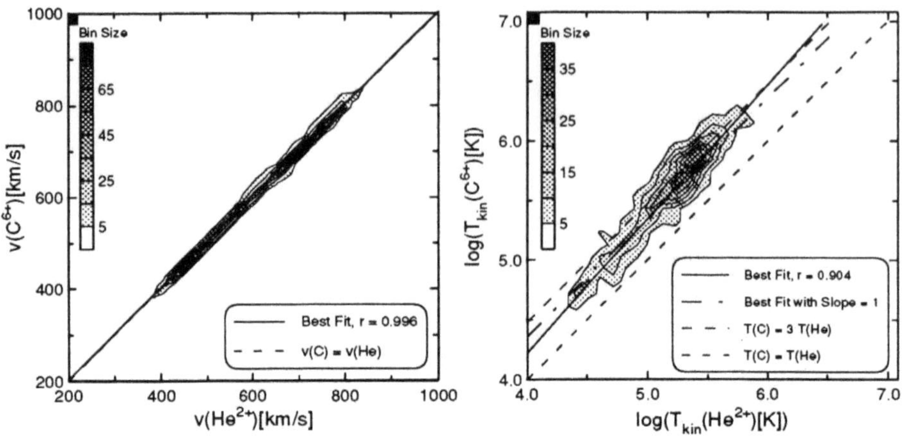

Fig. 2. Correlograms of bulk velocity (left) and kinetic temperature (right) of C^{6+} vs. He^{2+}. While the two velocities are equal over the full range of 400–800 km/s, the temperature of C^{6+} is enhanced over that of He^{2+} by almost a factor of three.

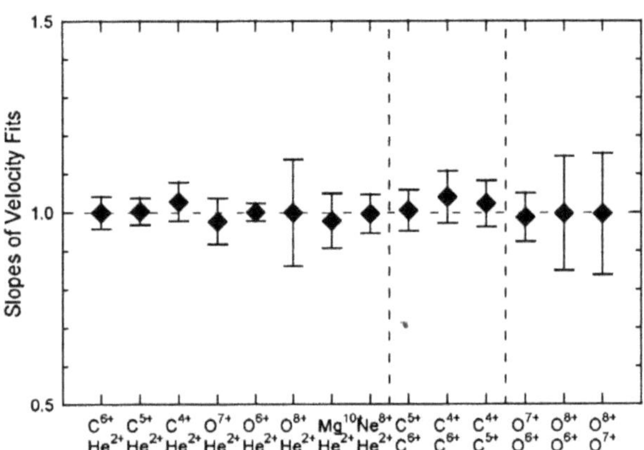

Fig. 3. Slopes of velocity fits as in Fig. 2 for all ion pairs investigated, as indicated on the abscissa. All heavy ion bulk velocities are found to be equal within the statistical errors.

proportional factor. This is, of course, not a new result by itself. Apart from the references cited in the Introduction, it has been found for a mixture of Fe charge states by Schmid *et al.* (1987), and for a mixture of Si charge states by Bochsler (1989), both using ICI/ISEE-3 data. However, our data spans a different range of solar wind velocities, from 400 to 800 km/s, as compared to 300–600 km/s covered by ICI.

In Fig. 3, we plot the slopes of the velocity fits of all ion pairs investigated (indicated on the abscissa). All slopes are found to be equal to unity, indicating that He and the heavy ions all flow with the same speed in the solar wind under all SW conditions encountered in our survey period, since the ordinate intercepts are all zero to within the error bars (not shown).

Figure 4 gives the fit parameters of the temperature fits, revealing the proportionality of T for all ion pairs. Exponents are between 0.9 and 1.1 (top panel),

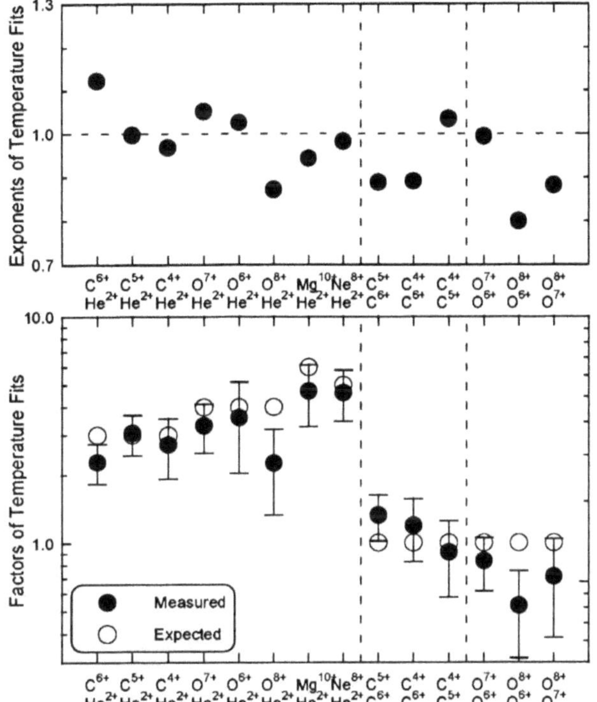

Fig. 4. Exponents of temperature fits as in Fig. 2 (top), and average temperature ratios (bottom) for all ion pairs investigated, as indicated on the abscissa. All heavy ion temperatures are found to be proportional to each other, but not equal. Their ratios agree with the mass ratios of the ions (indicated by open dots) within the statistical errors (see text for the case of O^{8+}).

and the average temperature ratios, calculated as the factor of the temperature fit with exponent fixed to one, are all found to agree with the mass ratios within the error bars (bottom panel), as expected from Eq. (1). The bad correspondence for the ratios including O^{8+} is probably caused by the difficulty to identify this ion even with SWICS, since it may be contaminated by the neighbouring and more abundant C^{6+}.

4. Discussion

We have demonstrated that the well-known isotachic property of the solar wind heavy ions is obeyed to a very high degree of accuracy for solar wind velocities between 400–800 km/s, and that the mass proportionality of their kinetic temperatures is obeyed also by ion charge states previously unavailable. These results are seemingly at variance with those obtained by ICI/ISEE-3 in two respects:

(1) Schmid *et al.* (1987), and Bochsler (1989) found that the heavier ions, Fe and Si, are lagging behind He at high bulk velocities (up to 600 km/s), while in our data no such difference is apparent up to the highest velocities of 800 km/s. A possible interpretation may be that the acceleration of the heavy species to the He speed is not yet fully developed at 1 AU (the distance of ISEE-3), but only at 5 AU (the distance of Ulysses), due to the longer expansion time. This

view is supported by the observation of a bimodal distribution of He–Si velocity differences by ICI, with a narrow peak at $\Delta v = 0$ superimposed to a broader distribution with $\Delta v > 0$. At 1 AU, the relative number of $\Delta v = 0$ observations increases with decreasing v, i.e. with increasing expansion time. It may thus be that, during the transition from 1 to 5 AU, the solar wind speed is further equalized, leading to more $\Delta v = 0$ observations at larger distances, and thus to a smaller $|\Delta v|$ on the average.

(2) Bochsler *et al.* (1985) found a deviation of the mass proportionality in T towards equal temperatures at low solar wind velocities. We could not observe such a deviation, but this is simply due to the fact that the lowest velocities in our data set are of the order of 400 km/s, while the isothermal solar wind was only observed at the extremely low velocity of $\sim$ 300 km/s by ICI. Our data set is thus not well suited to search for isothermal solar wind, which is limited to instances of very low velocity, combined with high density, resulting in enhanced Coulomb friction equalizing T.

In summary, we find that the solar wind velocities of He and heavy ions are equal and that their kinetic temperatures are mass proportional up to the highest bulk velocities observed. These rules are obeyed to a high degree of accuracy. Furthermore, previously unobservable ion species such as C^{6+} and Mg^{10+} fully support this picture. It is hoped that these results will trigger a new effort in theoretical modelling of the solar wind expansion, which eventually will result in understanding the cause of these simple observational rules.

Acknowledgements

This work was supported by the Swiss National Science Foundation, and by NASA/JPL contract 955460.

References

Bame, S. J., Goldstein, B. E., Gosling, J. T., Harvey, J. W., McComas, D. J., Neugebauer, M., and Phillips, J. L.: 1993, *Geophys. Res. Lett.* **20**, 2323–2326.

Bochsler, P.: 1989, *J. Geophys. Res. A* **94**, 2365–2373.

Bochsler, P., Geiss, J., and Joos, R.: 1985, *J. Geophys. Res. A* **90**, 10,779–10,789.

Gloeckler, G., Geiss, J. *et al.*: 1992, *A&AS* **92**, 267–289.

Goldstein, B. E., Neugebauer, M., Gosling, J. T., Bame, S. J., Phillips, J. L., and McComas, D. J.: 1994, *Space Sci. Rev.*, this issue.

Isenberg, P. A. and Hollweg, J. V.: 1983, *J. Geophys. Res. A* **88**, 3923–3935.

Marsch, E., Mühlhäuser, K.-H., Rosenbauer, H., Schwenn, R., and Neubauer, F. M.: 1982, *J. Geophys. Res. A* **87**, 35–51.

Neugebauer, M.: 1981, *Fundamentals of Cosmic Physics* **7**, 131–199.

Ogilvie, K. W., Bochsler, P., Geiss, J., and Coplan, M. A.: 1980, *J. Geophys. Res. A* **85**, 6069–6074.

Schmid, J., Bochsler, P., and Geiss, J.: 1987, *J. Geophys. Res. A* **92**, 9901–9906.

Schmidt, W. K. H., Rosenbauer, H., Shelley, E. G., Sharp, R. D., Johnson, R. G., and Geiss, J.: 1980, *Geophys. Res. Lett.* **7**, 697–700.

von Steiger, R.: 1992, *ESA SP* **348**, 309–314.

von Steiger, R., Christon, S. P., Gloeckler, G., and Ipavich, F. M.: 1992, *ApJ* **389**, 791–799.

RADIO OCCULTATION STUDIES WITH SOLAR CORONA SOUNDERS

M. PÄTZOLD and F.M. NEUBAUER
Institut für Geophysik und Meteorologie, Universität Köln, Köln, Germany
and
M.K. BIRD
Radioastronomisches Institut, Universität Bonn, Bonn, Germany

May 31, 1994

Abstract. Solar Corona Sounders (SCS), a mission designed to utilize the radio occultation technique for investigations of the inner heliosphere, was submitted to ESA in response to a call for new mission concepts. The SCS platforms are two small multi-frequency transmitters placed at the "anti-Earth" position (superior solar conjunction) for continuous radio sounding of the solar corona. Appropriately specifying certain orbital elements for the heliocentric trajectories of the spacecraft, their radially-aligned positions as seen from Earth appear to circle the solar disk over the course of a year. The two radio sources would be most effectively positioned at apparent solar distances inside and outside the nominal solar wind critical point, respectively, e.g., at $\sim$3 $R_\odot$ and $\sim$10 $R_\odot$. Radio parameters to be measured using the linearly polarized, coherent dual-frequency links to ground include the group time delay, signal amplitude, the phase (Doppler) shift, linewidth, and Faraday rotation. The link frequencies for coronal sounding observations this close to the Sun could be the interplanetary standards at S-band (2.3 GHz) and X-band (8.4 GHz). These measurements are used to derive both mean values and fluctuation spectra of such coronal parameters as the electron density, the solar wind velocity, and the magnetic field. The geometry afforded by the two radio ray paths from the SCS transmitters would provide unprecedented observations of the radial evolution of dynamic coronal events such as coronal mass ejections.

1. Introduction and Scientific Objectives

Over the last forty years, radio propagation experiments using both occulted natural (Interplanetary Scintillation IPS) and spacecraft signals (Radio Science) have probed the solar corona. These observations are one of the only viable means for obtaining information about plasma conditions in the corona at distances between 2 and 90 solar radii ($R_\odot$). The purpose of the Solar Coronal Sounder Program is to locate two small spacecraft at the "anti-Earth" position (superior conjunction) for continuous radio sounding of the solar corona at two frequencies over one year or even longer. As seen from Earth, the two spacecraft will appear to circle about the solar disk in radial alignment (see Figure 1).

The apparent distances to the Sun's center would vary between 3–5 $R_\odot$ and 10–13 $R_\odot$ for the inner and outer SCS probes, respectively (Figure 2). Noting that the critical point of solar wind acceleration is nominally located at $\simeq$5 $R_\odot$, the ray path from the inner probe would thus be well within the

Space Science Reviews **72**: 77–80.
© 1995 *Kluwer Academic Publishers.*

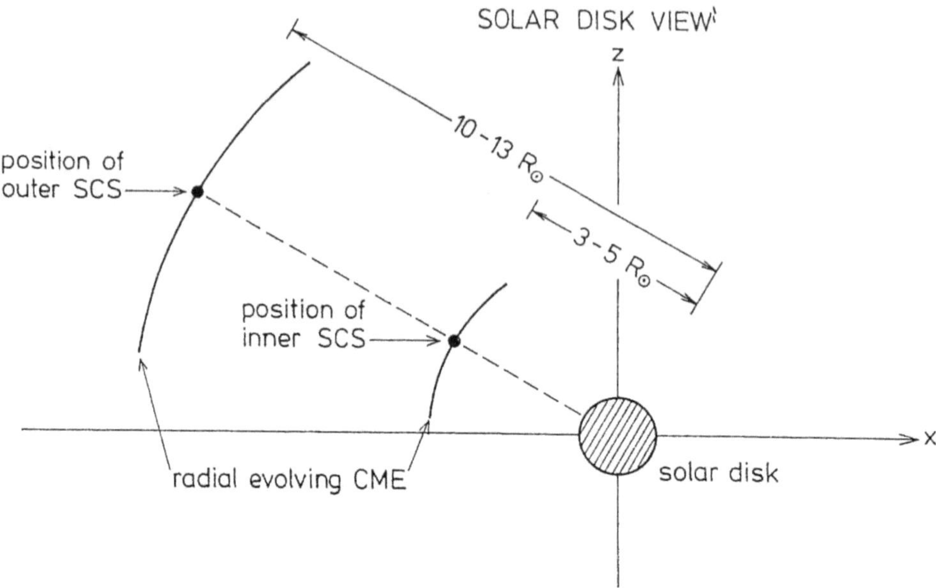

Fig. 1. Sketch of the geometrical constellation. Solar disk view as seen from the Earth: the radial alignment relative to the center of the Sun allows the observation of radial evolution of the coronal mass ejection. The shock front passes through the ray pass of the inner SCS (closest approach of the ray path 3 – 5 $R_\odot$ and then, approximately one hour later, through the ray path of the outer SCS (closest approach of the ray path 10 – 13 $R_\odot$).

region of subsonic solar wind speed at its solar proximate point. Likewise, the outer probe's signal would pass the Sun at a distance well within the region of supersonic speed.

It is then possible to monitor the temporal and spatial variations of the following physical quantities:

- coronal electron density using observations of the group time delays of range codes on two up- and downlink carriers (Pätzold et al., 1992)
- solar wind velocity and acceleration by observing the phase (Doppler) shifts of the two frequencies (Woo, 1978)
- coronal plasma fluctuations inferred from signal amplitude fluctuations (Woo and Armstrong, 1979)
- coronal magnetic field by measuring the Faraday rotation of the linear polarized signals (Pätzold et al., 1987)

- coronal mass ejections (CMEs) (Bird et al., 1985), specifically their radial evolution (as demonstrated in Figure 1).

It should also be noted that opportunities during solar oppositions would be available during cruise in the transfer orbit for the search for gravitational waves (Bertotti et al., 1992).

The Solar Coronal Sounders would represent an interplanetary complement to the SOHO mission or any other spacecraft located at L_1 or in interplanetary space monitoring solar wind parameters *in-situ*. In fact, data from other interplanetary monitoring stations should be used to facilitate the interpretation of the radio sounding data.

2. Conceptual Description

The two small and inexpensive Solar Corona Sounder spacecraft each carry an optimized radio system for radio science purposes. The primary components of this radio system are:

- one High Gain Antenna (HGA) of 1 meter diameter
- Very Stable Oscillators (VSO) and transponders for simultaneous, phase-coherent downlink transmissions at two frequencies (S- and X-band for coronal sounding; X- and K-band for the gravitational wave search)
- X- or K-band uplink capability for reducing uplink noise in two-way mode
- Linear polarized downlinks at S- and X-band for measurements of coronal Faraday rotation
- Spacecraft transmission power sufficient for receiving an adequate signal-to-noise ratio in the carrier signals at 30-meter class ground stations

The light weight of the SCS spacecraft would enable a tandem launch of both probes together with a second or third payload using Ariane. After launch the spacecraft are injected into a heliocentric transfer orbit. Upon arrival at the "anti-Earth" position, the spacecrafts are injected in an orbit with small eccentricity and inclination. As seen from Earth in the plane of the sky, the spacecraft circle in radial alignment around the solar disk. An entire rotation through all solar position angles is accomplished in one year (Figure 2).

References

Bertotti B. *et al.*: 1992, *Astron. Astrophys.*, **92**, 431.

Bird M.K. *et al.*: 1985, *Solar Phys.*, **98**, 341.

Pätzold M. *et al.*: 1987, *Solar Phys.*, **109**, 91.

Pätzold M. *et al.*: 1992, in *Solar Wind Seven*, Marsch E. and Schwenn R. (Ed.), Oxford.

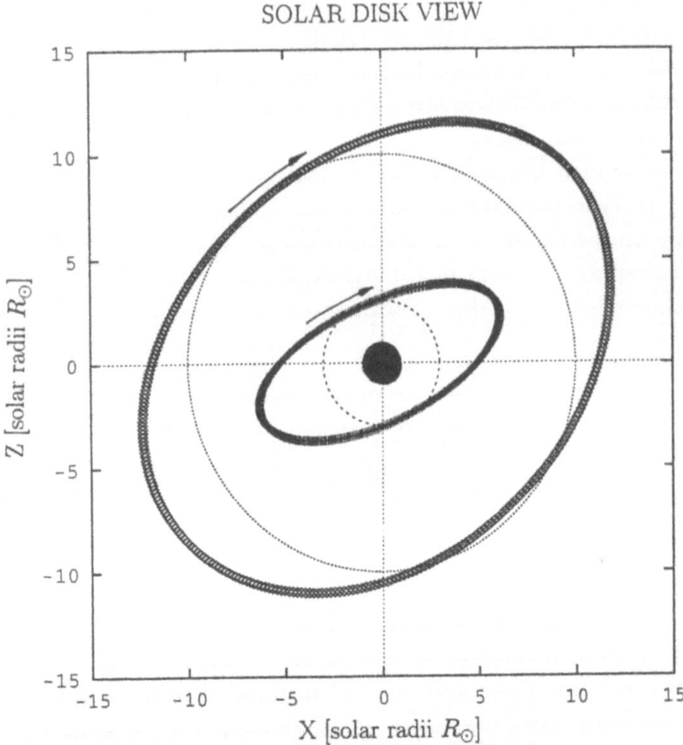

Fig. 2. Final constellation of both SCS probes circling about the solar disk. The inner dashed circle has a radius of 3 $R_\odot$; the outer dotted circle a radius of 10 $R_\odot$. The time step between positions is one day. The outer spacecraft reaches apparent distances between 10–13 $R_\odot$; the inner spacecraft distances between 3–7 $R_\odot$.

Woo R.: 1978, *Astrophys. Journ.*, **219**, 727.

Woo R. and Armstrong J.W.: 1979, *Journ. Geophys. Res.*, **84**, 7288.

SOHO:

THE SOLAR AND HELIOSPHERIC OBSERVATORY

V. DOMINGO AND B. FLECK
Space Science Department of ESA, ESTEC, 2200 AG Noordwijk, The Netherlands

and

A.I. POLAND
Laboratory for Astronomy and Solar Physics, NASA/Goddard Space Flight Center, Greenbelt, MD 20771 USA

Abstract. The Solar and Heliospheric Observatory (SOHO), together with the Cluster mission, constitutes ESA's Solar Terrestrial Science Programme (STSP), the first "Cornerstone" of the Agency's long-term programme "Space Science — Horizon 2000". STSP, which is being developed in a strong collaborative effort with NASA, will allow comprehensive studies to be made of the both the Sun's interior and its outer atmosphere, the acceleration and propagation of the solar wind and its interaction with the Earth. This paper gives a brief overview of one part of STSP, the SOHO mission.

Key words: solar physics, solar interior, solar corona, solar wind, SOHO

1. Scientific Objectives

The principal scientific objectives of the SOHO mission are a) to reach a better understanding of the structure and dynamics of the solar interior using techniques of helioseismology, and b) to gain a better insight into the physical processes that form and heat the Sun's corona, maintain it and give rise to its acceleration into the solar wind. To achieve these goals, SOHO carries a payload consisting of 12 sets of complementary instruments which are briefly described in Table I. Detailed instrument descriptions by the consortia involved can be found in the ESA volumes edited by Domingo and Guyenne (1989) or Mattok (1992).

2. Spacecraft and Orbit

The SOHO spacecraft (Fig. 1) is planned to be launched on an Atlas II-AS in July 1995. After a transfer phase of ca. 4 months it will be injected in a halo orbit around the L1 Sun-Earth Lagrangian point, about 1.5×10^6 km sunward from the Earth. The halo orbit will have a period of about 180 days and has been chosen because, 1) it provides a smooth Sun-spacecraft velocity change throughout the orbit, appropriate for helioseismology, 2) is permanently outside of the magnetosphere, appropriate for the "in situ" sampling of the solar wind and particles, and 3) allows permanent observation of the Sun, appropriate for all the investigations. The Sun-spacecraft velocity will be measured with an accuracy better than 0.5 cm/s.

Space Science Reviews **72**: 81–84.
© 1995 *Kluwer Academic Publishers.*

TABLE I
The SOHO Payload

Instrument	Measurements	Technique	Bit Rate (kb/s)
HELIOSEISMOLOGY			
GOLF	Global Sun velocity oscillations (ℓ=0-3)	Na-vapour resonant scattering cell, Doppler shift and circular polarization	0.160
VIRGO	Low degree (ℓ=0-7) irradiance oscillations and solar constant	Global Sun and low resolution (12 pixels) imaging, active cavity radiometers	0.1
MDI/SOI	Velocity oscillations, harmonic degree up to 4500	Fourier tachometer, angular resolution: 1.5 and 4"	5 (+160)
SOLAR ATMOSPHERE REMOTE SENSING			
SUMER	Plasma flow characteristics (T, density, velocity) chrom. through corona	Normal incidence spectrometer, 50-160 nm, spectral res. 20000-40000, angular res. 1.5"	10.5 (or 21)
CDS	Temperature and density: transition region and corona	Normal and grazing incidence spectrometers, 15-80 nm, spectr. res. 1000-10000, angular res.≈3"	12 (or 22.5)
EIT	Evolution of chromospheric and coronal structures	Images (1024 x 1024 pixels in 42' x 42') in the lines of He I, Fe IX, Fe XII and Fe XV	1 (or 26.2)
UVCS	Electron and ion temp., densities, velocities in corona (1.3-10 $R_\odot$)	Profiles and/or intensity of several EUV lines between 1.3 and 10 $R_\odot$	5
LASCO	Structures evolution, mass, momentum and energy transp. in corona (1.1-30 $R_\odot$)	1 internal and 2 externally occulted coronagraphs Spectrometer for 1.1-3 $R_\odot$	4.2 (or 26.2)
SWAN	Solar wind mass flux anisotropies and its temporal variations	Scanning telescopes with hydrogen absorption cell for Ly-α	0.2
SOLAR WIND 'IN SITU'			
CELIAS	Energy distribution and composition (mass, charge) of ions (0.1-1000 keV/e)	Electrostatic deflection, time-of-flight measurements, solid state detectors	1.5
COSTEP	Energy distribution of ions (p, He) 0.04-53 MeV/n and electrons 0.04-5 MeV	Solid state and plastic scintillator detector telescopes	1.01
ERNE	Energy distribution and isotopic composition of ions (p - Ni) 1.4-540 MeV/n and electrons 5-60 MeV	Solid state and scintillator crystal detector telescopes	1.01

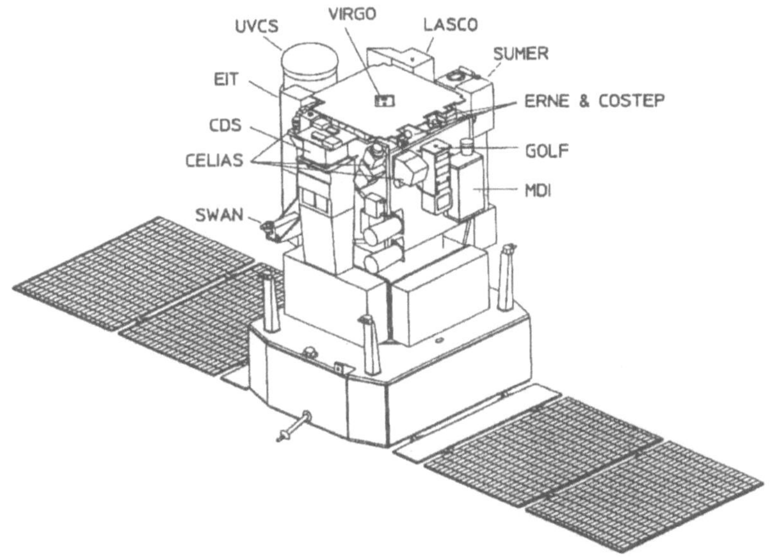

Fig. 1. SOHO spacecraft schematic view.

SOHO will be three-axis stabilized and point to the Sun within an accuracy of 10 arc sec. Pointing stability will be better than 1 arcsec per 15 minutes interval. The spacecraft will consist of a payload module which accommodates the instruments and a service module carrying the spacecraft subsystems and the solar arrays. The total mass will be about 1850 kg and 1150 W power will be provided by the solar panels. The payload will weigh about 640 kg and will consume 450 W in orbit.

SOHO is being designed for a lifetime of two years, but will be equipped with sufficient on-board consumables for an extra four years.

3. Operations

The SOHO Experiment Operations Facility (EOF), to be located at NASA's Goddard Space Flight Center, will serve as the focal point for mission science planning and instrument operations. At the EOF, experiment PI representatives will receive real-time and playback flight telemetry data, process these data to determine instrument commands, and send commands to their instruments, both in near real-time and on a delayed execution basis. They will be able to perform data reduction and analysis, and have capabilities for data storage.

The EOF has the following functions:

- Provides the means with which the PI teams participating in the SOHO programme can monitor and control their instruments on-board the spacecraft.

- Is the center where the solar atmosphere investigators of SOHO will coordinate and plan the near real-time operation of their instruments, and will be the focal point on the one hand, for the overall SOHO science operations planning, and on the other, for coordinating science studies through the organisation of campaigns and workshops.
- Provides electronic interfaces with the appropriate data bases and networks to provide the necessary input from ground stations and other spacecraft data for the planning of the SOHO science operations.
- Provides data storage for science, engineering and housekeeping data.

SOHO operations and telemetry is done using the Deep Space Network (DSN) which will receive S/C telemetry during three short (1.6 hrs) and one long (8 hrs) station pass per day. The SOHO payload will produce a continuous stream of science data of 40 kbits/s. Science data acquisition during non-station pass periods will be stored on-board and played back during the short station passes. The MDI high data rate stream (+ 160 kbits/s) will be transmitted only during the long station passes. For 2 consecutive months per year continuous data transmission, including MDI high data rate, will be supported by DSN. Whenever there is data transmission, the basic science data (40 kbits/s) will be available in near real-time at the EOF.

4. Coordinated Research

The SOHO payload has been conceived as an integrated package of complementary instruments, such that their measurements would produce a complementary set of data for the study of the phenomena in the solar atmosphere, in the solar wind, and in the solar interior. Therefore, to achieve the scientific aims of the mission, it is essential to operate the instruments in a coordinated programme and analyse the data in a correlative and cooperative spirit. Further, to maximize the scientific return of the mission, it is essential to have a close collaboration with ground-based solar observers (both radio and optical) and other space-based solar researchers (Yohkoh, Coronas, etc.). Finally, it should be emphasized that SOHO is part of STSP and that there is a large fraction of space science and plasma physics that may benefit from a coordinated approach between "in situ" observations such as obtained by Cluster, Ulysses, Geotail, Interball, Wind, or Polar and the remote and "in situ" measurements taken by SOHO.

References

Domingo, V., Guyenne, T.D. (eds.): 1989, The SOHO mission — scientific and technical aspects of the instruments, ESA SP-1104

Mattok, C. (prep.): 1992, Coronal streamers, coronal loops, and coronal and solar wind composition, ESA SP-348

SESSION 2
LARGE-SCALE HELIOSPHERIC STRUCTURE

LARGE-SCALE SOLAR WIND STREAM STRUCTURE AT HIGH HELIOLATITUDES

IGOR S.VESELOVSKY
Institute of Nuclear Physics, Moscow State University
119899, Moscow, Russia

Abstract. The tilted rotator model is considered using the collisionless supersonic and superalfvenic approximation. The results show that the maximal and minimal solar wind corotating stream velocities appear at the intermediate heliospheric latitudes depending on the tilt angle. The high-speed streams with the strongest velocity difference between maximal and minimal values are expected to be observed at the middle heliolatitudes near 45 degrees. Other consequences of the model are indicated and may be checked during the rapid pole-to-pole Ulysses passage in 1994-1995.

Key words: solar wind, high heliolatitudes, stream structure, tilted rotator

1. Introduction

A three-dimensional large-scale solar wind stream structure at high heliolatitudes is determined by the heliomagnetic organization at the source region. We consider theoretically the tilted rotator model which was first suggested by Zhao and Hundhausen (1981) to describe solar wind plasma properties. Based on this approach we introduce the simplest angular dependence on the source surface for the plasma flow velocity. The radial evolution is described by the collisionless supersonic and superalfvenic approximation. This approximation is valid in the region where characteristics do not cross each other. Kinematic effects play the dominant role in this region around the Sun and may be calculated using an expansion in inverse Mach and Mach-Alfven numbers of governing fluid or kinetic equations for ions. Both fluid (Matsuda and Sakurai, 1972; Shabansky and Shister, 1973; Sakurai, 1983) and gas-kinetic (Veselovsky, 1980, 1981) approaches give the same zero order results for lowest momenta (density and velocity) in a maxwellian plasma.

A tilted dipole two-sector solar wind MHD numerical simulations have been performed by Pizzo (1983, 1991, 1994) to investigate 2-D and purely 3-D dynamic processes in alternate corotating interaction regions mainly near the heliographic equator.

We use the gas-kinetic approximation for obtaining the global heliolatitudinal and heliolongitudinal velocity pattern based on the tilted rotator model.

© 1995 *Kluwer Academic Publishers*

2. Simple Model

The solar wind bulk velocity on the source surface $r = r_0$ is assumed to be radial $V(\vartheta, \varphi, t)$ and can be expanded on spherical harmonics $Y_{lm}(\vartheta, \varphi)$ as the function of the helio-colatitude ϑ and heliolongitude φ

$$V_0(\vartheta, \varphi, t) = \sum_{l=0}^{\infty} \sum_{m=-l}^{+l} \int_{-\infty}^{+\infty} V_{lm}(\omega) Y_{lm}(\vartheta, \varphi) e^{-i\omega t} \, d\omega. \tag{1}$$

The function $V(\vartheta, \varphi, t)$ is known when all spectral amplitudes $V_{lm}(\omega)$ are given.

The ion distribution function obeys the equation

$$\frac{\partial f}{\partial t} + \mathbf{v} \frac{\partial f}{\partial \mathbf{r}} = 0. \tag{2}$$

We assume the ion distribution function $f(\mathbf{r}, \mathbf{v}, t)$ to be locally Maxwellian at $r = r_0$ with given boundary conditions for density, velocity and temperature. In this case the solution of the governing equation (2) may be constructed along characteristics

$$\mathbf{r} = \mathbf{r}_0 + \int_{t_0}^{t} \mathbf{V}_0(\vartheta, \varphi, t') \, dt'. \tag{3}$$

The solar wind bulk velocity of corotating streams is equal to

$$V(r, \vartheta, \varphi, t) = V_0(\vartheta, \varphi_0), \tag{4}$$

where

$$\varphi_0 = \varphi + \Omega \left(t - \frac{r - r_0}{V_0} \right) \tag{5}$$

in the case $\frac{\partial V_0}{\partial t} = 0$, $\Omega = \frac{2\pi}{T}$ is angular velocity of the solar rotation.

The boundary value $V_0(\vartheta, \varphi_0)$ at the initial surface $r = r_0$ near the Sun is known very poor. Following Zhao and Hundhausen (1981) we retain in Formula (1) only two terms: a monopole (point source) and an axial quadrupole:

$$V_0(\vartheta, \varphi_0) = V(1 + a \cos^2 \vartheta_m), \tag{6}$$

where V and a are constants, ϑ_m is the heliomagnetic colatitude,

$$\cos \vartheta_m = \cos \vartheta \cos \alpha + \sin \vartheta \sin \alpha \cos \phi, \tag{7}$$

$\phi = \varphi_0 - \varphi_\alpha$ is the azimuthal phase angle, α is the magnetic axis polar angle, φ_α is the magnetic axis azimuth.

The Formula (6) represents the simplest case for symmetric northern and southern polar coronal holes when the dipole term is absent in the

Formula (1). This Formula can be considered as a rough approximation, which will be corrected after the rapid pole-to-pole Ulysses passage in 1994-1995. Not only numerical values in Formula (6) will be determined, but the additional (dipole) term proportional to $\cos\vartheta_m$ and characterizing the north-south asymmetry in the solar wind source region can be estimated for the first time from in situ solar wind measurements.

3. Corotating Stream Structure

Let us consider the large-scale corotating solar wind stream structure given by Equations (4-7) as it would be observed onboard Ulysses spacecraft. The parameters V, a, α and spacecraft position $\vartheta(t)$ are slowly varying functions of the time. These variations of the parameters V, a with characteristic time scales of the order of years are poorly known. Faster changes of $\alpha(t)$ during several solar rotations are possible. All these variations may be considered as rather slow amplitude modulations superimposed on the main solar rotation period.

Taking into account only this fastest phase ϕ variation the maximal and the minimal velocities in streams can be found for fixed V, a, α, ϑ parameters during each solar rotation. These extremal velocities are determined by the condition $\frac{\partial V}{\partial \phi} = 0$ combined with the Equations (4-7) which results in the Equation

$$\cos\vartheta_m \sin\vartheta \sin\alpha \sin\phi = 0. \tag{8}$$

The solutions of the Equation (8) are

$$\vartheta_m = \frac{\pi}{2}, \tag{9}$$

$$\vartheta = 0, \pi, \tag{10}$$

$$\phi = 0, \pi, \ \vartheta \text{ is arbitrary} . \tag{11}$$

It is seen from Formulae (9-11) that extremal velocities are reached at two perpendicular planes: 1) the magnetic equatorial plane (9) with minimal velocities; 2) the meridional plane (11) containing the rotation axis and the magnetic axis of the Sun. The extrema in the plane (11) are maxima for $0 < \vartheta < \pi/2 + \alpha$, $\varphi = 0$ and $\pi/2 - \alpha < \vartheta < \pi$, $\varphi = \pi$ or minima for $\pi/2 + \alpha < \vartheta < \pi$, $\varphi = 0$ and $0 < \vartheta < \pi/2 - \alpha$, $\varphi = \pi$.

There exists one maximum and one minimum in the corotating solar wind stream velocities per each solar rotation at near polar regions $0 < \vartheta < \alpha$ and $\pi - \alpha < \vartheta < \pi$. There are two maxima and two minima per each solar rotation in the near equatorial region $\pi/2 - \alpha < \vartheta < \pi/2 + \alpha$. This pattern corresponds to the well known one- or two sector IMF and corotating solar wind stream structure. It is usually observed during declining and minimum solar cycle phases.

From Equations (4-7) we obtain for maximal velocities

$$V_{max} = V\left[1 + a\cos^2(\vartheta - \alpha)\right], \quad \begin{array}{ll} 0 < \vartheta < \pi/2 + \alpha\ , & \varphi = 0; \\ \pi/2 - \alpha < \vartheta < \pi\ , & \varphi = \pi. \end{array} \tag{12}$$

and for minimal velocities

$$V_{min} = V\begin{cases} 1 & , \vartheta_m = \pi/2 \\ 1 + a\cos^2(\vartheta + \alpha) & , \begin{array}{l} \pi/2 + \alpha < \vartheta < \pi\ , \quad \varphi = 0; \\ 0 < \vartheta < \pi/2 - \alpha\ , \varphi = \pi. \end{array} \end{cases} \tag{13}$$

It is evident that the maximum maximorum velocities are reached at $\vartheta = \alpha$, $\varphi = 0$ and $\vartheta = \pi/2 - \alpha$, $\varphi = \pi$ on the projection of the magnetic axis $\vartheta_m = 0, \pi$. The minimum minimorum velocities are reached in the heliosphere on the projection of the magnetic equator along Archimedean spirals (5) starting from the source surface $r = r_0$. Weaker minima exist in the polar regions during the each solar rotation on the side opposite to the maxima for the fixed ϑ values.

It is easy to calculate from Equations (12, 13) the amplitude $\Delta V = V_{max} - V_{min}$ of the corotating solar wind streams as a function of the heliolatitude $\lambda = \pi/2 - \vartheta$

$$\Delta V = Va\begin{cases} \sin^2(\lambda + \alpha) & , 0 < |\lambda| < \alpha, \\ \sin 2\lambda \sin 2\alpha & , \alpha < |\lambda| < \pi/2. \end{cases} \tag{14}$$

Formula (14) shows that the amplitude of the corotating solar wind streams is growing with the heliolatitude in the near equatorial region till the maximal value reached at the intermediate latitudes and than is gradually diminishing at higher heliolatitudes and vanishing at the heliographic poles. In the case $\alpha \leq \pi/4$ the maximal amplitude

$$max\,\Delta V = V\,a\,\sin 2\alpha \tag{15}$$

is reached when $\lambda = \pi/4$. In the case $\alpha \geq \pi/4$ the maximal amplitude

$$max\,\Delta V = V\,a \tag{16}$$

is reached when $\lambda = \pi/2 - \alpha$.

The approximation (6, 7) holds for the period of the declining solar cycle with the tilt α being minimal during the solar cycle minimum. The tilt was approximately 29° in April 1993 (Bame et al., 1993). It may be expected that the tilt will decline according to the present solar cycle phase in the analogy with the previous solar cycle (Hoeksema, 1991).

4. Discussion and Conclusions

Kinetic breaking starts near crossing characteristics at heliocentric distances $r = r_0 + \frac{V}{\Omega a}\left(1 + a \cos^2 \vartheta_m\right)^2 \left(2 \cos \vartheta_m \sin \alpha \sin \vartheta \sin \varphi_0\right)^{-1}$. Interpenetrating solar wind streams and corotating shocks are formed here (Matsuda and Sakurai, 1972; Shabansky and Shister, 1973; Veselovsky, 1981).

More complicated and realistic approximations are possible instead Formula (6). This will change details in calculations. Qualitative results are the same for models with different heliomagnetic organizations considered by Zhao and Hundhausen (1981).

According to the simple model considered here the strongest velocity difference between maximal and minimal values in the corotating solar wind streams is expected to be observed at the middle latitudes near 45 . Other consequences of the model: 1) The strength and the number of corotating shocks are expected to be relatively low near the heliographic poles; 2) Ion distribution functions with interpenetrating solar wind streams along a nearly radial magnetic field are expected to be more frequent phenomena at intermediate heliolatitudes; 3) Tangential discontinuities occurrence frequency may have two broad maxima as a function of the heliolatitude because of the "Maunder's butterfly" type boundary conditions.

The validity of the simple model may be checked during the rapid pole-to-pole Ulysses passage in 1994-1995.

Acknowledgements

The author's participation in the 28th ESLAB Symposium "High latitude heliosphere" has been supported by the Organizing Committee chaired by Dr. R.G.Marsden. The author thanks Dr. A.R.Moszhukhina for her help in the preparation of the text.

References

Bame, S.J., Goldstein, B.E., Gosling J.T., Harvey, J.W., McComas, D.J., Neugebauer, M. and Phillips, J.L.: 1993,'Ulysses observations of a recurrent high speed solar wind stream and the heliomagnetic streamer belt', *Geophys.Res.Lett.*,**20**, 2323.

Hoeksema, J.T.: 1991, in E.Marsch and R.Schwenn (eds.),'Large scale structure of the heliospheric magnetic field:1979-1991', *Solar Wind Seven*, Pergamon Press, Oxford, 191.

Matsuda, T. and Sakurai, T.: 1972, 'Dynamics of the azimuthally dependent solar wind', *Cosmic Electrodynamics*,**3**, 97.

Pizzo, V.J.: 1983, 'Quasi-steady solar wind dynamics',*Solar Wind Five*, NASA CP-2280, 675.

Pizzo, V.J.: 1991, 'The evolution of corotating stream fronts near the ecliptic plane in the inner solar system. II. Three-dimensional tilted-dipole fronts',*J.Geophys.Res.*,**96**, 5405.

Pizzo, V.J.: 1994, 'Global, quasi-steady dynamics of the distant solar wind. 1. Origin of north-south flows in the outer heliosphere', *J.Geophys.Res.*,**99**, 4173.

Sakurai, T.: 1983, 'Analytical method of characteristics applied to the azimuthally dependent solar wind', *Mon. Not. Roy. Astr. Soc.*, **203**, 1187.
Shabansky, V.P. and Shister, A.R.: 1973, 'Supersonic plasma outflow from the rotating source', *Doklady USSR Academy of Sciences*, **211**, 825 (in Russian).
Veselovsky, I.S.: 1980, 'Kinetic theory of weak inhomogeneities in the solar wind', *Geomagnetism and Aeronomy*, **20**, 769 (in Russian).
Veselovsky, I.S.: 1981, 'Modelling of nonstationary processes in the solar wind', *Geomagnetism and Aeronomy*, **21**, 968 (in Russian).
Zhao, X.-P. and Hundhausen, A.J.: 1981, 'Organization of solar wind plasma properties in a tilted heliomagnetic system', *J.Geophys.Res.*, **86**, 5423.

ULYSSES SOLAR WIND OBSERVATIONS TO 56° SOUTH

D.J. McCOMAS, J.L. PHILLIPS, S.J. BAME, J.T. GOSLING
Los Alamos National Laboratory, Los Alamos, NM USA

B.E. GOLDSTEIN and M. NEUGEBAUER
Jet Propulsion Laboratory, Pasadena, CA USA

Abstract. In the 25 months since Jupiter flyby, the Ulysses spacecraft has climbed southward to a heliolatitude of 56°. This transit has been marked by an evolution from slow, dense coronal streamer belt solar wind through two regions where the rotation of the Sun carried Ulysses back and forth between streamer belt and polar coronal hole flows, and finally into a region of essentially continuous fast, low density solar wind from the southern polar coronal hole. Throughout these large changes, the momentum flux normalized to 1 AU displays very little systematic variation. In addition, the bulk properties of the polar coronal hole solar wind are quite similar to those observed in high speed streams in the ecliptic plane at 1 AU. Coronal mass ejections and forward and reverse shocks associated with corotating interaction regions have also been observed at higher heliolatitudes, however they are seen less frequently with increasing southern heliolatitude. Ulysses has thus far collected data from 20° of nearly contiguous solar wind flows from the polar coronal hole. We examine these data for characteristic variations with heliolatitude and find that the bulk properties in general show very little systematic variation across the southern polar coronal hole so far.

1. Introduction

The Ulysses spacecraft has been returning essentially continuous measurements of the interplanetary particles and fields environment since turn-on shortly after launch in October 1990. Ulysses will achieve its highest southern heliolatitude of 80° S in September 1994. At this writing, Ulysses plasma data are available through 11 March 1994 when Ulysses was at a heliolatitude of 55.6° S.

The solar wind plasma instruments on Ulysses are curved plate electrostatic analyzers which independently measure the distributions of electrons and ions as functions of spatial look direction and energy per charge (E/q) (Bame et al., 1992). The ion instrument covers an energy range from 255 eV/q to 34.4 keV/q over all look directions that the super-Alfvenic solar wind beam can come from. The electron instrument measures particles incident from >95% of 4π sr continuously from 1.6 to 862 eV.

Two previous studies of the Ulysses plasma observations have examined the latitudinal evolution of bulk solar wind properties (Bame et al., 1993; Phillips et al., 1994), In brief, Ulysses has transited from observing primarily slow, dense solar wind from the coronal streamer belt to fast, low density solar wind from the southern polar coronal hole. In this study we examine and extend the analysis of this evolution in the observed solar wind and provide a more detailed analysis of the 20° of observations since Ulysses entered the polar coronal hole.

Space Science Reviews **72**: 93–98.
© 1995 *Kluwer Academic Publishers.*

2. Observations

Figure 1 displays 6-hour averages of the solar wind proton speed, proton density, proton flux, and proton mass flux/amu observed by Ulysses between Jupiter and 56° S. The three bottom panels are normalized by the heliocentric distance squared to adjust to 1 AU equivalent values. The triangles along the top axis of the top panel indicate the heliocentric distance of the Ulysses spacecraft, while those along the bottom indicate the heliolatitude in 5° increments from 6°S at Jupiter to 56°S.

The solar wind speed and density display significant variations with heliolatitude; these variations can be characterized by four distinct regions. At low heliolatitudes (<13° S) Ulysses remained primarily in higher density, slow solar wind flow (<500 km s^{-1}) from near the solar coronal streamer belt. From 13° to 29° S Ulysses oscillated in each solar rotation between observing a low density, high speed solar wind from the southern polar coronal hole and the high density, slow solar wind (Bame et al., 1993). Since it is heliomagnetic latitude and not heliolatitude that orders coronal structures, such periodic variations are undoubtedly due to the effects of solar rotation with an offset magnetic axis. The change at 13° S to recurrent coronal hole flows was probably caused by a change in the solar corona and consequently in the solar wind structure (Bame et al., 1993). From 29° to 36° S Ulysses continued to oscillate between the high and low speed solar wind, but penetrated less deeply into the streamer belt flow as evidenced by higher minimum speeds. Finally, from 36° to 56° Ulysses has remained essentially immersed in the high speed solar wind flow from the southern polar coronal hole.

In addition to streamer belt and polar hole solar flows, Ulysses observed a number of coronal mass ejections (CMEs) and forward and reverse shocks in the solar wind. Many of these are not clearly evident in the six-hour averages shown in Figure 1, however, several can be seen, primarily as large decreases in proton density and flux (e.g., near 20°, 32°, and 35° S). Gosling et al. (1994a) documented six CMEs at mid latitudes, between 31° and 41.3° S; since then two additional CMEs have been identified at a higher heliolatitudes (52.3° and 54.3° S). Forward and reverse shocks associated with corotating stream interaction regions are also present in the Ulysses data; however, fewer and fewer of the forward shocks and then the reverse shocks are seen with increasing heliolatitude (Gosling et al., 1993). The highest heliolatitude forward and reverse shocks documented previously were at 33.6° and 45.6° S, respectively (Phillips et al., 1994). Since then, both a forward-reverse shock pair (at 54.1° and 54.5° S, respectively) was observed associated with the most recent CME (54.3° S); these were probably caused by the expansion of the CME as described for a CME at 32.5° S (Gosling et al., 1994b). Finally another reverse shock was observed on 10 March 1994 (55.5° S).

While the proton density displays roughly a factor of two decrease on average from ~7 cm^{-3} near the ecliptic to 3.4 cm^{-3} at high heliolatitudes, the proton flux displays only an ~30-40% decrease, and the proton momentum flux/amu displays little, if any, characteristic variation with heliolatitude. There are, however, large amplitude, short period variations in these six hour averages of all three of these parameters with each successive parameter displaying smaller variations. Finally, the variability of all the

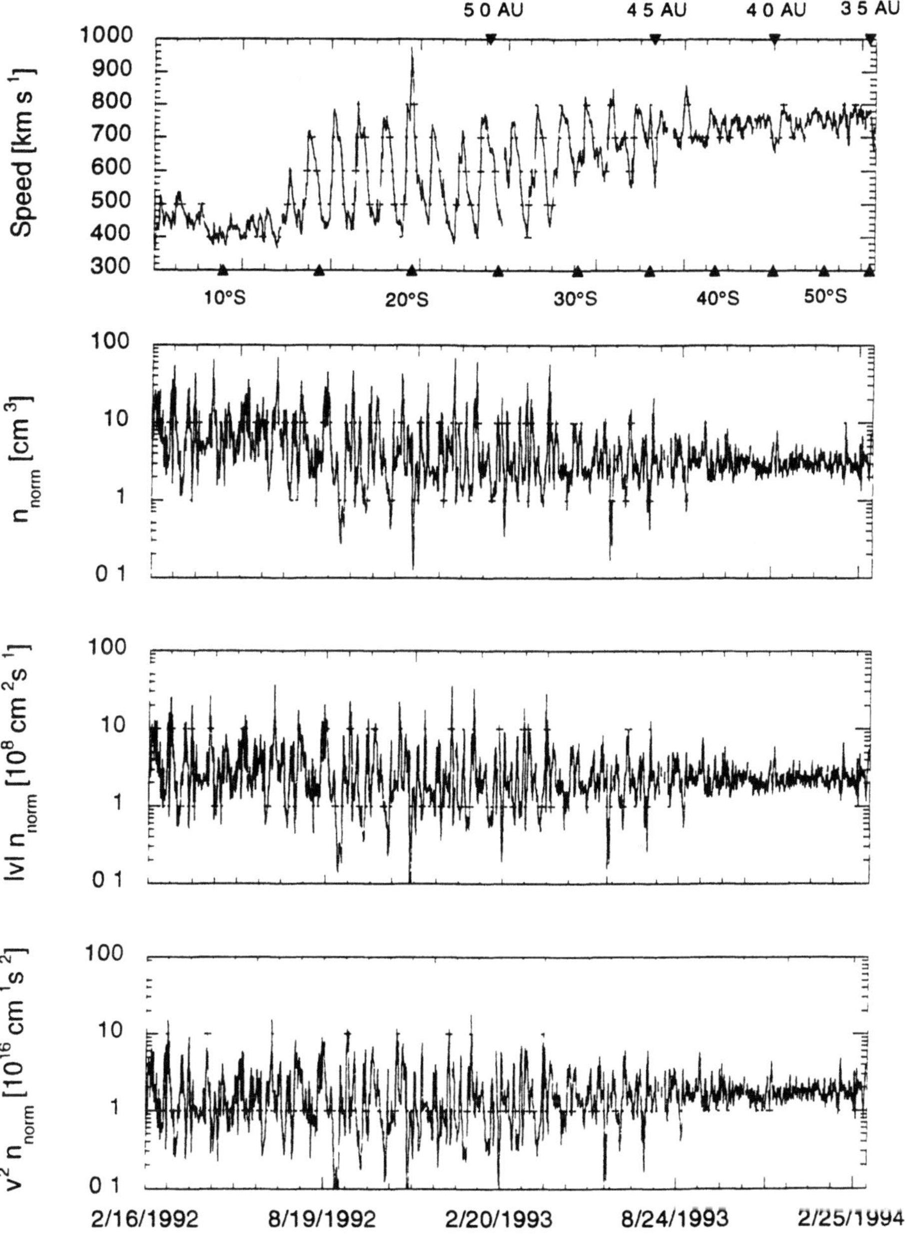

Figure 1 Six-hour averaged solar wind speed, proton density, proton flux, and proton momentum flux/amu observed with the Ulysses solar wind ion experiment. The latter three parameters have been normalized to 1 AU Four general epochs of observations are separated by vertical lines: streamer belt (SB), streamer belt/polar hole (SB/PH), partial streamer belt/ polar hole (pSB/PH), and polar hole (PH)

parameters decreased substantially after Ulysses entered the polar hole flow.

As pointed out above, Ulysses began periodically observing solar wind from the polar coronal hole at a southern heliolatitude of only ~13° and has been intermittently observing such flows for at least a portion of each solar rotation since. These intervals correspond to the periodic peaks in solar wind speed and drops in solar wind density in Figure 1. Other than the anomalously high speed peak at ~20° S, associated with a very fast CME, the polar hole intervals have relatively constant peak speeds between 700 and 850 km s^{-1}. These observations suggest that there has been little variation in the solar wind bulk flow from the southern polar hole over the past 19 months. Solar wind observations since the last Ulysses update (Phillips et al., 1994) indicate that, other than a few transient CMEs, Ulysses has remained immersed in solar wind from the southern coronal polar hole for nearly 20° of heliolatitude, from 36° to 56° S.

Figure 2 displays 1-hour averages of selected solar wind parameters over the polar hole region from 36° to 56° S. From top to bottom these are proton speed, proton density (normalized to 1 AU as in Figure 1), normalized mass flux, normalized proton momentum flux/amu, proton temperature, and relative helium abundance (n_α/n_p). Linear fits to the data are superposed on each of the curves. The average solar wind speed appears to increase gradually with heliolatitude over this interval while the other five parameters display less characteristic changes compared to their levels of variability.

Table 1 summarizes the results of the linear fit analysis. For comparison, the first column gives mean values observed in high speed streams (HSSs) in the ecliptic plane at 1 AU (Feldman et al., 1977). The remaining columns give the average value of the parameter observed over from 36° to 56° S, the slope (i.e., change in parameter per degree of heliolatitude south), the slope normalized to the average value (i.e., percent change per degree of heliolatitude south), and the extrapolated value at 80° S (September 1994). Note that all six of the solar wind parameters given show less than or about 0.5% change per degree of heliolatitude and that of all the parameters given, the proton momentum flux/amu, $v^2\, n_{norm}$, is the most constant over the polar hole region just as it was extremely constant, on average, over the transition from low to high heliolatitudes. The only major difference between the parameters observed in the southern polar coronal hole flows and high speed streams observed in the ecliptic is in the proton temperature which has not been corrected for heliocentric distance.

Table 1.

	Mean in HSSs at 1 AU	Average in PH	Slope	$\frac{\text{Slope}}{\text{Average}}$	Extrapolated to 80°
$\lvert v\rvert$ [km s^{-1}]	702	740	1.9	0.3%	810
n_{norm} [cm^{-3}]	3.9	3.4	-0.017	0.5%	2.8
$\lvert v\rvert\, n_{norm}$ [10^8 $cm^{-2}s^{-1}$]	2.7	2.5	-0.0048	0.2%	2.3
$v^2\, n_{norm}$ [10^{16} cm^{-1} s^{-2}]	1.9	1.8	0.0020	0.1%	1.9
T_p [10^5 K]	2.3	1.5	-0.0064	0.4%	1.3
n_α/n_p	0.048	0.043	-0.00013	0.3%	0.039

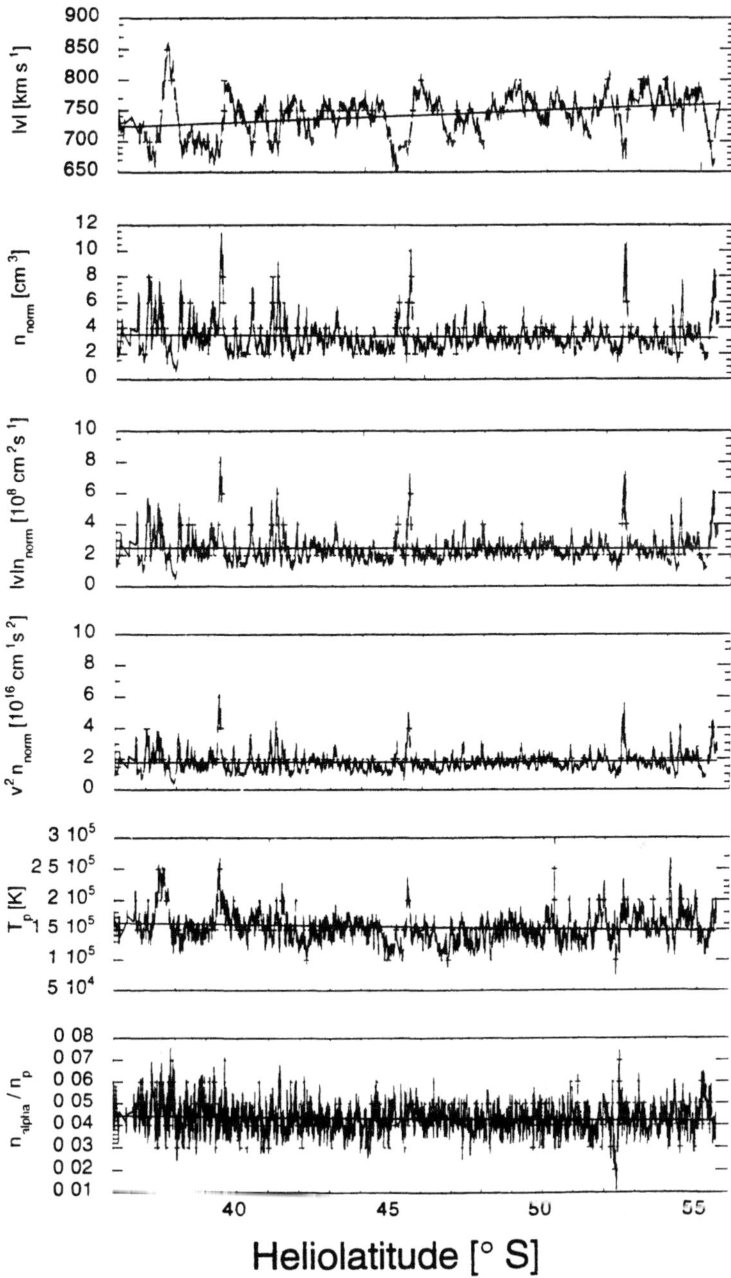

Figure 2 Observations of selected solar wind plasma parameters since Ulysses became continuously immersed in the polar hole flow at a heliolatitude of 36° S Linear fits to these parameters are overplotted on each curve; fit parameters are given in Table 1

4. Conclusions

The Ulysses solar wind instruments are fulfilling their primary objective of charting the solar wind properties out of the ecliptic plane, over the poles of the Sun. Thus far in the mission, Ulysses has achieved a heliolatitude of 56° S and has transited from slow, dense solar wind flows from the coronal streamer belt to fast, less dense solar wind from the southern polar coronal hole. Throughout these changes in average speed and density, the momentum flux was quite constant on average. Ulysses observed two other regions at intermediate heliolatitudes where it repeatedly sampled both higher and lower heliomagnetic latitudes as the rotation of the Sun carried both streamer belt and polar hole types of solar wind back and forth across the spacecraft. By examining the peak vaules in the polar hole portions of these rotations, we demonstrated that the solar wind speed from the southern polar coronal hole appears to have been relatively constant over the 19 months that Ulysses has been at least intermittently sampling it. In addition, the bulk properties of the solar wind from the polar coronal hole are generally quite similar to those normally observed in high speed streams in the ecliptic plane at 1 AU.

A more detailed analysis of the last 20° of essentially continuous polar hole flow indicates that the bulk properties of the solar wind from this region are relatively constant and do not display strong variations with increasing heliolatitude; this is particularly true of the proton flux which displays the least average percentile change over the polar hole observations thus far. Nonetheless, small characteristic variations, such as the nearly 2 km s^{-1} deg^{-1} increase in the average bulk solar wind speed, may be real. If the solar wind continues to increase with heliolatitude at this rate, an average solar wind speed of over 800 km s^{-1} would be observed at the maximum Ulysses heliolatitude of 80° S in September 1994. The next several months of Ulysses observations hold the key to what the polar solar wind is really like.

Acknowledgements. This work was carried out under the auspices of the United States Department of Energy with support from the National Aeronautics and Space Administration.

References

Bame, S.J., *et al.*: 1992, 'The Ulysses solar wind plasma experiment', *Astron. and Astrophys., Suppl. Ser.*, **9 2**, 221.

Bame, S.J., *et al.*: 1993, 'Ulysses observations of a recurrent high speed solar wind stream and the heliomagnetic streamer belt', *Geophys. Res. Lett.*, **2 0**, 2323.

Feldman, W.C., *et al.*: 1977, Plasma and magnetic fields from the Sun, The solar output and its variation, ed., O.R. White, Col. Assoc. Univ. Press, Boulder.

Gosling, J.T. *et al.*: 1993, Latitudinal variation of solar wind corotating interaction regions: Ulysses, *Geophys. Res. Lett.*, **2 0**, 2789.

Gosling, J.T. *et al.*: 1994a, The speeds of coronal mass ejections in the solar wind at mid heliographic latitudes: Ulysses, submitted to *Geophys. Res. Lett.*

Gosling, J.T. *et al.*: 1994b, A forward-reverse shock pair in the solar wind driven by over-expansion of a coronal mass ejection:Ulysses observations, *Geophys. Res. Lett.*, **2 1**, 237.

Phillips, J.L., *et al.*: 1994, 'Ulysses at 50° south: constant immersion in the high-speed solar wind', submitted to *Geophys. Res. Lett.*

SOLAR WIND COROTATING STREAM INTERACTION REGIONS OUT OF THE ECLIPTIC PLANE: ULYSSES

J.T. GOSLING[1], S.J. BAME[1], D.J. MCCOMAS[1], J.L. PHILLIPS[1], V.J. PIZZO[2], B.E. GOLDSTEIN[3], and M. NEUGEBAUER[3]

[1]Los Alamos National Laboratory, Los Alamos, New Mexico USA

[2]San Juan Institute, San Juan Capistrano, California USA

[3]Jet Propulsion Laboratory, Pasadena, California USA

Abstract. Ulysses plasma observations reveal that the forward shocks that commonly bound the leading edges of corotating interaction regions (CIRs) beyond ~2 AU from the Sun at low heliographic latitudes nearly disappeared at a latitude of S26°. On the other hand, the reverse shocks that commonly bound the trailing edges of the CIRs were observed regularly up to S41.5°, but became weaker with increasing latitude. Only three CIR shocks have been observed poleward of S41.5°; all of these were weak reverse shocks. The above effects are a result of the forward waves propagating to lower heliographic latitudes and the reverse waves to higher latitudes with increasing heliocentric distance. These observational results are in excellent agreement with the predictions of a global model of solar wind flows that originate in a simple tilted-dipole geometry back at the Sun.

The solar wind at low heliographic latitudes is often organized into alternating streams of high and low speed flows, particularly on the declining phase of the solar activity cycle. While each parcel of solar wind plasma flows nearly radially out from the Sun, the stream patterns tend to be quasi-stationary in time and appear to corotate with the Sun. With increasing heliocentric distance the high-speed streams steepen and overtake slower solar wind ahead forming compressive, corotating interaction regions (CIRs) on the rising speed portions of the streams (Again, it is the pattern of interaction regions that corotates.) A CIR is a region of high pressure; its leading edge is a forward wave that propagates into the slow solar wind ahead, while the trailing edge is a reverse wave that propagates backward into the high-speed flow. These waves commonly steepen into forward and reverse shocks that bound the CIR at heliocentric distances beyond ~2 AU (e.g., Hundhausen and Gosling, 1976; Smith and Wolfe, 1976). Here we (1) use Ulysses plasma observations to extend to much higher latitudes our previous work (Gosling *et al.*, 1993) on the latitudinal evolution of CIRs, and (2) relate the Ulysses observations to a global 3-dimensional simulation of corotating solar wind flows.

Since July 1992 when the spacecraft was at ~S13°, the plasma experiment on Ulysses has detected a single, broad high-speed stream with peak speed greater than 700 km s^{-1} during each solar rotation. Minimum speeds in the troughs between the reappearances of the stream were ~400 km s^{-1} until Ulysses passed poleward of ~S28° in April 1993, at which time the minimum speed rose to ~550 km s^{-1}. The combined plasma and

Space Science Reviews **72**: 99–104.
© 1995 *Kluwer Academic Publishers.*

field observations from Ulysses have been used to infer that the coronal streamer belt and its embedded heliospheric current sheet were tilted at ~29° to the solar equatorial plane in April 1993 (Bame *et al.*, 1993; Smith *et al.*, 1993). Since July 1993 Ulysses has been immersed continually in high speed flows greater than ~650 km s^{-1} (Phillips *et al.*, 1994).

Figure 1 provides a summary of Ulysses shock observations from July 1992 when the spacecraft was at ~S12° through April 9, 1994 when the spacecraft was at ~S59°. The figure shows as a function of heliographic latitude a plot of the ratio of downstream to upstream density minus one (a measure of shock strength) for each observed shock. Forward shocks are plotted above the horizontal axis and reverse shocks are plotted below the axis. Most of the forward shocks without matching reverse shocks equatorward of S26° appear to have been associated with interplanetary disturbances driven by fast coronal mass ejections (CMEs). The relatively weak shock pair observed near S32.5° was associated with over-expansion of a CME (Gosling *et al.*, 1994) as was the considerably stronger shock pair observed more recently near S54°. The remainder of the shocks were associated with CIRs.

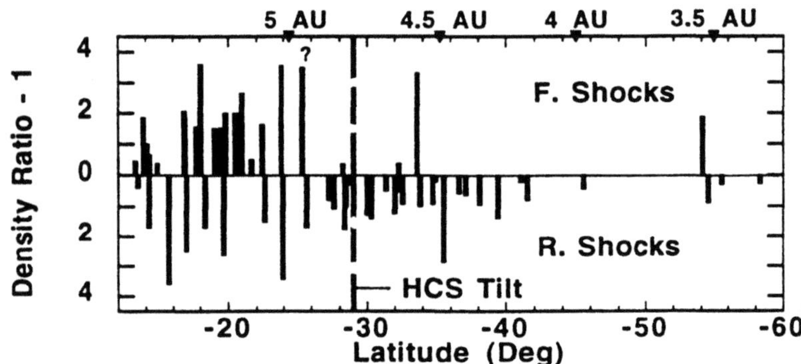

Fig 1. Bar diagram of shock "strength" vs latitude for all shocks observed by the plasma experiment on Ulysses from S12° to S59°. Heliocentric distance of the observations is indicated by the triangles at the top of the plot.

Several interesting features are evident in Figure 1. At latitudes below ~S26° the CIRs were usually bounded by strong forward-reverse shock pairs. Above that latitude, which corresponded roughly to the tilt of the streamer belt in April 1993, Ulysses encountered only two CIR forward shocks. No CIR forward shocks have been detected since Ulysses passed S33.6° in late June 1993. On the other hand, CIR reverse shocks continued to be observed on a regular basis up to a latitude of ~S41.5°, ~12.5° poleward of the tilt of the streamer belt. These shocks were almost always weaker than those observed at lower latitudes. Only three CIR reverse shocks have been observed poleward of S41.5°.

The left panel of Figure 2 shows selected plasma parameters encompassing a CIR on the leading edge of a stream observed when Ulysses was at S19.5°. Since both the

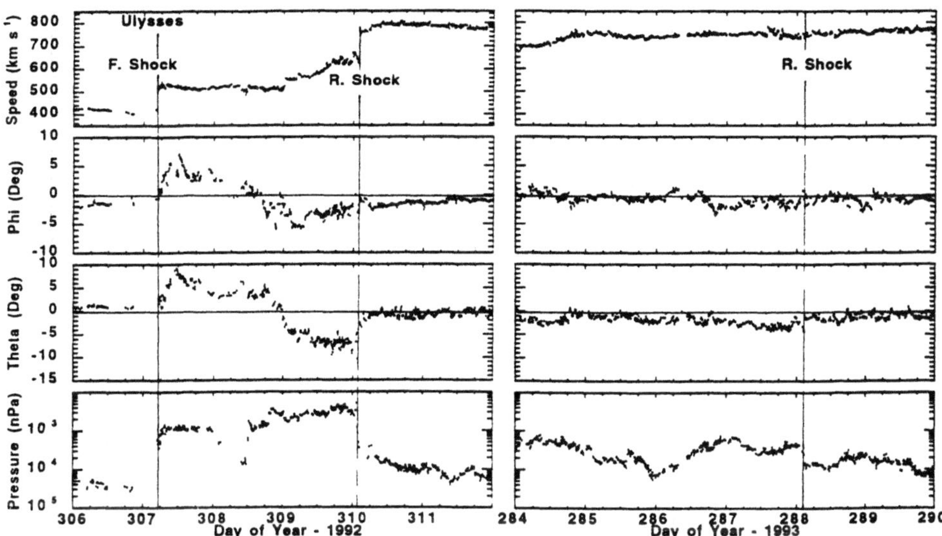

Fig. 2. (Left) A corotating interaction region observed at 5.12 AU and S19.5°. Parameters plotted from top to bottom are solar wind flow speed, azimuthal and meridional flow angles (positive for flow in the direction of planetary motion about the Sun and equatorward, respectively), and proton thermal pressure. (Right) A weak reverse shock observed at 4.18 AU and S41.5°. Parameter scaling is as on the left.

forward and reverse shocks were convected away from the Sun by the supersonic flow of the solar wind, Ulysses sampled the region downstream of the forward shock after its passage and the region downstream of the reverse shock prior to its passage. The flow was deflected equatorward and westward (i.e., in the sense of planetary motion about the Sun) as it crossed the forward shock and poleward and eastward as it crossed the reverse shock. Deflections of a similar nature were observed for most of the other CIRs encountered at latitudes equatorward of S26° (Gosling *et al.*, 1993). Such deflections provide important information on the orientations of the CIRs and the propagation directions of the bounding waves. In particular, they indicate that, relative to an observer moving with the solar wind, the forward waves preferentially were propagating antisunward, westward, and equatorward while the reverse waves were preferentially propagating sunward, eastward, and poleward.

The reverse shocks tended to get weaker as Ulysses moved poleward. The right panel of Figure 2 shows plasma parameters for an interval encompassing the weak reverse shock observed at S41.5°. Not only was there no accompanying forward shock, but the reverse shock itself was relatively poorly distinguished in the data (compare with the event on the left). A small poleward deflection of the flow was observed downstream (earlier in time) of the shock. Similar or larger deflections were measured downstream of all but two of the reverse shocks observed poleward of S26°, indicating that most of the reverse shocks at high latitudes were propagating poleward, as at lower latitudes.

The sketches shown in Figure 3 provide a qualitative explanation for these effects. On the left a band of slow solar wind is shown girding the Sun at a modest tilt relative to the solar equatorial plane. This band is surrounded on either side by regions of fast solar wind flow. We nominally associate the band of slow wind with the coronal streamer belt with its embedded heliospheric current sheet and the fast wind regions with coronal holes. The overall tilt of the slow solar wind band is caused by the tilt of the Sun's magnetic dipole axis relative to its rotation axis. As the Sun rotates the fast wind overtakes the slow wind along interfaces in interplanetary space that are also tilted relative to the equator. The tilts of these interaction regions are opposed in opposite hemispheres. With increasing heliocentric distance in both hemispheres the forward waves bounding the interaction regions propagate westward and equatorward and eventually across the equator into the opposite hemisphere, while the reverse waves propagate eastward and poleward and eventually to latitudes above the band of slow solar wind flow.

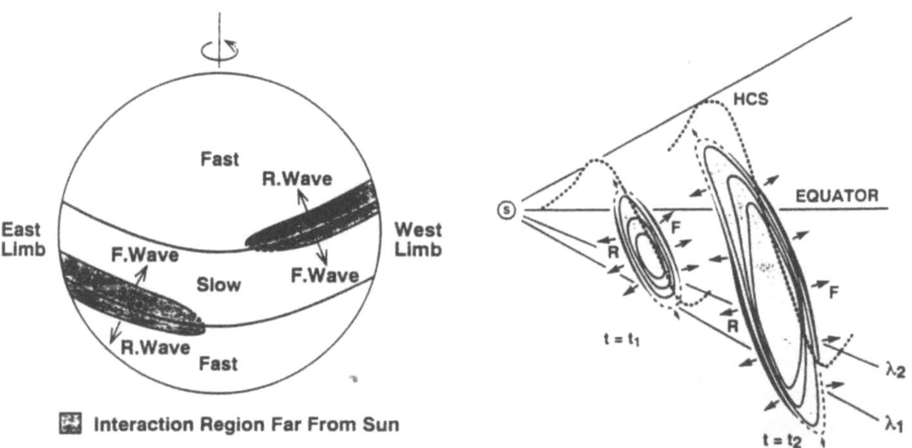

Fig. 3. (Left) A sketch illustrating the origin of tilted CIRs in interplanetary space in terms of a tilted-dipole stream structure back at the Sun. See text. (Right) Snapshots of the southern CIR at two different times. The CIR is tilted in the same sense as the heliospheric current (HCS). The solid line of varying thickness surrounding the interaction region marks the forward (F) and reverse (R) shocks, with the thickness of the line indicating the strength of the shocks. Solid lines within the CIR are pressure contours. Both sketches adapted from Gosling *et al.* (1993).

The right portion of Figure 3, which shows a meridional cut through the southern CIR at two different times, helps illustrate the evolution of the CIRs with increasing heliocentric distance. The sketches are based upon simulations to be described. We note that the CIR has roughly the same tilt in a meridional plane as the heliospheric current sheet since both originate from the same geometry. The tilt illustrated arises because the high latitude portion of a CIR arrives in any meridional plane before the low latitude portion. With increasing heliocentric distance the CIR expands in the direction transverse to the tilt, with the reverse wave propagating to higher latitudes and the forward wave propagating to lower latitudes and across the equator. The shocks are strongest at latitudes where the direct interaction between the high and low speed flows actually occurs; they weaken as they propagate to higher and lower latitudes.

The geometry illustrated in the left portion of Figure 3 has been used to specify the inner boundary conditions for a 3-D MHD simulation of corotating solar wind flows using a numerical code previously described (Pizzo, 1994a,b). The input flow configuration is that of a dipole tilted 30° to the solar spin axis, with fast, hot, low-density flow emanating radially from regions centered on the dipole axis and slow, cold, dense flow emanating from a belt girding the dipole equator. Parameter values are chosen at the inner boundary of the calculation at 0.15 AU to produce a 310 - 750 km s^{-1} speed range at 1 AU, and the calculation is carried out over a latitude range of +/- 64° and out to a heliocentric distance of 10 AU (Pizzo and Gosling, 1994).

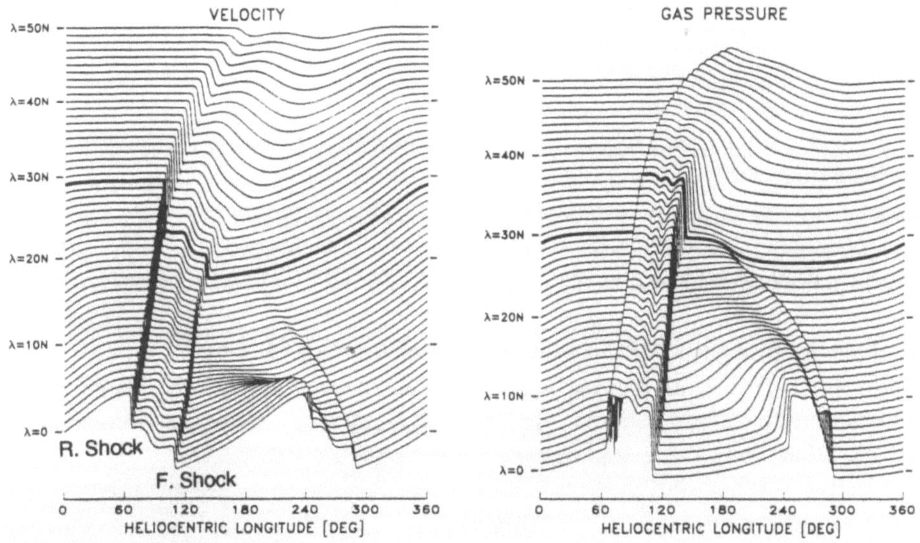

Fig 4. Solar wind speed and gas pressure at 5 AU vs longitude in one degree intervals in latitude predicted by the 3-D stream model discussed in the text. The heavy trace at 29° indicates the inferred tilt of the current sheet in April 1993. The longitudinal traces from right to left mimic the time-profiles that a stationary spacecraft would measure at various latitudes at 5 AU and are similar to those actually measured by Ulysses during its initial transit to high heliographic latitudes. Adapted from Pizzo and Gosling (1994).

Figure 4 shows resulting longitudinal traces of flow speed and gas pressure at 5 AU in one degree intervals in latitude (λ) from the equator up to λ = N50° (the pattern is identical at southern latitudes although shifted in longitude). Since the pattern rotates with the Sun, these longitudinal traces from right to left mimic the time-profiles that a spacecraft would measure at various latitudes at this distance from the Sun. The simulation qualitatively reproduces the basic observational effect shown in Figure 1. In particular, the interaction region for the stream from the northern polar region (the one near a longitude of 90°) is bounded by a strong forward- reverse shock pair up to a latitude near that associated with the tilt of the dipole axis to the solar spin axis (λ = 30°). However, poleward of that point the forward shock decays rapidly into a broad forward wave, while the reverse shock persists with ever weakening intensity to latitudes 10-15° higher. Moreover, although not shown here, the simulation also accurately predicts the sense and magnitude of the flow deflections observed downstream of the shocks (see Pizzo and Gosling, 1994).

In conclusion, the Ulysses plasma observations have demonstrated systematic variations in the structure of CIRs with increasing heliographic latitude. Our ability to distinguish the pattern of these variations has been aided by the relatively simple and persistent nature of coronal and solar wind structure at the time of these observations. The observed variations are in excellent agreement with the predictions of a global model of solar wind flows that originate in a simple tilted-dipole geometry back at the Sun. The overall 3-D structure of CIRs and their associated shocks in the outer heliosphere and at latitudes well removed from the ecliptic plane is considerably more complex than was generally appreciated prior to the Ulysses mission. It is worth emphasizing that the reverse shocks associated with CIRs continue to propagate to high latitudes beyond 5 AU. Thus field lines intercepted by Ulysses at high latitudes can be connected directly to these reverse shocks at larger heliocentric distances even when Ulysses does not itself directly encounter the shocks. The implication of this effect for energetic particle measurements made by experiments on Ulysses should be obvious.

We thank A. Balogh and his Ulysses magnetometer team for helping with the shock identifications. This work was performed under the auspices of the U.S. Department of Energy with financial support from the National Aeronautics and Space Administration (NASA). Work done at JPL was supported under contract with NASA.

References

Bame, S. J. *et al.*: 1993, *Geophys. Res. Lett.*, **20**, 2323.

Gosling, J. T. *et al.*: 1993, *Geophys. Res. Lett.*, **20**, 2789.

Gosling, J. T. *et al.*: 1994, *Geophys. Res. Lett.*, **21**, 237.

Hundhausen, A. J. and Gosling, J. T.: 1976, *J. Geophys. Res.*, **81**, 1436.

Phillips, J. L. et al.: 1994, *Geophys. Res. Lett.*, in press.

Pizzo, V. J. and Gosling, J. T.: 1994, *Geophys. Res. Lett.*, in press.

Pizzo, V. J.: 1994a, *J. Geophys. Res.*, **99**, 4173.

Pizzo, V. J.: 1994b, *J. Geophys. Res.*, **99**, 4185.

Smith, E. J. and Wolfe, J. H.: 1976, *Geophys. Res. Lett.*, **3**, 137.

Smith, E. J. *et al.*: 1993, *Geophys. Res., Lett.*, **20**, 2327.

LATITUDINAL VARIATIONS IN THE SOLAR WIND ELECTRON HEAT FLUX

EARL E. SCIME, SAMUEL J. BAME, AND JOHN L. PHILLIPS
Los Alamos National Laboratory, Los Alamos, NM 87545, U.S.A.

ANDRE BALOGH
Blackett Laboratory, Imperial College, London, U.K.

Abstract. Ulysses measurements of the solar wind electron heat flux as a function of heliographic latitude are presented. The latitudinal variations in the electron heat flux presented have been normalized by the radial gradient in the electron heat flux obtained during the in-ecliptic phase of the Ulysses mission ($q_e \sim R^{-3.0}$). We find no significant variation in electron heat flux with latitude.

1. Introduction

The Ulysses mission to the solar poles has provided an opportunity to study the field-aligned electron heat flux, q_e, in a solar wind environment significantly different from previous in-ecliptic investigations [Feldman *et al.*, 1975; Pilipp *et al.*, 1990; Scime *et al.*, 1994]. Those studies have shown that the electron heat flux is actively dissipated from 0.3 to 5.0 AU and that the electron heat flux is smaller than expected for a collisional heat flux. Suggested dissipation mechanisms range from complicated coupled ion-electron collision operators [Cuperman *et al.*, 1988] to heat flux limiting instabilities [Gary *et al.*, 1975]. The mechanism responsible for the heat flux regulation remains unknown, but a recent comparison of the observations with the predictions of a whistler heat flux instability model is encouraging [Scime *et al.*, 1994].

As Ulysses moves poleward the magnetic field is expected to become less tightly wrapped. Since the heat flux is carried by the field aligned halo electrons [Scime *et al.*, 1994], the changing magnetic field geometry permits a comparison between the electron heat flux at identical heliocentric distances carried by electrons which have significantly different path lengths from the Sun to the observer. In addition, Ulysses has moved into a region of relatively constant high speed solar wind [Phillips *et al.*, 1994].

In this paper we report the variations in solar wind electron heat flux as a function of heliographic latitude and discuss the implications of the observations for heat flux dissipation models.

2. Instrument descriptions

The electron heat flux was measured with the Solar Wind Plasma Experiment 3D electron spectrometer aboard Ulysses [Bame *et al.*, 1992]. The electron spectrometer is a hemispherical electrostatic analyzer with seven discrete polar detectors, an energy range of 1.6 to 862 eV, and uses the spacecraft spin to map 95% of 4π steradians of solid angle.

Space Science Reviews **72**: 105–108.
© 1995 *Kluwer Academic Publishers.*

The electron distributions are split into core and halo parts using inflections in the energy spectra. Velocity moments of the distribution, such as q_e, are calculated via numerical integrations of the core, halo, and total distributions. The Ulysses magnetometer consists of both a vector helium magnetometer and a fluxgate magnetometer [Balogh *et al.*, 1992]; 32 s averaged data are used in this investigation.

3. Latitudinal variations in the solar wind electron heat flux

Figure 1 shows the total electron heat flux versus heliographic latitude normalized by the radial gradient ($R^{-3.0}$) obtained for the in-ecliptic portion of the Ulysses mission [Scime *et al.*, 1994]. The most distinctive features in Figure 1 are 40% decrease in total electron heat flux as Ulysses traveled from 5° to 20° in heliographic latitude and the reduced level of heat flux fluctuations after Ulysses passed 35°. Coincident with the decrease in electron heat flux, a plasma density decrease of approximately 40% was observed [Bame *et al.*, 1993]. Although a collisional heat flux is independent of density, the maximum electron heat flux permitted by the whistler heat flux instability [Gary *et al.*, 1994] and the heat flux carried by free streaming electrons are proportional to electron density. Previous studies have shown that Ulysses left the heliospheric current sheet at 35° heliographic latitude [Bame *et al.*, 1993] and entered a region of nearly steady high speed solar wind due to the southern coronal hole. The decrease in the heat flux variability is well correlated with the decrease in solar wind speed variability.

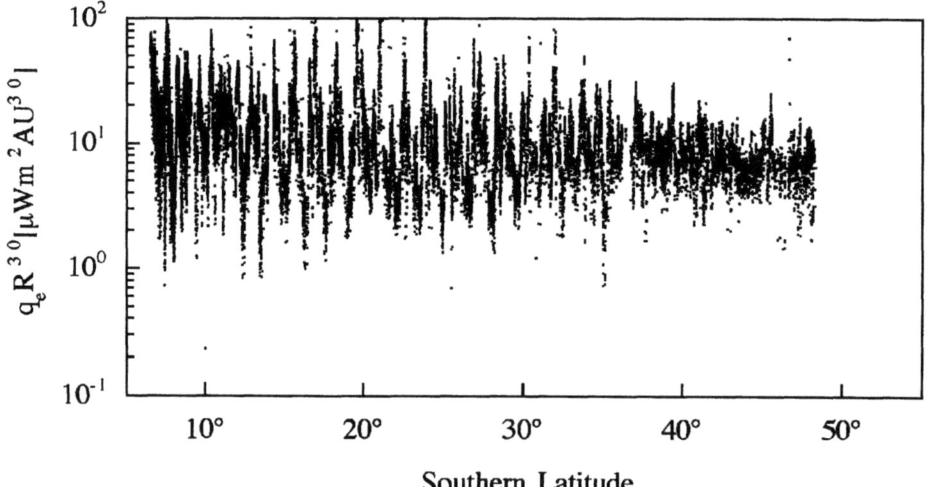

Figure 1. Normalized total electron heat flux versus heliographic latitude.

The median total, core, and halo electron heat fluxes binned by solar rotation and normalized by the respective core and halo heat flux in-ecliptic radial scalings of Scime *et al.* [1994] are shown in Figure 2. As the core, halo, and total electron heat fluxes are each calculated in the frame of their respective distributions, the sum of the core and halo heat fluxes does not necessarily equal the total (see Scime *et al.,* 1994). The component heat fluxes are essentially a measure of the asymmetry of that component distribution. As seen in the unbinned total electron heat flux, the total heat flux decreases slightly. The halo electron heat flux shows no trend with latitude and the core electron heat flux decreases by nearly 80%. It is believed that the core heat flux is a by-product of the acceleration of the core electrons, through some coupling with the solar wind ions, to nearly the solar wind speed [Feldman *et al.*, 1975]. The substantial decrease in the core heat flux occurs over the entire latitude range and this suggests that the ion-electron coupling is more a function of latitude and solar wind speed than simply plasma density.

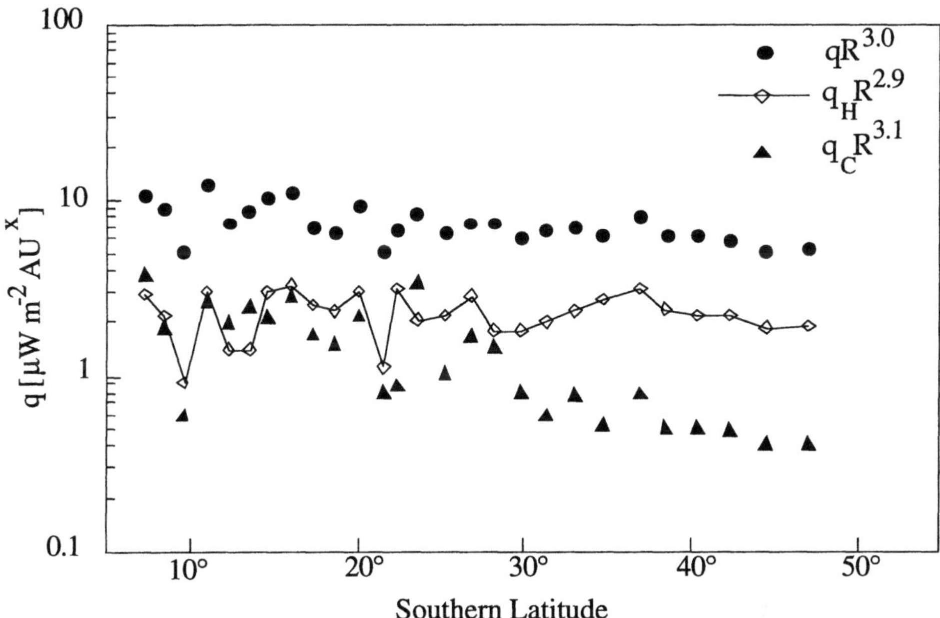

Figure 2. Normalized median total (circles), core (triangles), and halo (diamonds) electron heat fluxes versus heliographic latitude binned by solar rotation.

As shown in Scime *et al.*, [1994], the magnitude of the observed electron heat flux is directly proportional to the magnetic field magnitude. To eliminate this effect we show the total electron heat flux divided by the magnetic field magnitude, normalized by the in-ecliptic results, versus heliographic latitude in Figure 3. The slight decrease in total electron heat flux from 5° to 20° visible in Figure 1 is not evident in Figure 3 as the magnetic field strength has also decreased over the same interval. The most striking aspect of Figure 3 is the very small latitudinal variation in the electron heat flux. The very slight increasing trend with latitude apparent in Figure 3 suggests that the electron heat flux is less dissipated at the higher latitudes. If this trend continues, it would demonstrate that the heat flux dissipation is dependent upon the path length traveled by the halo electrons along the interplanetary magnetic field. It is worth noting, however, that the predicted (Parker) interplanetary magnetic field spiral is only slightly less wound at Ulysses' current location (~66°) than at the same in-ecliptic distance (~75°). Thus significant magnetic field configuration effects may not be observable until Ulysses' moves further south in latitude and closer to the Sun.

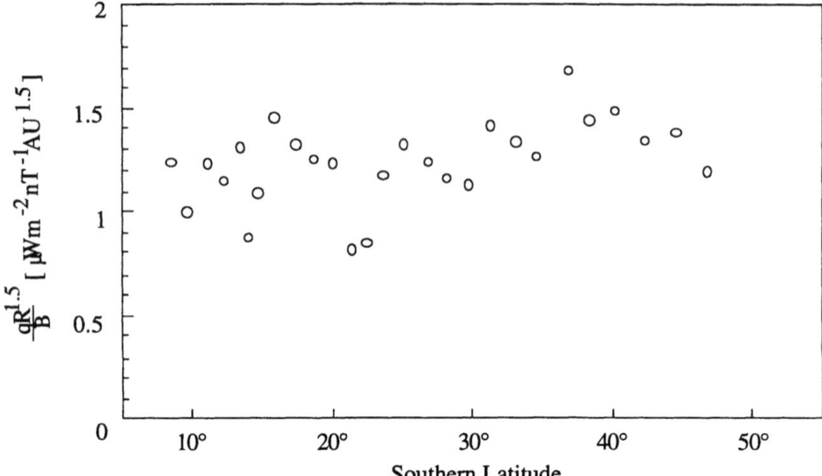

Figure 3. Median total heat electron heat flux binned by solar rotation and normalized by magnetic field magnitude and in-ecliptic scaling. Note that vertical scale is linear.

4. Conclusions

The correlated decrease in total electron heat flux and plasma density support the conclusion that collisional processes are not responsible for the electron heat flux regulation. The lack of any substantial variation in the electron heat flux as Ulysses has traversed 50° of heliographic latitude suggests that the heat flux regulatory mechanism is not extraordinarily sensitive to the magnetic field configuration, i.e. the distance traversed by the halo electrons along the magnetic field spiral.

Acknowledgments

This work was performed under the auspices of the United States Department of Energy with financial support of NASA. One of the authors (EES) was also supported by the Department of Energy Distinguished Post-Doctoral Research Program

References

Balogh, A., et al.:1992, *Astro. Astrophys. Supp. Ser.,* **92,** 221.
Bame, S.J., et al.: 1992, *Astro. Astrophys. Supp. Ser.,* **92,** 266.
Bame, S. J., et al.: 1993, *Geophys. Res. Lett.*, **20**, 2323.
Cuperman. S.,, et al.: 1988, *Astrophys. J.,* **330,** 466.
Feldman, W. C.:, et al. 1975, *J. Geophys. Res.,* **80,** 4181.
Feldman, W. C., et al.: 1979, *J. Geophys. Res.,* **81**, 2377.
Gary, S. P.: 1975, *J. Geophys. Res.*, **80,** 4197.
Gary, S. P., et al.: 1994, *J. Geophys. Res.*, in submission.
Pilipp, W. G., et al. 1990, *J. Geophys. Res.*, **95,** 6305.
Phillips, J. L., et al.: 1994, *Geophys. Res. Lett.* in submission.
Scime, Earl E., et al.: 1994, *J. Geophys. Res.*, in submission.

RADIAL AND MERIDIONAL TRENDS IN SOLAR WIND THERMAL ELECTRON TEMPERATURE AND ANISOTROPY: ULYSSES

J. L. PHILLIPS, S. J. BAME, S. P. GARY, J. T. GOSLING, E. E. SCIME
Los Alamos National Laboratory, Los Alamos, New Mexico 87545, USA

R. J. FORSYTH
Imperial College, London, UK

Abstract. Ulysses plasma measurements from 1.15 to 5.31 AU and from S6.4° to S48.3° solar latitude are used to assess the trends in the solar wind thermal electron temperature and anisotropy. Improved spacecraft potential corrections and data products have been incorporated. The radial temperature gradient is steeper than in previous determinations, but flatter than adiabatic. When normalized to 1 AU, temperature decreases with increasing latitude. Little change in the average thermal anisotropy has been seen during the mission.

1. Introduction

Solar wind electron distributions include thermal or "core" and suprathermal or "halo" populations. Measurements from various spacecraft have been used to assess the radial gradient in core temperature, with little consensus emerging except that the gradient is flatter than the adiabatic value of $T \propto R^{-4/3}$. Voyager/Mariner results from 0.45 to 4.76 AU yielded a power-law exponent of -0.37 (Sittler and Scudder, 1980), Helios measurements yielded a range of exponents from -0.48 to -0.84 (Pilipp *et al.*, 1990), and a preliminary Ulysses study yielded an exponent of -0.56 (Phillips *et al.*, 1993). Electron temperature measurements were previously unavailable for a wide range of solar latitude; however, Phillips and Gosling (1990) predicted a temperature increase and a change in distribution anisotropy with increasing latitude for the high-speed wind.

2. Observations

We use 3-d measurements from the electron spectrometer of the Ulysses solar wind plasma experiment (Bame *et al.*, 1992), plus 32-sec averaged magnetic field observations from the dual magnetometer experiment (Balogh *et al.*, 1992). Improvements in data analysis relative to a previous study (Phillips *et al.*, 1993) include: (1) incorporation of a vector correction for spacecraft potential (Scime *et al.*, 1994); (2) calculation of core temperatures via numerical integration rather than bi-Maxwellian fitting; (3) determination of parallel and perpendicular temperatures based on the measured field; (4) deletion of spectra altered by the active RF sounder on Ulysses. Data were obtained from November 22, 1990 through the end of 1993, were binned by solar rotation, and were truncated to avoid partial rotations. The resulting data sets are: (1) in-ecliptic, 25,591 spectra from 1.15 to 5.31 AU, 17 solar rotations; (2) out-of-ecliptic, 32,158 spectra from 5.40 AU at S6.4° to 3.83 AU at S48.3°, 27 rotations.

Space Science Reviews **72**: 109–112.
© 1995 *Kluwer Academic Publishers.*

The anisotropy of the core distributions, defined here as $T_{\parallel}/T_{\perp}$, provides insight into the overall plasma transport processes. Figure 3 shows $T_{\parallel}/T_{\perp}$ vs. R for selected in-ecliptic data (left panel) and vs. heliographic latitude for selected out-of-ecliptic data (right). The prevailing anisotropy shows a slight upward trend with increasing R for the in-ecliptic data. This is in contrast to the model of Phillips and Gosling (1990), which predicted a reduction in the prevailing $T_{\parallel}/T_{\perp}$ with increasing heliocentric distance, at least for the high-speed wind. The discrepancy suggests that non-adiabatic processes preferentially increase $T_{\parallel}$. The out-of-ecliptic data (right) show considerable variation in anisotropy (note the interval near 30° latitude where ~50% of the observations have $T_{\parallel}/T_{\perp} < 1$) but on average do not show a consistent meridional trend. The existence of distributions with $T_{\parallel}/T_{\perp} < 0.8$ is striking when compared with 1 AU ecliptic data, where only very small $T_{\perp} < T_{\parallel}$ anisotropies have been noted (Phillips et al., 1989). Southward of 33°, the prevailing anisotropy increases slightly with increasing latitude; this trend is in qualitative agreement with the Phillips and Gosling prediction.

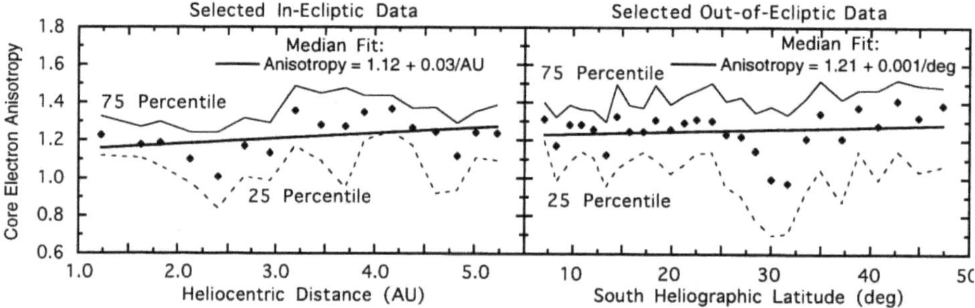

Fig. 3. Median (diamonds), plus 25 and 75 percentile core anisotropy, $T_{\parallel}/T_{\perp}$, binned by solar rotation, vs. heliocentric distance (left panel) and heliographic latitude (right). Heavy traces and equations represent power-law fits to medians.

In order to examine some of the overall characteristics of the core temperature and anisotropy without the need for normalization, we have chosen the Ulysses mission phase during which the spacecraft was between 5.0 and 5.4 AU. This interval was over 13 months long and included a solar latitude range of S5.8° to S24.5°; the resulting selected data set comprises 14,070 3-d spectra. Figure 4 shows medians of core electron density (top panel), $T_{\parallel}$ (bottom, solid diamonds), and $T_{\perp}$ (bottom, circles), binned by $T_{\parallel}/T_{\perp}$. The top panel indicates that Coulomb collisions probably play a significant role in regulating the distribution shapes: the nearly isotropic distributions have, on average, the highest densities, while anisotropies of either sense are accompanied by lower densities. The bottom panel shows that $T_{\parallel}$ is substantially more variable than is $T_{\perp}$. Similar trends in ISEE-3 measurements at 1 AU motivated the development of the Phillips and Gosling (1990) model, in which the core distributions are controlled by a combination of collisional isotropization and expansion in a spiral magnetic field. One difference between the 1 AU and 5+ AU data is that the $T_{\perp} > T_{\parallel}$ spectra had the highest observed densities at 1 AU but had intermediate densities at 5+ AU; this qualitative trend was predicted by the model (Phillips *et al.*, 1989; Phillips and Gosling, 1990).

Figure 1 (left panel) shows percentiles of core temperature for the in-ecliptic measurements, binned by solar rotation. Medians of temperature and heliocentric distance (R) have been fit to a power law, yielding $T \propto R^{-0.82}$. The right panel is an identical display for in-ecliptic data without shocked intervals (*e.g.*, Balogh *et al.*, this issue), coronal mass ejections, magnetic connections to the Jovian bow shock (Moldwin et al., 1993), and the extended electron foreshocks typical for corotating shocks (Gosling et al., 1993). Data thus chosen will be referred to henceforth as "selected" data. For this selected subset of 16,229 spectra, the power law exponent is -0.91 and the temperature variability (range between 25 and 75 percentiles) is generally less than for the entire in-ecliptic set (left).

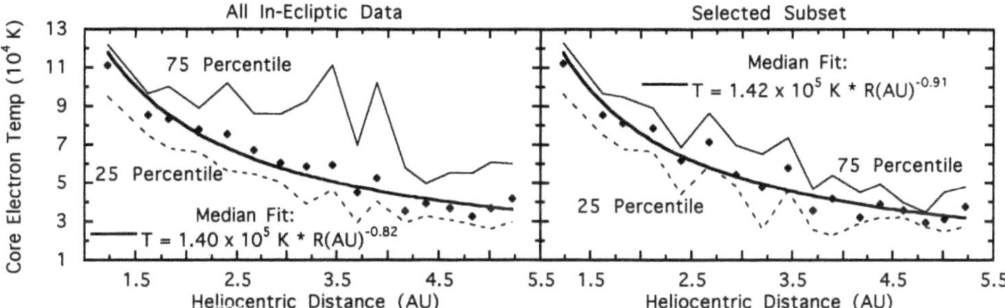

Fig. 1. Median (diamonds), plus 25 and 75 percentile core electron temperature vs. heliocentric distance, binned by solar rotation, for all in-ecliptic data (left) and a subset described in text (right). Heavy traces and equations represent power-law fits to medians.

Figure 2 (left) shows percentiles of core temperature vs. south heliographic latitude, with a linear fit to the medians, for selected out-of-ecliptic data from Jupiter encounter in February 1992 through the end of 1993. This data subset includes 19,318 spectra. Figure 2 (right) shows the same data normalized to 1 AU using the $T \propto R^{-0.91}$ fit from the in-ecliptic data. Note that the measured temperature (left) appears to have no consistent latitude dependence, while the normalized temperature (right) declines with increasing latitude as a result of the decreasing heliocentric distance of the spacecraft.

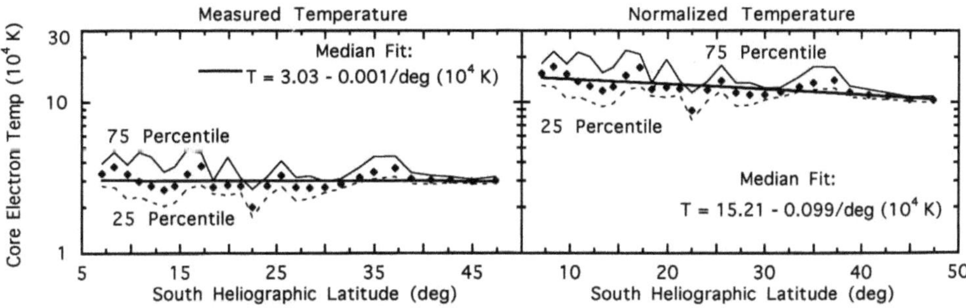

Fig. 2. Median (diamonds), plus 25 and 75 percentile measured core temperature vs. heliographic latitude, binned by solar rotation, for selected out-of-ecliptic data through 1993 (left panel); right panel shows temperature normalized to 1 AU.

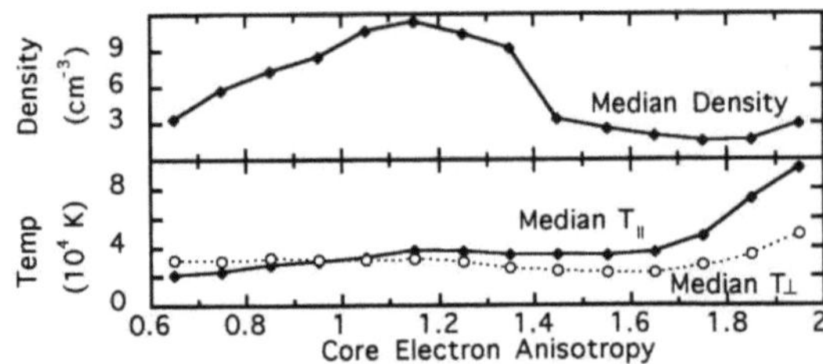

Fig. 4. Median density (top), and parallel and perpendicular temperature components (bottom), binned by core anisotropy, for a selected subset of electron spectra beyond 5 AU.

3. Conclusions

During the declining phase of the solar cycle, the Ulysses solar wind electron data yielded the following findings. (1) Median core T for selected data from 1.15 to 5.31 AU varied as $R^{-0.91}$, steeper than previous results but flatter than the adiabatic value of $R^{-1.33}$. (2) Temperature, normalized to 1 AU, decreased with increasing latitude. (3) Prevailing $T_{\|}/T_{\perp}$ increased slightly with increasing R, and with increasing latitude in the latest data. (4) Characteristics of T and $T_{\|}/T_{\perp}$ beyond 5 AU resemble those at 1 AU, though intervals of very low $T_{\|}/T_{\perp}$ first appear at greater R. Assessment of the Phillips and Gosling (1990) model produced mixed results. Noteworthy disagreements were a non-adiabatic radial gradient in T, a lack of decrease in $T_{\|}/T_{\perp}$ with increasing R, and a poleward cooling trend. Although it is impossible to ensure that the measured gradients represent spatial and not temporal variations, the relatively short (3-year) observational interval, and the data selection used, should minimize temporal effects.

Acknowledgments

Work at Los Alamos was carried out under the auspices of the U.S. Department of Energy with support from NASA. We thank C.M. Hammond and L.A. Weiss for help in solar wind event identification, and R.K. Sakurai for identifying sounderized spectra.

References

Balogh, A. *et al:* 1992, *Astron. Astrophys. Suppl. Ser.*, **92**, 221.
Bame, S.J. *et al.*: 1992, *Astron. Astrophys. Suppl. Ser.*, **92**, 237.
Gosling, J.T. *et al.*: 1993, *Geophys. Res. Lett.*, **20**, 2335.
Moldwin, M.B. *et al.*: 1993, *Planet. Space Sci.*, **41**, 795.
Phillips, J.L. and Gosling, J.T.: 1990, *J. Geophys. Res.*, **95**, 4217.
Phillips, J.L. *et al.*: 1989, *J. Geophys. Res*, **94**, 13,377.
Phillips, J.L. *et al.*: 1993, *Adv. Space Res.*, **13**, (6)47.
Pilipp, W.G. *et al.*: 1990, *J. Geophys. Res.*, **95**, 6305.
Scime, E.E. *et al.*: 1994, *J. Geophys. Res.*, in press.
Sittler, E.C. and Scudder, J.D.: 1980, *J. Geophys. Res.*, **85**, 5131.

ULYSSES OBSERVATIONS OF SOLAR WIND PLASMA PARAMETERS IN THE ECLIPTIC FROM 1.4 TO 5.4 AU AND OUT OF THE ECLIPTIC

B. E. GOLDSTEIN, M. NEUGEBAUER
Jet Propulsion Laboratory, California Institute of Technology, Pasadena, CA, 91109-8099, U.S.A.

J. T. GOSLING, S. J. BAME, J. L. PHILLIPS, D. J. MCCOMAS
Los Alamos National Laboratory, Los Alamos, NM, 87545, U.S.A.

A. BALOGH
Imperial College, London, England

ABSTRACT We report observations of radial and latitudinal gradients of Ulysses plasma parameters. The solar wind velocity increased rapidly with latitude from 0° to 35°, then remained approximately constant at higher latitudes. Solar wind density decreased rapidly from 0° to 35° of latitude, and also was approximately constant beyond that latitude. The mass flux similarly decreased away from the equator (but less than the density), whereas the momentum flux was relatively constant. The radial gradient of the entropy at high latitude indicated a value for the polytrope index of about 1.72 (close to adiabatic); the in-ecliptic estimates of radial gradients for temperature and entropy may be biased by temporal variation. A striking increase in the alpha particle-proton velocity difference with latitude is found.

1. Observations

As Ulysses climbed in latitude it encountered a recurrent high-speed stream arising from an equatorward extension of the South Polar Coronal Hole (Bame et al., 1993); followed by steady immersion in the high speed stream (Phillips et al. 1994) We display in Fig. 1 the solar wind proton velocity, for data obtained since June 30, 1992, as a

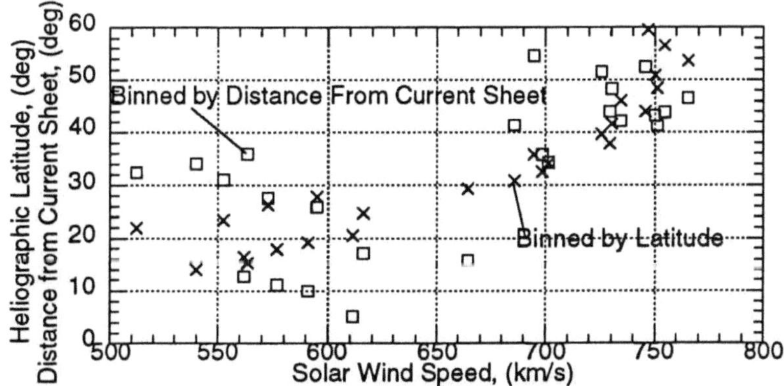

function of heliospheric latitude, and as a function of angular distance from the current sheet in degrees, both averaged over solar rotations. The angular distance from the current sheet was estimated by determining the location of the current sheet from the

Space Science Reviews **72**: 113–116.
© 1995 *Kluwer Academic Publishers.*

Stanford source surface data, and tracing back to the Sun using the measured velocity. The data are similar at the highest latitudes, but the current sheet distance binned averages do not show the low velocities at low latitudes that are apparent in the heliolatitude averages. Significant errors can occur from the constant velocity assumption, and, the source surface model exagerrates the distance of the current sheet from the equatorial plane (Schulz, 1994). Heliographic coordinates are used in the rest of this paper.

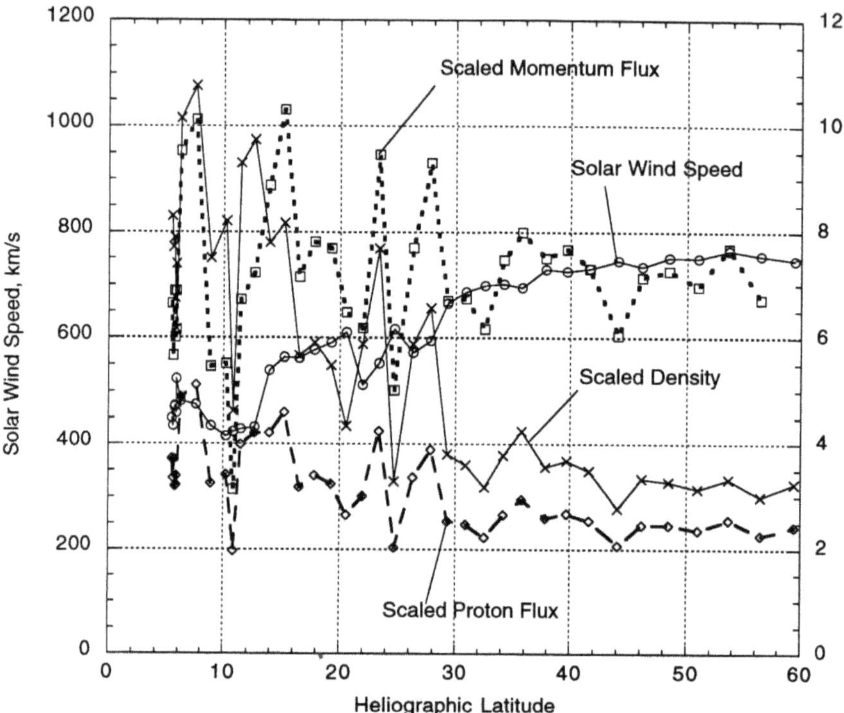

Fig. 2 shows data from Sept. 12 of 1991, at which time Ulysses was about 4.15 AU from the Sun, to the end of currently available data, March 26, 1994, when the distance was 3.4 AU. Shown are proton bulk speed, V_p, (km/s), scaled (to 1 AU) density, $r^2 n_p$, ($/cm^3$), scaled number flux, $r^2 n_p V_p$, (km/sec) $/cm^3$ 10^{-3}) , and scaled momentum flux (including alpha particles, magnetic terms, and electron pressure, dynes cm^{-2} $2x10^8$), all averaged over solar rotations. The velocity axis is on the left, on the right are density, mass flux divided by 1000 and momentum flux divided by 50000 . The in-ecliptic velocity of the solar wind is ~450 km/sec, increasing rapidly over the range from 5° to 35° to ~700 km/sec. From 35° to 55° the velocity increases from ~700 km/sec to ~750 km/sec, and by 55° the velocity seems to be no longer increasing in these rotation-binned data (but McComas et al. 1994, find a modest gradient examining these data with higher latitudinal resolution). The scaled density drops precipitously from an in-ecliptic value of about 8.6 /cc to a value of 3.7 /cc. The proton flux decreases with latitude by about 35% from 0° to 40°, which is less than the decrease in density. Similarly, the decrease in radial momentum flux is even less, ~10%.

The general tendency of density and velocity to be anticorrelated is typical of what is observed at low latitudes in the solar wind.

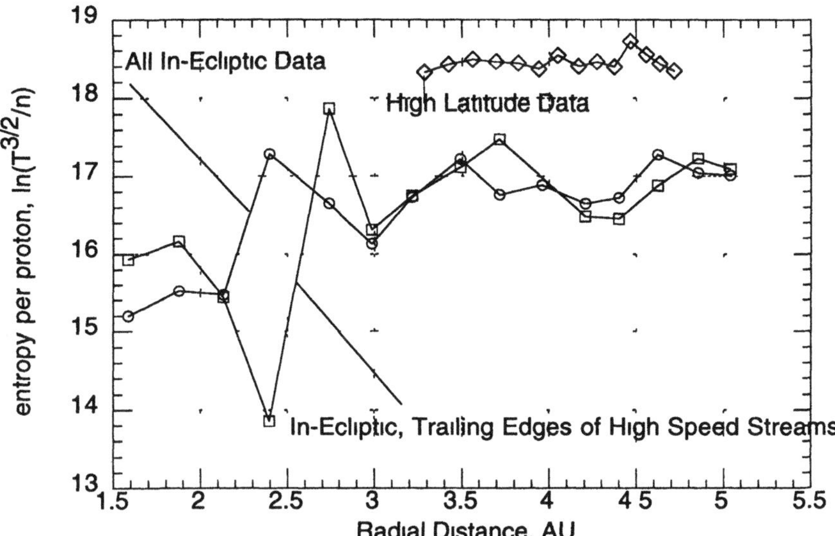

The solar wind proton distribution is much hotter at high latitudes than at low latitudes. The entropy per proton, $\ln(T^{3/2}/n)$, averaged over each solar rotation as seen from Ulysses, is shown in Fig. 3 which compares high latitude observations and two selections of in-ecliptic data. The in-ecliptic data period for Fig. 3 is from Dec. 18, 1990, when nutation ceased, to January 20 , 1992 (ending before Jupiter encounter, a range of 1.4 to 5.2 AU). The high latitude observations were obtained at latitudes of from 30° to 57° while the spacecraft traveled from 4.76 to 3.4 AU. The in-ecliptic data shown are for two cases, the full data set and a data set restricted to the trailing edges of high speed streams. Since coronal mass ejections have different thermal properties, they were excluded. The entropy at high latitudes is very much greater than in the ecliptic; there are large latitudinal entropy and temperature gradients. At high latitudes the entropy does not appear to be increasing with distance from the Sun as is seen at lower latitudes; a logarithmic fit to the data provides an equivalent polytrope index of 1.72 (close to adiabatic expansion). If all the in-ecliptic observations are used, there is apparently a large increase of entropy with distance from the Sun, correspondingly, the temperature (not shown) decreases with distance from the Sun much more slowly than predicted by adiabatic expansion. Heating by shocks or other effects at corotating interaction regions should not affect trailing edges of high speed streams, yet, this data has an entropy increase with distance similar to that of the full data set. From a logarithmic fit in r, and assuming a r^{-2} dependence of density upon distance from the Sun, the resulting polytrope indices, γ, are: 1.20 (all in-ecliptic data), 1.16 (trailing edge in-ecliptic data). However, examining the temperature variations indicates that temporal effects may be as important as spatial effects.

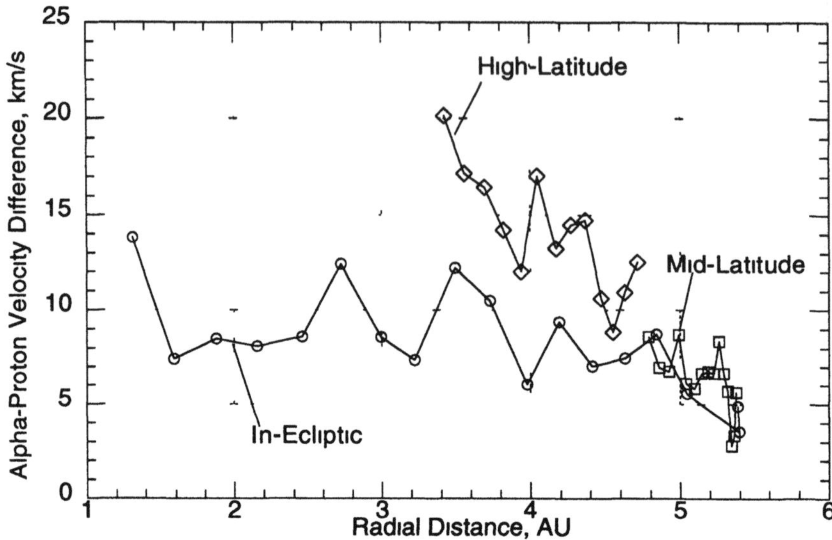

In the solar wind at 1 AU, relative streaming between alpha-particles and protons is typically observed in high speed streams. However, Nuegebauer et al. (1994) found that in the Ulysses plasma observations such relative streaming occurs less frequently beyond 1 AU. Further, when relative streaming does occur it is not associated with high speed streams, but rather occurs in the vicinity of solar wind shocks. Fig. 4 shows the alpha-proton velocity difference for three different regimes, in-ecliptic, 10° to 30° (mid-latitude), and $>30^{\circ}$ (high latitude). As Ulysses enters the high latitude region, there is a large increase in the relative streaming with the velocity difference being relatively constant and about half the Alfvén speed. Such large relative streaming was not observed in the ecliptic plane, and might be due to reduced Coulomb collisions.

References

Bame, S J., et al : 1993, Ulysses and IMP 8 Solar Wind Observations in the Heliosphere: A Recurrent High Speed Stream and the Heliomagnetic Streamer Belt, *Geophys. Res. Letts.*, **20**, 2323.

Neugebauer, M., et al.: 1994, Ulysses Near-Ecliptic Observations of Differential Flow Between Protons and Alphas in the Solar Wind, *J. Geophys. Res.*, **99**, 2505.

McComas, D et al · 1994, Ulysses Solar Wind Observations to 56° South, Space Science Reviews, this issue

Phillips, J. L , et al 1994, Ulysses at 50° South· Constant Immersion in the High-Speed Solar Wind, submitted to *Geophys. Res. Letts.*

Schulz, M 1994, Fourier Parameters of Heliospheric Current Sheet and Their Significance, Space Science Reviews, this issue

PIONEER AND VOYAGER OBSERVATIONS OF LARGE-SCALE SPATIAL AND TEMPORAL VARIATIONS IN THE SOLAR WIND

P. R. GAZIS *, A. BARNES and J. D. MIHALOV
NASA Ames Research Center, Moffett Field, CA 94035, USA

Abstract. The Pioneer 10, Pioneer 11, and Voyager 2 spacecraft were launched in 1972, 1974, and 1977, respectively. While these three spacecraft are all at comparatively low heliographic latitudes compared with Ulysses, their observations span almost two solar cycles, a range of heliocentric distances from 1 to 57 AU, and provide a unique insight into the long-term variability of the global structure of the solar wind. We examine the spatial and temporal variation of average solar wind parameters and fluxes. Our observations suggest that the global structure of the outer heliosphere during the declining phase of the solar cycle at heliographic latitudes up to 17.5°N was characterized by two competing phenomena: 1) a large-scale increase of solar wind density, temperature, mass flux, dynamic pressure, kinetic energy flux, and thermal energy flux with heliographic latitude, similar to the large-scale latitudinal gradient of velocity seen in IPS observations, 2) a small-scale decrease in velocity and temperature, and *increase* in density near the heliospheric current sheet, which is associated with a band of low speed, low temperature, and high density solar wind similar to that observed in the inner heliosphere.

1. Introduction

In situ observations of the inner heliosphere (Rhodes and Smith, 1981; Schwenn et al., 1978; Mitchell et al., 1981) and IPS measurements (Sime and Rickett, 1981; Kakinuma and Kojima, 1986) have shown that solar wind stream structure varies solar cycle and heliographic latitude. The heliospheric current sheet (HCS), which changes in inclination over the course of a solar cycle (Hoeksema et al., 1983), appears to be associated with a region of low speed, low temperature, and high density solar wind (Zhao and Hundhausen, 1981; Miyake et al., 1989; Gazis, 1993). A similar association of low speed and low temperatures with the HCS has been observed in the outer heliosphere (Gazis 1994; Bame et al., 1993). In this paper we examine solar cycle variability in the outer heliosphere.

2. Observations and Conclusions

Comparison of data from Pioneer 10, Pioneer 11, and Voyager 2 spacecraft suggests that velocities and temperatures measured at these three spacecraft are comparable. Density is more difficult to measure, and the average densities measured at Pioneer 10 were 20-80% greater than the average densities measured at Voyager 2 even when all three spacecraft observed the same

* affliated to San Jose State University Foundation

Space Science Reviews **72**: 117–120.
© 1995 *Kluwer Academic Publishers.*

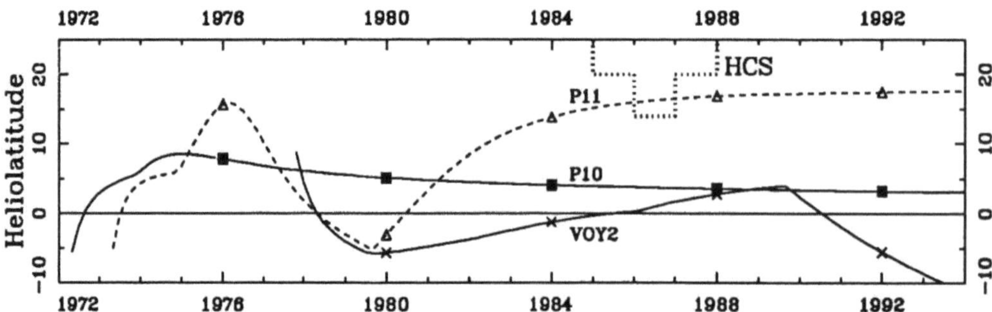

Fig. 1. Heliographic latitudes of Pioneer 10, Pioneer 11, Voyager 2, and the approximate inclination of the HCS.

solar wind. This may be indicative of the callibration uncertainties of the existing Pioneer and Voyager density measurements.

Figure 1 shows the heliographic latitudes of the Pioneer 10 (solid line, filled symbols), Pioneer 11 (dashed line), and Voyager 2 (solid line, open symbols), plotted versus time. While Pioneer 10 and Voyager 2 have spent most of their time near the solar equator, Pioneer 11 has been steadily moving north since 1980, passed 10° N in 1982, and is now near 17°. Figure 1 also shows the average inclination of the HCS. Near the time of the 1986 solar minimum, the HCS dropped below the latutude of Pioneer 11 while Pioneer 10 and Voyager 2 spent much of their time in or near the HCS (Gazis, 1989).

Figure 2 shows twelve-solar-rotation averages of solar wind dynamic pressure, mass flux, temperature, density, and velocity, respectively. Pioneer 11 and Voyager 2 data have been time-shifted to account for solar wind travel time and the longitudinal separation of these spacecraft from Pioneer 10. Densities and fluxes have been multiplied by the square of the heliocentric distance to account for the presumed radial expansion.

The three spacecraft observed similar solar wind speeds, except near the time of the 1986 solar minimum, when the inclination of the HCS was low and the solar wind stream structure varied with latitude, which caused the average solar wind speed to increase at Pioneer 11 and decrease at Pioneer 10 and Voyager 2 (Gazis, 1989). Pioneer 11 has observed higher temperatures than Pioneer 10 and Voyager 2 ever since 1981 (Gazis, 1994).

The average solar wind density, mass flux, and dynamic pressure observed at Pioneer 11 were consistently higher than the values observed at Voyager 2, and were comparable to or higher than the values observed at Pioneer 10 throughout most of the last solar cycle, which suggests that the density and fluxes *increased* with latitude between the solar equator and 17° N. This gradient, if it exists, cannot be associated with the velocity gradient that was observed near the time of the 1986 solar minimum, for it began

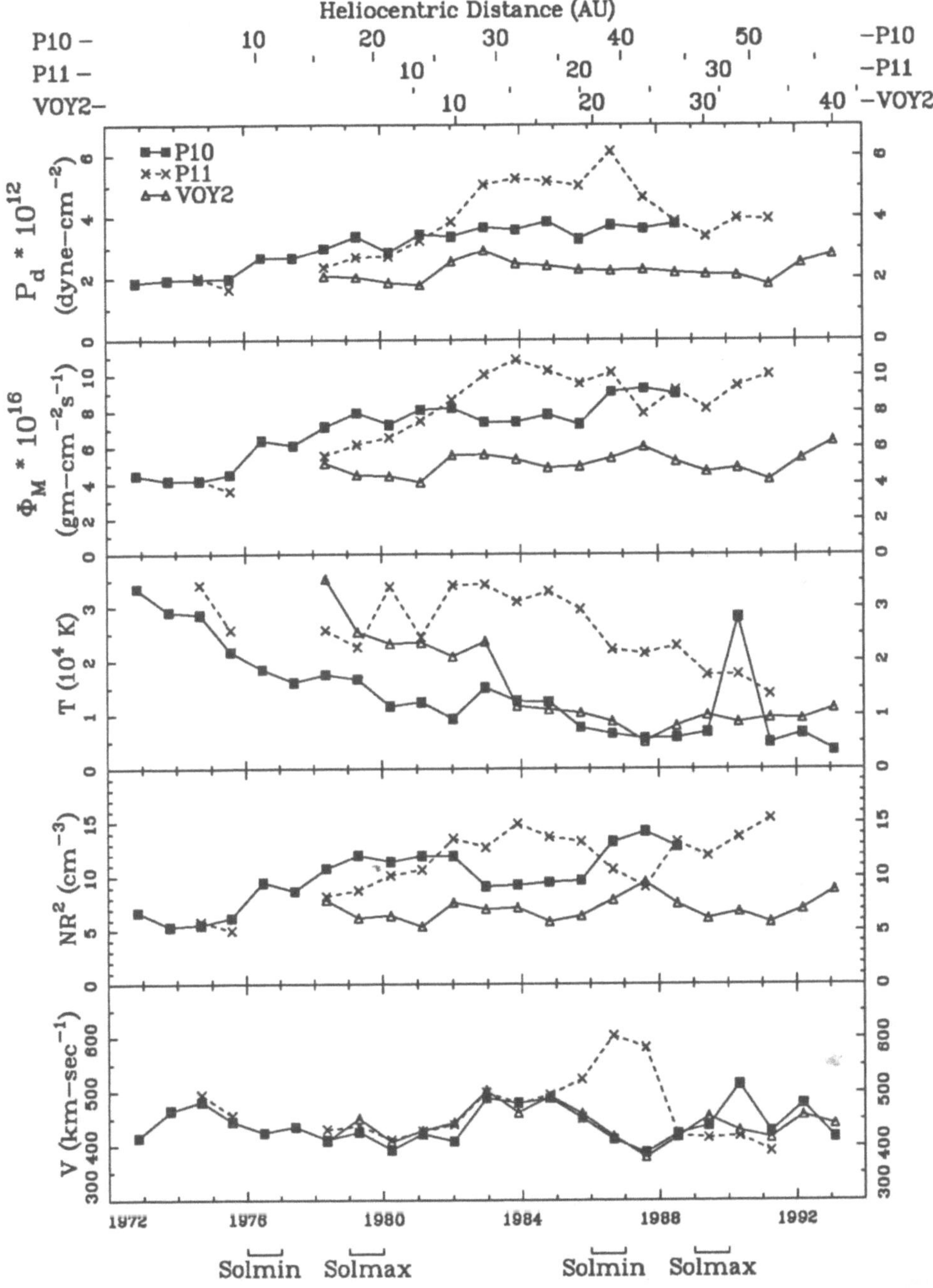

Fig. 2. Twelve-rotation averages of solar wind parameters observed at Pioneer 10, Pioneer 11, and Voyager 2, plotted versus time. From the top, panels show dynamic pressure, mass flux, temperature, density, and velocity, respectively. Densities and fluxes have been multiplied by R^2. Heliocentric distances of the spacecraft are shown above the top panel.

more than four years earlier. Recent observations from Ulysees (Schwenn, 1994) show that the density, mass flux, and dynamic pressure remained the same or *decreased* with heliographic latitude between the solar equator and 55° S. This is not necessarily inconsistent with our observations, for the Ulysees measurement were made over a wider latitude range, a shorter time interval, in the opposite hemisphere, during a different solar cycle, and at a significantly smaller distances from the sun. Still, it is important to note that the latitude gradient we seem to observe is comparable to the calibration uncertainties of the existing Pioneer and Voyager density measurements.

The solar cycle variations observed at Pioneer 11 were much much larger than the variations observed at Pioneer 10 and Voyager 2. While the average density, mass flux, and dynamic pressure at Pioneer 10 and Voyager 2 remained roughly constant throughout the last solar cycle, except for a slight rise near the time of the 1986 solar minimum when these two spacecraft spent most of their time in or near the HCS the density and fluxes varied strongly at latitude of Pioneer 11. The dynamic pressure varied the most, and increased by more than a factor of two between 1979 and 1985, increased furthur in 1986 when the inclination of the HCS dropped below the latitude of this spacecraft, then dropped during the ascending phase of the most recent solar cycle. This variation cannot be attributed to the velocity changes near the time of the 1986 solar minimum, for persisted throughout the entire solar cycle, nor can it be attributed to intercallibration problems, for it is apparent in measurements from a single spacecraft.

Acknowledgements

The authors would like to thank A. J. Lazarus and J. W. Belcher for use of the Voyager 2 data. This work was supported at NASA Ames by Contract NCA2-479 (San Jose State University Foundation).

References

Bame, S. J., et al.: 1993, *Geophys. Res. Lett.* **20**, 2323
Gazis, P. R.: 1993, *J. Geophys. Res.* **98**, 9391
Gazis, P. R.: 1994, *J. Geophys. Res.* in press,
Gazis, P. R., et al.: 1989, *Geophys. Res. Lett.* **16**, 223
Hoeksema, J. T., et al.: 1983, *J. Geophys. Res.* **88**, 9910
Kakinuma, T., and Kojima, M.: 1986, *Proc. Res. Inst. Atmos. Nagoya Univ.* **33**, 1
Mitchell, D. G., et al.: 1981, *J. Geophys. Res.* **86**, 165
Miyake, W., et al.: 1989, *J. Geophys. Res.* **94**, 15,359
Rhodes, E. J., and Smith, E. J.: 1981, *J. Geophys. Res.* **86**, 8877
Schwenn, R.: 1994, *28th Symposium of the ESLAB* ,
Schwenn, R., et al.: 1978, *J. Geophys. Res.* **83**, 1011
Sime, D. G., and Rickett. B. J.: 1978, *J. Geophys. Res.* **83**, 5757
Zhao, Xue-Pu and Hundhausen, A. J.: 1981, *J. Geophys. Res.* **86**, 5423

NUMERICAL 3–D TIME–DEPENDENT MHD MODEL OF THE SOLAR WIND

A. V. USMANOV
Institute of Physics, University of St. Petersburg, St. Petersburg 198904, Russia

Abstract. A fully three-dimensional, time-dependent MHD model of the solar corona and the solar wind is developed. The proposed model is an upgrading of the steady-state model (Usmanov, 1993a,b). A numerical self-consistent solution for 3–D MHD equations is constructed for the region between the solar photosphere and 1.5 AU. The unique features of the proposed model are: (i) uniform coverage and self-consistent treatment of the regions of subsonic/sub-Alfvénic and supersonic/super-Alfvénic flows, (ii) as a background for transients, a steady-state solution for the ambient solar wind is used. The parameters of the initial pulses are prescribed in terms of the near-Sun shock velocities (as inferred from the metric Type II radio burst observations) relative to the preshock steady-state flow parameters.

1. Introduction

Numerous studies dealing with the dynamics of the solar wind have been based upon numerical solutions of time-dependent hydrodynamic (HD) or magnetohydrodynamic (MHD) equations (see reviews by Dryer, Smith, and Wu, 1988; Dryer, 1994). The published models have been drastically improved from simple one- and two-dimensional HD simulations (e.g., Hundhausen and Gentry, 1969; De Young and Hundhausen, 1971) to one- and two-dimensional MHD models (e.g., Hirshberg, Nakagawa, and Wellck,1974; Wu, Dryer, and Han, 1983; Steinolfson and Dryer, 1984) and even up to fully three-dimensional MHD models in which all the spatial components and derivatives are taken into account (Han, Wu, and Dryer, 1988; Wu and Wang, 1991; Steinolfson, 1992).

All the models referred above have two important drawbacks. Firstly, none of them covers the full range of heliocentric distances between the solar surface where disturbances originate (about 1 R_s where R_s is the solar radius) and 1 AU where the bulk of solar wind measurements has been obtained by Earth-orbiting satellites. The second drawback is that they use some simple models rather than observational data for inferring the structure of ambient (steady-state) solar wind.

To study the interplanetary dynamics on a global scale, an initial state describing the ambient solar wind should be determined for a computational domain extending from the solar surface to, at least, 1 AU. As the initial state, we apply the steady-state 3-D MHD model (Usmanov, 1993a,b). This model is a generalization of streamer simulation studies (Steinolfson, Suess, and Wu, 1982; Linker, Van Hoven, and Schnack, 1990) in combination with supersonic/super-Alfvénic steady-state flow calculations (Pizzo, 1982).

Space Science Reviews **72**: 121–124.
© 1995 *Kluwer Academic Publishers.*

Our purpose is to upgrade the simulation of steady solar wind (Usmanov, 1993a,b) for incorporation of transient dynamics.

2. Simulation Procedure

Being constructed as consisting of two regions, the steady model includes an outer supersonic/super–Alfvénic flow region where stationary MHD equations for a polytropic single–fluid flow with negligible dissipation are solved and an inner region of transonic/trans–Alfvénic flow where a steady–state solution for the time–dependent, fully 3–D system of MHD equations is sought as a result of time-relaxation process. The same time–dependent system of equations should be solved in order to find steady–state solution from relaxation and to simulate non–stationary processes; thus, the propagation of transient disturbances can be studied by using the same numerical algorithm.

The simulation domain is subdivided into two regions: $1 \leq r \leq 9.85\ R_s$ and $9.85\ R_s \leq r \leq 329\ R_s$. The radial distribution of mesh points in the inner region is taken to be $r_i = (R_s + \Delta r)^{i-1}$, where $\Delta r = 0.1\ R_s$. In the outer region, the radial spacing is chosen to increase as $r^{1/2}$ and the mesh points are located at $r_i = r_{i-1} + \sqrt{r_{i-1}/r_1}$, where $r_1 = 9.85\ R_s$. The mesh points are uniformly distributed from $\theta = 6°$ to $\theta = 174°$ and from $\phi = -12°$ to $\phi = 360°$ with the spacing of 12° (the two auxiliary points at $\phi = -12°$ and 360° are added in order to conveniently prescribe periodic boundary conditions). Thus, the computations are carried out on a mesh of 15 positions in θ–, 32 positions in ϕ–direction, and 25 (96) positions in radial direction in the inner (outer) region. As in the case of steady-state calculations, we use different values for the polytropic index γ, 1.05 and 1.17, for the inner and outer simulation region, respectively. However, unlike the steady–state case, the numerical procedure of solving the MHD equations is essentially the same for both regions. The values of dependent variables obtained at the outflow boundary from the simulation in the inner region are used to prescribe the inflow boundary condition for the outer region.

The outline of our approach is as follows. The simulation begins with computation of a background steady-state solar wind. Then, the flare-associated disturbances are introduced into the background flow by changing the properties of the flow near the observed flare sites at the inner boundary. The shock velocity and the steady-state parameters at the boundary mesh point closest to the observed flare site are used to infer the postshock parameters from the Rankine-Hugoniot shock jump conditions. These parameters are used as peak values in specifying the input pulse imitating the solar flare at the point. The temporal profile of pulse intensity is chosen to be a ramp function that varies linearly between preshock and postshock parameters during the rising phase and vice versa during the falling phase.

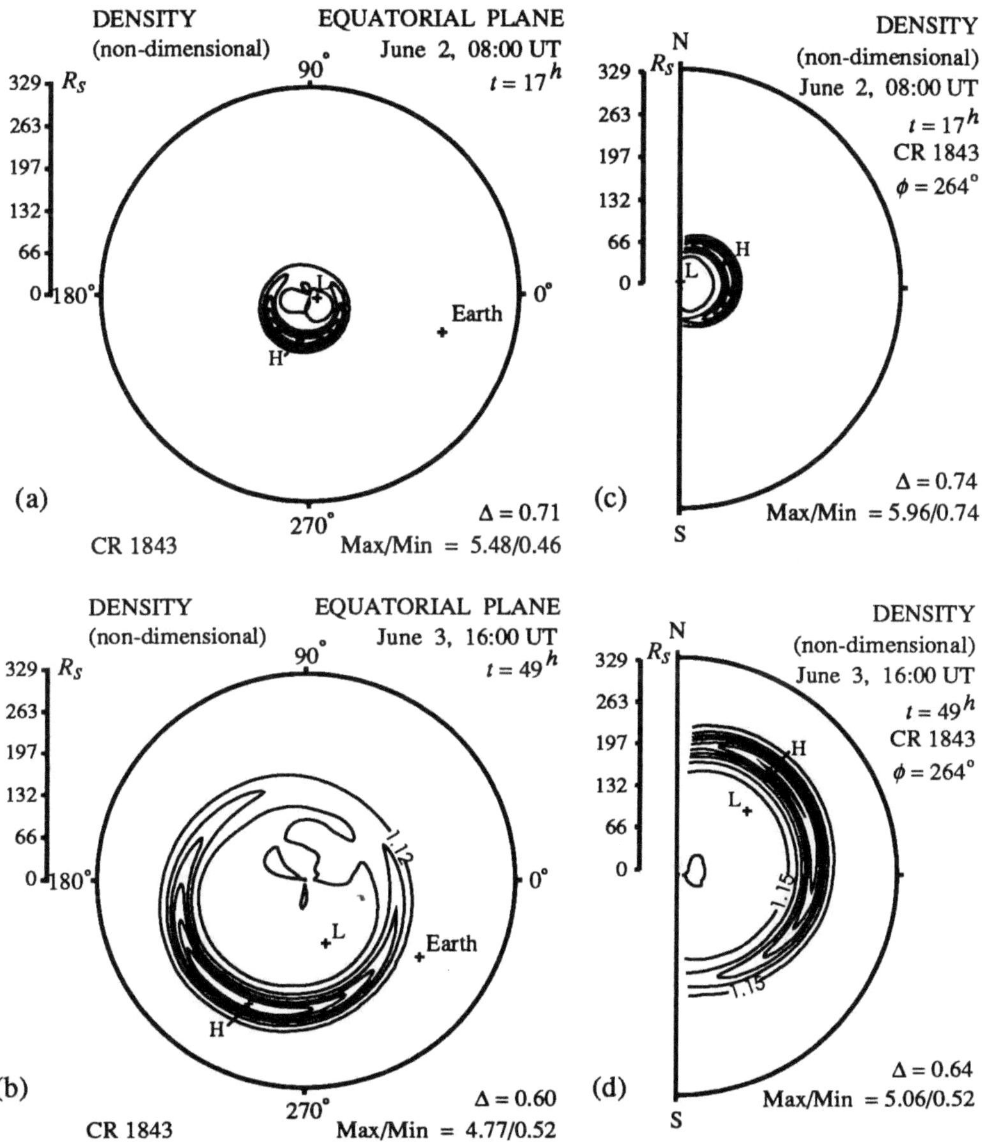

Fig. 1. Time evolution of disturbances associated with solar flare of June 1 presented by pairs of simulated contour maps of plasma density in the equatorial plane (a, b) and in the $\phi = 264°$ meridional plane (c, d) for $t = 17$ and 65^h. The density is referred to the respective values in the initial state. Max/Min are the peak values denoted as **H/L** on the plots and Δ is the contour spacing.

3. Simulation Example

An example of simulation is presented in Figure 1 where the contour maps of constant density are shown as a function of the time elapsed since the initiation of pulse imitating the X12 solar flare of 1991, June 1, 15:00 UT. The pulse peak parameters are correspond to the shock velocity of 1400 $km{\cdot}s^{-1}$. Figure 1 displays density distributions in the flare central meridian plane and those in the solar equatorial plane. The successive frames at times $t = 17$ and 49^h after the initiation of the pulse demonstrate the evolution of the large–scale density structure in the transient. The compression region followed by a rarefaction region is clearly seen in these plots. Note the location of Earth in Figure 1.

We have developed a global fully 3–D, time–dependent MHD model of the solar wind allowing a self–consistent simulation of heliospheric dynamics on a global scale. The input parameters for the model are: (i) a synoptic map of the observed line-of-sight magnetic field on the solar surface, (ii) the data on solar flares: location, importance, variation of soft X–ray emission, characteristics of Type II radio bursts (if associated). The output from the model is the solar wind and interplanetary magnetic field parameters (the velocity and magnetic field vectors, the thermodynamic characteristics: the density, pressure, temperature) at any point of interplanetary space (inside, of course, the computational domain, the upper boundary of which can be chosen as far as the interest goes and the computer resources allow, while the inner boundary is at 1 R_s).

Acknowledgements

The author wishes to thank the SCOSTEP project, SOLTIP, SOLar connection with Transient Interplanetary Processes, and the Russian Foundation for Fundamental Research for partial support.

References

De Young, D. S. and Hundhausen, A. J.: 1971, *J. Geophys. Res.*, **76**, 2245.
Dryer, M., Smith, Z. K., and Wu, S. T.: 1988, *Astrophys. Space Sci.*, **144**, 407.
Dryer, M.: 1994, *Space Sci. Rev.* (in press).
Han, S. M., Wu, S. T., and Dryer, M.: 1988, *Computers and Fluids*, **16**, 81.
Hirshberg, J., Nakagawa, Y., and Wellck, R. E.: 1974, *J. Geophys. Res.*, **79**, 3726.
Hundhausen, A. J. and Gentry, R. A.: 1969, *J. Geophys. Res.*, **74**, 2908.
Linker, J. A., Van Hoven, G., and Schnack, D. D.: 1990, *Geophys. Res. Letters*, **17**, 2281.
Pizzo, V. J.: 1982, *J. Geophys. Res.*, **87**, 4374.
Steinolfson, R. S., Suess, S. T., and Wu, S. T.: 1982, *Astrophys. J.*, **255**, 730.
Steinolfson, R. S. and Dryer, M.: 1984, *Astrophys. Space Sci.*, **104**, 111.
Steinolfson, R. S.: 1992, *J. Geophys. Res.*, **97**, 10811.
Usmanov, A. V.: 1993a, *Solar Phys.*, **146**, 377.
Usmanov, A. V.: 1993b, *Solar Phys.*, **148**, 371.
Wu, S. T., Dryer, M., and Han, S. M.: 1983, *Solar Phys.*, **84**, 395.
Wu, S. T., and Wang, A.-H.: 1991, *Adv. Space Res.*, **11(1)**, 187.

THE THREE-DIMENSIONAL EXTENT OF A HIGH SPEED SOLAR WIND STREAM

R.J. MACDOWALL [1], M.D. DESCH [1], M.L. KAISER [1], R.G. STONE [1], R.A. HESS [2], A. BALOGH [3], S.J. BAME [4], and B.E. GOLDSTEIN [5]

[1] *Code 695, NASA/GSFC, Greenbelt, MD 20771 USA*

[2] *Hughes STX, Lanham, MD 20706 USA*

[3] *The Blackett Laboratory, Imperial College, London SW7 2BZ UK*

[4] *Los Alamos National Laboratory, Los Alamos, NM 87545 USA*

[5] *Jet Propulsion Laboratory, Pasadena, CA 91109 USA*

Abstract. A primary goal of the Ulysses mission is to study the 3–dimensional structures making up the interplanetary medium, an example of which is the high speed solar wind stream observed *in situ* by Ulysses beginning in July 1992. In order to study the longitudinal extent of this stream as a function of Ulysses' increasing heliographic latitude, a second point of reference is required to separate spatial and temporal variations. Such a reference point is provided at Jupiter by a class of Jovian radio bursts, whose occurrence rate varies in a predictable way with solar wind speed. Using the *in situ* and remote observations from Ulysses, the extent of the high speed stream at ~5 AU is mapped and compared to the associated coronal hole boundary on the Sun.

1. Introduction

The Ulysses spacecraft provides the first opportunity for direct study of the solar wind at high ecliptic latitudes. In this paper, a cross-sectional mapping of a high speed solar wind stream as a function of longitude and latitude is presented. Such a mapping provides an indication of the evolution of the high speed stream (hereafter HSS) as it propagates from the coronal hole source on the Sun through the interplanetary medium.

The Ulysses observations consist of *in situ* data from the Ulysses plasma instrument and remote observations derived from the Ulysses radio instrument. The latter data are the occurrence rates of a particular class of Jovian radio burst, which provide a proxy for solar wind data in the ecliptic. During the time interval for which data are presented (January 1992 through July 1993), the Ulysses-Sun distance ranged from 5.40 to 4.48 AU, and its heliographic latitude, from −5.8 to −35.8° . The out-of-ecliptic trajectory was initiated by the Jupiter flyby on 8 February 1992.

The Jovian radio bursts used in this analysis are readily identifiable by their morphological characteristics and their recurrence with a 40–50 min periodicity. They were discovered by the Ulysses spacecraft (MacDowall *et al.*, 1993; Desch, 1994) and named QP-40 bursts because of the 40 min quasiperiodicity. As

Space Science Reviews **72**: 125–128.
© 1995 *Kluwer Academic Publishers.*

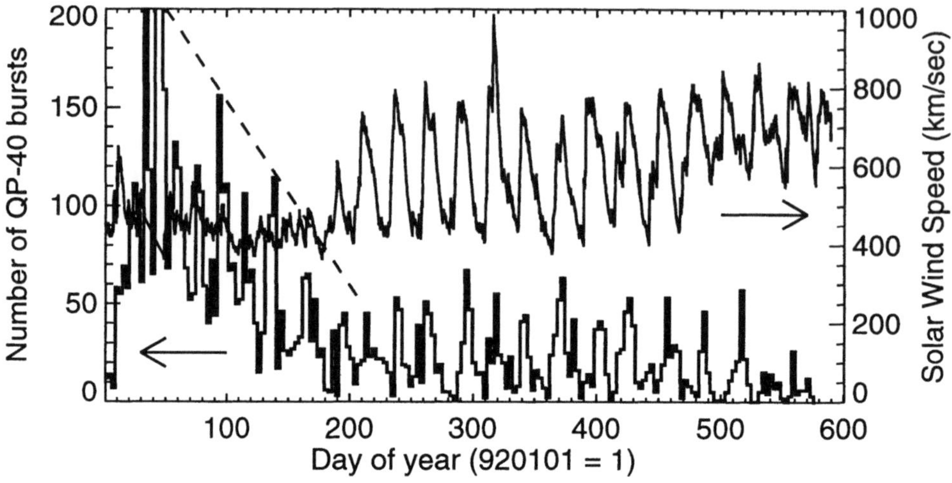

Fig. 1. Plots of solar wind speed at Ulysses and occurrence rate of Jovian QP-40 radio bursts in 1992–3. The dashed line shows the trend of decreasing numbers of QP-40 bursts with distance from Jupiter, until about day 200. At about this time, the extension of the southern solar coronal hole moved equatorward and intercepted Ulysses and Jupiter.

shown in Figure 1, after approximately day 200 of 1992, the occurrence rate of QP-40 bursts varies with a 25 to 30 day period, correlated with solar wind speed (MacDowall *et al.*, 1993; Desch, 1994).

2. Temporal Evolution of the High Speed Stream

Beginning in mid-1992, Ulysses detected a recurring high-speed solar wind stream, evident in the solar wind speed plotted in Figure 1. This feature corresponds to the extension of a coronal hole from high southern solar latitudes towards the solar equator, as described by Bame *et al.* (1993). In Figure 2a, the successive occurrences of high speed solar wind are plotted, assuming a recurrence period of 25.38 days. This period, which is the defined sidereal period of solar rotation at the heliographic equator, corresponds to the synodic period seen from Ulysses, which is in a high inclination orbit after the Jupiter encounter. The rotation period of the solar wind structure appears to change in late-1992. This variation will be shown to be primarily a time-dependent phenomenon; nevertheless, Ulysses' increasing southern heliographic latitude may also introduce latitude-dependent effects, including differential rotation.

Determining the extent of any latitude dependence requires measurements at a second location, preferably near the heliographic equator and at a similar distance from the Sun. As discussed in MacDowall *et al.* (1993), the good correlation of solar wind speed and QP-40 burst activity provides a method of determining when the HSS impacts the Jovian magnetosphere. The QP-40

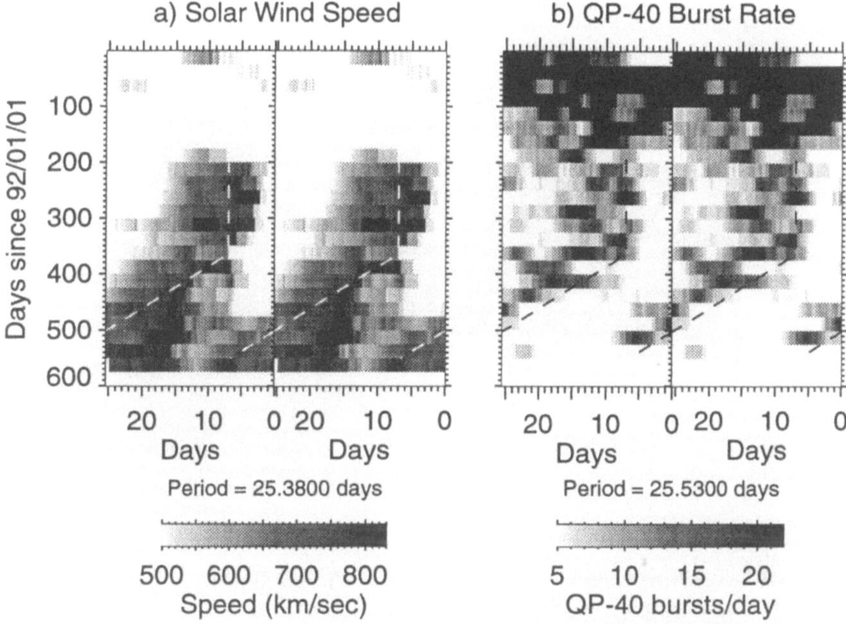

Fig. 2. Gray-scale plots of the recurrence of a) high speed solar wind and b) occurrence rate of QP-40 bursts. The dashed line, plotted in the same location on each panel, provides an estimate of the arrival times of the HSS at Jupiter, based on the onsets of enhanced QP-40 burst activity. Note that each parameter is cycled using the appropriate synodic period.

burst activity is plotted in Figure 2b. Although the QP-40 burst onsets exhibit more variability, the trends are similar to those evident in Figures 2a. The QP-40 bursts provide strong evidence that the recurrence period of the HSS at Jupiter changed from 25.5 days to ~29 days in late 1992, as indicated by the dashed reference line. This change corresponds to a shift in the position of the southern coronal hole, evident in the solar Fe XIV 5303Å synoptic charts, which indicate the boundaries of coronal holes because of their lower intensity (Solar Geophysical Data, 1992–1993). Furthermore, the times of the QP-40 burst onsets diverge from those of the HSS in 1993, which will be shown to result from latitudinal variation in the longitudinal extent of the HSS.

3. Three-dimensional shape of the High Speed Stream

Although it is evident from Figure 2 that the HSS drifts in heliographic longitude with time, it is still possible to construct a time-independent map from the Ulysses data under the assumption that the cross-section itself is time-independent. (Such a major assumption is required in order to combine the data obtained from only two locations — Jupiter and Ulysses.) To construct the map, the solar wind data for 1993 are shifted by the appropriate amount of time to

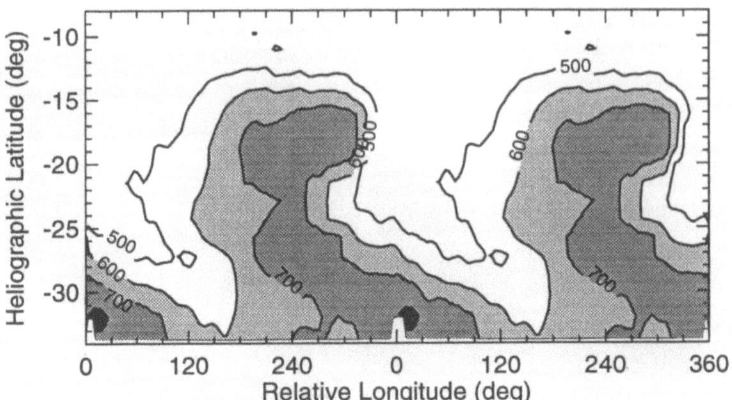

Fig. 3. Contour plot of solar wind speed as a function of Ulysses' heliographic latitude. The data were organized as a function of longitude as described in the text.

convert the diagonal dashed reference line (Figure 2a) into a vertical line. This compensates for the moving HSS by using the QP-40 burst onsets to determine when the leading edge of the stream reached a known location (Jupiter) in the ecliptic (heliographic latitude from -5.9 to $-4.8°$). The adjusted data are smoothed and contoured, yielding Figure 3.

The contours in Figure 3 are closer together at the leading edge of the HSS than at the trailing edge. This is the expected result of the dynamic evolution of the HSS with distance from the Sun (e.g., Pizzo, 1982). The other dominant feature of this map is that the HSS at low latitudes lags the HSS at high latitudes by $\sim 120°$ of longitude. This trend as a function of latitude does not correspond to any structure seen in the Fe XIV synoptic charts of the coronal hole boundaries. (Note that a trend of this form would have resulted no matter what (reasonable) reference line had been chosen for the QP-40 onsets in Figure 2b.) Furthermore, it has the opposite sense of the result expected from differential rotation. Some of this latitudinal variation results because the lower latitude regions of the HSS interact with the slower, higher density solar wind comprising the streamer belt. Further analysis of this dataset, such as detailed comparison with Yohkoh x-ray images and comparison to 3–D MHD models, is required to understand any additional causes.

References

Bame, S. J., Goldstein, B. E., Gosling, J. T., Harvey, J. W., McComas, D. J., Neugebauer, M., and Phillips, J. L.: 1993, *Geophys. Res. Lett.* **20**, 2323–2326.

Desch, M. D.: 1994, *Ap. J. Suppl. Series* **90**, 541–546.

MacDowall, R. J., Kaiser, M. L., Desch, M. D., Farrell, W. M., Hess, R. A., and Stone, R. G.: 1993, *Planet. Space Sci.* **41**, 1053–1066.

Pizzo, V. J.: 1982, *J. Geophys. Res.* **87**, 4374–4394.

Solar Geophysical Data, 1992–1993, U.S. Dept. of Commerce, Boulder.

RECONNECTION ON OPEN FIELD LINES AHEAD OF CORONAL MASS EJECTIONS

D.J. McCOMAS, J.T. GOSLING, C.M. HAMMOND,
M.B. MOLDWIN, J.L. PHILLIPS
Space and Atmospheric Sciences Group,
Los Alamos National Laboratory, Los Alamos, NM USA

R.J. FORSYTH
Space & Atmospheric Physics,
Imperial College, Blackett Laboratory, London, England

Abstract. Plasma and magnetic field signatures from 29 November 1990 indicate that the Ulysses spacecraft passed through a series of interplanetary structures that were most likely formed by magnetic reconnection on open field lines ahead of a coronal mass ejection (CME). This reconnection changed the magnetic topology of the upstream region by converting normal open interplanetary magnetic field into a pair of regions: one magnetically disconnected from the Sun and the other, a tongue, connected back to the Sun at both ends. This process provides a new method for producing both heat flux dropouts and counterstreaming suprathermal electron signatures in interplanetary space. In this paper we expand upon the 29 November case study and argue that reconnection ahead of CMEs should be less common at high heliolatitudes.

The bulk of solar wind observed at 1 AU displays evidence of "open" field lines that extend outward from the Sun and presumably close somewhere far out in the heliosphere. These regions can be identified in interplanetary space by the presence of a unidirectional beam or "strahl" (Rosenbauer et al., 1977) of hot halo electrons, with energies above ~80 eV, directed outward from the million-degree corona along the IMF. Other solar wind halo electron distributions display a pair of counterstreaming electron beams (Gosling et al., 1987). Such distributions are indicative of magnetic structures that are connected back to the hot coronal source at both ends (magnetic tongues) or that are detached plasmoids in which the counterstreaming populations are trapped on closed field lines (e.g., Gosling, 1990).

In still other electron distributions, named heat flux dropouts or HFDs, the strahl is absent (McComas et al., 1989). These authors suggested that HFDs could represent regions that are magnetically disconnected from the Sun and connected to the outer heliosphere at both ends instead. They suggested that such detached U-shaped structures could be formed by reconnection in the corona across the heliospheric current sheet above helmet streamers. Subsequently, coronagraph observations indicated that such

Space Science Reviews **72**: 129–132.
© 1995 *Kluwer Academic Publishers.*

reconnection does occur, at least on occasion (McComas et al., 1991). An analysis of more energetic electrons for HFD events indicated that, while a few of the HFD intervals probably represent disconnected regions, others probably represent some other sort of phenomena (Lin and Kahler, 1992).

An unusual sequence of solar wind structures was observed by Ulysses on 29 November 1990 in the ecliptic plane near 1.25 AU (McComas et al., 1994). Over a day and a half interval the speed increased from <400 km s^{-1} to >600 km s^{-1}, indicating that faster moving material was gradually overtaking the slower solar wind ahead of it. Figure 1 displays the halo electron density, electron heat flux, relative alpha particle abundance (n_α/n_p), proton temperature, and magnetic field magnitude for the nine hour interval containing the solar wind structures. Five distinct regions can be identified based on the electron distributions (not shown here, see McComas et al. (1994)) and plasma and field properties displayed in Figure 1. Prior to 1:18 UT (A) and after 6:15 UT (E) the halo distribution has a unidirectional strahl indicating open IMF lines. A heat flux dropout (B) indicated by the strong depression in the measured heat flux and an absence of a beam in the halo distributions is observed from 1:18 to 1:42 UT. Regions C (1:42 to 2:53 UT) and D (2:53 to 6:15 UT) both have counterstreaming suprathermal electron distributions and substantial enhancements in the halo electron density indicative of counterstreaming intervals. However, these two counterstreaming intervals represent quite different plasma regimes. In region D the alpha abundance is higher, while the proton temperature is lower than in region C or any of the other surrounding regions. The enhanced alpha abundance in region D is characteristic of many CMEs while the reduced plasma temperature is characteristic of a closed field region which has expanded and cooled more than the surrounding solar wind (e.g., Gosling, 1990). The enhanced and smooth magnetic field magnitude and the slow, even rotation of the field direction (not shown) in region D are characteristic of CMEs that are also magnetic clouds (Klein and Burlaga, 1982) or flux ropes (Gosling, 1990). The plasma and field properties of region C resemble those of the open IMF (A and E) and the HFD (B) far more than they do the other counterstreaming region (D).

Thus, observations from 29 November 1990 show that Ulysses passed in sequence through a heat flux dropout, a non-CME region of counterstreaming halo electrons, and a CME. These observations suggest that the magnetic topology of a region of normal open IMF was altered by a faster moving CME. In particular, reconnection upstream of the CME between previously open field lines created a pair of new magnetic topologies: a region magnetically disconnected from the Sun with field lines connected to the outer heliosphere at both ends and one connected back to the solar corona at both ends.

Figure 2 shows a simplified "flat heliosphere" representation of a region of IMF that illustrates how reconnection can change the magnetic topology in interplanetary space. The top panel shows a CME rising up (moving from left to right) behind a region of normal open IMF. It is not important for the upstream processes described here whether the CME is a detached plasmoid (dashed lines) or a connected tongue (dotted lines). If the swept-up IMF contains a magnetic field reversal, then conditions are right for reconnection. In the bottom panel, reconnection has converted the open IMF into a new pair of structures: a magnetic tongue and a disconnected U-shaped structure. Previous discussions of the formation of detached structures have emphasized reconnection in the corona. The 29 November 1990 example provides direct evidence for reconnection in interplanetary space driven by the relative motion between a CME and the slower plasma ahead leading to the modification of previously open IMF into new magnetic topologies (McComas et al., 1994).

In the ecliptic plane the rotation of the Sun causes the IMF to wrap up into a roughly planar Arcimedean spiral. At higher heliolatitudes, the IMF should, on average, spiral less tightly along conical surfaces. This variation of the global IMF configuration with heliolatitude should effect the likelihood of the sort of reconnection examined above. In particular, the wrapping-up of the IMF in and near the ecliptic

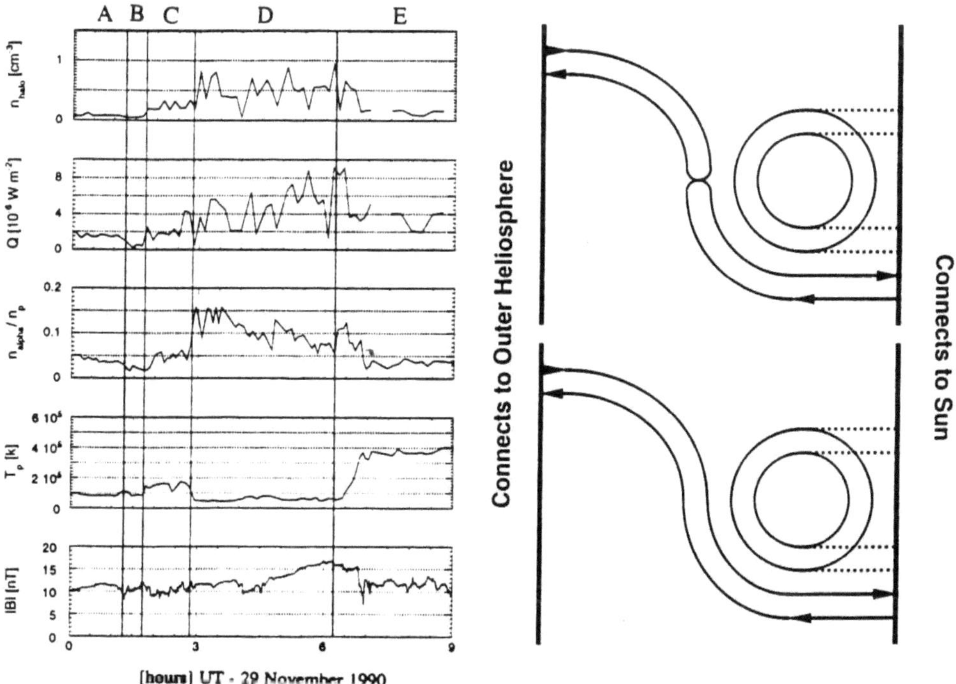

Figure 1 (left). Selected solar wind plasma and magnetic field parameters for the first nine hours of 29 November 1990.

Figure 2 (right). Schematic diagram of a CME overtaking normal, open IMF (top). In the bottom panel reconnection has converted the magnetic topology of the open IMF ahead of the CME into a pair of regions: a tongue, connected back to the Sun at both ends, and a disconnected U-shaped region, which closes in the outer heliosphere at both ends.

plane provides a favorable geometry for reconnection where the solar wind flow is largely perpendicular to the field and consequently can sweep it up and compress it. At higher heliolatitudes, within several AU of the Sun, we would expect the geometry to be less favorable for compression of the IMF and reconnection because the IMF should be less tightly wound-up. In addition, we would not expect to find the heliospheric current sheet at high heliolatitudes (Suess et al., 1993). Without swept-up regions of field reversal, conditions are not appropriate for reconnection.

Another factor which could effect the likelihood of reconnection in the solar wind is the rate of CMEs, in general, and of CMEs which are faster than the surrounding solar wind, in particular. Recent work by Gosling et al. (1994) shows that the high latitude CMEs observed so far by Ulysses were all quite fast (average speed ~740 km s^{-1}) compared to typical ecliptic plane CME speeds. However, these high speeds are also characteristic of essentially all high latitude solar wind flows observed by Ulysses so far (Phillips et al., 1994). In addition, there are fewer CMEs at high than at low heliolatitudes, particularly away from times of solar maximum (see, for example, Figure 6 of Hundhausen (1993)). To the extent that CMEs plowing out through slower solar wind play a role in interplanetary reconnection, there should be far less at higher heliolatitudes.

Magnetic reconnection on open field lines in interplanetary space can alter the magnetic topology of the IMF. This study suggests that while the conditions are sometimes right for reconnection to occur in the ecliptic plane, and that at least one good example has already been observed with Ulysses, the conditions are less appropriate for reconnection at higher heliolatitudes. We predict that comparable examples of reconnection in interplanetary space will not be observed over the polar pass portions of the Ulysses mission.

This work was carried out under the auspices of the United States Department of Energy with financial support from the National Aeronautics and Space Administration.

References

Gosling, J.T.: 1990, *Geophys. Mono.* **58**, AGU, 343.
Gosling, J.T. *et al.*: 1987, *J. Geophys. Res.*, **92**, 8519.
Gosling, J.T. *et al.*: 1994, submitted to *Geophys. Res. Lett.*
Hundhausen, A.J.: 1993, *J. Geophys. Res.*, **98**, 13177.
Klein, L.W. and L.F. Burlaga: 1982, *J. Geophys. Res.*, **87**, 613.
Lin, R.P. and S.W. Kahler: 1992, *J. Geophys. Res.*, **97**, 8203.
McComas D.J. *et al.*: 1989, *J. Geophys. Res.*, **94**, 6907.
McComas, D.J. *et al.*: 1991, *Geophys. Res. Lett.*, **18**, 73.
McComas, D.J. *et al.*: 1994, in press in *Geophys. Res. Lett.*
Phillips, J.L. *et al.*: 1994, in press in *Geophys. Res. Lett.*
Rosenbauer, H. *et al.*: 1977, *J. Geophys.*, **42**, 561.
Suess, S.T. *et al.*: 1993, *Geophys. Res. Lett.*, **20**, 161.

CORONAL MASS EJECTIONS AT HIGH HELIOGRAPHIC LATITUDES: ULYSSES

J. T. GOSLING[1], S. J. BAME[1], D. J. MCCOMAS[1], J. L. PHILLIPS[1], A. BALOGH[2], and K. T. STRONG[3]

[1]*Los Alamos National Laboratory, Los Alamos, New Mexico USA*

[2] *Imperial College, London England*

[3]*Lockheed Palo Alto Research Laboratory, Palo Alto, California USA*

Abstract. Nine coronal mass ejections (CMEs) have been detected in the solar wind by the Ulysses plasma experiment between 31° and 61° South. One of these events, which was also a magnetic cloud, was directly associated with an event observed by the soft X-ray telescope on Yohkoh in which large magnetic loops formed in the solar corona directly beneath Ulysses. This association suggests that the flux rope topology of the magnetic cloud resulted from reconnection between the "legs" of neighboring magnetic loops within the rising CME. The average CME speed (~740 km s^{-1}) at these latitudes was comparable to that of the normal solar wind there and is much greater than average CME speeds observed either in the solar wind in the ecliptic plane or in the corona close to the Sun. We suggest that the same basic acceleration process applies to both slow CMEs and the normal solar wind at any latitude.

We have scanned color-coded plots of solar wind electron angular distributions obtained by the Los Alamos plasma experiment on Ulysses to identify the counterstreaming electron events characteristic of coronal mass ejections (CMEs) in the solar wind (e.g., Gosling, 1990). At least 9 CMEs have been identified in the data poleward of S31°. All but three of these events were observed equatorward of S41.3°; the most recent events were observed at S52.3°, S54.3°, and S60.5°.

The left panel of Figure 1 shows 20-min averages of selected solar wind plasma and magnetic field parameters from Ulysses for the interval June 7 - 15, 1993. A distinct counterstreaming suprathermal electron event characteristic of CMEs occurred roughly in the center of this interval, as indicated by the dashed vertical lines. The combined measurements indicate that the CME, which had expanded to a radial width of ~1.1 AU, was also a magnetic cloud. The measured speed within the center of the CME was ~740 km s^{-1}, which is greater than the speeds of 98% of all CMEs observed in the ecliptic plane at 1 AU; however, its speed was typical of the other high-latitude CMEs identified in the present survey.

There were two shock waves associated with the CME - a forward shock that passed the spacecraft ~17 hours before the arrival of the CME and a reverse shock that passed the spacecraft ~27 hours after the end of the CME. These shocks could not have been produced as a result of the CME overtaking slower plasma ahead, because the plasma

Space Science Reviews **72**: 133–136.
© 1995 *Kluwer Academic Publishers.*

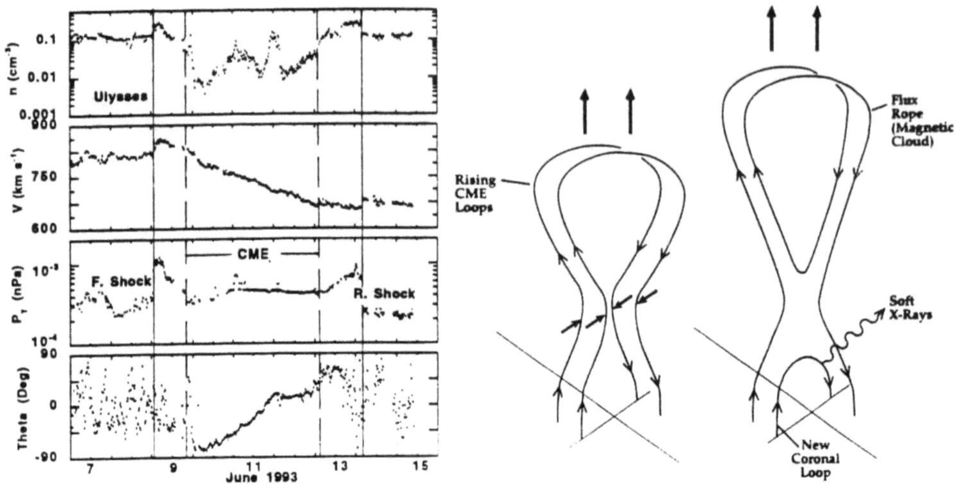

Fig. 1. (Left) An expanding coronal mass ejection (CME) and associated forward-reverse shock pair observed in the solar wind by Ulysses at 4.64 AU and S32.5°. The parameters plotted from top to bottom are the proton density, bulk flow speed, total pressure (protons, electrons, and magnetic field), and magnetic field polar angle in a spherical coordinate system centered on the Sun. (Right) Sketches illustrating the pinching off (reconnection) of magnetic loops in a rising CME whose legs are sheared relative to one another. Such reconnection produces new coronal loops (compare with Figure 2) and a magnetic flux rope extending into interplanetary space. Adapted from Gosling *et al.* (1994a).

ahead was moving faster than the CME. Rather, the roughly symmetrical appearance of the overall total pressure profile and the declining speed across the CME suggest that the shocks were the result of an over-expansion of the CME driven by a high initial internal pressure. This suggestion is consistent with simple numerical simulations of the event (Gosling *et al.*, 1994a); two additional forward-reverse shock pair events associated with over-expansion of CMEs have been observed recently at high latitudes.

The field topology of a magnetic cloud is that of a magnetic flux rope characterized by helical field lines wrapped about a central axis (e.g., Burlaga, 1991). We have previously suggested (Gosling, 1990; 1993) that the flux rope topology of a magnetic cloud is a natural consequence of reconnection occurring within the "legs" of the outward-moving magnetic loops that constitute a CME. As illustrated in the right panel of Figure 1, such reconnection produces both a magnetic flux rope topology within the CME and new magnetic loops in the corona below the CME.

Assuming a constant outward speed, the CME shown in Figure 1 left the Sun near midday on May 31. At this time Ulysses was located almost directly above the southeast limb of the Sun as viewed from Earth. Figure 2 shows two images of the Sun obtained by the soft X-ray telescope on Yohkoh on May 31. These images show that large coronal loops formed on and probably slightly behind the east solar limb almost directly beneath Ulysses between 08:26 and 17:51 UT near the time when we

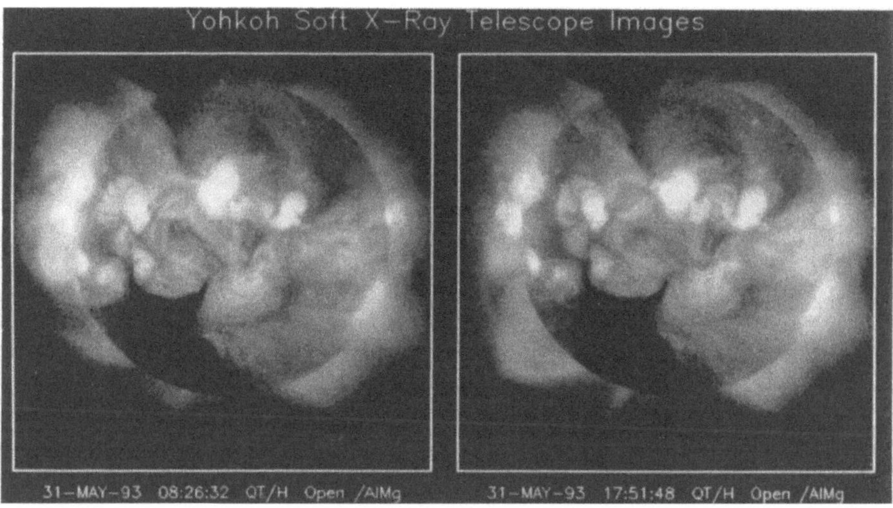

Fig. 2. Images of the Sun taken by the Soft X-ray Telescope on Yohkoh on May 31, 1993 at 08:26:32 and 17:51:48 UT. Note the large coronal loops that formed on the southeast limb directly beneath Ulysses in the interval between the two snapshots.

calculate that the CME should have departed from the Sun. The combined Ulysses and Yohkoh observations thus provide substantial support for the idea that the flux rope topology of magnetic clouds is a result of reconnection between the magnetic loops embedded within CMEs as illustrated in Figure 1. Statistical support for this flux rope formation mechanism can be found in the fact that magnetic loops visible in X-rays form in the corona after ~1/3 of all CMEs (Sheeley *et al.*, 1983) while, similarly, ~1/3 of all CMEs in the solar wind appear to be magnetic flux ropes (Gosling, 1990).

A striking aspect of our observations is that all of the CMEs observed at high latitudes had speeds comparable to the rest of the solar wind there (Gosling *et al.*, 1994b). The event in Figure 1 is typical in this respect; its speed of ~740 km s^{-1} is about the same as the average flow speed associated with the large coronal hole that has dominated the high-latitude Ulysses observations (Phillips *et al.*, 1994). Average speeds for the other high-latitude events were 650, 840, 750, 700, 740, 740, 760, and 750 km s^{-1} respectively. All of these events were observed at times of declining or nearly constant flow speed. Thus the high-latitude CMEs do not appear to have achieved their high speeds by virtue of being "pushed" outward by faster plasma from behind. They simply seem to have experienced the same basic acceleration as the rest of the solar wind at high latitudes.

It is useful to compare the above result with observations of CME speeds in the corona and in the ecliptic plane at 1 AU. Coronagraph observations show that CMEs close to the Sun seldom have speeds comparable to those measured by Ulysses at high latitudes. For example, less than 12% of all CMEs observed in the corona by the SMM coronagraph over a 7-yr interval had speeds greater than 700 km s^{-1} (Hundhausen *et al.*, 1994). On the other hand, ~50% of all CMEs observed by SMM had speeds less than 300 km s^{-1}, and many had speeds less than 100 km s^{-1}. Such low speeds are not observed in the solar wind either in the ecliptic plane or at high latitudes, even though there is good evidence that these slow CMEs do escape from the Sun into the interplanetary medium. In the ecliptic plane the minimum CME speed observed (~280 km s^{-1}) is virtually identical to that of the rest of the solar wind there (Gosling *et al.*, 1994b). A similar situation prevails at high heliographic latitudes, but the minimum solar wind speed is much higher (~650 km s^{-1}). At all latitudes most CMEs do not appear to have been pushed up to their observed speeds by faster plasma from behind. They simply seem to have received the same minimum acceleration as the rest of the solar wind at their respective latitudes.

It is usually assumed that the forces that accelerate CMEs outward from the Sun are different from those responsible for accelerating the normal solar wind (see, e.g., Chen and Garren, 1993). Our results, on the other hand, suggest that slow CMEs experience the same basic acceleration as does the normal solar wind. It may be, for example, that an instability in the corona is responsible for the initial release of a slow CME, but thereafter the material within the CME is subject to the same forces responsible for accelerating the surrounding solar wind (e.g., Low, 1993). The fact that the magnetic topology of CMEs is quite different from that of the normal solar wind ("closed" versus "open" field lines) suggests that observations of CME accelerations might be used to discriminate between solar wind acceleration models. Furthermore, coronagraph observations of CME speed versus height may provide the best optical means of determining the acceleration profile of the normal solar wind close to the Sun.

This work was performed under the auspices of the U.S. Department of Energy with financial support from the National Aeronautics and Space Administration under W-15847. K.T.S. was supported by NASA contract NAS8-37336 and the Lockheed Independent Research Program

References

Burlaga, L.F.: 1991, in *Physics of the Inner Heliosphere II*, edited by R. Schwenn and E. Marsch, Springer-Verlag, Berlin, 1-22.

Chen, J. and Garren, D.A.: 1993, *Geophys. Res. Lett.*, **20**, 2319.

Gosling, J.T.: 1990, in *Physics of Magnetic Flux Ropes, Geophys. Monogr. 58*, edited by C. T. Russell, E. R. Priest, and L. C. Lee, AGU, Washington D.C., 343-364.

Gosling, J.T.: 1993, *J. Geophys. Res.*, **98**, 18,937.

Gosling, J.T. *et al.*: 1994a, *Geophys. Res. Lett.*, **21**, 237.

Gosling, J.T. *et al.*: 1994b, *Geophys. Res. Lett.*, in press.

Hundhausen, A.J. *et al.*: 1994, *J. Geophys. Res.*, 99, 6543.

Low, B.C.: 1993, *Adv . Space Res.*, **13**, (9)63.

Phillips, J.L. *et al.*: 1994, *Geophys. Res. Lett.*, in press.

Sheeley, N.R. *et al.*: 1983, *Astrophys. J.*, **272**, 349.

THE LARGE-SCALE STRUCTURE OF THE HELIOSPHERIC CURRENT SHEET DURING THE ULYSSES EPOCH

J. TODD HOEKSEMA
Center for Space Science and Astrophysics
Stanford, CA 94305-4055 USA
JTHoeksema@SOLAR.Stanford.EDU

June 16, 1994

Abstract. Ulysses is traversing the Sun's polar regions for the first time a year or two before solar minimum. If the heliospheric magnetic field behaves as we expect, the heliospheric current sheet (HCS) during this stage of the sunspot cycle should be quite stable and lie nearly flat, close to the equator. The high latitude solar fields should be unipolar and nearing their maximum strength. The overlying polar coronal holes should be well developed, producing a nearly uniform high-speed solar wind. Because the Sun's polar fields live longer than any other solar magnetic phenomenon, Ulysses will remain within a single coronal hole for an extended period and probe its structures in a unique way.

Of course everything will almost certainly not turn out to be as quiet and well-ordered as we expect. We know now that the photospheric field strength in the northern polar cap is less than the south. What will this mean for the solar wind speed and the magnitude of the interplanetary magnetic field? Solar and coronal observations suggest that the first magnetic signatures of the next solar cycle may already be emerging at high latitudes. Will Ulysses sense these fields "leaking" into the heliosphere or is the interplanetary magnetic structure completely dominated by the large-scale field as the models currently predict? How will the sources of variations in field, velocity, density, and composition fit into our conceptual picture?

Ulysses' rapid dash from south to north provides a unique opportunity to determine the latitudinal width of the equatorial region influenced by the HCS. This rapid change in latitude over a short interval during which the structure should be stable, should provide a definitive measurement of the latitudinal gradients of various quantities relative to the heliomagnetic equator.

In the next several years Ulysses will gradually decrease in latitude, eventually meeting the HCS as it gradually expands away from the equator during the rising part of the cycle. It will then race the HCS to the poles as maximum approaches. The situation in six years, at solar maximum, will be radically different. There will be no unipolar high latitude field, the HCS will extend to the poles, existing structures will be less stable, and there will likely even be multiple current sheets. Ulysses will have another opportunity to confirm or disprove our understanding of the high latitude heliosphere in a much different environment. The extended mission is essential to advance our understanding of the heliospheric field at the extremes of the solar cycle.

Key words: Heliospheric Current Sheet – Ulysses

1. Modeling the Heliospheric Current Sheet

Attempts to model coronal and interplanetary magnetic fields from photospheric observations have been made since the late 1960's (Schatten et al., 1969; Altschuler and Newkirk, 1969). Such models make simplifying assumptions about the physics to compute the magnetic field configuration

Space Science Reviews **72**: 137–148.
© 1995 *Kluwer Academic Publishers.*

near the Sun. At some height, often called the source surface, one assumes that the magnetic field configuration becomes frozen into the accelerating coronal plasma. The neutral line, dividing regions of opposite radial field at the source surface, is swept radially outward by the solar wind to define the 3-dimensional surface called the heliospheric current sheet (HCS). Fig. 1 shows the computed location of the HCS for the last solar cycle.

While the field configuration predicted at the base of the solar wind compares favorably with coronal data (Wilcox and Hundhausen, 1983; Bruno et al., 1984; Sime and McCabe, 1990) and with in-ecliptic spacecraft measurements of the interplanetary magnetic field (IMF) polarity (Hoeksema et al, 1982, 1983), it is generally difficult to verify the HCS location at high latitudes. The disappearance of the IMF sector structure during Pioneer 11's sojourn at just 16° north in 1976 demonstrated that the Sun's polar field dominates the heliospheric field configuration at solar minimum (Smith et al., 1978). Pioneer and Voyager measurements at high latitudes (and large heliocentric distances) confirmed the flattening of the sector structure again at the following minimum (Gazis, et al, 1989). However, making a detailed comparison between observations and predictions in the inner heliosphere and the outer heliosphere is very difficult because of dynamic processes acting in the solar wind (Behannon et al., 1989). Until Ulysses, sporadic cometary observations have provided the only *in situ* confirmation that the HCS extends to high latitudes (Neidner and Brandt, 1978; see also Farnham and Meech, 1994, and references therein). Ulysses passed below the maximum latitude of the current sheet enroute to the south pole in April, 1993 at about 30° S (Smith, et al., 1993).

The potential field model postulates that the coronal fields are effectively current free between the photosphere and a larger concentric shell called the source surface. Observations constrain the field at the photosphere, while at the source surface the model enforces a purely radial field configuration. These conditions are enough to determine a solution that reproduces fairly well what we know about the coronal field. The height of the source surface is fixed by physical considerations and optimized to match observations. Until recently the model apportioned the measured line-of-sight photospheric field to both a radial and a meridional component. In this case the fields in the Sun's polar caps were relatively weak. Observations of the annual variation of the polar fields (Svalgaard et al., 1978) and the 16° Pioneer observations (Smith et al., 1978) indicated that the Sun's strong polar fields must be included in the model. Setting the source surface radius at 2.5 R_s and adding a very sharply peaked 10 G polar field enabled the calculations of the ecliptic polarity to match the daily measurements about 80% of the time over the solar cycle (Hoeksema et al., 1982, 1983).

More recently Wang and Sheeley (1992) proposed treating the surface fields as purely radial, a very reasonable assumption at that height in the

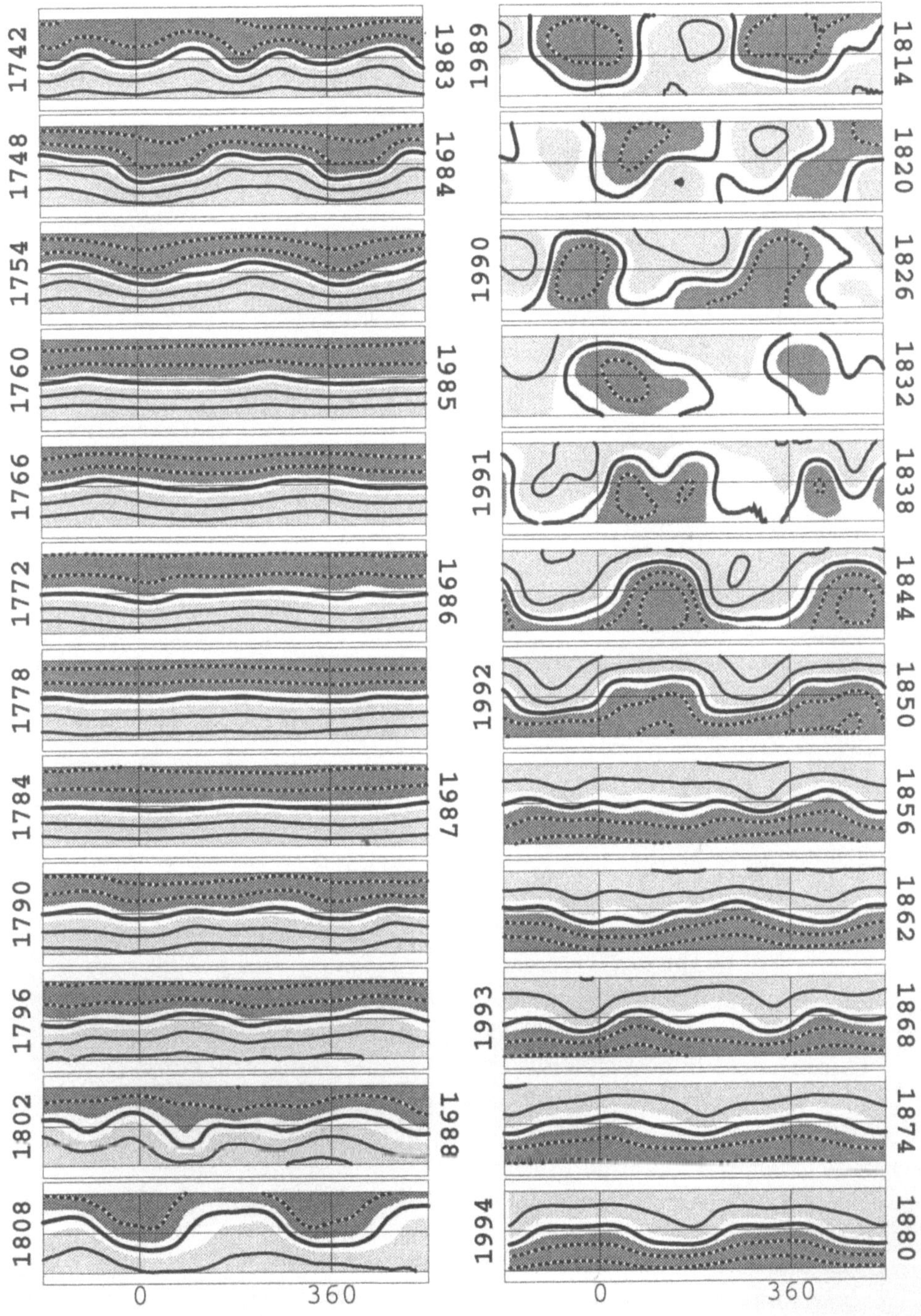

Fig. 1. Each panel shows the source surface field in a window two rotations wide centered on the indicated Carrington rotation. The neutral line is bold and negative regions are more darkly shaded. This shows every 6th rotation for the last 11 years (the first rotation within each year is indicated in the inner margin).

photosphere. This effectively increases the polar field strength and obviates the need for the indirect polar field correction. Except as noted, results in this paper use this radial inner boundary condition and a source surface radius of 3.25 R_s. With these conditions the daily predicted IMF polarity matches the *in situ* data with a correlation coefficient of about 0.59 in the interval from 1976 to 1988. The correlation ranges from 0.52 at solar minimum in 1976-8 to 0.71 in the declining phase, 1982-4. The overall in-ecliptic agreement depends relatively little on the source surface radius except at solar minimum. The HCS computed with the radial surface field is generally flatter and agrees better with the timing of Ulysses' last sector observations (Phillips et al., 1994). Potential field models do not correctly determine the IMF strength, but one such effort is discussed below.

2. The Source of the Heliospheric Field

The field lines at the source surface are open, by definition. Following open field lines to the photosphere shows that essentially all the IMF comes from coronal holes (Levine, 1982, Wang & Sheeley, 1992). Fig. 2 shows how the photospheric field connects to the open flux in the corona in March 1993, just before Ulysses passed south of the HCS. The footpoints in Panel 3 are traced from an evenly spaced grid on the source surface. In the declining and minimum phases nearly all of the heliospheric field originates in the polar coronal holes; the rest comes from low-latitude extensions of the polar hole.

The divergence of the field is inversely related to the solar wind speed (Levine, 1978; Wang & Sheeley, 1990, 1991; Sheeley et al, 1994). Low divergence in the lower panel of Fig. 2 indicates expected regions of high solar wind speed, e.g. in the northern polar region and at mid northern latitudes near 330° and at low southern latitudes near 60°. The HCS forms over closed coronal features that constitute the streamer belt. The HCS always lies over areas of large field line divergence that generate slow solar wind.

While the large-scale structure of the coronal field is stable for long periods of time, there are occasions when the field rapidly rearranges itself. Fig. 3 shows a dramatic example of such an episode in May - July, 1992, when over just a couple rotations the long-standing coronal pattern consisting of an extension of the southern polar hole around 60° was replaced by a totally new structure with the polarity of the northern hole. The north-south arcade at 90° was replaced by an east-west arcade in the south (left column). The rearrangement of the open field regions is shown better on the right. Panels for CR 1850, 1856, and 1862 in Fig. 1 help to put this in context. This reorganization was reflected in a major shift in the IMF sector structure pattern observed by Ulysses (Balogh et al., 1993).

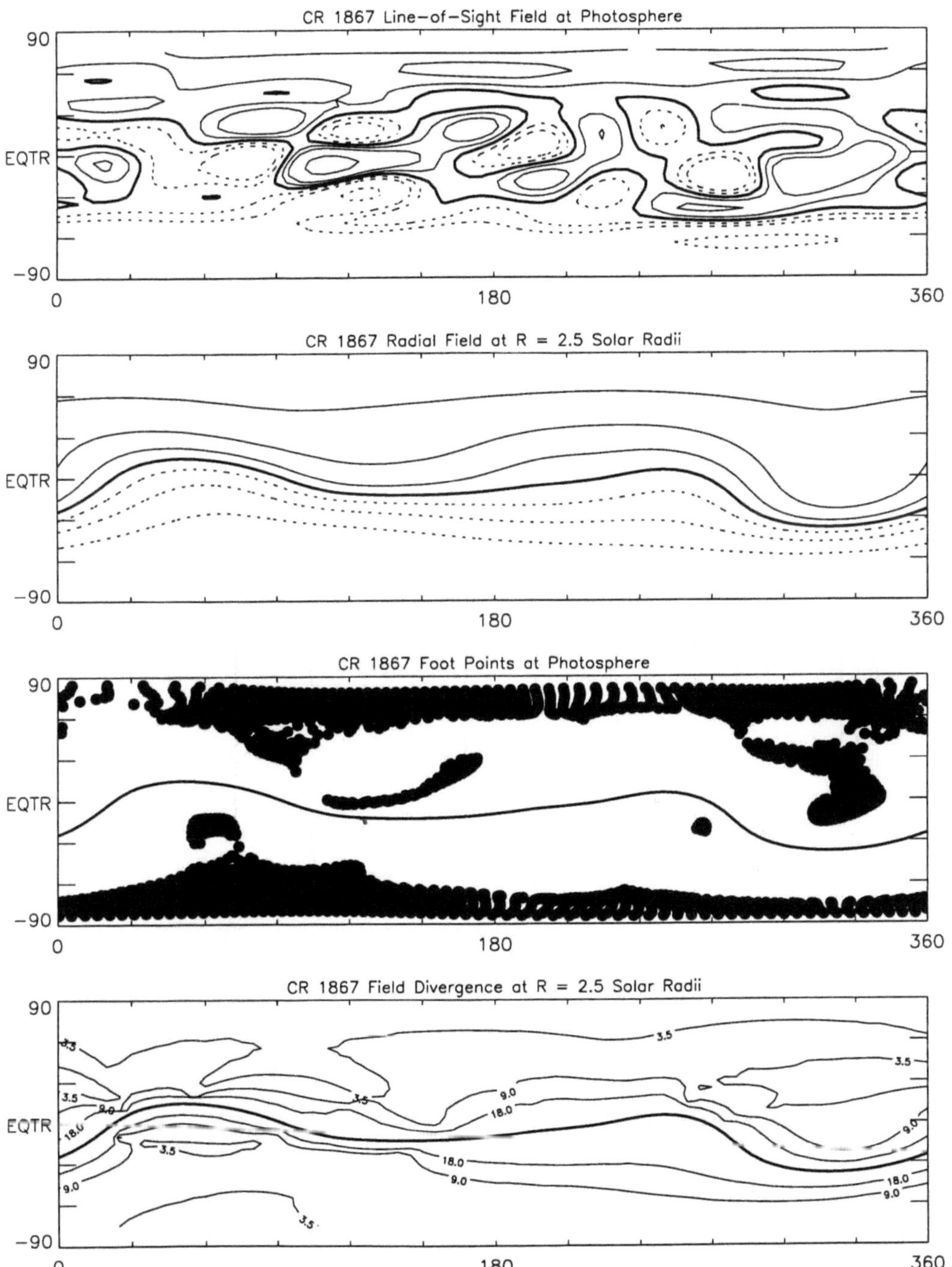

Fig. 2. Panel 1 (top) shows a Carrington map of the photospheric magnetic field in March 1993. Contours are at 0, 1, 2, 5 and 10 G. Dashed contours show negative field. The second panel shows the field calculated with the source surface at 2.5 R_s. Contours are at 1% of the photospheric level. Dots in the third panel show the foot points of open fields mapped back to the photosphere. The divergence of flux tubes is related to the solar wind velocity (bottom panel). Bold lines in panels 3 and 4 are the source surface neutral line.

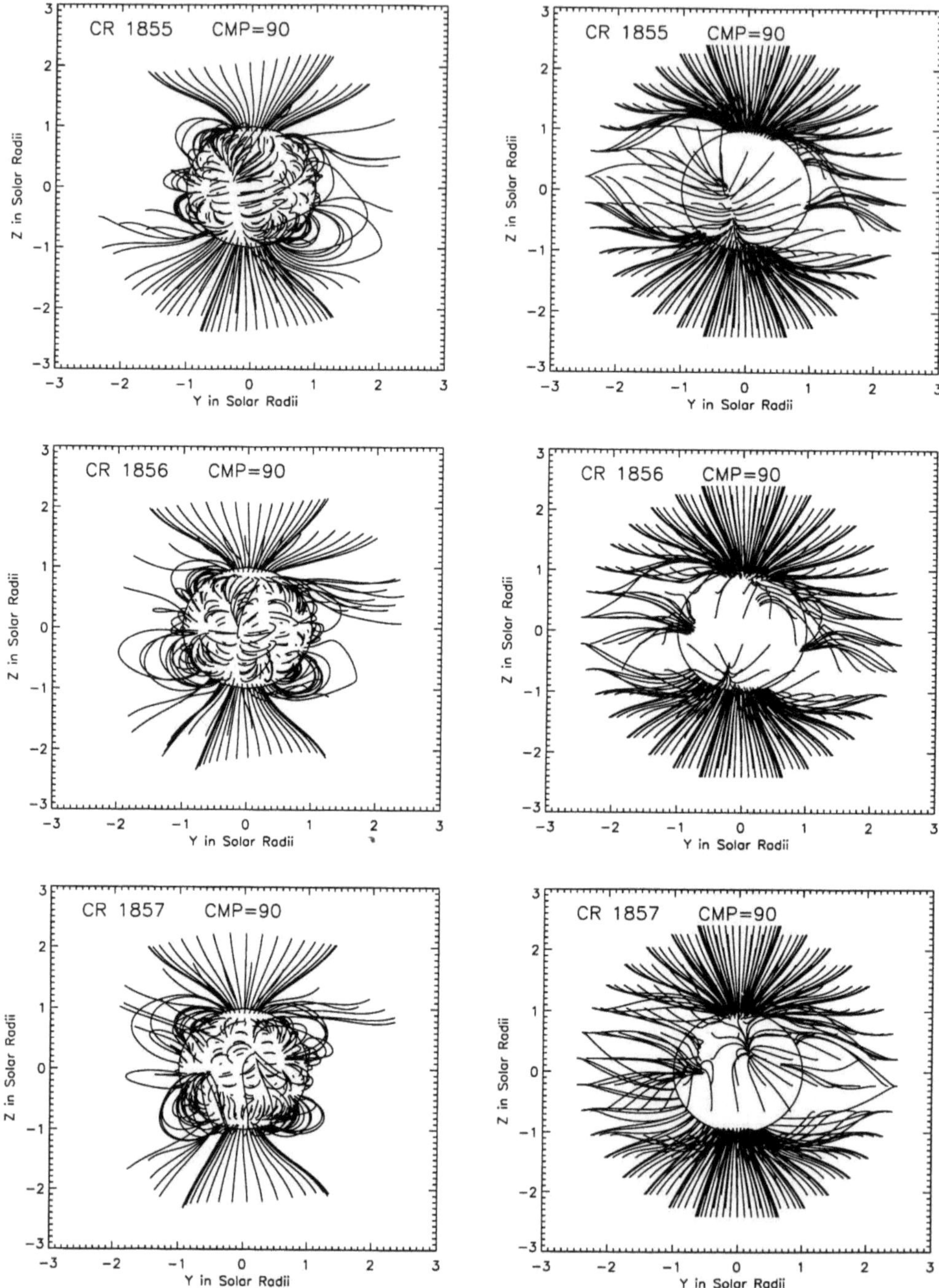

Fig. 3. The coronal field changed its configuration in the spring of 1992. The left panels show the field lines originating on a regular grid in the photosphere for CR 1855, 1856, & 1857. Only open field lines traced downward from 2.5 R_s are shown on the right. Notice the decay of the vertical arcade near CMP 90° and the development of a horizontal structure in the south. The northward extension of the negative southern coronal hole in CR 1855 is replaced by low latitude open positive field regions.

3. The High-Latitude HCS in the Ulysses Epoch

Stability characterizes the coronal and heliospheric field during the declining and minimum phases of the solar cycle, particularly at high latitudes. The strong polar fields confine the HCS to the equatorial region for several years around solar minimum. Near minimum the ecliptic and the HCS are never far apart, explaining the corresponding minimum in average solar wind velocity near Earth. In the declining phase of Cycle 22, even more than in Cycle 21, the (sometimes tilted) dipole field dominates the higher order contributions to the HCS shape everywhere except near the "heliomagnetic equator." Fig. 4 (top) shows that the computed HCS currently reaches to about 25° in the south and about 10° in the north. The predicted latitudes are up to 20° higher when using the standard potential field model (light, solid curve). The maximum latitude and complexity of the HCS structure vary through the solar cycle (see also Fig. 1 and from one cycle to the next.

Ulysses will be immersed in the polar coronal holes for several months. The photospheric sources of the polar holes should change little, so the solar wind should be uniform at high latitudes. Inhomogeneities should arise from variations inherent to the acceleration processes rather than the evolution of the photospheric fields. Ulysses' unique perspective at solar minimum allows this kind of coronal hole monitoring for the first time.

The photospheric field does not remain completely static. In fact the first precursors of Cycle 23 may be erupting at high latitudes now. For example, Harvey (1992) shows that ephemeral regions of the new cycle emerge at high latitudes several years before sunspot minimum, as do Ca^+ plage regions. Evidence for an extended high-latitude solar cycle is also seen in coronal brightness observations (Altrock, 1988). Effects of such photospheric field irregularities may find a way to "leak" out into the high latitude heliosphere.

The polar fields are not strictly antisymmetric, each hemisphere evolves semi-independently. The lower panel of Fig. 4 shows that field strength in the region from about 55° to the pole is currently as much as 60% larger in the south than the north. This was true near the minimum in 1986 as well. The polar fields were roughly equal in 1976. Note that the large annual variation of the line-of-sight polar fields is caused by the strongly peaked distribution of the polar field and the inclination of the ecliptic to the solar equator. This imbalance of north and south results in differences between the hemispheres. One would expect differences in high-latitude IMF strength and solar wind velocity. The HCS reaches to higher northern latitudes as well. The differences between Ulysses' two polar passes should tell us some interesting things about the dependence of coronal holes and solar wind acceleration on absolute field strength. Another difference to look for will be the rotation rate of the magnetic field pattern; Hoeksema and Scherrer (1987) showed that the computed coronal field pattern rotated about 4%

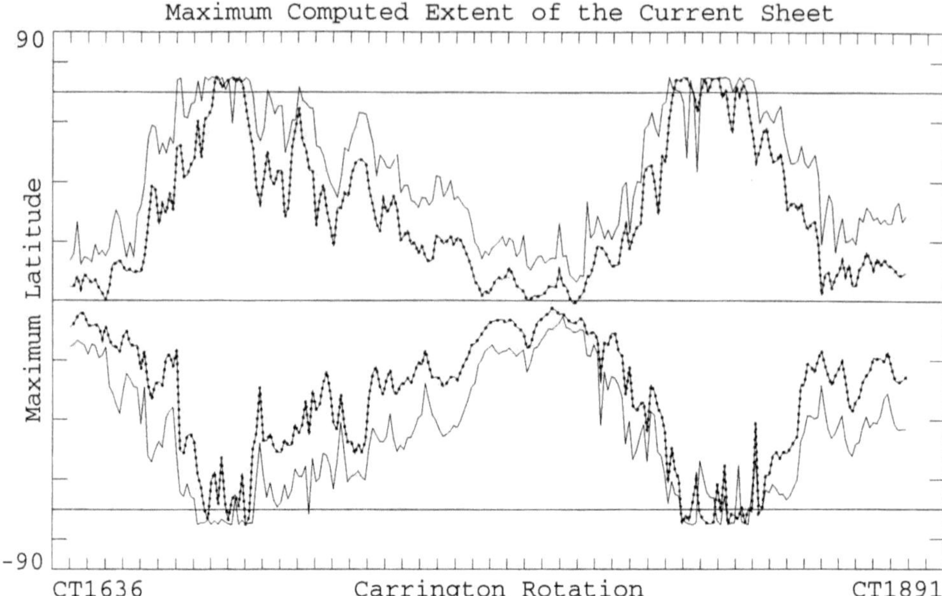

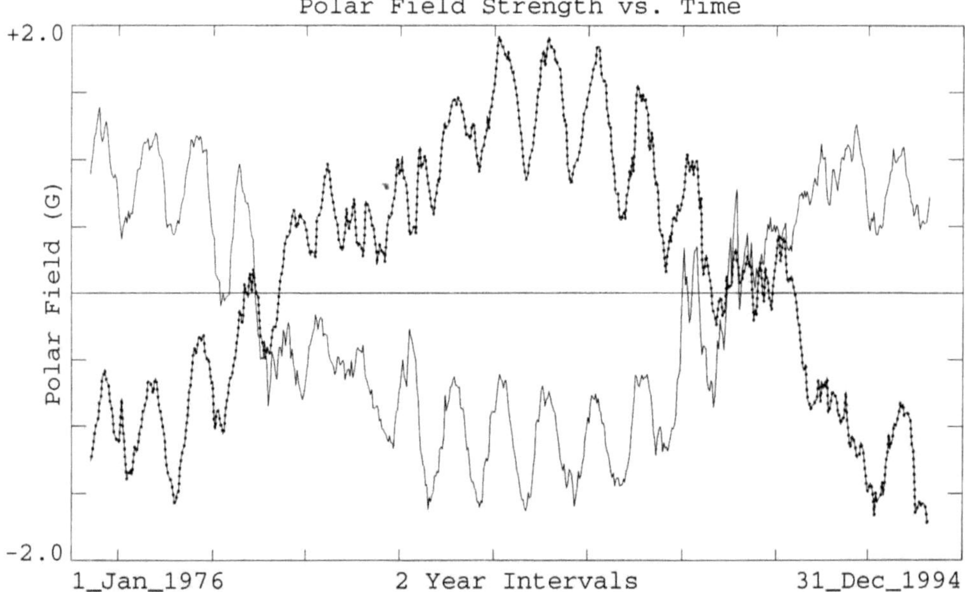

Fig. 4. The upper panel shows the maximum latitude reached in each hemisphere by the computed heliospheric current sheet from 1976 to 1994. The solid line represents the standard potential field model with $R_s s$=2.5 R_s; the dotted line shows the better radial model result with $R_s s$=3.25 R_s. The measured line-of-sight polar region field strength has a large annual variation due to the 7.25° inclination of the ecliptic to the solar rotation axis. The south polar field was much stronger than the north during two most recent solar minima. They were roughly equal in 1976.

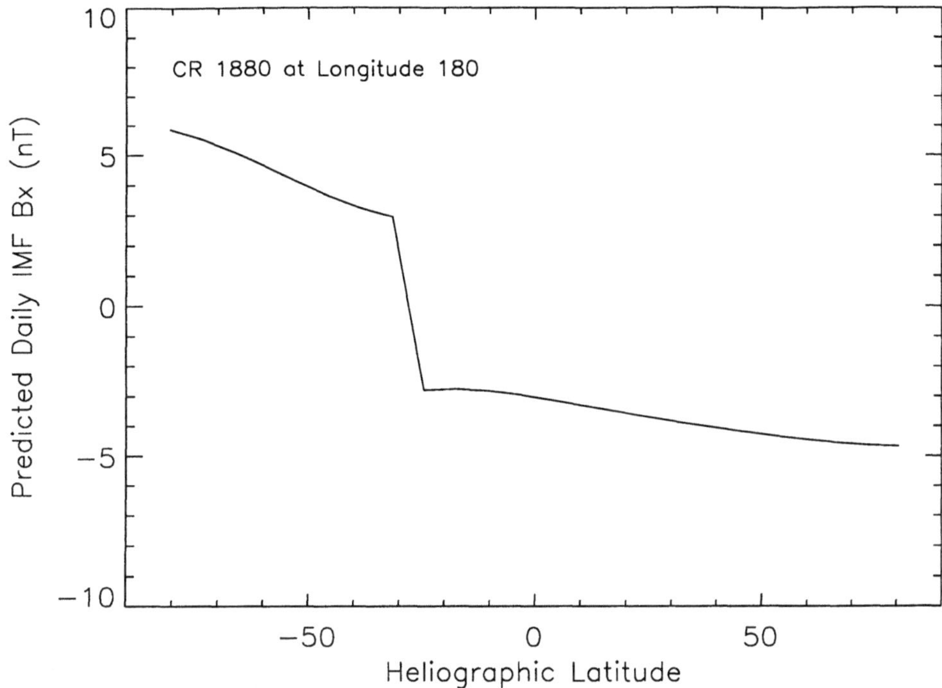

Fig. 5. The composite coronal field model predicts the sunward component of the IMF versus latitude for longitude 180 of Carrington Rotation 1880. The south is stronger than the north and more sharply peaked. Note that Bx is -Br. (From Zhao and Hoeksema, 1994a)

slower in the south during Cycle 21.

Recently Zhao & Hoeksema (1994b) have developed a composite model that combines many of the best features of the potential field model, Bogdan and Low's magnetostatic model that includes horizontal currents (1986) and Schatten's current sheet model (1971). Besides being more physically realistic, the model makes reasonable predictions of the in-ecliptic radial IMF magnitude and direction. Ulysses will provide the high-latitude test of the prediction shown in Fig. 5. At the computed Carrington longitude the HCS lies between 25° and 30° S and the field strength is sharply peaked, particularly in the south.

4. The Transit Phase and the Extended Mission

Between the southern and northern polar passes Ulysses will move rapidly through the equatorial region. Traditionally the solar magnetic field is extremely stable during this part of the cycle, thus the spacecraft should get a fairly clear reading of how various heliospheric phenomena vary with lata fairly clear reading of how various heliospheric phenomena vary with latitude, independent of temporal evolution. Depending on the longitudinal phase of the HCS structure with respect to the trajectory, it should be possible to detect differences in latitudinal extent of the HCS in the two hemispheres, though the neutral line will be fairly flat. It will also be interesting to learn about the size of the belt where the velocity is affected by the HCS, not to mention the gradients in cosmic ray intensity. Previous probes have been limited to relatively low latitudes and have been forced to compare measurements made at greatly different heliocentric distances and/or times.

The extended mission will encounter in very different circumstances. After the northern flyby in 1995 Ulysses will slowly head toward lower latitudes, eventually meeting the HCS again near the equator just after solar minimum. Then the race toward the poles will begin again, this time as the Sun approaches its maximum in 2000-2001. Both Ulysses and the HCS will move quickly poleward, neck and neck at first, but the HCS will probably arrive at the poles first. While Fig. 4 shows that extrapolations from one cycle to the next may be unreliable in detail (see also Suess et al., 1993), Fig. 1 clearly indicates that the heliospheric current sheet will extend to high latitudes. There may, in fact, be multiple current sheets in the heliosphere near solar maximum.

Even if we develop confidence in our understanding of the relatively simple inner heliospheric field at minimum from the presently available data, exploring the high latitude heliosphere at solar maximum will lead to important discoveries because conditions are expected to be completely different. The polar fields will likely have mixed polarities and will almost certainly be weak. In a certain sense this will make the high latitude heliosphere more like the ecliptic at maximum, but the region will be much less influenced by active regions.

5. Conclusions

The heliospheric current sheet will remain fairly flat for the next several years, with only small warps apparent at low latitudes to distort it from a simple slightly tilted dipole configuration. The maximum latitudinal extent of the structure will continue to decrease through solar minimum and only begin to rise a couple of years later. The coronal and heliospheric structure will continue to remain stable, though episodes of large-scale field rearrangement can occur in the corona.

Tradition suggests that the solar wind in the polar regions will be fairly homogeneous, as far as the magnetic field is concerned. Ulysses will cross the HCS only when it travels near the solar equator from now until the late 1990's. If there are nonuniformities in the solar wind that are not due to dynamic effects caused by lower latitude activity, they may be the result of high-latitude precursors of Solar Cycle 23.

There are significant differences between the northern and southern polar fields. The southern cap is currently stronger; therefore solar wind conditions over the north pole should be different. Because the stable polar fields are sharply peaked near minimum, models predict that the radial component of the IMF will increase with latitude; more in the south than in the north. Such will likely not be the case at solar maximum. Rotation rates of the magnetic source pattern may also differ between north and south, as it has in past cycles.

The solar wind velocity will remain high over the stable polar coronal holes, though the highest solar wind velocity is predicted for mid-latitudes over coronal holes that extend toward the equator. The rapid south-north transit in 1995 will be ideal for determining latitudinal variations in solar wind velocity and magnetic field strength.

The extended mission offers exciting possibilities for learning about the high-latitude heliosphere under radically different conditions. Experience gained from the current solar minimum will help place the more complex structures expected to be observed at solar maximum into some kind of reasonable perspective.

Contributions from other spacecraft, such as SOHO, will be very important for constructing a complete and consistent picture of the heliosphere throughout the solar cycle.

6. Acknowledgements

This work and the Wilcox Solar Observatory have been supported by the National Aeronautics and Space Administration, the National Science Foundation, and the Office of Naval Research.

References

Altrock, R.C.: 1988, 'Variation of Solar Coronal Fe XIV 5303A Emission During Solar Cycle 21' in R.C. Altrock, ed(s)., *Solar and Stellar Coronal Structure and Dynamics*, NSO/Sacramento Peak: Sunspot, New Mexico, 414.

Altschuler, M.D. and Newkirk, G. Jr.: 1969, 'Magnetic Fields and the Structure of the Solar Corona', *Solar Phys.* **9**, 131.

Balogh, A., Erdos, G., Forsyth, R.J., and Smith, E.J.: 1993, 'The Evolution of the Interplanetary Sector Structure in 1992', *Geophys. Res. L.* **20**, 2331.

Behannon, K.W., Burlaga, L.F., Hoeksema, J.T., and Klein, L.W.: 1989, 'Spatial Variation and Evolution of Heliosphere Sector Structure', *J. Geophys. Res.* **94**, 1245.

Bogdan, T.J. and Low, B.C.: 1986, 'The Three-Dimensional Structure of Magnetostatic Atmospheres II. Modeling the Large-scale Corona', *Astrophys. J.* **306**, 271.

Bruno, R., Burlaga, L.F., and Hundhausen, A.J.: 1984, 'K-Coronameter Observations and Potential Field Model Comparison in 1976 and 1977', *J. Geophys. Res.* **89**, 5381.

Farnham, T.L. and Meech, K.J.: 1994, 'Comparison of the Plasma Tails of Four Comets: P/Halley, Okazaki-Levy-Rudenko, Austin, and Levy', *Astrophys.J.* **91**, 419.

Gazis, P.R., Mihalov, J.D., Barnes, A., and Lazarus, A.J., and Smith, E.J.: 1989, 'Pioneer and Voyager Observations of the Solar Wind at Large Heliocentric Distances and Latitudes', *Geophys. Res. Let.* **16**, 223.

Harvey, K.L.: 1992, 'The Cyclic Behavior of Solar Activity' in K. Harvey, ed(s)., *The Solar Cycle*, A.S.P Conf. Proc. 27: San Francisco, CA, 335.

Hoeksema, J.T., and Scherrer, P.H.: 1987, 'Rotation of the Coronal Magnetic Field', *Astrophys.J.* **318**, 428.

Hoeksema, J.T., Wilcox, J.M., and Scherrer, P.H.: 1982, 'Structure of the Heliospheric Current Sheet in the Early Portion of Sunspot Cycle 21', *J. Geophys. Res.* **87**, 10331.

Hoeksema, J.T., Wilcox, J.M., and Scherrer, P.H.: 1983, 'The Structure of the Heliospheric Current Sheet: 1978-1982', *J. Geophys. Res.* **88**, 9910.

Levine, R.H.: 1982, 'Open Magnetic Fields and the Solar Cycles, I', *Solar Phys* **79**, 203.

Niedner, M.B. and Brandt, J.C.: 1978, 'Interplanetary Gas XXIII: Plasma Tail Disconnection Events in Comets: Evidence for Magnetic Field Line Reconnection at Interplanetary Sector Boundaries?', *Astrophys.J.* **223**, 655.

Phillips, J.L., Balogh, A., Bame, S.J., Goldstein, B.E., Gosling, J.T., Hoeksema, J.T., McComas, D.J., Neugebauer, M., Sheeley, N.R., and Wang, Y.-M.: 1994, 'Ulysses at 50°S: Constant Immersion in the High-speed Solar Wind', *Geophys. Res. L.* , in press.

Schatten, K.H.: 1971, 'Current Sheet Magnetic Model for the Solar Corona', *Cosmic Electrodynamics* **2**, 232.

Schatten, K.H., Wilcox, J.M., and Ness, N.F.: 1969, 'A Model of Interplanetary and Coronal Magnetic Fields', *Solar Phys* **6**, 442.

Sheeley, N.R. Jr., Wang, Y.-M., and Phillips, J.L.,: 1994, 'Near-Sun Magnetic Fields and the High-Latitude Solar Wind', *Space Sci. Rev.* , this volume.

Sime, D.G., and McCabe, M.K.: 1990, 'The Structure of the White-light Corona and the Large-scale Solar Magnetic Field', *Solar Phys* **126**, 267.

Smith, E.J., Neugebauer, M., Balogh, A., Bame, S.J., Erdos, G., Forsyth, R.J., Goldstein, B.E., Phillips, J.L., and Tsurutani, B.T.: 1993, 'Disappearance of the Heliospheric Sector Structure at Ulysses', *Geophys. Res. Lett.* **20**, 2327.

Smith, E.J., Tsurutani, B.T., and Rosenberg, R.L.: 1978, 'Observations of the Interplanetary Sector Structure up to Heliographic Latitudes of 16°: Pioneer 11', *J. Geophys. Res.* **83**, 717.

Suess, S.T., McComas, D.J. and Hoeksema, J.T.: 1993, 'Prediction of Heliospheric Current Sheet Tilt: 1992-1996', *Geophys. Res. L.* **20**, 161.

Svalgaard, L., Duvall, T.L. Jr., and Scherrer, P.H.: 1978, 'The Strength of the Sun's Polar Fields', *Solar Phys* **58**, 225.

Wang, Y.-M.: 1994, 'Latitude and Solar-Cycle Dependence of Radial IMF Intensity', *Space Sci. Rev* , this volume.

Wang, Y.-M., and Sheeley, N.R. Jr.: 1990, 'Magnetic Flux Transport and the Sunspot-Cycle Evolution of Coronal Holes and Their Wind Streams', *Astrophys.J.* **365**, 372.

Wang, Y.-M., and Sheeley, N.R. Jr.: 1991, 'Why Fast Solar Wind Originates from Slowly Expanding Coronal Flux Tubes', *Astrophys.J.* **372**, L45.

Wang, Y.-M., and Sheeley, N.R. Jr.: 1992, 'On Potential Field Models of the Solar Corona', *Astrophys.J.* **392**, 310.

Wilcox, J.M., and Hundhausen, A.J.: 1983, 'Comparison of Heliospheric Current Sheet Structure Obtained from Potential Magnetic Field Computations and from Observed Maximum Coronal Brightness', *J. Geophys. Res.* **88**, 8095.

Wolfson, R.: 1985, 'A Coronal Magnetic Field Model with Volume and Sheet Currents', *Astrophys. J.* **288**, 769.

Zhao, X.P. and Hoeksema, J.T.: 1994a, 'Prediction of the Interplanetary Magnetic Field Strength', *Space Sci. Rev* , this volume.

Zhao, X.P. and Hoeksema, J.T.: 1994b, 'Prediction of the Interplanetary Magnetic Field', *J. Geophys. Res.* , submitted.

FOURIER PARAMETERS OF HELIOSPHERIC CURRENT SHEET AND THEIR SIGNIFICANCE

MICHAEL SCHULZ
Lockheed Palo Alto Research Laboratories, Palo Alto, California 94304, U.S.A.

Abstract. If the path of the neutral line on the coronal source surface is expressible as a single-valued function (colatitude θ vs longitude φ), then Fourier analysis of ctn θ with respect to φ leads to a simple algorithm for realistically mapping the neutral line outward to model the heliospheric current sheet (HCS) at distances $r \gtrsim 1$ AU. To be compatible with MHD, the source surface used for this mapping should be prolate (aligned with dipole axis) rather than spherical. Orientation of the Sun's magnetic-dipole moment is indicated by the $m = 1$ Fourier amplitude ($a_1 \sin \varphi + b_1 \cos \varphi$) of ctn θ on the source surface. Physical features (including the neutral line) on a prolate source surface intrinsically map to lower dipole latitudes at $r \gtrsim 1$ AU in the heliosphere, and Ulysses observations of a unipolar field at latitudes beyond 30° S (when the neutral line on the source surface still reached 39° S) confirm the expected geometry.

1. Background

As observed at heliocentric distances $r \gtrsim 0.3$ AU, the solar wind flows almost radially and with a velocity that seems to increase with angular separation from the heliospheric current sheet (HCS). The field configuration is more complicated near solar maximum, but otherwise the HCS divides the heliosphere into a quasi-northern hemisphere of one magnetic polarity and a quasi-southern hemisphere of the opposite polarity. Traversals of the HCS account for events previously called encounters with "sector boundaries," until Pioneer-11 observations by Smith *et al.* (1978), indicating absence of "sector structure" at heliographic latitudes $\gtrsim 15°$ during the 1976 solar minimum, clarified the global structure of the heliospheric magnetic field (HMF). Ulysses' voyage toward ever-higher southern latitudes offers a new *in situ* perspective of the heliosphere during the declining phase of the present solar cycle.

Sector structure disappeared from Ulysses' HMF observations around May 1993, when Ulysses was at a heliographic latitude $\approx 30°$ S (Smith *et al.*, 1993). The HCS must not have extended southward of 30° latitude then (nor presumably since, as this has been the declining phase of the solar cycle). Smith *et al.* (1993) found the absence of "sector structure" at latitudes $\approx 30°$ S difficult to reconcile with the usual source-surface model of the heliosphere, as the neutral line on the spherical source surface at $r = 2.5 r_\odot$ (hence presumably at all $r > 2.5 r_\odot$) in May 1993 reached a latitude $\approx 39°$ S (WDC-A, 1993).

Such a discrepancy in HCS position, between the standard source-surface model and Ulysses observations, lends support to the idea (Schulz *et al.*, 1978) that the source surface used for mapping from the solar corona to $r \gtrsim 1$ AU should have a non-spherical shape in order best to simulate the magnetohydrodynamic (MHD) aspects of the mapping. The premise here is that the coronal magnetic field (and also the solar-wind

Space Science Reviews **72**: 149–152.
© 1995 *Kluwer Academic Publishers.*

velocity in the reference frame that corotates with the Sun) should be perpendicular to the source surface, and that the **B** field from the source surface out to the boundary of the heliosphere should be determined (under the "frozen-in" approximation) purely by geometry. The solar wind presumably follows rectilinear streamlines except near interplanetary shocks, which require a separate treatment anyway. The above construction leads essentially to the spiral-field model of Parker (1958) if the source surface is spherical and to an appropriately generalized spiral field (Levine *et al.*, 1982) if the source surface is not spherical.

The source surface compatible with MHD results of Pneuman and Kopp (1971), who treated the simplified case of a dipolar solar **B** field, is prolate and aligned with the dipole axis. Schulz *et al.* (1978) characterized it somewhat schematically as an isogauss of the dipole field, since this specification could be generalized easily to more complicated solar fields (*e.g.*, Levine *et al.*, 1982). Prolateness of the source surface makes little difference to the modeling of magnetic structures inside the corona. However, field lines crossing a prolate source surface at mid-latitudes map to significantly lower magnetic latitudes at $r \gtrsim 1$ AU. This result is desired, as it makes $|B_r|$ nearly independent of heliospheric latitude λ at $r \gtrsim 1$ AU and thus makes the HCS appropriately thin (corresponding to an abrupt change in the direction of **B**) at such distances (Schulz *et al.*, 1978).

2. Source Surface

The source surface is conveniently regarded as a surface of constant $\tilde{F} \equiv r^{-k}\tilde{B}$, where $\tilde{B}$ ($\equiv |\tilde{\mathbf{B}}|$) is the absolute value of the magnetic field $\tilde{\mathbf{B}}$ produced by currents inside the Sun and k (≥ 0) is an adjustable parameter. The limit $k \rightarrow \infty$ leads back to the usual spherical source surface (Schatten *et al.*, 1969), which was intended primarily for modeling the corona (not the entire heliosphere). The choice $k = 0$ leads to a source surface of constant $\tilde{B}$ (Schulz *et al.*, 1978). For general k and dipolar $\tilde{\mathbf{B}}$, the unit vector normal to the source surface is given in spherical coordinates by

$$\hat{\mathbf{n}} \equiv -\left|\nabla\tilde{F}\right|^{-1}\nabla\tilde{F} = \frac{3\hat{\boldsymbol{\theta}}\cos\theta\sin\theta + (k+3)\hat{\mathbf{r}}(1+3\cos^2\theta)}{[9\cos^2\theta\sin^2\theta + (k+3)^2(1+3\cos^2\theta)^2]^{1/2}}, \tag{1}$$

where θ is the magnetic colatitude. The direction of $\hat{\mathbf{n}}$ is easier to visualize in cylindrical coordinates ($\rho \equiv r\sin\theta$, $z \equiv r\cos\theta$), since the value of

$$\frac{dz}{d\rho} = \frac{[k+3(k+4)\cos^2\theta]\cos\theta}{[(k+3)+3(k+4)\cos^2\theta]\sin\theta} \tag{2}$$

along $\hat{\mathbf{n}}$ at the source surface is equal to the tangent of the heliomagnetic latitude λ to which the corresponding field line would map asymptotically. This $dz/d\rho$ should well approximate the value of $\tan\lambda$ along the field line at $r \sim 1.1$–5.4 AU (the range spanned by Ulysses). The dipolar MHD simulation of Pneuman and Kopp (1971) suggests $k = 1.4$ as the best choice (cf. Levine *et al.*, 1982) for the shape parameter k, in which case a source surface of equatorial radius $r_0 = 2.33r_\odot$ would enclose the same volume as a sphere of radius $2.5r_\odot$.

3. Application

The solid curve in Figure 1 shows the neutral line (θ vs φ) on the standard (spherical) source surface of radius $2.5 r_{\odot}$ for Carrington Rotation 1869, during which Ulysses had clearly ceased to observe sector structure in the HMF. Fourier analysis of digitized neutral-line coordinates from WDC-A (1993) suggests a representation of the form

$$\operatorname{ctn} \theta = \sum_{m=0}^{\max m} [a_m \sin m\varphi + b_m \cos m\varphi] , \tag{3}$$

with $b_0 = -0.05072$, $a_1 = +0.25817$, $b_1 = -0.06409$, $a_2 = +0.49407$, $b_2 = -0.36479$, $a_3 = -0.06856$, $b_3 = +0.15113$, and a_0 indeterminate. The value of b_0 corresponds to a mean southward offset of the neutral line by 2.9° on the Sun-centered source surface. The values of a_1 and b_1 suggest that the dipole moment is tilted by 14.9° (in the half-plane φ = 283.9°) relative to the Sun's rotation axis. The dashed sinusoidal curve in Figure 1 reflects these three Fourier coefficients. The straight dashed line reflects b_0 only.

The reason for choosing ctn θ (rather than some other function of θ) for Fourier analysis is that ctn θ would vary sinusoidally with φ along the neutral line on the source surface (whether spherical or prolate) if the Sun's magnetic field were entirely the field of a centered but tilted dipole. In this case the heliographic latitude of any point on the HCS would be equal (even if the source surface is not spherical) to the heliographic latitude of the corresponding point on the source surface (the connection being mapped along a spiral field line). The remaining (largely quadrupolar) deformation of the neutral line on the source surface in Figure 1 would map roughly in accordance with (2) if θ is now interpreted as the polar angle measured relative to the dipole axis (rather than relative to the Sun's rotation axis). The dotted curve in Figure 1 represents the resulting

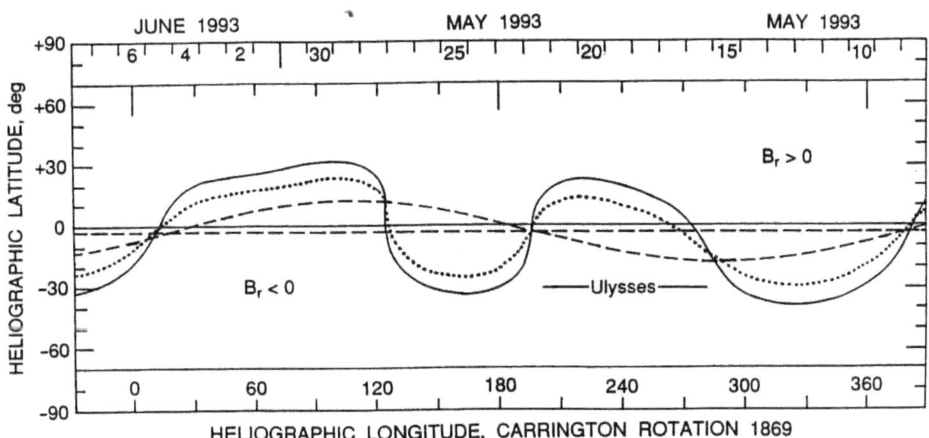

Fig. 1. Analysis (as described in text) of the neutral line (solid curve) on the source surface and its mapping to the distant heliospheric current sheet (dotted curve) for the month during which sector structure ceased to be observed at Ulysses (latitude ≈ 30° S). Dashed line and curve represent Fourier components (m = 0, 1) of neutral line on source surface.

HCS configuration at asymptotic distances ($r \gtrsim 1$ AU). The model source surface for this mapping is Sun-centered and dipole-oriented; its symmetry axis is tilted by 14.9° relative to the Sun's rotation axis. The equation of the source surface for $k = 1.4$ is $r \approx (1 + 3\cos^2\theta)^{5/44} r_0$, with θ measured from the dipole axis. As mapped from this surface, the HCS reaches extrema of +23.2° heliographic latitude at $\varphi \approx 100°$, –24.9° latitude at $\varphi \approx 160°$, +13.4° latitude at $\varphi \approx 220°$, and –29.2° latitude at $\varphi \approx 320°$. The last of these maps back to a heliographic latitude of –38.9° on the source surface. Representative Fourier coefficients of the HCS at $r \gtrsim 1$ AU are $b_0 = -0.05642$, $a_1 = +0.21238$, $b_1 = -0.04318$, $a_2 = +0.31025$, $b_2 = -0.22948$, $a_3 = -0.06652$, and $b_3 = +0.10184$.

These results account well for the HCS configuration inferred by Smith *et al.* (1993), notably that the HCS at $r \gtrsim 1$ AU must have extended only to a latitude ≈ 30° S in May 1993. The model source surface requires a prolate shape (preferably with $k \approx 1.4$) in order to replicate MHD simulation results of Pneuman and Kopp (1971). Mid-latitude features (including the neutral line) on the source surface thereby map to significantly lower latitudes at $r \gtrsim 1$ AU. The present exposition facilitates a rough estimate of the mapping but lacks some of the internal consistency provided by Levine *et al.* (1982). Ideally, for example, the quadrupolar and higher components of the Sun's **B** field, which have clearly influenced the shape of the neutral line in Figure 1, should influence the shape of the source surface also. An even better (but purely numerical) approach to modeling the global HCS configuration, of course, would be a three-dimensional MHD simulation of the entire heliosphere (*e.g.*, Kóta, 1992; Usmanov, 1993; Pizzo, 1994).

Acknowledgements

Data used in this study were provided by the Wilcox Solar Observatory of Stanford University through World Data Center A for Solar-Terrestrial Physics, NOAA E/GC2, 325 Broadway, Boulder, Colorado 80303, U.S.A. The study itself was supported by JPL Contract 959569 under the Ulysses interdisciplinary scientist (IDS) program. Work described here has been inspired by scientific discussions at several Science Working Team (SWT) meetings held by the Ulysses project under NASA and ESA sponsorship.

References

Kóta, J.: 1992, in *Solar Wind Seven*, E. Marsch and R. Schwenn (eds.), pp. 205–208, Pergamon, Oxford (U.K.).

Levine, R. H., Schulz, M., and Frazier, E. N.: 1982, *Solar Phys.* **77**, 363–392.

Parker, E. N.: 1958, Astrophys. J. **128**, 664–675.

Pizzo, V. J.: 1994, *J. Geophys. Res.* **99**, 4173–4183; **99**, 4185–4191.

Pneuman, G. W., and Kopp, R. A.: 1971, *Solar Phys.* **18**, 258–270.

Schatten, K. A., J. M. Wilcox, and N. F. Ness: 1969, *Solar Phys.* **6**, 442–455.

Schulz, M., Frazier, E. N., and Boucher, D. J., Jr.: 1978, *Solar Phys.* **60**, 83–104.

Smith, E. J., Tsurutani, B. T., and Rosenberg, R. L.: 1978, *J. Geophys. Res.* **83**, 717–724.

Smith, E. J., Neugebauer, M., Balogh, A., Bame, S. J., Erdös, G., Forsyth, R. J., Goldstein, B. E., Phillips, J. L., and Tsurutani, B. T.: 1993, *Geophys. Res. Lett.* **20**, 2327–2330.

Usmanov, A. V.; 1992, *Solar Phys.* **146**, 377–396.

WDC-A for Solar-Terr. Phys.: 1993, *Solar-Geophysical Data, Part I*, **587**, 41; **588**, 39; **589**, 37.

THE HIGH LATITUDE HELIOSPHERIC MAGNETIC FIELD

R. J. FORSYTH
The Blackett Laboratory, Imperial College, London SW7 2BZ, U.K.

Abstract. As the Ulysses spacecraft approaches its first pass under the south pole of the sun, it is an appropriate time to review our current knowledge and predictions regarding the three dimensional behaviour of the heliospheric magnetic field, in particular at high heliographic latitudes. Optical techniques for measuring the photospheric magnetic field and observations of coronal brightness structures provide indications of the behaviour of the source of the heliospheric field in the corona. As the coronal fields are carried out into the heliosphere by the solar wind, from Parker's model we would expect that the spiral field observed in the equatorial plane should gradually unwind with latitude leading to open, approximately radial, field lines over the polar regions. Predictions of departures from, and models extending this simple picture are discussed. Both the Pioneer and Voyager spacecraft have spent brief periods in the regions above the maximum latitude of the heliospheric current sheet - relevant results from these missions are reviewed as well as results from the early stages of the out-of-ecliptic phase of the Ulysses mission. The configuration of the coronal magnetic field exhibits a strong dependence on the phase of the solar activity cycle. While the forthcoming Ulysses polar passes take place near to solar minimum, the different conditions which might be encountered on a second orbit of the sun at solar maximum are described.

1. Introduction

Up to the present day, observational data on the behaviour of the heliospheric magnetic field (HMF) by spacecraft has been confined to heliographic latitudes close to the ecliptic plane. One of the aims of the Ulysses mission (Wenzel *et al.*, 1992) is to expand our knowledge of the HMF into a full three dimensional picture.

In the equatorial regions of the heliosphere, the model first proposed by Parker (1958) has been found to give a good first order description of the observed magnetic field configuration by, for example, Thomas and Smith (1980). From the assumptions that the magnetic field lines are frozen in to the highly conducting solar wind and that the field lines are radial at their source on the sun, the rotation of the sun leads to the field lines being wound up into the Archimedean spiral.

The first spacecraft observations showed that the heliospheric field was systematically divided into regions where the field lines were directed along the spiral alternately towards and away from the sun (Wilcox and Ness, 1965). Analysis of data from the Pioneer spacecraft by Thomas and Smith (1981) supported the idea proposed by Schulz (1973) that this sector structure arose from the dominant northern and southern hemisphere solar magnetic field polarities being carried out into interplanetary space by the solar wind. These polarity regions are separated by a wavy heliospheric

Space Science Reviews **72**: 153–163.
© 1995 *Kluwer Academic Publishers.*

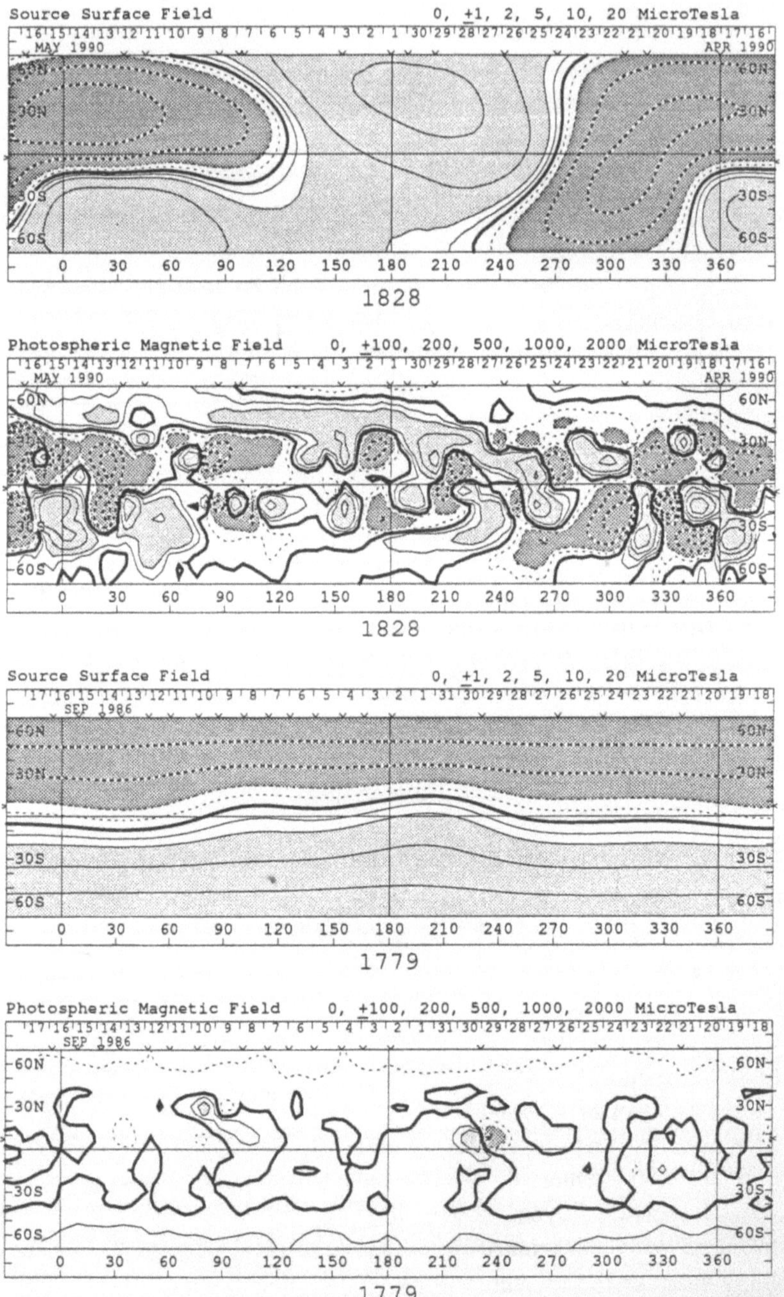

Fig. 1. Coronal and photospheric fields at solar maximum (top two panels) and at solar minimum (lower panels), plotted against heliographic longitude on the horizontal axes and heliographic latitude on the vertical axes. The heavy solid line is the neutral line separating opposite polarity regions, darker shaded regions representing inward polarity. From Hoeksema (1991b).

current sheet which can be thought of as the outward projection of the sun's magnetic equator.

The purpose of this paper is to summarise our current knowledge on the extension of the equatorial view of the heliospheric magnetic field into three dimensions, and in particular to high heliographic latitudes, to put into perspective the results that should soon be forthcoming from Ulysses. A review of the large scale heliospheric magnetic field was given at a previous ESLAB symposium by Balogh (1986). A more recent review by Smith (1989) covers the knowledge gained from the spacecraft which have explored the heliosphere in and near the ecliptic plane. Prior to Ulysses only two spacecraft have made significant excursions away from the heliographic equator. Pioneer 11 reached 16° heliographic latitude in 1975 and 1985 while Voyager 1 reached 31° in 1987. Ulysses will reach 80° in both solar hemispheres.

2. Photospheric and Coronal Magnetic Fields

The heliospheric magnetic field can be viewed as the outward extension of the magnetic field in the solar corona, which in turn is determined by the surface fields in the photosphere. Thus it is to the photosphere and corona that we must first look if we wish to understand the field configurations that Ulysses will measure at high latitudes. Here we have the advantage that using optical observations from the Earth we can gain a three dimensional picture of the solar magnetic field patterns. Photospheric fields can be measured by spectroscopic techniques. Hoeksema (1991a) gives a summary of the main features of the global photospheric magnetic fields thus obtained.

However, the coronal magnetic field is difficult to measure directly, although a qualitative idea of the field configuration can be gained from optical or X-ray observations of coronal brightness structures. Much of the photospheric field does not reach the outer corona but returns to the photosphere in closed loops. Large regions of open field lines are observed, especially at solar minimum, these are the coronal holes where the majority of heliospheric magnetic field lines are believed to originate (Bohlin, 1977). The most common method of quantitatively determining coronal fields is to use the potential field model first developed by Altschuler and Newkirk (1969) which introduces the concept of a source surface for the HMF in the corona. At the source surface (commonly taken to be at 2.5 solar radii) the solar wind is assumed to have been accelerated to the point where the thermal energy dominates the magnetic field energy, hence it is the surface beyond which the field lines are frozen in to the radially outflowing solar wind. An inner boundary condition is provided by the measured photospheric fields and an outer boundary condition by the constraint that the field should be radially directed at the source surface. If the region between is assumed to be current free, then Laplace's equation can be solved to determine the field configuration throughout, and also to determine the polarity and strength of the field at the source surface. Hoeksema *et al.* (1982) applied this technique to photospheric field observations made at Stanford Solar Observatory and validated it by comparing the polarity at the source surface to the sector structure observed at 1 AU. Hoeksema (1991b) has studied the behaviour of the large scale coronal magnetic fields deduced in this way over 14 years.

Figure 1, taken from Hoeksema (1991b), shows examples of the photospheric and deduced coronal source surface fields at solar maximum in the top two panels and at solar minimum in the lower two panels. Ulysses will sample heliospheric fields originating from a coronal configuration much like that shown for solar minimum. At this time the neutral line on the source surface is close to the equator and the field looks very much like an axial dipole. Strong fields are present in the polar regions of the corona. On the other hand, at solar maximum the neutral line stretches right up to the polar regions, there are no strong polar fields and the overall field pattern is much more complicated. At this time the polarity of the fields in the polar regions reverse. Although the picture can be approximated by imagining that the dipole has rotated to now lie with axis in the equatorial plane, analysis of the components and strength of the dipole from the potential field calculations shows the variation to be more complex (Hoeksema, 1991b). Observationally, the polar fields are the least well measured and are determined for the potential field model calculations by interpolation from the higher latitude measured fields.

3. High Latitude Heliospheric Fields: Observations and Model Predictions

3.1 Strength of the HMF and its components

If we know the radial magnetic field strength spatial variation over the source surface in the corona, Parker's model tells us how this pattern will be reproduced in the heliosphere. Modelling work by Wang and Sheeley (1988) has extended the potential field model approach to determining the coronal source surface field with the aim of relating the photospheric observations to heliospheric magnetic field strength. They tested their model by comparing the predicted fields at 1 AU with those observed by spacecraft but had to scale up photospheric field strengths to obtain agreement. They found that the behaviour of the radial component of the heliospheric field could be approximated from the dipole component of the photospheric field. Further work by Wang (1993) improved the fit by including the effects of a current sheet in the plane of the dipole equator and discusses the predicted behaviour of the radial component of the HMF at high heliographic latitudes. The prediction, shown in Figure 2, is that Ulysses should encounter relatively strong fields compared to the ecliptic during the high latitude passes close to solar minimum. The model predicts that at solar maximum the latitudinal gradients in the field strength should be less pronounced.

Luhmann *et al.* (1988) found observational evidence of latitudinal variation of the source surface field strength by considering how the ratio of the radial components of the heliospheric field observed at two spacecraft located near Venus (0.7 AU) and Earth (1.0 AU) varied with heliolatitude separation of the two spacecraft. Their result was

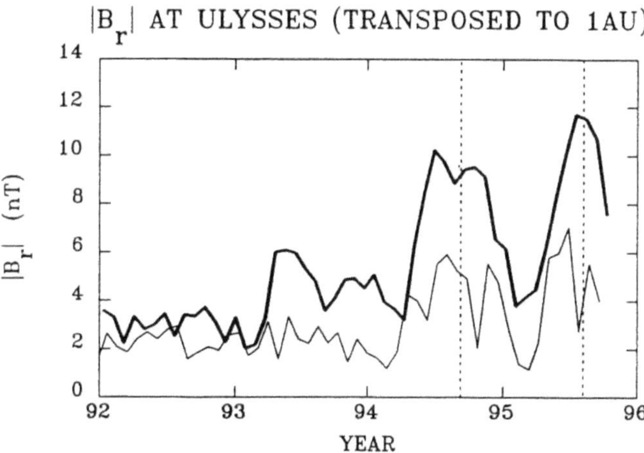

Fig. 2. The predicted value of the radial component of the magnetic field, normalised to 1 AU that Ulysses should measure during the polar passes, from the model of Wang (1993).

consistent with a north - south asymmetry in field strength.

Burlaga and Ness (1993) made use of the passage of Voyager 1 to a latitude of 31°N to look for evidence of latitudinal variations in magnetic field strength. They noted that the field strength measured by Voyager 1 was lower than that measured by Voyager 2 which remained close to the equator and, although solar wind speed measurements were unavailable for Voyager 1, they were able to conclude that a latitudinal variation of the source field at the sun was the most probable cause.

Particularly with the Pioneer and Voyager spacecraft travelling out to very large radial distances efforts have been made to look for systematic spatial deviations from Parker's model in the field magnitude and components. Winterhalter *et al.* (1990) analysed Pioneer 11 data between 1 and 24 AU and reported evidence of a magnetic flux deficit of the order of 1%/AU. On the other hand Burlaga and Ness (1993) analysing data from both Pioneer spacecraft and both Voyager spacecraft between 1 and 19 AU found no evidence of a flux deficit with radial distance within the statistical uncertainties in the data. In this type of study it is very important to have a good baseline spacecraft elsewhere in space, usually near the Earth, so that temporal changes in the heliospheric field can be removed from the analysis. This will be equally true in the analysis of Ulysses data for latitudinal gradients in the field strength and components.

Suess *et al.* (1985) and Pizzo and Goldstein (1987) have proposed models that could explain a flux deficit with radial distance if one does exist. These attempt to provide mechanisms that would give a meridional pressure gradient transporting magnetic flux away from equatorial latitudes. Suess *et al.* examined the differential spiralling with latitude that is inherent in Parker's model, while Pizzo and Goldstein invoked the pressure gradient that would arise from the interaction of solar wind streams of differing velocity, the higher speed streams from coronal holes tending to be at higher latitudes.

3.2 HMF DIRECTION

In Parker's model the direction of the heliospheric magnetic field does not depend on the source surface field strength. As the field is assumed to be radial at the source surface, the direction at any point in interplanetary space depends on the velocity with which the field lines are transported outwards by the solar wind and on the twist imposed by the solar rotation. The simplest extension of this picture to high heliographic latitudes implies that the spiral configuration observed in the ecliptic plane should gradually unwind with latitude until over the poles open radial field lines should be observed. Alternatively, at any colatitude θ, the field lines can be viewed as being wrapped round the surface of a cone of half angle θ.

Tests of Parker's model have shown that it predicts well the long term average direction of the heliospheric field when averaging intervals of at least a month are used. However, distributions of the angles describing the field direction have a significant width indicating the shorter term directional variability. The theory predicts that the radial component of the field should fall off with distance r according to $1/r^2$. Jokipii and Kóta (1989) point out that any large-scale random transverse component present in the photospheric field that survives beyond the source surface will decay with distance only as $1/r$ and hence at suitably large heliocentric distances will come to dominate the radial component. This effect is predicted to be greatest over the solar poles. The effect would be that at a large enough distance the polar heliospheric fields would have a randomly oriented transverse direction although if the vector field was averaged it would still produce the on average radial field predicted by Parker's model. Jokipii and Kóta suggest around 40 AU as the distance where this behaviour will become dominant. Whether any evidence of this will be detected by Ulysses during its polar passes at around 2 AU is open to some debate. There is also little observational evidence on the magnitude of any random transverse components in the coronal field although their existence is certainly possible. A further similar study by Hollweg and Lee (1989) predicted the same effect but slightly weaker.

Smith and Bieber (1991) have studied the variation of the spiral angle in the ecliptic over the solar cycle and found that the spiral had a tendency to be more tightly wound at solar maximum than at minimum, effects first noted by Svalgaard and Wilcox (1974). They also found that the spiral was generally more tightly wound than could be accounted for by the observed solar wind velocity and solar rotation rate. On this basis they conjectured that there might be a systematic azimuthal component in the magnetic field at the source surface. Extending this to higher heliographic latitudes would also lead to more tightly wound heliospheric field lines over the polar regions. In a similar vein to the above study Smith and Bieber (1993) have reported that the spiral angle north of the heliospheric current sheet is more tightly wound than that south of the current sheet independent of whether the northern hemisphere was in the positive or negative polarity state.

TABLE I

List of occasions when spacecraft have passed above the maximum latitude of the heliospheric current sheet.

Spacecraft	Years	Latitude	Solar activity	Reference
Pioneer 11	1975-76	16°N	Minimum	Smith *et al.* (1978)
Pioneer 11	1985-86	16°N	Minimum	Smith (1989)
Voyager 1	1986-88	27.8-30.8°N	Rising	Ness and Burlaga (1993)
Ulysses	1993-	30°S	Declining	Smith *et al.* (1993)

3.3 THE HELIOSPHERIC CURRENT SHEET

Probably the most thoroughly studied three dimensional feature in interplanetary space is the heliospheric current sheet. Of particular interest in out-of-ecliptic observations is the latitude at which a spacecraft passes permanently above the current sheet, a phenomena which is strongly dependent on the phase of the solar cycle. Table I lists the four occasions where this has occurred.

These provide rough estimates of the inclination of the current sheet at the appropriate phase of the solar activity cycle which can be compared with the behaviour of the neutral line at the source surface previously described. From such studies there is growing evidence that the maximum latitude of the current sheet at any particular time decreases with radial distance (Smith, 1989; Smith *et al.*, 1993). This is supported by a study by Behannon *et al.* (1989) which utilising simultaneous observations by spacecraft at a variety of radial distances found that four sector structures observed near 1 AU had evolved to two sector structures far out in the heliosphere. We may find that some high latitude effects seen by Ulysses will be ordered by heliomagnetic latitude, measured with respect to the latitude of the current sheet, rather than by heliographic latitude. At solar minimum the current sheet should be close to the heliographic equator. It will be interesting to see what the inclination of the current sheet is when Ulysses makes its rapid transit from south to north poles, crossing the equator around the beginning of 1995. Perhaps of greater interest still, there are no occasions where spacecraft have passed above the current sheet near solar maximum - an opportunity to study what happens to the current sheet in interplanetary space at high heliographic latitudes near solar maximum will only be given by a second orbit of the sun by Ulysses.

3.4 HMF RESPONSE TO SOLAR WIND STREAM STRUCTURE

The phenomena of interaction regions in the heliosphere arise from the interaction of solar wind streams of differing velocity. Plasma originating from different sources on the sun cannot interpenetrate and hence regions of compression develop where fast solar wind catches up with slower wind ahead and expansion regions develop behind (for example, Smith and Wolfe, 1976). In magnetic field observations the interaction regions reveal themselves as intervals of enhanced magnetic field strength corresponding to regions where the solar wind plasma has been compressed together. Beyond about 1 AU these regions are often bounded by forward and reverse shock waves.

It has become accepted that the fast solar wind streams originate from coronal holes on the sun (Sheeley and Wang, 1991, and references therein). At solar minimum the coronal holes cover much of the polar caps and stretch down to low heliographic latitudes leading to stable patterns of interaction regions recurring with the solar rotation period. Clearly there is going to be a strong dependence with latitude on how the interaction regions manifest themselves in HMF data. This has been exemplified in Ulysses data from 1993. Figure 3, taken from Smith *et al.* (1993), shows data from six solar rotations from January to May of 1993. During this time the heliographic latitude of Ulysses increased from 23° to 32°S. The upper panel displays the field azimuth angle ϕ which is approximately 90° for the southern hemisphere polarity and approximately 270° for the northern hemisphere. The second panel shows the field magnitude and the third shows solar wind velocity. The velocity plot is dominated by a solar wind stream emanating from a northward extension of the southern polar coronal hole (Bame *et al.*, 1993). The field magnitude shows the region of compressed magnetic field (labelled d) associated with the interaction region created by the fast solar wind. Note that in the last appearance of the interaction region in the plot, once Ulysses is above the maximum latitude of the current sheet, the lowest speed solar wind associated with the streamer belt is no longer observed and the compression of the magnetic field is much

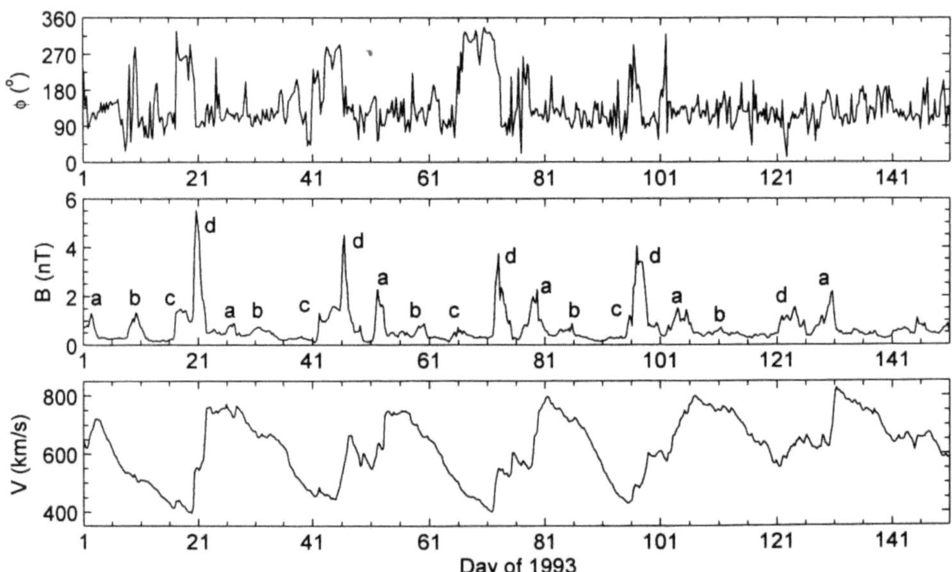

Fig. 3. Ulysses observations of corotating interaction regions during the first five months of 1993. From Smith *et al.* (1993).

diminished. Since then Ulysses has gone on to become more and more immersed in solar wind originating from the polar coronal hole (Phillips *et al.*, 1994) and the structured magnetic field associated with the corotating interaction regions has gradually disappeared. Hence, at solar minimum over the poles we expect the magnetic field to be free of large scale structure due to interacting solar wind streams.

At solar maximum the current sheet and associated streamer belt reach to high latitudes into the polar regions of the corona. The large stable polar coronal holes are no longer present, the coronal holes are smaller and more broken up. In the ecliptic regions most of the large scale field variability at this time is associated with transient solar wind streams such as coronal mass ejections. These are thought to primarily originate from the streamer belt regions (Hundhausen, 1993), hence at solar maximum this behaviour should extend to all latitudes. (Coronal mass ejections also occur at solar minimum but are less frequent and should be confined to low latitude regions.) Hence, both fast (predominantly transient) and slow solar wind streams should be present in the polar regions at this time. However, in the same way that the solar rotation induced spiral field is expected to unwind over the polar regions, one would expect the compression of streams against each other to be less and that streams of differing velocity should be more able to slip past each other. Clearly any interaction between streams and resulting effects on the magnetic field is going to be of a very different nature from that at low latitudes.

Thus during the current polar passes, Ulysses will be sampling the heliospheric magnetic field in the least complex state that it is likely to be in during the 11 year solar cycle. During a second orbit of the sun by Ulysses, the field over the poles would be expected to be in a much more complex state and the results should very much complement those that will be obtained over the next year.

3.5 HMF FLUCTUATIONS

There have been many studies of the short time scale, less than 1 day, variability in the heliospheric magnetic field. It has been established that the variances of the field magnitude and the components decline slowly with radial distance when measured relative to average field values (for example, Klein *et al.*, 1987). Klein *et al.* (1991) report on the evolution with radial distance of the shape of the spectrum of magnetic field fluctuations while Bieber *et al.* (1993) have studied variations of the field spectra with solar cycle reporting greater amplitudes at solar maximum compared to solar minimum. However, the author is not aware of any studies that have looked at how the variability of the magnetic field changes with heliographic latitude. Such studies will undoubtedly be undertaken with the Ulysses data. The forthcoming polar pass will be spent almost certainly sampling field lines originating within the polar coronal holes and will thus allow us to determine whether the nature of the fluctuations in the plasma from this source differ significantly from those observed in the ecliptic to date.

4. Summary

With the forthcoming polar passes of Ulysses in mind, an attempt has been made to collect together information that puts into context the observations that will be made of the high latitude heliospheric magnetic field. Inherent in the discussion has been the distinction between the conditions prevailing during the present polar passes at solar minimum and the conditions at solar maximum which would be encountered if a further two polar passes were made on a second orbit of Ulysses around the sun. In contrast to the simple coronal hole fields that will be sampled at minimum, a much more complex source field would be expected with the heliospheric current sheet and transient solar wind streams present at all latitudes. It is very much hoped that the opportunity will be given to us to study the polar heliospheric regions in both of these states.

Acknowledgements

I wish to thank A. Balogh and E.J. Smith for commenting on the original manuscript of this paper and to thank J.T. Hoeksema for assistance in reproducing Figure 1. Ulysses related work at Imperial College, London is supported by the U.K. Particle Physics and Astronomy Research Council.

References

Altschuler, M.D. and Newkirk, G., Jr.: 1969, *Solar Phys.* **9**, 131.

Balogh, A.: 1986, in R.G. Marsden (ed.), *The sun and the heliosphere in three dimensions*, D. Reidel, Dordrecht, pp255-266.

Bame, S.J., Goldstein, B.E., Gosling, J.T., Harvey, J.W., McComas, D.J., Neugebauer, M. and Phillips, J.L.: 1993, *Geophys. Res. Lett.*, **20**, 2323.

Behannon, K.W., Burlaga, L.F., Hoeksema, J.T. and Klein, L.W.: 1989, *J. Geophys. Res.*, **94**, 1245.

Bieber, J.W., Chen, J., Matthaeus, W.H., Smith, C.W. and Pomerantz, M.A.: 1993, *J. Geophys. Res.*, **98**, 3585.

Bohlin, J.D.: 1977, in J. Zirker (ed.), *Coronal holes and high speed solar wind streams*, Colorado Associated University Press, Boulder, pp27-69.

Burlaga, L.F., and Ness, N.F.: 1993, *J. Geophys. Res.*, **98**, 3539.

Hoeksema, J.T.: 1991a, *J. Geomag. Geoelectr.*, **43** Suppl., 59.

Hoeksema, J.T.: 1991b, *Adv. Space Res.*, **11**, (1)15.

Hoeksema, J.T., Wilcox, J.M. and Scherrer, P.H.: 1982, *J. Geophys. Res.*, **87**, 10,331.

Hollweg, J.V. and Lee, M.A.: 1989, *Geophys. Res. Lett.*, **8**, 919.

Hundhausen, A.J.: 1993, *J. Geophys. Res.*, **98**, 13,177.

Jokipii, J.R. and Kóta, J.: 1989, *Geophys. Res. Lett.*, **16**, 1.

Klein, L.W., Burlaga, L.F. and Ness, N.F.: 1987, *J. Geophys. Res.*, **92**, 9885.

Klein, L.W., Matthaeus, W.H., Roberts, D.A. and Goldstein, M.L.: 1992, in E. Marsch and R. Schwenn (eds.), *Solar Wind Seven*, Pergamon, Oxford, pp. 197-200.

Luhmann, J., Russell, C.T. and Smith, E.J.: 1988, in V.J. Pizzo, T.E. Holzer and D.G. Sime (eds.), *Proceedings of the sixth international solar wind conference*, Rep. NCAR/TN-306, Natl. Cent. for Atmos. Res., Boulder, pp. 323-327.

Ness, N.F. and Burlaga, L.F.: 1993, (Abs.), *Eos, Trans. AGU*, **74** Suppl., 241.

Parker, E.N.: 1958, *Astrophys. J.*, **128**, 664.

Phillips, J.L., Balogh, A., Bame, S.J., Goldstein, B.E., Gosling, J.T., Hoeksema, J.T., McComas, D.J., Neugebauer, M., Sheeley, N.R., Jr. and Wang, Y.-M.: 1994, *Geophys. Res. Lett.*, in press.

Pizzo, V.J. and Goldstein, B.E.: 1987, *J. Geophys. Res.*, **92**, 7241.

Schulz, M.: 1973, *Astrophys. Space Sci.*, **24**, 371.

Sheeley, N.R., Jr. and Wang, Y.-M.: 1991, *Solar Phys.*, **131**, 165.

Smith, C.W. and Bieber, J.W.: 1991, *Astrophys. J.*, **370**, 435.

Smith, C.W. and Bieber, J.W.: 1993, *J. Geophys. Res.*, **98**, 9401.

Smith, E.J.: 1989, *Adv. Space Res.*, **9**, (4)159.

Smith, E.J. and Wolfe, J.H.: 1976, *Geophys. Res. Lett.*, **3**, 137.

Smith, E.J., Tsurutani, B.T., and Rosenberg, R.L.: 1978, *J. Geophys. Res.*, **83**, 717.

Smith, E.J., Neugebauer, M., Balogh, A., Bame, S.J., Erdös, G., Forsyth, R.J., Goldstein, B.E., Phillips, J.L. and Tsurutani, B.T.: 1993, *Geophys. Res. Lett.*, **20**, 2327.

Suess, S.T., Thomas, B.T. and Nerney, S.F.: 1985, *J. Geophys. Res.*, **90**, 4378.

Svalgaard, L. and Wilcox, J.M.: 1974, *Science*, **186**, 51.

Thomas, B.T. and Smith, E.J.: 1980, *J. Geophys. Res.*, **85**, 6861.

Thomas, B.T. and Smith, E.J.: 1981, *J. Geophys. Res.*, **86**, 11,105.

Wang, Y.-M. and Sheeley, N.R., Jr.: 1988, *J. Geophys. Res.*, **93**, 11,227.

Wang, Y.-M.: 1993, *J. Geophys. Res.*, **98**, 3529.

Wenzel, K.-P., Marsden, R.G., Page, D.E. and Smith, E.J.: 1992, *Astron. Astrophys. Suppl. Ser.*, **92**, 207.

Wilcox, J.M. and Ness, N.F.: 1965, *J. Geophys. Res.*, **70**, 5793.

Winterhalter, D., Smith, E.J., Wolfe, J.H. and Slavin, J.A.: 1990, *J. Geophys. Res.*, **95**, 1.

ULYSSES OBSERVATIONS OF LATITUDE GRADIENTS IN THE HELIOSPHERIC MAGNETIC FIELD: RADIAL COMPONENT AND VARIANCES

E. J. SMITH[1], M. NEUGEBAUER[1], A. BALOGH[2], S. J. BAME[3], R. P. LEPPING[4], AND B. T. TSURUTANI[1]

[1]*Jet Propulsion Laboratory, California Institute of Technology, Pasadena, CA 91109-8099*
[2]*Imperial College of Science & Technology, The Blackett Laboratory London, England SW7 2BZ*
[3]*Los Alamos National Laboratory, Los Alamos, NM 87545*
[4]*Goddard Space Flight Center, Greenbelt, MD 20771*

Abstract. The radial component of the magnetic field at Ulysses, over latitudes from -10° to -45° and distances from 5.3 to 3.8 AU, compares very well with corresponding measurements being made by IMP-8 in the ecliptic at 1AU. There is little, if any, evidence of a latitude gradient. Variances in the field, normalized to the square of the field magnitude, show little change with latitude in variations in the magnitude but a large increase in the transverse field variations. The latter are shown to be caused by the presence of large amplitude, long period Alfvénic fluctuations. This identification is based on the close relation between the magnetic field and velocity perturbations including the effect of anisotropy in the solar wind pressure. The waves are propagating outward from the Sun, as in the ecliptic, but variance analysis indicates that the direction of propagation is radial rather than field-aligned. A significant long-period component of ≥ 10 hours is present.

1. Introduction

The radial component of the heliospheric magnetic field, B_R, merits special attention. It is representative, after extrapolation back to the Sun, of the magnetic field in the source region of the solar wind. On the other hand, the Parker (1963) theory shows that the tangential component, B_T, derives not only from the radial component but is also dependent on the solar wind speed and the angular velocity of the field line at the source. The latitude gradient in B_R, of interest here, contains implicit information about the character of the magnetic field at the solar wind source surface (Hoeksema, 1986).

The time dependent field changes superposed on the average field are quantified by their variances. Power associated with the various contributors, including shocks, waves, discontinuities, and turbulence, are all included. The study of latitude gradients in the variances shows how "rough" or irregular the magnetic field becomes as the observations depart from the solar equatorial zone. The nature of the underlying fluctuations can have important implications for the origin of the solar wind, e.g., the contribution made by wave pressure to its momentum (Alazraki and Couturier, 1971; Belcher 1971), and to the access of galactic cosmic rays to the polar heliosphere (Jokipii and Kóta, 1989).

Space Science Reviews **72**: 165–170.
© 1995 *Kluwer Academic Publishers.*

2. Radial Field Component

The radial component in solar heliospheric (SH or RTN) coordinates, B_R, is positive when outward from the Sun. Although less strongly influenced by stream interactions and CMEs, it is affected by the sector structure and ever-present irregularities. The fields in magnetic sectors tend to cancel, so a simple average of B_R is not appropriate. If $|B_R|$ is averaged, the background fluctuations can contribute power to the average. The safest procedure is to use data from a single sector. Ulysses passed below the heliospheric current sheet (HCS) in April-May 1993 (Smith et al., 1993). Accordingly, we use values of B_R in negative sectors both at Ulysses and at 1AU in the ecliptic by IMP-8. The corresponding B_R values are then averaged over several successive solar rotations. Values at Ulysses are multiplied by r^2 equivalent to extrapolating them back to 1AU. The amplitude distributions of B_R showed reasonable-looking distributions in good agreement at the two locations.

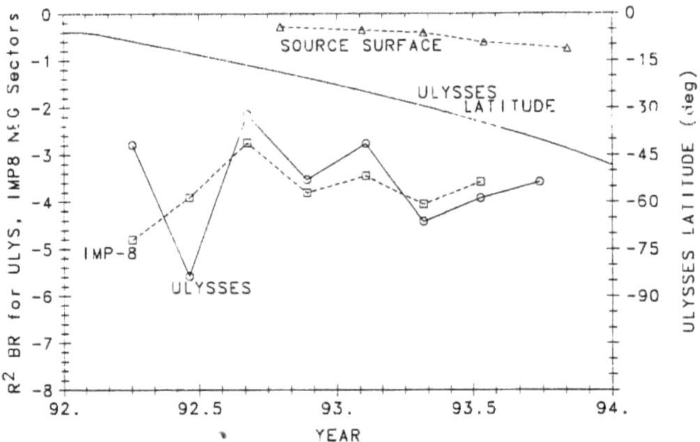

Figure 1. B_R at Ulysses, IMP-8 and the Solar Wind Source Surface. The average values at Ulysses (circles) and IMP-8 (squares) are from negative sectors only. The solid curve is the heliolatitude of Ulysses (whose scale is shown to the right). The field at the source surface, at the latitude of Ulysses, averaged over a solar rotation, appears at the top (triangles).

The averages of B_R are shown in figure 1. The overall long-term averages are both approximately equal to -3.5 nT, the "canonical" average of B_R at 1AU corresponding to a field of 5 nT oriented along a Parker spiral of 45°. The close correspondence between B_R at IMP and Ulysses implies the absence of a significant latitude gradient from 0° to ≈ -45°.

The latitude variation in B_R has also been compared with the radial field at the solar wind source surface using the field contours regularly published by Stanford University. Solar rotation averages along the latitude trace of Ulysses are shown in figure 2. They have also been extrapolated to 1AU. The field magnitudes at the source surface are too low. The model also contains a definite gradient with the field strength increasing with latitude. Increasing the field strength by a factor of 1.8 to allow for solar magnetograph saturation still leaves too small a field value (cf. Hoeksema et al., 1982). Thus, although such a model leads to a reasonable topology for the neutral sheet and predicts the sector structure quite well, it does poorly at reproducing the magnitude of B_R.

Y-M Wang (1993 and these proceedings) and Zhao and Hoeksema (these proceedings) have developed models which include the heliospheric current sheet. A significant prediction of the models is the relative absence of a gradient below ≈ 50° latitude in agreement with figure 2, followed by a large increase in B_R up to the south pole. It will be of obvious importance to see if this latter prediction is confirmed.

3. Variances

Variances in the field magnitude, σ_B^2, and in the three components, $\sigma_R^2, \sigma_T^2, \sigma_N^2$, have been computed over intervals of 1 hour. (In SH coordinates, the R axis is radially outward, $\hat{T}$ is perpendicular to $\hat{R}$ and the solar rotation axis, $\hat{H}$, and is positive for prograde motion, while $\hat{N}$ is orthogonal to $\hat{R}$ and $\hat{T}$ and is positive northward). One-hour intervals appear to be a reasonable compromise that distinguishes between waves and/or turbulence and longer period solar wind structure. The sum of the variances in the components, $\sigma_S^2 = \sigma_R^2 + \sigma_T^2 + \sigma_N^2$, representing the power in the fluctuations principally in direction or transverse to the field, has then been calculated. Both σ_S^2 and σ_B^2 have been normalized by dividing by the square of the average field magnitude, B^2, over the corresponding one-hour intervals. Both at 1AU and in the outer heliosphere, as observed by Pioneer 10, 11 and Voyager, the normalized variances tend to be nearly constant when averaged over intervals of one or more solar rotations.

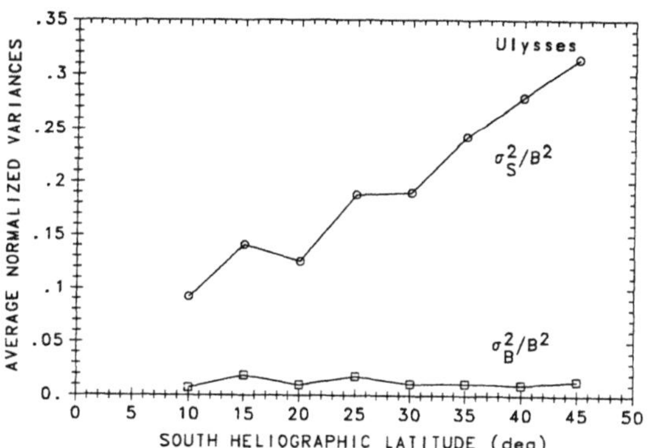

Figure 2. Average Variances at Ulysses as a Function of Latitude. The two curves are hourly values of σ_S^2/B^2 and σ_B^2/B^2 averaged over successive latitude intervals of 5° each. There is a large growth in power in the components but none in the field magnitude.

The normalized variances observed by Ulysses are shown in figure 2 averaged over successive latitude intervals of 5°. σ_B^2/B^2 is essentially independent of latitude, whereas, σ_S^2/B^2 exhibits a strong dependence, increasing approximately linearly with latitude.

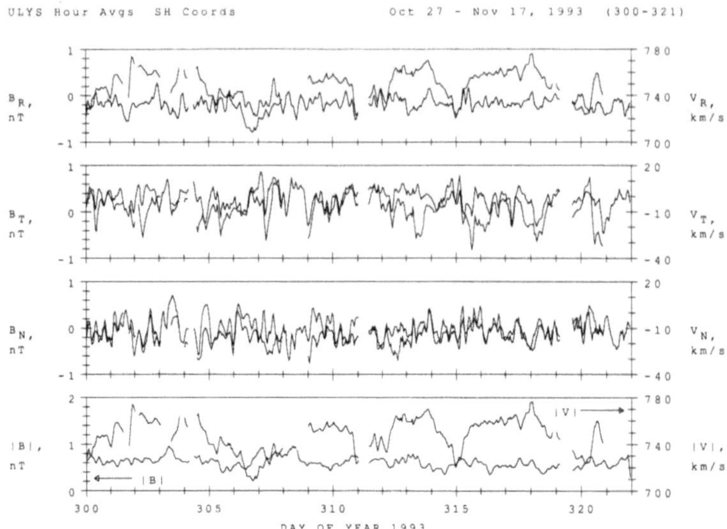

Figure 3. Field and Solar Wind Velocity Components at -45° Latitude. Data from October 27 to November 17, 1993, (22 days) are shown. The upper three panels are superposed plots of hourly averages of the field and velocity components. The lower panel shows the field magnitude and the solar wind speed. The scales used to plot the components have been adjusted to enhance their correlation which is particularly striking for the two transverse components.

What is the cause of these different dependences? Figure 3 contains one-hour averages of the three components of the magnetic field and its magnitude over a 21-day interval in late 1993. The solar wind velocity components, V_R, V_T, V_N, and the speed, $|V|$, are superposed on the field plots.

The bottom panel shows that only slight variations are present in V ($\Delta V \leq \pm 35$ km/s) and B($\Delta B \leq \pm 0.3$ nT). This behavior is typical of the data from the Fall of 1993 onward when the solar wind structure all but disappeared, attributable to Ulysses remaining continuously within flow from the south polar coronal hole.

There is a good correlation between $\vec{B}$ and $\vec{V}$. This relationship is especially evident in the transverse components. The scales have been adjusted to enhance the correlation resulting in a figure corresponding to the well-known Alfvén wave figure 3 of Belcher and Davis (1971), hereafter BD.

We have tested the large amplitude, long period fluctuations apparent in figure 3 to see if they are Alfvén waves. We have followed the analysis first carried out by BD. The wave identification is based on the well-known relationship between the perturbations in velocity and magnetic field:

$$\delta V / \delta B = \pm(4\pi\rho)^{-\frac{1}{2}} \eta^2.$$

In the above, ρ is the solar wind mass density (including alpha particles) and $\eta = 1 - (p_{11} - p_{\perp})/(B^2/4\pi)$ is associated with the pressure anisotropy in the solar wind. This factor is the usual criterion for the firehose instability which operates for $\eta \geq o$.

As is evident from figure 3, V_T - B_T and V_N - B_N are in-phase. Given the inward-directed background magnetic field, the positive sign implies that the waves are propa-

As is evident from figure 3, V_T - B_T and V_N - B_N are in-phase. Given the inward-directed background magnetic field, the positive sign implies that the waves are propagating outward (away from the Sun). This direction is characteristic of Alfvenic fluctuations in the ecliptic.

The wave identification has been further tested by comparing the values of $\delta V/\delta B$ and $\eta^2/(4\pi\rho)^{1/2}$. Correlograms of $\delta V/\delta B$ produce values of $\delta V_N/\delta B_N = 31$ and $\delta V_T/\delta B_T =$ 42 km/s·nT. The average value of $\eta^2/(4\pi\rho)^{1/2}$ over the same 21-day interval is 34.4 km/s·nT which represents as good an agreement as is customarily achieved in the ecliptic. The average value of η is approximately 0.7.

Other aspects of the Alfven waves have been investigated and reveal some important differences as compared to the waves studied by BD. The amplitude and phase of the coherency (between V_N and B_N)have been determined as a function of frequency. The two parameters are in-phase throughout the frequency range while the degree of correlation is high ($\approx$ 0.75). There is, however, no evidence of a low frequency cutoff such as that found by BD at periods of $\approx$ 2.5 hours. Thus, the waves have significantly longer periods (T $\gtrsim$ 10 hours) and wavelengths at the higher latitudes. Inspection of figure 3 (hourly averages) shows that there is significant power at these longer periods.

Another major difference is the direction of minimum variance, customarily identified with the k vector. A variance analysis of the magnetic data in figure 3 leads to an eigenvector that lies within 10° of the radial direction. The analysis by BD led to the minimum variance direction being parallel to $\vec{B}$ rather than $\hat{R}$. Theoretically, waves originating near the Sun and propagating outward are expected to have their k vectors parallel to $\hat{R}$.

4. Discussion

A direct comparison of the latitude gradient with the Stanford source surface contours results in two disagreements. The source surface fields are too weak, implying $|B_R| \approx \lesssim$ 1 nT at 1AU whereas the observed value is $\approx$ 3.5 nT. Further, the source surface model implies a significant gradient (increase in field) at latitudes up to 50°. However, a model which includes the contribution to B_R of the HCS reconciles the field strengths and implies little, if any, observable gradient along the Ulysses trajectory at this phase of the solar cycle.

Obviously, the Ulysses observations have the potential to subject available models to stringent testing and possible improvements. Knowledge of the polar cap magnetic fields, expected to be forthcoming this Fall, should also contribute greatly to our understanding of the polar fields in the photosphere as well as in the source region of the solar wind.

The study of the field variances show that the field irregularities increase markedly with latitude compared to the average field. Inspection of the waveforms suggests that the dominant contribution is from Alfvenic fluctuations, a hypothesis that has been tested successfully by comparison with the in-ecliptic study of Belcher and Davis (1971).

Several differences between the high latitude and in-ecliptic Alfven waves are noteworthy. The waves are continuously present at Ulysses rather than being observed intermittently as in the ecliptic. This difference is presumably because Ulysses is continuously sampling flow from a polar coronal hole rather than just when solar wind is observed from an equatorward extension that is limited in longitude. In addition, the direction of propagation, although anti-sunward, is nearly radial at 50° latitude rather than along the average field direction.

Another important difference is the presence of very long period waves ($T \geq 10$ hours) at Ulysses in contrast to a low frequency cutoff at ≈ 2.5 hours for the in-ecliptic waves. In view of the high solar wind speed of ≈ 750 km/s, the corresponding wavelengths at high latitude extend to 0.2-0.3 AU, i.e., a considerable fraction of an astronomical unit. The waves may represent those postulated by Jokipii and Kóta (1989) caused by motions of the field lines at the Sun.

The long periods are likely to have important implications. Hollweg (1973) considered the effect of the waves on both the linear momentum (wave pressure) and angular momentum of the solar wind. His analysis revealed a strong dependence of both effects on wavelength so that both issues should probably be revisited.

The waves may also have important consequences for galactic and anomalous cosmic rays. Protons resonant with waves having periods between 1 and 10 hours have energies between 10 MeV and 1 GeV, an energy range over which solar modulation effects are well known and well studied (McKibben, 1990; Jokipii, 1989). The waves may inhibit the anticipated easier entry of the cosmic rays over the polar caps (associated with the more radial global fields) and influence their basic properties such as the latitude gradient in flux or intensity.

Acknowledgments. John Phillips is the Principal Investigator of the Ulysses Solar Wind Experiment. Joyce Wolf prepared the plots and carried out much of the computer analyses described in this article. This work was supported by the Jet Propulsion Laboratory, California Institute of Technology, under contract with the National Aeronautics and Space Administration. Work performed at the Los Alamos National Laboratory was carried out under the auspices of the US Department of Energy with support under Contract S-04039.

References

Alazraki, G. and Couturier, P.: 1971, *Astron. Astrophys.*, **13**, 380.
Belcher, J.W.: 1971, *Astrophys. J.*, **168**, 509.
Belcher, J.W. and Davis, Jr., L.: 1971, *J. Geophys. Res.*, **76**, 3534.
Hoeksema, J.T., Wilcox, J.M. and Scherrer, P.H.: 1982, *J. Geophys. Res.*, **87**, 10,331.
Hollweg, J.V.: 1973, *J. Geophys. Res.*, **78**, 3643.
Jokipii, J.R.: 1989, *Adv. Space Res.*, **9**, (12)105.
Jokipii, J.R. and Kóta, J.: 1989, *Geophys. Res. Lett.*, **16**, 1.
McKibben, R.B.: 1990, in S. Grzedzielski and D.E. Page (eds.), *Physics of the Outer Heliosphere*, Pergamon Press PLC, London, pp. 107-118.
Parker, E.N.: 1963, *Interplanetary Dynamical Processes*, Wiley-Interscience, New York.
Smith, E.J., et al: 1993, *Geophys. Res. Lett.*, **20**, 2327.
Wang, Y.-M.: 1994, *ESLAB Symposium Proceedings*, this volume.
Wang, Y.-M.: 1993, *J. Geophys. Res.*, **98**, 3529.
Zhao, X.P. and Hoeksema, J.T.: 1994, this volume.

INTERPLANETARY SHOCK WAVES: ULYSSES OBSERVATIONS IN AND OUT OF THE ECLIPTIC PLANE

A. BALOGH, J.A. GONZALEZ-ESPARZA, R.J. FORSYTH
The Blackett Laboratory, Imperial College, London, U.K.

M.E. BURTON, B.E. GOLDSTEIN, E.J. SMITH
Jet Propulsion Laboratory, California Institute of Technology, Pasadena, Ca., U.S.A.

S.J. BAME,
Los Alamos National Laboratory, Los Alamos, NM, U.S.A.

Abstract. Between its launch in October 1990 and the end of 1993, approximately 160 fast collisionless shock waves were observed in the solar wind by the Ulysses space probe. During the in-ecliptic part of the mission, to February 1992, the observed shock waves were first caused mainly by solar transient events following the solar maximum and the reorganisation of the large scale coronal fields. With the decay in solar activity, relatively stable Corotating Interaction Regions (CIRs) were observed betwen 3 and 5.4 AU, each associated with at least one forward-reverse shock pair. During the out-of-ecliptic phase of the orbit, from February 1992 onwards, CIRs and shock pairs associated with them continued to dominate the observations. From July 1992, Ulysses encountered the fast solar wind flow from the newly developed southern polar coronal hole, and from May 1993 remained in the unipolar magnetic region associated with this coronal hole. At latitudes beyond 30°, CIRs were associated almost exclusively with reverse shocks only. A comprehensive list of shock waves identified in the magnetic field and solar wind plasma data from Ulysses is given in Table 1. The principal characteristics were determined mainly from the magnetic field data. General considerations concerning the determination of shock characteristics are outlined in the Introduction.

1. Introduction

Collisionless shock waves in the solar wind result from the interaction of plasma flows of different velocities, densities and temperatures. When the speed difference between interacting plasma flows is greater than the characteristic velocity at which information about dissipation processes at the interface can propagate away from it, a shock wave develops from the non-linear steepening of waves generated in the dissipation process. In the classical view of MHD shock waves, the interaction between the two plasma flows takes place over a very short, in fact negligible distance, and at a sufficient distance away from the shock surface, the two solar wind regimes are related by the Rankine-Hugoniot jump relations, based on the conservation of mass,

Space Science Reviews **72**: 171–180.
© 1995 *Kluwer Academic Publishers.*

momentum and energy flow across the shock surface. Furthermore, boundary conditions derived from Maxwell's equations also hold at the shock. Characteristics of heliospheric shock waves are normally derived from the MHD formulation.

However, when examined more closely, shock waves encompass a wide variety of complex plasma phenomena, involving wave-particle interactions which can have a significant spatial extent, affecting the properties of the upstream as well as the downstream solar wind flow. The geometry of the magnetic field with respect to the shock surface plays a key role in determining the nature of the dissipation processes. The angle θ_{Bn}, defined as the acute angle between the normal to the shock surface and the upstream magnetic field, is used as the basis of the major classification of shocks into quasi-perpendicular and quasi-parallel categories which correspond to very different upstream and downstream plasma characteristics. In quasi perpendicular shocks, plasma particles interacting with the shock are unable to escape far upstream, and therefore most of the dissipation occurs over a short spatial distance when measured along the shock normal. The shock transition in this case tends to be abrupt, and the upstream and downstream plasma remains relatively undisturbed. By contrast, quasi-parallel shocks are characterised by pre- and post-shock disturbances over considerably longer spatial scales, in particular in the pre-shock region, where particles outrunning the shock wave can generate non-linear waves which in turn interact with the particles and form a complex region leading to the shock transition which tends to be less well defined than in the quasi-perpendicular case. In extreme cases, normally only encountered at the Earth's bow shock, the quasi-parallel shock is unstable and can be reformed in the upstream region. The term "oblique shock" is somewhat ambiguous, but is often used to describe the wide range of shock transition characteristics which are observed when the shock wave is neither nearly perpendicular nor nearly parallel.

The shock strength is normally represented by the ratio B_d/B_u of the magnitudes of the downstream and upstream magnetic field vectors. The shock velocity is characterised by the sonic Mach number, the ratio of the shock velocity and the magnetosonic speed characteristic of the upstream solar wind. Above a critical value of the Mach number, which depends not only on the shock velocity but also on upstream plasma conditions and θ_{Bn}, dissipation at the shock front cannot be supplied by resistivity alone, and other processes, such as the reflection of particles is required contribute to the dissipation process. The reflected particles, in turn, lead to instabilities and wave generation which disturb considerably the upstream medium.

These brief and general remarks on collisionless shock waves in the heliosphere are intended to show that actually observed shock waves rarely, if ever, conform to a simple MHD picture. Shock waves propagate through a medium in which discontinuities and waves are already present, and which interact with the disturbances generated by the shock wave. The unambiguous identification of the characteristics of the two states of the solar wind plasma upstream and downstream of the shock wave to be used in the classical jump equations is rarely, if ever possible. Nevertheless, several techniques have been developed to derive "best fit" values of θ_{Bn} and other characteristics of shock waves. These normally combine solar wind plasma and magnetic field data and search, often iteratively, for the best fit values satisfying the Rankine-Hugoniot jump conditions and the boundary conditions at the shock front. However, the time resolutions of the plasma and magnetic field data often differ by as much as a factor 100 or more (on Ulysses, the highest resolution of the solar wind data is 4 minutes, the corresponding resolution of the magnetic field data is 1 second). This leads to a difficult choice, whether to average the magnetic field data, and if so, over what interval, for use in the determination of shock characteristics in combination with the solar wind data. A somewhat related and important question concerns the estimation of the uncertainties with which these characteristics can be determined. Recognising that it is impossible to determine the characteristic shock parameters to a very high accuracy, it is important to obtain an estimate of the likely uncertainties.

2. Shock Waves Observed by Ulysses

The characteristics of shock waves observed by the Ulysses magnetometer (Balogh *et al.*, 1992) and solar wind plasma detector (Bame *et al.*, 1992) are listed in Table 1. In particular, the times associated with the shock waves have been determined from the magnetic field data. Shock normals have been determined by a new technique based on the coplanarity theorem. The basic technique (Colburn and Sonett, 1966, Abraham-Schrauner, 1971) takes advantage of the coplanarity of the upstream ($\mathbf{B}_u$) and downstream ($\mathbf{B}_d$) magnetic field vectors and the shock normal vector. However, when applying this technique, the values used for $\mathbf{B}_u$ and $\mathbf{B}_d$ are critical. Normally, suitable upstream and downstream averages are used, but it is often difficult to know to what extent, and in what sense these averages represent the values on either side of the idealised shock wave. The new technique uses a large number of upstream-downstream magnetic field vector pairs to determine a large ensemble of θ_{Bn} values. Given that disturbances associated with the shock are basically wave-like, the ensemble of θ_{Bn} values has, in general, a bell-shaped distribution, with a well defined maximum, and a width which is representative of the uncertainties in determining the shock normal for a given shock. The uncertainties indicated in Table 1 for both θ_{Bn} and the shock strength B_d/B_u have been determined using this technique. The sonic Mach numbers (M_s) shown in Table 1 were determined by using two solar wind plasma data points, one on either side of the shock transition, and hence are very rough estimates only.

In general, shock waves are easily identifiable phenomena. Inevitably, however, there is a size distribution reaching to very weak shocks, and at the limit their identification becomes difficult, even questionnable. This is particularly acute for quasi-parallel shock waves which are associated with significant wave fields, upstream as well as downstream. In this paper, we have chosen to be comprehensive rather than restrictive in selecting shock waves for inclusion in Table 1. For a few shocks in Table 1 either the magnetic signature or the plasma signature was unclear, and these are specially indicated. It is unlikely that a more definitive yet complete shock list is possible. Any statistical or case study involving shock waves observed by Ulysses needs to take into account the necessary limitations of the present list.

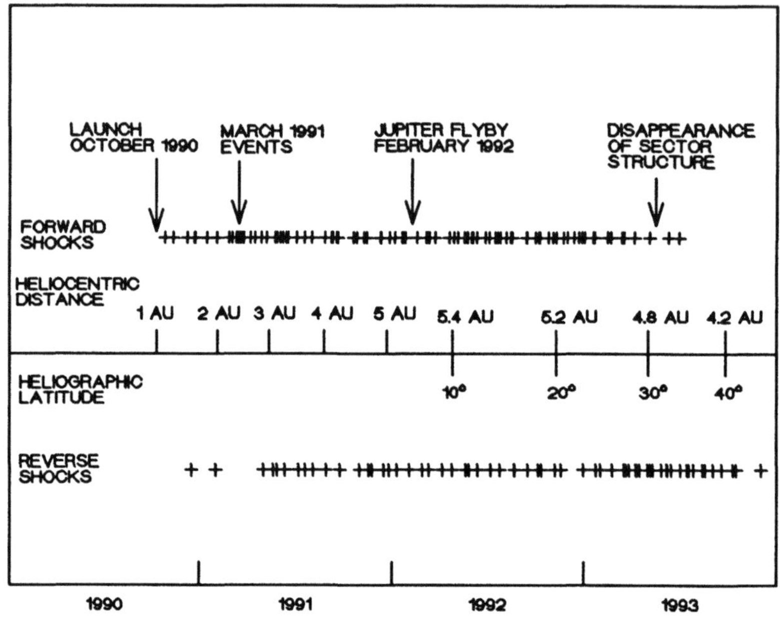

Fig. 1: Overview of shock waves seen by Ulysses from 1990 to 1993.

The list contains in total 160 shock waves from instrument switch-on in October 1990 to the end of 1993. Of these, 97 have been identified as fast forward shocks and 63 as fast reverse shocks. No special search of the data set has been carried out for slow shocks, but none has been identified up to now. A summary plot of the occurence of shock waves as a function of time is shown in Figure 1. In the early phase of the mission, up to July 1991, most shock waves were caused by solar transients, following the maximum in solar activity a year earlier (Balogh *et al.* 1993a). Nine fast forward shock waves were observed in March 1991 alone (Phillips *et al.* 1992) and a total of 27 in the period March to June 1991. The development of a more stable pattern of recurrent high speed streams from mid-1991 led to the formation of Corotating Interaction Regions (CIRs) which were bounded by fully developed forward-reverse shock pairs, similar to those observed on the Pioneer mission in this heliocentric distance range (Smith and Wolfe 1976). In the case of the Ulysses observations, there is no obvious dependence of the development of the CIR-associated shock waves on heliocentric distance, as the very active state of the Sun clearly prevented the formation of a stable, recurrent solar wind flow pattern earlier, when Ulysses was travelling from 1 to about 3 AU. The Ulysses observations in the ecliptic plane therefore represent the state of the medium in the two heliocentric distance ranges, from 1 to 3 AU and 3 to 5

AU, at different solar activity levels. It would not be reasonable to conclude that shock waves generated by solar transient do not propagate beyond 3 AU (clearly they do) nor that CIR-associated shock pairs only develop beyond 3 AU; the decrease in solar activity and the fast solar wind streams associated with the developing solar coronal holes resulted in the formation of large scale recurrent interaction regions with well defined shock waves between the Sun and the location of Ulysses.

The shock waves observed in the period up to July 1992, past the Jupiter flyby and including the early phase of the out-of-ecliptic trajectory of Ulysses mostly fall into the category of CIR-associated shocks. A reorganisation of coronal fields at that time, caused by the the magnetic evolution of the Sun towards solar minimum activity (Balogh *et al.* 1993b), exposed Ulysses to the fast solar wind stream from the expanding southern coronal hole, leading to the formation of a new series of CIRs (Bame *et al.* 1993). This series of CIRs was associated with forward-reverse shock pairs, similar to those encountered in the ecliptic. However, the evolving geometry of the large southern solar coronal hole, combined with the southbound trajectory of Ulysses, led, in early 1993, to a change in the geometry of encounters with the CIRs. This change is characterised by a reduction in the numbers of forward shocks and an increase, in absolute and relative terms, in the number of reverse shocks, as illustrated in Figure 2.

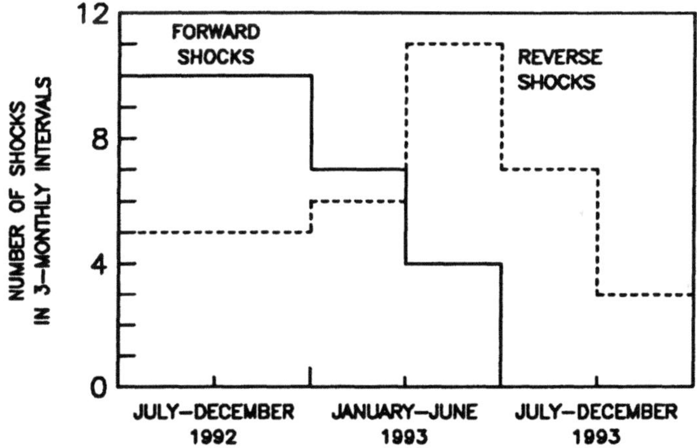

Fig. 2: Statistics of fast forward and reverse shocks observed by Ulysses between July 1992 and December 1993. The disappearance of forward shocks and the significant increase in the number of reverse shocks in 1993 has been associated with the large scale geometry and evolution of Corotating Interaction Regions at mid-latitudes.

This feature has been noted by Gosling et al. (1993) who explained it using the three-dimensional models of CIRs put forward by Pizzo (1994). In this model, forward shocks leading CIRs propagate preferentially towards lower latitudes, while reverse shocks propagate to higher latitudes; in the case of Ulysses, this means towards higher southern latitudes. The change in the relative frequency of forward and reverse shocks is well explained by the model; further investigations (such as those by MacDowall *et al.* 1994) are needed to use the Ulysses observations for determining the overall extent and latitude-dependent propagation characteristics of CIRs. One particular feature of CIRs is that they are not always bounded by a single forward-reverse shock pair, but that near the ecliptic there may be two forward shocks associated with their leading edge and that at the higher latitudes explored by Ulysses, more than one reverse shock has been observed at the trailing edge of CIRs.

A number of transient shocks were also observed in association with the high latitude Coronal Mass Ejections in 1993. A particularly noteworthy example, a forward-reverse shock pair associated with an over-expanding CME has been described by Gosling *et al.* (1994). The origin and propagation characteristics of high latitude CMEs will require further investigation.

References

Abraham-Schrauner, B.: 1972, 'Determination of magnetohydrodynamic shock normals', *J. Geophys. Res.* **77**, 736.

Balogh, A. *et al.*: 1992, 'The magnetic field investigation on the Ulysses mission: Instrumentation and preliminary scientific results', *Astron. Astrophys. SOuppl. Ser.* **92**, 221.

Balogh, A. *et al.*: 1993a, 'The interplanetary magnetic field from 1 to 5 AU: Ulysses observations', *Adv. Space Res.*, **13**, 6(15).

Balogh, A. *et al.*: 1993b: 'The evolution of the interplanetary sector structure in 1992', *Geophys. Res. Lett.* **20**, 2331.

Bame, S.J. *et al.*: 1992, ' The Ulysses solar wind plasma experiment', *Astron. Astrophys. Suppl. Ser.* **92**, 237.

Bame, S.J. *et al.*: 1993, 'Ulysses observations of a recurrent high speed solar wind stream and the heliomagnetic streamer belt', *Geophys. Res. Lett.* **20**, 2319.

Colburn, D.C. and Sonett, C.P.: 1966 'Discontinuities in the solar wind', *Space Sci. Rev.*, **5**, 439.

Gosling, J.T. *et al.*: 1993, Latitudinal variation of solar wind corotating stream interaction regions: Ulysses', *Geophys. Res. Lett.* **20**, 2789.

Gosling, J.T. *et al.*: 1994, 'A forward-reverse shock pair in the solar wind driven by over-expansion of a coronal mass ejection: Ulysses observations', *Geophys. Res. Lett.* **21**, 237.

MacDowall, R.J. *et al.*: 1994, 'The three-dimensional extent of a high speed solar wind stream' *Space Sci. Rev.* (this issue)

Pizzo, V.J.:1994, 'Global, quasi-steady dynamics of the distant solar wind. 1. Origin of north-south flows in the outer heliosphere' *J. Geophys. Res.* **99**, 4173.

Phillips, J.L. *et al.*: 1992, 'Ulysses plasma observations of coronal mass ejections near 2.5 AU' *Geophys. Res. Lett.* **19**, 1239.

Smith, E.J. and Wolfe, J.H.: 1976, 'Observations of interaction regions and corotating shocks between one and five AU: Pioneers 10 and 11', *Geophys. Res. Lett.* **3**, 137.

No.	Date	Time	F/R	θ_{Bn}	B_d / B_u	M_s	NB	No.	Date	Time	F/R	θ_{Bn}	B_d / B_u	M_s	NB
1	1990 299	14:13:08	F	28°±10°	1.40 ± 0.05	*		21	1991 097	04:46:02	F	75°±07°	2.77 ± 0.27	2.79	
2	1990 299	16:15:35	F	09°±09°	1.42 ± 0.12	*		22	1991 105	19:54:01	F	47°±09°	1.20 ± 0.05	(0.76)	
3	1990 315	20:35:30	F	26°±10°	1.56 ± 0.08	*		23	1991 118	06:16:47	F	75°±07°	1.93 ± 0.02	1.97	
4	1990 343	19:16:32	F	61°±16°	1.77 ± 0.11	1.98		24	1991 121	07:28:23	R	65°±10°	1.94 ± 0.15	1.55	
5	1990 350	07:52:12	R	31°±20°	1.38 ± 0.03	1.43		25	1991 122	14:21:59	R	43°±12°	1.51 ± 0.05	1.21	
6	1990 356	00:39:30	F	17°±13°	1.24 ± 0.07	1.05		26	1991 128	10:12:49	F	68°±10°	1.86 ± 0.09	1.46	
7	1990 358	15:59:50	F	66°±09°	1.56 ± 0.07	1.64		27	1991 140	21:58: ..	R	* *	* *	1.42	
8	1991 014	07:55:09	F	42°±12°	1.58 ± 0.07	1.49		28	1991 144	14:22: ..	F	* *	* *	*	
9	1991 031	04:12:37	R	60°±05°	1.87 ± 0.07	1.48		29	1991 144	18:41:24	F	69°±12°	1.41 ± 0.02	1.01	
10	1991 034	12:39:30	F	59°±07°	2.14 ± 0.08	1.52		30	1991 147	04:08:16	F	55°±16°	2.34 ± 0.05	1.50	
11	1991 056	22:04:56	F	55°±15°	1.30 ± 0.01	1.41		31	1991 147	19:04:39	R	45°±07°	1.63 ± 0.07	2.58	
12	1991 062	23:49:15	F	60°±16°	1.34 ± 0.02	1.59		32	1991 154	17:05:54	F	60°±15°	2.01 ± 0.01	2.30	
13	1991 069	00:43:39	F	43°±15°	1.28 ± 0.02	1.30		33	1991 158	15:15:14	F	71°±10°	1.74 ± 0.06	(0.98)	
14	1991 070	22:13:09	F	81°±02°	1.78 ± 0.01	1.66		34	1991 158	15:45:19	F	43°±13°	1.84 ± 0.17	1.16	
15	1991 073	01:56: ..	F	* *	* *	1.93		35	1991 162	09:14:52	R	39°±14°	1.46 ± 0.09	1.52	
16	1991 074	05:40:07	F	72°±04°	1.92 ± 0.06	2.12		36	1991 165	01:53:11	F	76°±04°	1.66 ± 0.04	1.82	
17	1991 077	13:08:07	F	32°±05°	1.66 ± 0.01	1.17		37	1991 168	22:58:15	F	62°±04°	1.89 ± 0.11	1.33	
18	1991 078	22:25:15	F	43°±16°	1.31 ± 0.05	2.79		38	1991 185	02:18:42	F	62°±13°	1.66 ± 0.07	1.71	
19	1991 082	15:39:43	F	74°±06°	2.24 ± 0.07	1.96		39	1991 188	16 :34:06	R	81°±04°	3.23 ± 1.00	2.93	
20	1991 084	05:52:11	F	47°±17°	1.59 ± 0.05	1.51		40	1991 199	16:49:34	F	53°±10°	1.25 ± 0.02	*	

np = not confirmed by Ulysses plasma data
nb = not confirmed by Ulysses magnetometer data (? doubtful)

TABLE 1: Characteristic parameters of heliospheric shock waves observed by Ulysses

No.	Date	Time	F/R	θ_{Bn}	B_d / B_u	M_s	NB	No.	Date	Time	F/R	θ_{Bn}	B_d / B_u	M_s	NB
41	1991 200	20:26:00	R	25°±19°	1.75 ± 0.20	1.48		61	1991 345	15:20:26	F	72°±08°	1.44 ± 0.04	(0.94)	
42	1991 213	03:33:38	F	76°±05°	3.22 ± 0.28	3.68		62	1991 349	16:53:48	R	39°±22°	1.62 ± 0.20	2.73	
43	1991 216	13:22	R	* *	* *	*	nb?	63	1991 359	08:47:10	R	35°±17°	1.88 ± 0.24	1.74	
44	1991 238	13:07:13	F	69°±06°	2.32 ± 0.14	3.38		64	1991 360	19:45:22	F	66°±10°	1.76 ± 0.24	(0.99)	
45	1991 238	19:50:59	F	78°±04°	2.70 ± 0.13	2.37		65	1992 006	01:58:36	F	54°±12°	2.45 ± 0.34	3.33	
46	1991 240	21:22:51	R	30°±16°	1.82 ± 0.21	1.61		66	1992 010	10:20:58	R	52°±16°	2.34 ± 0.18	2.27	
47	1991 251	08:26:41	F	64°±04°	1.53 ± 0.03	1.61		67	1992 019	00:12:33	F	43°±13°	1.36 ± 0.13	*	
48	1991 253	05:45	F	* *	* *	*	nb	68	1992 022	09:07:31	F	43°±12°	1.14 ± 0.04	(0.89)	np
49	1991 261	07:10	F	* *	* *	*	nb?	69	1992 026	15:42:16	F	56°±16°	1.32 ± 0.15	1.17	
50	1991 263	15:22:12	F	69°±05°	2.14 ± 0.06	2.55		70	1992 033	13:44:51	R	42°±07°	1.39 ± 0.06	1.58	
51	1991 266	15:36:02	R	35°±14°	1.51 ± 0.08	1.31		71	1992 048	22:02:22	F	68°±14°	1.92 ± 0.12	1.91	
52	1991 292	04:40:21	F	50°±11°	2.50 ± 0.13	2.53		72	1992 058	14:15:22	R	20°±20°	1.32 ± 0.19	3.61	
53	1991 299	22:03:22	F	44°±10°	2.74 ± 0.40	2.20		73	1992 065	16:09:26	F	32°±22°	1.56 ± 0.10	2.21	
54	1991 304	10:45:31	R	46°±18°	2.03 ± 0.14	1.33		74	1992 069	03:00:06	F	61°±02°	2.16 ± 0.03	2.18	
55	1991 313	08:21:12	F	50°±17°	2.42 ± 0.18	2.23		75	1992 070	16:58:50	R	75°±10°	1.22 ± 0.01	1.23	
56	1991 318	03:51:09	F	52°±08°	1.45 ± 0.07	1.48		76	1992 071	00:15:16	F	21°±16°	1.11 ± 0.01	2.22	
57	1991 318	12:40:37	F	85°±03°	1.79 ± 0.03	1.98		77	1992 082	17:57:11	F	61°±08°	2.00 ± 0.09	2.00	
58	1991 321	05:26:30	R	* *	* *	*		78	1992 096	17:11:00	R	* *	1.5/1.6	1.42	
59	1991 327	17:07:37	R	63°±11°	1.30 ± 0.05	1.65		79	1992 109	15:09: ..	F	* *	* *	3.50	
60	1991 343	04:38:26	F	67°±06°	1.21 ± 0.01	(0.97)	np	80	1992 115	02:45:45	R	18°±06°	1.58 ± 0.08	*	

np = not confirmed by Ulysses plasma data
nb = not confirmed by Ulysses magnetometer data (? doubtful)

TABLE 1 (cont'd): Characteristic parameters of heliospheric shock waves observed by Ulysses

No.	Date	Time	F/R	θ_{Bn}	B_d / B_u	M_s	NB	No.	Date	Time	F/R	θ_{Bn}	B_d / B_u	M_s	NB
81	1992 118	13:42:28	F	26°±16°	1.46 ± 0.22	1.64		101	1992 229	01:02:12	F	73°±09°	1.11 ± 0.01	1.51	np
82	1992 126	17:46:37	F	36°±21°	1.55 ± 0.06	1.70		102	1992 233	12:13:43	R	52°±08°	1.59 ± 0.13	1.91	np
83	1992 138	01:38:10	F	62°±07°	1.28 ± 0.04	1.62		103	1992 234	04:48:02	R	43°±15°	1.62 ± 0.10	2.12	
84	1992 141	16:57:12	R	64°±07°	1.53 ± 0.06	1.24		104	1992 256	01:21:54	F	65°±10°	3.41 ± 0.39	4.51	
85	1992 143	18:26:53	F	41°±12°	1.36 ± 0.04	(0.93)	np	105	1992 257	13:52:48	F	51°±14°	1.44 ± 0.09	2.37	
86	1992 144	03:10:55	F	74°±08°	1.31 ± 0.03	4.10		106	1992 259	10:26:35	R	64°±10°	3.73 ± 0.84	4.14	
87	1992 147	23:12:06	R	57°±13°	1.63 ± 0.11	1.38		107	1992 273	11:35:26	F	10°±07°	1.87 ± 0.11	3.38	
88	1992 152	11:48:55	F	34°±11°	1.24 ± 0.04	*		108	1992 278	20:19: ..	R	* *	2.4 / 2.5	*	
89	1992 155	18:33:04	F	35°±22°	1.91 ± 0.15	1.51		109	1992 279	15:38:11	F	24°±17°	1.75 ± 0.18	1.91	
90	1992 162	01:04:04	F	69°±06°	1.17 ± 0.02	1.18	np	110	1992 282	06:16:00	F	75°±09°	1.35 ± 0.05	1.69	
91	1992 163	13:00	R	* *	* *	*	nb	111	1992 285	09:16:24	R	68°±11°	2.46 ± 0.25	5.21	
92	1992 178	04 59:55	F	57°±17°	1.79 ± 0.20	3.64		112	1992 298	17:42:53	F	72°±05°	2.01 ± 0.11	1.06	
93	1992 185	02:18:09	F	54°±15°	1.70 ± 0.14	1.35		113	1992 307	04:46:39	F	82°±05°	1.88 ± 0.09	3.03	
94	1992 188	13:00:31	R	77°±05°	1.73 ± 0.06	1.98		114	1992 310	02:00	R	* *	* *	*	nb?
95	1992 197	05:50:31	F	42°±15°	2.28 ± 0.19	1.87		115	1992 313	15:34:50	F	68°±11°	2.81 ± 0.11	2.85	
96	1992 202	05:45:34	F	59°±14°	2.17 ± 0.15	1.85		116	1992 322	19:17:29	R	49°±13°	1.31 ± 0.05	1.62	
97	1992 205	18:52:13	R	20°±16°	1.90 ± 0.17	2.12		117	1992 326	05:22:12	F	51°±18°	1.66 ± 0.16	1.58	
98	1992 206	03:50:47	F	56°±11°	1.97 ± 0.10	1.31		118	1992 334	02:54:21	F	42°±23°	1.83 ± 0.14	1.92	
99	1992 217	05:53: ..	F	32°±18°	1.53 ± 0.08	1.55		119	1992 347	09:50:40	F	72°±09°	1.39 ± 0.04	1.36	
100	1992 226	04:29:53	F	67°±08°	1.21 ± 0.01	*		120	1992 355	02:41:35	F	43°±14	1.30 ± 0.05	1.27	np

np = not confirmed by Ulysses plasma data
nb = not confirmed by Ulysses magnetometer data (? doubtful)

TABLE 1 (cont'd): Characteristic parameters of heliospheric shock waves observed by Ulysses

No.	Date	Time	F/R	θ_{Bn}	B_d / B_u	M_s	NB	No.	Date	Time	F/R	θ_{Bn}	B_d / B_u	M_s	NB
121	1992 361	19:38:04	F	25°±11°	1.56 ± 0.10	1.55		141	1993 124	03:05	F	* *	* *	*	nb
122	1992 364	23:44:44	R	70°±06°	2.45 ± 0.27	2.38		142	1993 125	21:38: ..	R	11°±08°	1.62 ± 0.12	1.42	
123	1993 001	15:33:46	F	72°±05°	1.49 ± 0.04	*	np	143	1993 130	19:17:58	R	59°±10°	2.07 ± 0.19	1.84	
124	1993 017	00:35:09	F	72°±08°	1.57 ± 0.05	1.61	np	144	1993 146	17:53:29	R	09°±04°	1.39 ± 0.04	(0.97)	
125	1993 020	04:59:49	F	69°±09°	2.94 ± 0.11	2.93		145	1993 156	01:05:27	R	36°±20°	1.99 ± 0.24	1.91	
126	1993 022	02:57:44	R	35°±13°	2.96 ± 0.46	2.51		146	1993 157	03:09:53	R	31°±16°	1.27 ± 0.05	*	
127	1993 031	23:19:37	R	30°±13°	1.24 ± 0.02	1.24		147	1993 160	01:26:53	F	32°±19°	1.16 ± 0.11	*	
128	1993 045	22:27:44	F	64°±11°	2.29 ± 0.11	2.02		148	1993 165	04:21:38	R	21°±11°	1.37 ± 0.04	(0.80)	
129	1993 051	15:42: ..	F	28°±17°	1.14 ± 0.05	1.21		149	1993 180	18:37:30	F	65°±08°	2.20 ± 0.09	2.04	
130	1993 053	07:10:10	R	72°±06°	2.73 ± 0.09	2.51		150	1993 183	09:18:21	R	19°±13°	2.29 ± 0.18	2.00	
131	1993 072	04:48:05	F	27°±21°	1.34 ± 0.11	1.35		151	1993 195	15:31:37	R	49°±15°	1.93 ± 0.16	1.86	
132	1993 075	03:26:56	R	51°±12°	1.32 ± 0.07	1.32		152	1993 199	18:53:54	R	58°±08°	1.36 ± 0.02	1.38	
133	1993 076	23:54:38	F	42°±09°	1.73 ± 0.13	1.77	np	153	1993 207	09:56:59	R	39°±23°	2.09 ± 0.25	2.25	
134	1993 079	18:34:25	R	57°±10°	1.63 ± 0.05	1.57		154	1993 224	23:24:16	R	42°±23°	1.22 ± 0.14	1.54	
135	1993 085	20:18:35	R	65°±06°	1.68 ± 0.03	1.66		155	1993 230	11:26:35	R	19°±16°	1.40 ± 0.25	*	
136	1993 096	05:35:36	F	51°±21°	1.44 ± 0.18	1.38		156	1993 244	06:54:57	R	46°±09°	1.50 ± 0.08	1.44	
137	1993 098	15:56:42	R	36°±16°	1.78 ± 0.36	*		157	1993 261	19:16:28	R	65°±21°	1.77 ± 0.36	1.78	
138	1993 100	19:17:00	R	68°±15°	1.49 ± 0.18	1.47		158	1993 282	16:33:43	R	51°±12°	1.35 ± 0.04	1.30	
139	1993 105	15:52:28	R	29°±19°	1.66 ± 0.19	2.13		159	1993 288	03:00:07	R	22°±15°	1.41 ± 0.07	1.42	
140	1993 121	18:44	R	* *	* *	*	nb	160	1993 335	22:16:27	R	46°±16°	1.38 ± 0.05	1.36	

np = not confirmed by Ulysses plasma data
nb = not confirmed by Ulysses magnetometer data (? doubtful)

TABLE 1 (cont'd): Characteristic parameters of heliospheric shock waves observed by Ulysses

INTERACTION BETWEEN WHISTLER-MODE WAVES AND ELECTRONS IN THE VICINITY OF INTERPLANETARY SHOCKS AS SEEN BY ULYSSES: A PRELEMINARY STUDY

J. SOLOMON[1,2], N. CORNILLEAU-WEHRLIN[1] and P. CANU[1]
[1] CETP, 92131 Issy-les-Moulineaux, France.
[2] IAS, 91405 ORSAY CEDEX, France.

D. LENGYEL-FREY
University of Mariland, College Park, MD 20742, USA.

S. J. BAME and E.E. SCIME
Los Alamos National Laboratory, Los Alamos, New Mexico 87545, USA.

A. BALOGH and R. FORSYTH
Imperial College, London, UK.

Abstract. We present a study of wave particle interaction in the vicinity of interplanetary shocks in the ecliptic plane, as observed by the Ulysses spacecraft. We focus here on some events, for which electromagnetic waves are observed, mainly down-stream of the shock. The whistler mode waves are studied by means of the Unified RAdio and Plasma wave experiment (URAP) and the electron distribution functions are obtained from the Ulysses' plasma experiment (SWOOPS). The results are discussed in the light of electron-cyclotron instabilities.

1. Introduction

Recent papers (Lengyel-Frey et al, 1992; 1993) have focused on the study of the emission of whistler mode waves at interplanetary shocks. In numerous shock events, whistler waves are observed in general downstream of the shock while no whistler waves (precursors) are observed upstream. Moreover, the wave emission can persist several hours after that the shock front has passed Ulysses position. A case study indicates that the waves are emitted at large angle with respect to the interplanetary magnetic field **B** (Lengyel-Frey et al, 1994). It is in general admitted that whistler wave amplification results from an electron-cyclotron instability due to anisotropic electron distribution functions in the halo energy range (100 eV-2keV). Several papers have dealt with this instability problem about the Earth's bow shock using data of the ISEE satellites (Tokar et al., 1984; Tokar and Gurnett, 1985). In this paper we intend to examine both whistler wave emissions and electron distribution functions (EDF) measured about interplanetary shocks on Ulysses, then to discuss the wave generation by an electron-cyclotron instability.

2. Data

3-D EDF are produced in a 2.5-minutes sweep by the Ulysses' plasma experiment (e. g., Bame et al., 1992)) in the energy range of 1.6 to 862 eV. Such 3-D distributions are produced about every 17 minutes. The wave data used for this study come from the

Space Science Reviews **72**: 181–184.
© 1995 *Kluwer Academic Publishers.*

URAP search coil. The wave form is analysed onboard to produce spectra, averaged over 64 seconds. Here we present data in the range 10-450 Hz, for the By component, which is in the spin plane of Ulysses.

3. Data analysis

In figure 1 (top panels), are displayed 2 EDFs observed before and after a shock passed Ulysses at 04:46 UT on April 7, 1991. After the shock, the EDF exhibits a flat-top shape. This effect has been observed for 2 other shock events, which we have studied (December 9, 1990 at 19:16 UT and May 27, 1991 at 04:08 UT). A first study of these EDF can be performed by estimating the differential temperature Tdiff(v) of the EDF (Pilipp et al, 1987). This last quantity is defined by the following formula, where k_B is the Boltzmann constant, m the electron mass and F(v) the EDF in a specified direction with respect to **B**:

$$\mathrm{Tdiff}(v) = -\left((2k_B/m)\ d\ \ln F(v)/dv^2\right)^{-1} \quad (1)$$

The results are shown in the two middle panels of figure 1 as the parallel temperatures T//+, T//– and the perpendicular temperature Tperp respectively (parallel and perpendicular with respect to the interplanetary magnetic field **B**). Concerning the parallel temperatures one has distinguished between the temperatures T//+ and T//– obtained for positive and negative parallel velocities with respect to the direction of **B**. Concerning the core of the EDF ($v \le 6$, in units of 10^3 km/s), the temperatures change slowly with v apart that the flat-top shape of the EDF after the shock crossing (right panel) results in large values of T//± for $v \le 2$. In the halo part, the temperatures increase steadily showing no plateau: this clearly indicates that a bimaxwellian cannot represent properly the halo part of the EDF (e. g. Pilipp et al, 1987). A clear change in the halo temperatures is seen on the plots: before the shock crossing Tperp is above T//± while Tperp is mostly below T//± after it. An average estimate gives Tperp/T//± ≈ 1.4 at 04:40 and 0.85 at 04:58. For the 2 other shock events studied, the tendency is partly the same for the halo: before the shock crossings Tperp/T//± >1 (Dec. 9, 90) and Tperp/T//+ >1 (May 27, 91); after the shock these temperature anisotropies decrease.

Concerning whistler wave emission, one observes on the spectra displayed in figure 1 (bottom left panel) a clear increase of the wave magnetic field below the electron cyclotron frequency f_{ce} at the shock crossing. Downstream of the shock (bottom right panel), the wave emission persists with even a slight reinforcement.

4 Discussion and conclusion

Amplification of the whistler waves results from an electron cyclotron instability (e. g. Kennel, 1966). One has for the resonant energy E_R in eV:

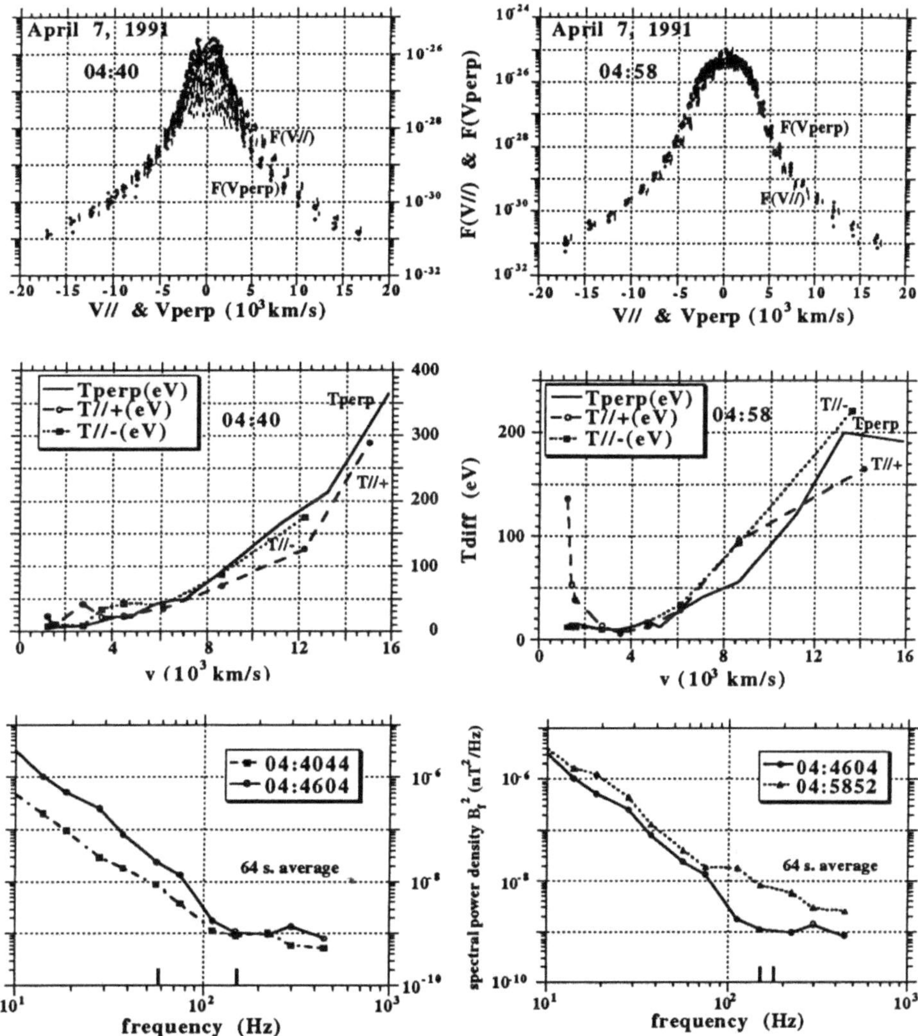

Fig 1 Top left and right panels cuts of the electron distribution function F(V) (in $cm^{-6}s^{-3}$) along the IMF **B** (V//) and perpendicular to **B** (Vperp), before and after a shock front passed Ulysses at 04 46 UT on April 7, 1991 Middle left and right panels differential temperatures obtained from the above distributions (see text) Bottom panels wave magnetic spectra obtained before the shock, at shock crossing and after the shock respectively The increasing values of the electron cyclotron frequency f_{ce} at the 3 considered times are 56 Hz, 142 Hz and 170 Hz respectively (those values are indicated by ticks on the frequency axis)

$$E_R \approx 2.6\ 10^5\ (f_{ce}/f_{pe})^2\ (1/x\ \cos\theta) \qquad (2)$$

where f, f_{ce}, f_{pe}, are the wave frequency, the electron cyclotron and the electron plasma frequency respectively. $x=f/f_{ce}$ is the reduced wave frequency and θ the wave normal angle with respect to **B**. For values of f_{pe}~10-20 kHz, f_{ce}~50-200 Hz, x~0.1-0.5, θ~0-60 deg., one finds that E_R~10 eV-2 keV. Therefore Ulysses EDF data should be sufficient to tackle the whistler wave generation problem around interplanetary shocks. The resonant velocity is $V_{//R}=(\omega+m\omega_{ce})/k_{//}$ where $\omega=2\pi/f$, $\omega_{ce}=2\pi/f_{ce}$ (positive quantity) and $k_{//}$ is the parallel wave vector with respect to **B**. One has to consider the cyclotron resonances m= ±1 and the Landau resonance m=0. For m= ±1, the sign of the wave growth rate $\gamma_{\pm 1}$ (> 0 for wave amplification) is given by :

$$-\,k_{//}\,(V_{//}\partial F/\partial V_{perp} - V_{perp}\partial F/\partial V_{//}) \sim -\,k_{//}V_{//R}(T_{perp}/T_{//}-1) \sim -m(T_{perp}/T_{//}-1) \qquad (3)$$

with $\omega \ll \omega_{ce}$ and by choosing a bimaxwellian $F(V_{//},V_{perp})$ with no drift velocity as an (incorrect) EDF model. By convention, $k_{//}$, $V_{//R}$ and $V_{//}$ are positive when oriented as **B**. Both cases, $T_{perp}/T_{//} > 1$ or $T_{perp}/T_{//} < 1$ can result in wave amplification, following the value of m that we consider. Note that the resonance m=1 only exists for a wave normal angle $\theta \neq 0$. For the Landau resonance, $\gamma_0 \sim k_{//}\partial F/\partial V_{//}$: a monotonous decreasing EDF can result in a damping of the wave propagating at an angle $\theta \neq 0$ with respect to **B**.

For the shock event of April 7, 1991, one should observe wave emission before the shock crossing due to the m= −1 resonance, which is not the case (bottom left panel of figure 1). It is possible that an accurate calculation would give a quite small value for γ_{-1} or that the wave spectrum is limited to a frequency range below 10 Hz, for which we have no data in the present case. Moreover, in order to explain wave emission at the shock crossing, it would be necessary to obtain at least in some cases, EDF measurements on a time range comprising the shock front. Another problem is the wave emission downstream of the shock (bottom right panel of figure 1): although the average temperature anisotropy $T_{perp}/T_{//}\pm \approx 0.85$ observed at 04:58 seems weak, the value of $\gamma_{\pm 1}$ could depend strongly on the detailed features of the EDF (middle right panel of figure 1). There is also the possibility for the waves of propagating downstream from the shock front. In any case an accurate model for the EDF is needed to properly estimate the wave growth rates γ_0 and $\gamma_{\pm 1}$.

References

Bame, S. J. et al.: 1992, *Astron. Astrophys. Suppl. Ser.*, **92**.
Kennel, C.:1966, *Phys. Fluids*, **9**, 2190.
Lengyel-Frey, D. et al.: 1991, *Proc. Solar Wind 7*, **3**, 477.
Lengyel-Frey, D. et al.: 1992, *ESA SP-346*, 71.
Lengyel-Frey, D. et al.: 1994, *J. Geophys. Res.*, to be published.
Pilipp, W. G. et al.: 1987, *J. Geophys. Res.*, **92**, 1075.
Tokar, R. L., Gurnett, D. A., and Feldman, W. C.: 1984, *J. Geophys. Res.*, **89**, 105.
Tokar, R. L., Gurnett, D. A.: 1985, *J. Geophys. Res.*, **90**, 105.

GLOBAL SOLAR MAGNETIC FIELD EVOLUTION INFERRED FROM GEOMAGNETIC VARIATIONS

D. I. PONYAVIN
Institute of Physics, St.Petersburg University, St.Petersburg, 198904, Russia

Abstract. I have analyzed geomagnetic disturbance index C9, mean solar magnetic field observed at Stanford Solar Observatory for the interval January 13, 1976 – December 30, 1993. It has been established a good correspondence between high-intensity geomagnetic recurrent and solar magnetic field patterns during whole period analyzed. A surprising thing is that the behavior of the solar mean field and interplanetary medium in the latest two solar cycles is very similar. Geomagnetic activity variations actually could serve as an ecliptic monitor of solar magnetic field structure and its evolution.

Key words: Geomagnetic activity – Solar Magnetic Fields

1. Introduction

It has been reliably established that solar magnetic fields determine a structure of the solar wind and interplanetary magnetic field (IMF). Severny et al. (1970) have found that the mean solar field shows the same polarity pattern as the IMF observed near the Earth's orbit. Sun-as-star integrated light measurements of the mean solar magnetic field are similar to coronal magnetic field determined by extrapolating the line-of-sight photospheric magnetic field to a source surface at several solar radii (Schatten, 1970). On the source surface, computed contour along which magnetic field changes its sign has been identified with heliospheric current sheet (HCS). The HCS manifests itself in the IMF sector boundary crossing the Earth. Sector boundaries separate usually high-speed recurrent solar wind streams. In this connection fairly good relation could appear between computed magnetic field strength at the source surface and the solar wind velocity (Burlaga et al., 1978; Hoeksema, 1984). Wang et al. (1990) have shown recently that, the fastest wind is in fact centered around the warps in the HCS.

It is widely recognized that the geomagnetic activity is almost entirely controlled by the solar wind and IMF variations. Therefore, it reflects a large-scale organization of interplanetary medium and solar magnetic fields. In this manner geomagnetic activity variations, especially in absence of *in situ* observations, could serve of use as probe for exploring solar magnetic field structure and its evolution (Ponyavin, 1993).

2. Observations and Discussion

For the present analysis the daily planetary index C9 was taken as representative of global geomagnetic activity. The historical record of C9 is

Space Science Reviews **72**: 185–188.
© 1995 *Kluwer Academic Publishers.*

continuous enough and usefully dense for the last 100 years. C9 data from St.Petersburg geomagnetic observations recorded since 1841 was added since 1890 to the available global data set (Zosimovich, 1981). It is significant that yearly or monthly averaging of original data often loss an important information about solar causes of long-term geomagnetic variations. A notable example is the origin of seasonal and annual variations in geomagnetic activity due to relatively stable IMF and solar wind structure during some periods of solar cycle (Ponyavin, 1992).

Bartels display of daily C9 geomagnetic disturbance index, mean solar magnetic field and sunspot numbers are presented in Figure 1. A technique of smoothing was applied to reveal recurrent pattern of original data for the interval January 13, 1976 – December 30, 1993. I have used the daily entries of mean solar magnetic field observations (plus four days to allow for an average Sun-Earth transit time) published in the Solar Geophysical Data series by Solar Stanford Observatory. Gaps in the data have been filled by linear interpolation technique. Only contours with C9 greater than four and mean field exceeding 25 microtesla are displayed in the Figure 1. Monthly sunspot numbers have been smoothed with a five point running mean.

An important point is that the sufficiently large mean solar magnetic field correspond in structure high-intensity geomagnetic recurrent pattern throughout whole period analyzed. The correlation coefficient as a function of time shows its maximum of 0.45 at a lag of four days. It is apparent that high-intensity geomagnetic variations of this type reflect high-intensity recurrent solar wind stream occurrence along the Earth's orbit. And in its turn solar wind streams originate from open intense magnetic fields at the source surface. As it can be seen from the Figure 1, mean solar field is much more stronger during the declining phase of solar cycles 21 and 22. A similar pattern of high-intensity streams is observed near the Earth.

Hoeksema (1991) pointed out that the solar magnetic field approximated as dipolar tends to diminish its strength and changes rapidly its orientation along the solar rotation axis in the vicinity of solar maximum, as it has been done in early 1980 and 1990. Just after axial polar field reversing near-equatorial large-scale stable magnetic field regions begin to characterize the declining phase corona in 1981 to 1984. A large increase in total strength seen in 1982 comes from equatorial dipole component, as evidenced by the absence of an increase in the axial dipole. Figure 1 shows also clearly that geomagnetic disturbance patterns have maintained a relatively high level of intensity at the same time.

While the details vary significantly from cycle 21 to cycle 22, large-scale patterns of mean solar field and geomagnetic recurrences remain the same. More likely that solar magnetic dipole determines in fact a quasi-stationary structure of interplanetary medium, i.e. HCS orientation and solar wind streams throughout the whole 22-year magnetic solar cycle. Actually, the

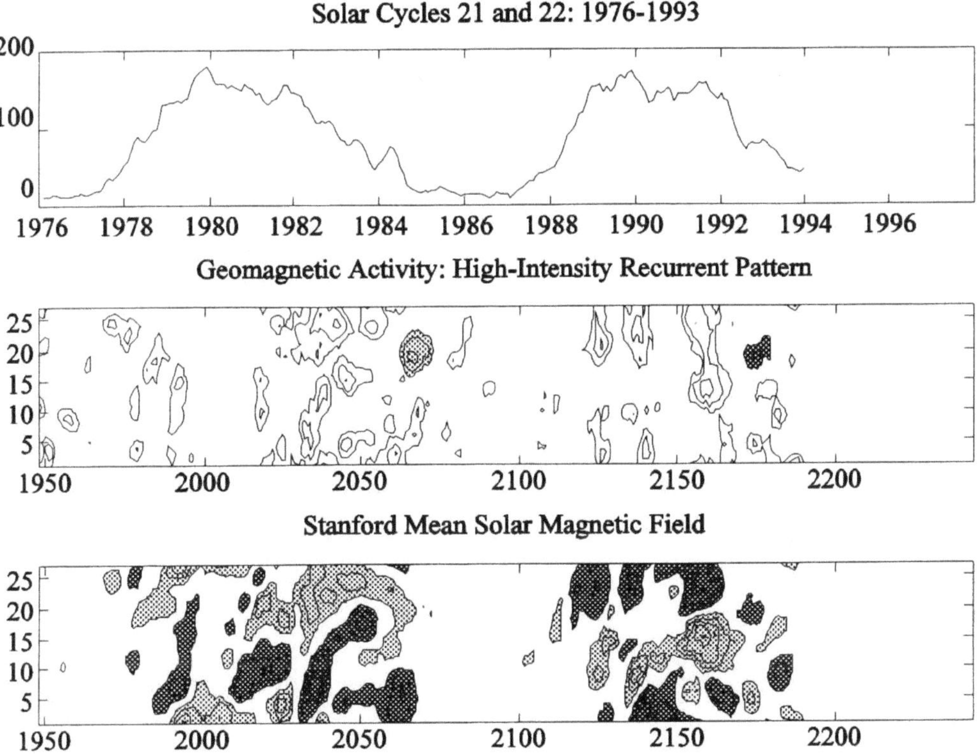

Fig. 1 Bartels display of smoothed C9, mean solar magnetic field, 1976–1993 (Bartels Rotations: 1948–2190). Contour plots of C9 are: 4,5,6, ... Mean solar field contours are: 25, 50, 75, ... microtesla. A top panel shows a smoothed sunspot numbers vs. time. Negative polarity is shaded more darkly High-intensity streams discussed in the paper are shaded also in the middle panel

corona acts a kind of filter that is most sensitive to the large-scale field pattern (Suess et al., 1993).

What is more, the recent Ulysses observation results discovered an excellent agreement with the predictions of a three-dimensional model of corotating solar wind flows that originate in a tilted dipole geometry back at the Sun (Gosling et al., 1993). In July 1992 while travelling south (Bame et al., 1993), Ulysses encountered a recurrent large-amplitude high-speed stream has been observed through June 1993. Polarity of the Ulysses's stream was inward, as well as polarity of associated south polar coronal hole. At the same period a succession of recurrent streams appeared at IMP 8, a 1 AU in-ecliptic monitor. As can be seen from the Figure 1, a high-intensity recurrent stream of negative polarity has been detected by the geomagnetic activity in the vicinity of Bartels day 20.

It is interesting to note, that this stream has appeared first during a time of major change in the heliospheric magnetic field structure (Balogh et al., 1993). This situation is very similar to solar wind stream occurrence just after IMF reorganization in 1984. Figure 1 shows that magnetic polarity of the stream was outward, and associated with flows from southern hemisphere as well (Ponyavin, 1993).

In May, 1993, the HCS ceased to be seen by the Ulysses a half year earlier and at heliocentric latitude lower than predicted on the basis of the persistence of the time profile of the neutral sheet tilt from solar cycle 21 to the next (Smith et al., 1993). As the cited source points out, a question is: will Ulysses now stay above the HCS all the way to the polar regions? The approach has been used by Suess et al. (1993), was to show that the behavior of the tilt in the past solar cycles is very similar.

As is evident from the foregoing analysis of geomagnetic activity, there is little likelihood that the HCS will extent to high latitudes at least within the next few years.

References

Balogh, A., Erdos, G., Forsyth, R.J., and Smith, E.J.: 1993, *Geophys.Res.Lett.* **20**, 2331

Bames, S.J., Goldstein, B.E., Gosling, J.T., Harvey, J.W., McComas, D.J., Neugebauer, M., and Phillips, J.L.: 1993, *Geophys.Res.Lett.* **20**, 2323

Burlaga, L.F., Behannon, K.W., Hansen, S.F., Pneuman, G.W., and Feldman, W.C: 1978, *J.Geophys.Res.* **83**, 4177

Gosling, J.T., Bames, S.J., McComas, D.J., Phillips, J.L., Pizzo, V.J., Goldstein, B.E., and Neugebauer, M.: 1993, *Geophys.Res.Lett.* **20**, 2789

Hoeksema, J.T.: 1984, *Structure and Evolution of the Large Scale Solar and Heliospheric Magnetic Fields*, CSSA-ASTRO-84-07, Ph.D.Diss.: Stanford University

Hoeksema, J.T.: 1991, *Adv. Space Res.* **11**, 15

Ponyavin, D.I.: 1992, *Biophysics* **37**, 499

Ponyavin, D.I.: 1993, '' in International Workshop on Artificial Intelligence Applications in Solar-Terrestrial Physics, ed(s)., *Joselyn, J., Lundstedt, H., Trolinger, J.*, NOAA SEL, U.S. Dept. of Commerce: Boulder, Colorado, 127

Schatten, K.H.: 1970, *Solar Phys.* **15**, 499

Severny, A., Wilcox, J.M., Scherrer, P.H., and Colburn, D.S.: 1970, *Solar Phys.* **15**, 3

Smith, E.J., Neugebauer, M., Balogh, A., Bames, S.J., Erdos, G., Forsyth, R.J., Goldstein, B.E., Phillips, J.L., and Tsurutani, B.T.: 1993, *Geophys.Res.Lett.* **20**, 2327

Suess, S.T., McComas, D.J., and Hoeksema, J.T.: 1993, *Geophys.Res.Lett.* **20**, 161

Wang, Y.-M., Sheeley, N.R.Jr., and Nash, N.G.: 1990, *Nature* **347**, 439

Zosimovich, I.D.: 1981, *Geomagnetic Activity and Corpuscular Field Stability of the Sun*, Nauka: Moscow

MODELING THE OUT-OF-ECLIPTIC INTERPLANETARY MAGNETIC FIELD IN THE DECLINING PHASE OF SUNSPOT CYCLE 22

XUEPU ZHAO and J. TODD HOEKSEMA
Center for Space Science and Astrophysics, Stanford University Stanford, CA, USA

Abstract. We have developed a new model of the coronal and interplanetary magnetic field. The model includes the effects of large-scale horizontal electric currents flowing in the inner corona, of the warped heliospheric current sheet, and of heliospheric volume currents in the super-Alfvenic solar wind. The model determines the interplanetary magnetic field (IMF) strength as well as its polarity from measurements of the photospheric magnetic field. A detailed comparison between the observed and calculated in-ecliptic IMF Bx in Cycles 22, confirms the fitness of the optimal set of free parameters inferred using data in Cycle 21. We can predict the latitudinal gradient of Bx in the declining phase of Cycle 22 and the temporal variation of the amplitude of the radial component of the IMF at various latitudes. The calculated IMF polarity and Bx strength agree best with the in-ecliptic observations when the photospheric field (measured with a $\lambda 5250 \mathring{A}$ magnetograph) is scaled up by a factor of two. Ulysses may provide the critical data to improve the model and check these inferences.

1. Comparison Between Observed and Calculated In-Ecliptic IMF B_x

The newly developed coronal and interplanetary magnetic field model (Zhao and Hoeksema, 1993) builds on a magnetostatic atmosphere (Bogdan and Low, 1986) and uses Schatten's techniques (Schatten et al., 1969; Schatten, 1971). The model includes three parameters, a, R_{cp}, and R_{ss}, characterizing, respectively, 1) the height scale of the large-scale horizontal current in the corona, 2) the height at which the heliocentric current sheet is rooted, and 3) the height beyond which the solar wind plasma totally controls the magnetic field. The three parameters were inferred on the basis of the OMNItape data in Cycle 21. We selected a 27-day interval in each year from 1977 to 1986 when most of the measured IMF B_x (Here the 'x' direction is in the radial direction toward the Sun) values at 1 AU were less than 5 nT and the sector structure was well defined. We then determined two sets of parameters, *viz.*, Set 1: $a = 0.2, R_{cp} = 2.25, R_{ss} = 5.00$ and Set 2: $a = 0.2, R_{cp} = 1.4, R_{ss} = 3.25$, that reproduce the rotation-averaged values of the unsigned daily IMF B_x with correlation coefficients greater than 0.85. In the calculations using parameter set 1, the photospheric field measured at $\lambda 5250 \mathring{A}$ with the Wilcox Solar Observatory magnetograph was scaled up by a factor of two, as suggested by Wang and Sheeley (1988).

In order to evaluate the two sets of parameters, we calculate the rotation-averaged B_x in Cycle 22 with each set and determine how well the predictions match the observations. We also calculate the fraction of successful predictions of the daily IMF polarity in each Carrington rotation using the two sets of parameters

Space Science Reviews **72**: 189–192.
© 1995 *Kluwer Academic Publishers.*

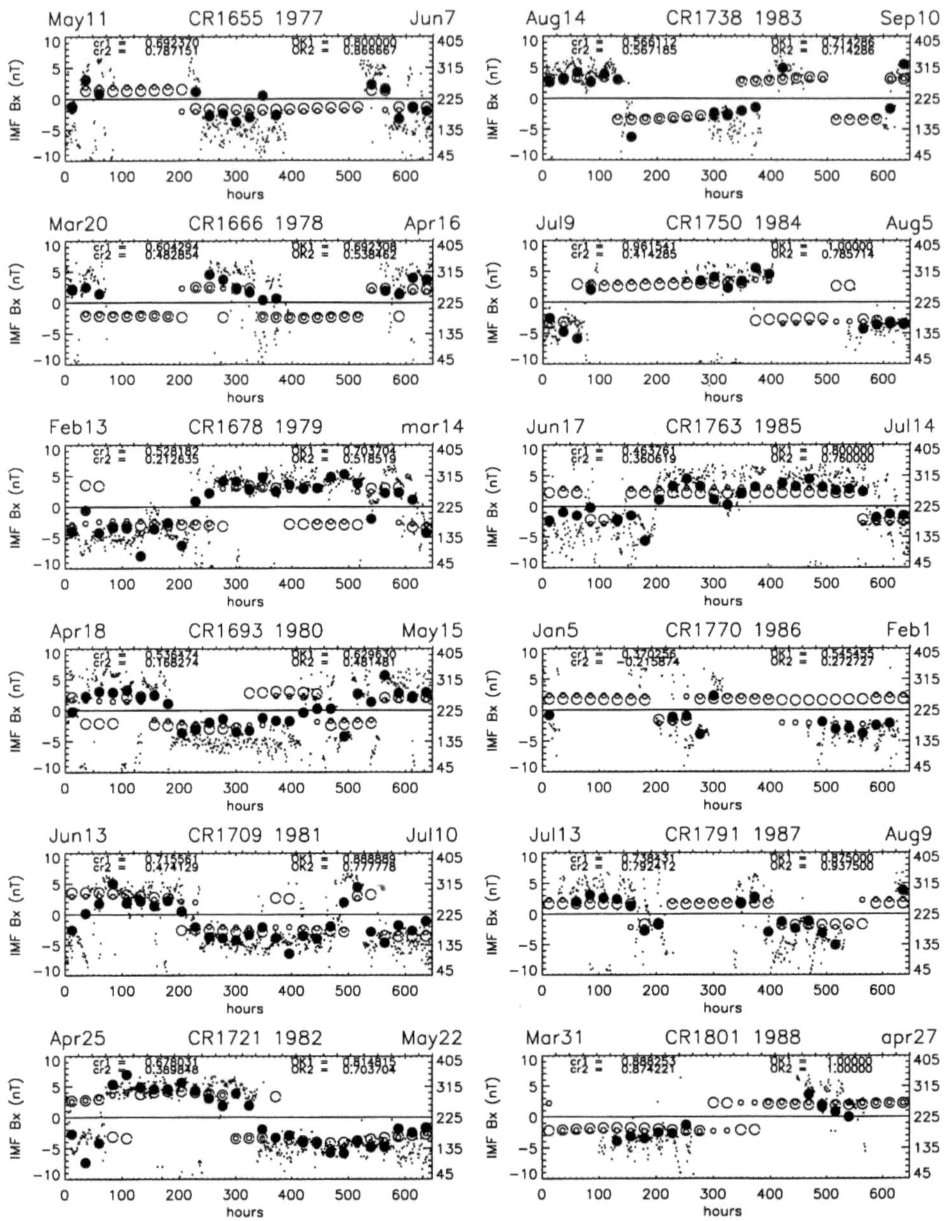

Figure 1. Comparison of the observed (filled circles) and calculated (open circles) daily IMF B_x at 1 AU. The small and large open circles denote B_x calculated using Set 1 and Set 2 of parameters. The dots denote the observed hourly azimuthal angles scaled at the right ordinate. The fraction of successful polarity prediction (ok1, ok2) and the correlation coefficient (co1, co2) for each rotation are displayed in appropriate panels.

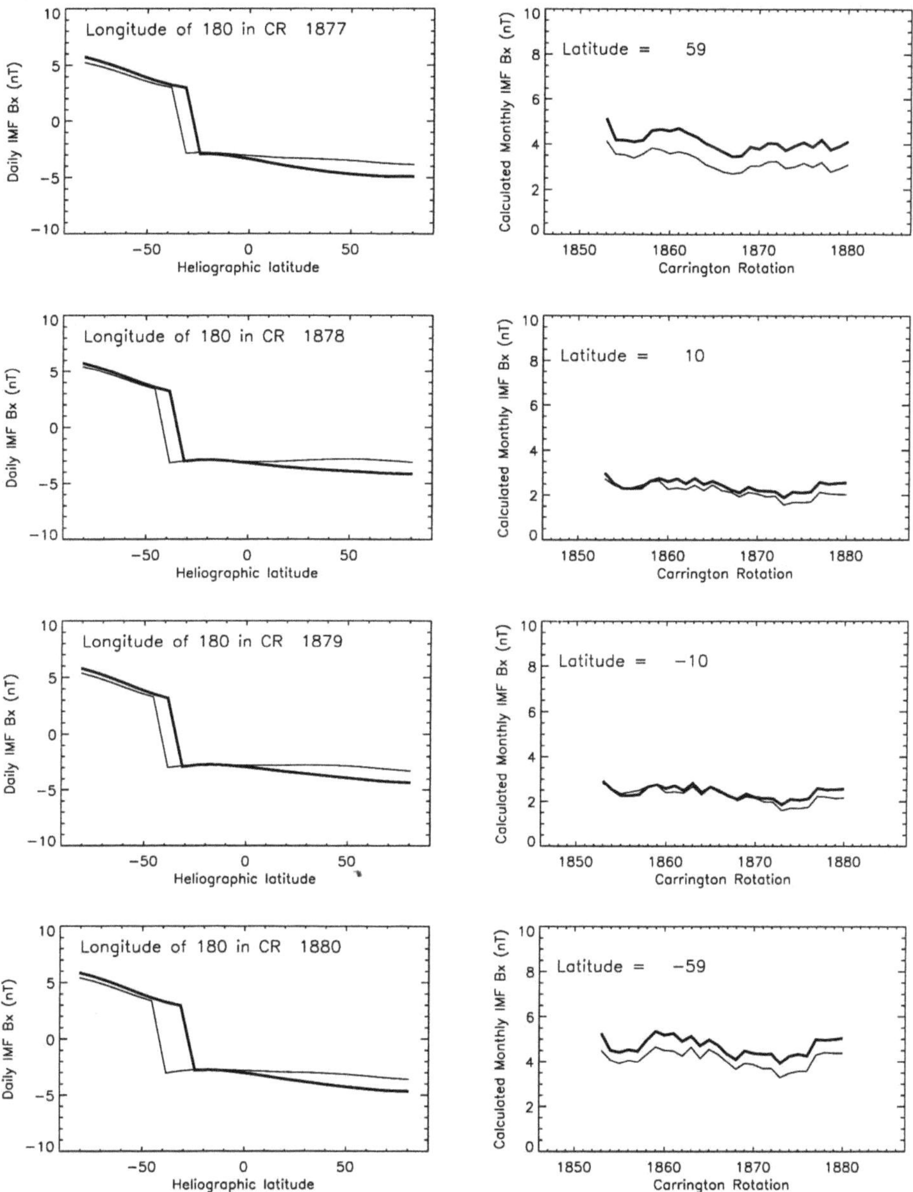

Figure 2. The latitudinal variation of IMF B_x (left column) and temporary variation of rotation-averaged values of unsigned IMF B_x (right column), predicted using Set 1 (thick lines) and Set 2 (thin lines) of parameters for Carrington Rotations 1978 – 1881 (19 December 1993 – 7 April 1994).

(see Figure 1). The fraction is also calculated using the potential field source surface model ((Hoeksema, et al., 1982). The results obtained for Set 1 and for the source surface model are basically identical. For comparison we also show the correlation coefficient between the observed and predicted, signed daily B_x for each rotation. Figure 1 shows that the values calculated with Set 1 reproduce both the strength and polarity of the daily IMF B_x better than Set 2, though the two sets determine the rotation-averaged B_x almost equally well. Detailed comparison of the observed and calculated in-ecliptic signed IMF Bx shows that when Parameter Set 1 is used together with doubled photospheric magnetic field values, the new model reproduces both the IMF Bx strength and the IMF polarity quite well.

2. Prediction of the Out-of-Ecliptic IMF B_x in Declining and Minimum Phases of Cycle 22

The left column of Figure 2 displays the latitudinal variation of the predicted IMF B_x at 1 AU at a Carrington longitude of 180° for Rotations 1877–1880 (Dec. 19 1993 – Apr. 7, 1994). The predicted latitudinal gradient of IMF B_x using Set 1 is greater than Set 2.

The right column of Figure 2 displays the temporal variation of the predicted rotation-averaged B_x at 1 AU and at low and moderate heliographic latitudes in the declining phase of Cycle 22 (1992–1994). The predicted rotation-averaged values using Set 1 are consistently greater than using Set 2.

Figures 2 shows that The daily IMF Bx observed by Ulysses would be expected to increase as the heliospheric latitude increases due to a nearly constant latitudinal gradient. The rotation-averaged values of the magnitude of Bx at middle and high latitudes would undergo a stronger temporal variation than near the ecliptic.

3. Conclusions and Discussion

The latitudinal gradient and rotation-averaged values predicted using Set 1 are greater than those using Set 2. Ulysses data will provide a unique opportunity to determine the optimum set of parameters for accurately predicting the IMF B_x not only near the Earth's orbit but also at higher latitudes.

We plan to use the model to determine the flux tube expansion factor and thus model the solar wind speed. We expect this will do a better job than the potential field source surface model.

References

Bogdan,T.J. and Low,B.C: 1986, *Astrophys. J.*, **306**, 271.

Hoeksema, J.T., Wilcox,J.M. and Scherrer,P.H.: 1982, *J. Geophys. R.*, **87**, 10331.

Schatten,K.H: 1971, *Cosmic Electrodynamics*, **2**, 232.

Schatten,K.H., Wilcox,J.M. and Ness,N.F: 1969, *Solar Phys.*, **9**, 442.

Wang,Y.-M. and Sheeley,N.R., Jr.:1988, *J. Geophys. Res.*, **93**, 11227.

Zhao, X. P. and Hoeksema,J.T.:1993, *EOS*, **74**, No. 43, 484.

LATITUDE AND SOLAR-CYCLE DEPENDENCE OF RADIAL IMF INTENSITY

Y.-M. WANG
Code 7672W, E. O. Hulburt Center for Space Research, Naval Research Laboratory, Washington, DC 20375-5352

Abstract. I describe a simple procedure for extrapolating the observed solar magnetic field into the heliosphere, which averages the asymptotic fields computed using the standard "source surface" and "current sheet" models. The resultant field is characterized by strong latitudinal gradients (maintained by volume currents outside the source surface) and by abrupt reversals in direction at the current sheets. The model yields good agreement with the observed long-term variation of the radial IMF component in the ecliptic, and is used to predict the variation of $|B_r|$ along the latitudinal trajectory of Ulysses during 1990–1994. As found in earlier studies, the magnitude of B_r at any latitude is determined largely by the strength and relative orientation of the Sun's dipole moment.

Key words: heliosphere – latitude effects – magnetic fields – solar cycle – Ulysses

1. MODELING THE IN-ECLIPTIC IMF INTENSITY

The interplanetary magnetic field (IMF) intensity measured near Earth undergoes surprisingly little long-term variation (King, 1979; Slavin, Jungman, and Smith, 1986). Figure 1*a* compares 3-month running means of the observed $|B_r|$ with the corresponding values obtained by applying the standard "source surface" model (Schatten, Wilcox, and Ness, 1969) to solar magnetic measurements during 1970–1994. In this model, the magnetic field is assumed to be potential between $r = R_\odot$ and a spherical source surface $r = R_{ss}$, where the field becomes purely radial; no current sheets are present, but the latitudinal gradients in the radial field beyond R_{ss} implicitly require the presence of *volume* currents. Although the SS model reproduces the observed IMF variation near sunspot maximum, it yields values that are much too low around sunspot minimum (1976, 1986), when the neutral sheet (where B_r gradually approaches 0) lies very close to the ecliptic.

In a very different prescription (Schatten, 1971), the magnetic field is assumed to be potential everywhere except at current sheets, where B_r suddenly reverses its sign. In this current sheet (CS) model, the asymptotic field is radial and constant in magnitude (independent of latitude and longitude). Applying this model to the solar magnetic data, we see (Fig. 1*b*) that it yields values of $|B_r|$ that are too *high* around sunspot minimum, in contrast to the behavior obtained with the SS model. The CS model essentially takes all of the open magnetic flux in the SS model and distributes it uniformly at large radii. The variation is thus proportional to the total amount of open flux on the Sun, which is smallest at sunspot maximum.

Space Science Reviews **72**: 193–196.
© 1995 *Kluwer Academic Publishers.*

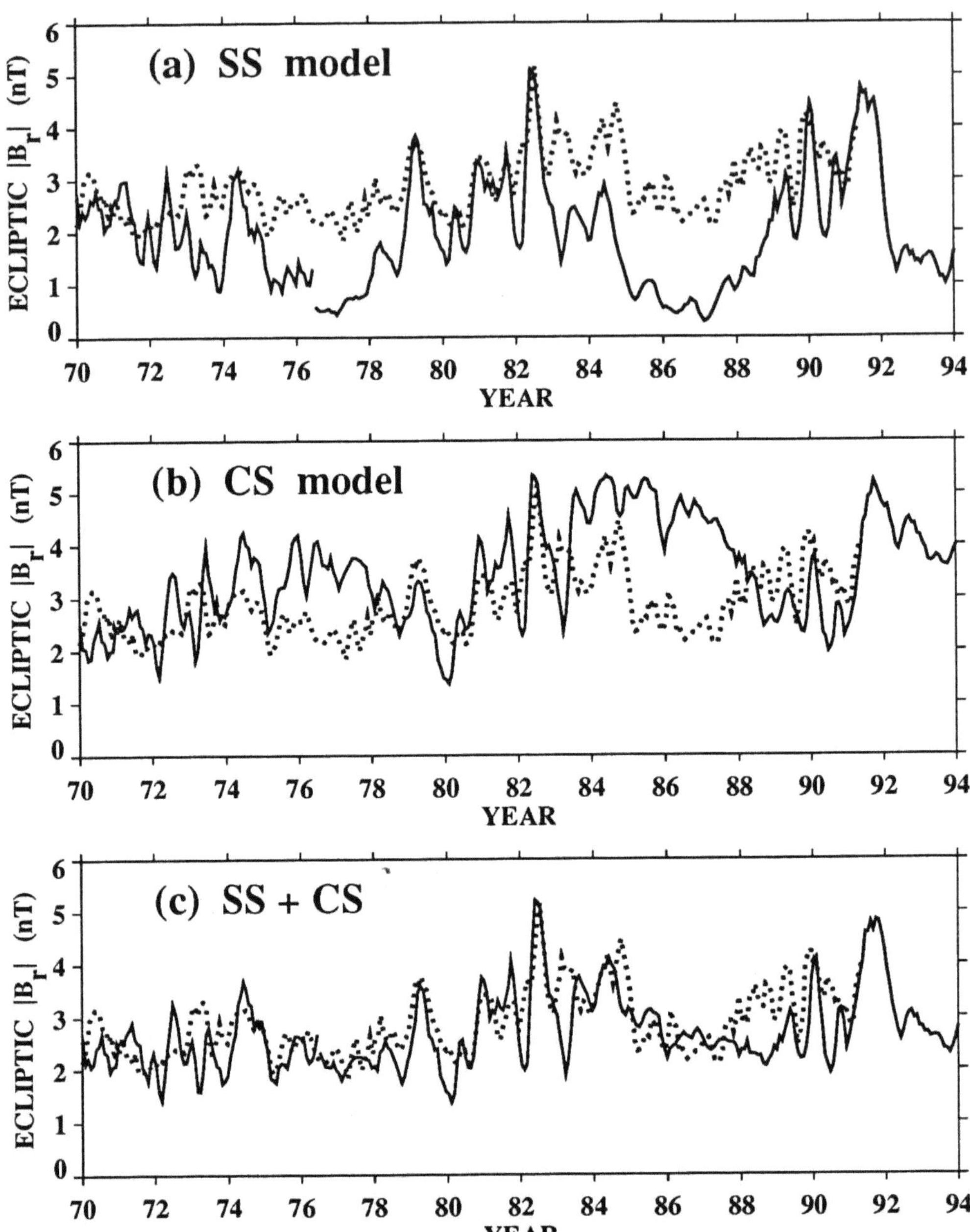

Fig. 1. Three-month running means of $|B_r|$ over two sunspot cycles, as measured by near-Earth spacecraft (*dotted lines*) and as calculated using the three models described in the text (*solid lines*). The models were applied to photospheric magnetic data from Mount Wilson Observatory (1970–1976) and from the Wilcox Solar Observatory (1976–1994); the magnetograph measurements were scaled upward by a factor of 2×1.8, as discussed in Wang (1993). The source surface and inner edge of the current sheets were located at $r = 2.35R_\odot$.

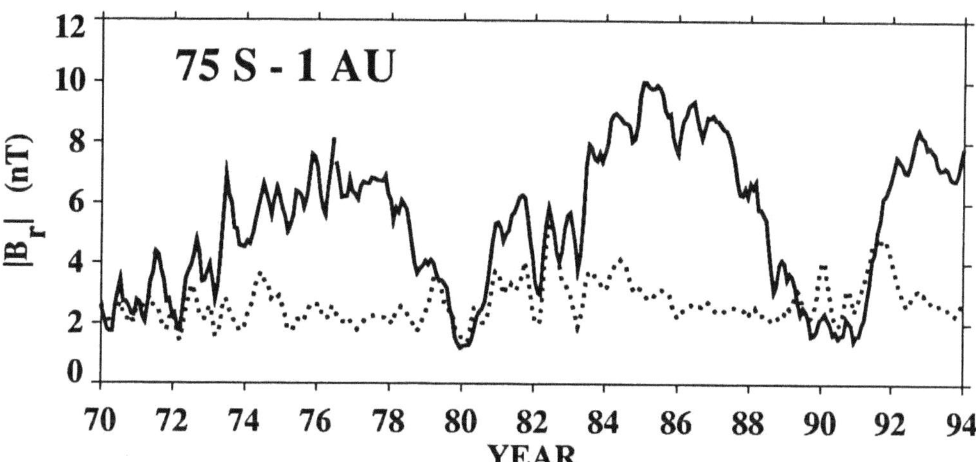

Fig. 2. Three-month running means of $|B_r|$ at heliographic latitude 75°S and distance 1 AU (*solid line*), as inferred by applying the SS + CS model to solar magnetic data during 1970–1994. For comparison, the calculated in-ecliptic variation is shown by the *dotted line*.

The failure of the SS and CS models individually to reproduce the observed IMF variation suggests that both volume and sheet currents affect the asymptotic flux distribution. Figure 1*c* shows that, by simply averaging the fields computed with the SS and CS models, we obtain a remarkably good fit to the observed in-ecliptic $|B_r|$. In this combined "SS + CS" model, volume and sheet currents are of equal importance, a characteristic also of the MHD model of Pneuman and Kopp (1971); these currents are required to maintain transverse pressure balance in the solar wind plasma.

2. PREDICTING THE HIGH-LATITUDE IMF INTENSITY

We now use the combined model to predict the out-of-ecliptic behavior of $|B_r|$. As an example, Figure 2 shows the calculated variation of $|B_r|$ at heliographic latitude 75°S and heliocentric distance 1 AU during 1970–1994. Unlike the behavior in the ecliptic, the high-latitude intensity varies by a factor of order 10, attaining its maximum at sunspot minimum and its minimum at sunspot maximum.

As a second example, Figure 3 (*solid line*) shows the predicted variation of $|B_r|$ along the latitudinal trajectory of Ulysses, as averaged over successive Carrington rotations and normalized to 1 AU assuming an r^{-2} fall-off. For comparison, the *dotted line* indicates the computed variation in the ecliptic. The SS + CS model predicts that $|B_r|$ should have been rising steeply by the time Ulysses reached 50°S at the beginning of 1994. The peak in $|B_r|$ seen in the ecliptic during 1991 was associated with the development of a strong solar equatorial dipole component, just as occurred during the 1982

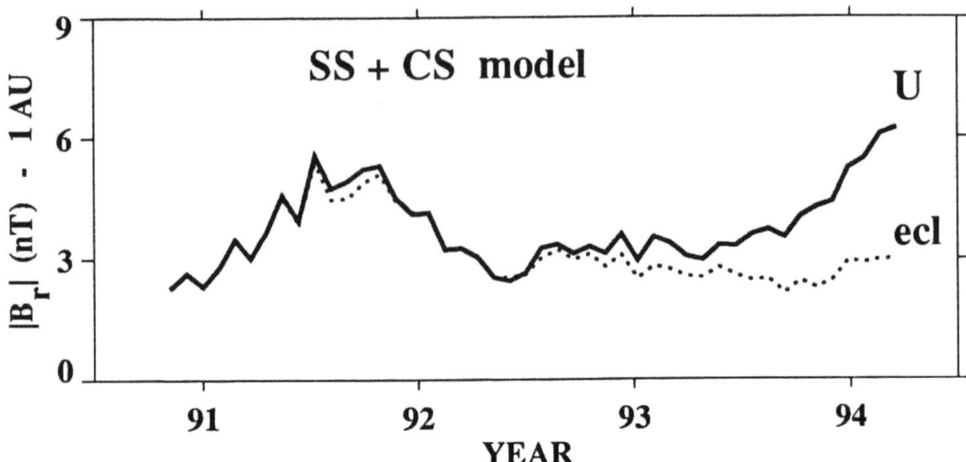

Fig. 3. Variation of $|B_r|$ along the latitudinal trajectory of Ulysses (transposed to 1 AU), as predicted by applying the SS + CS model to Wilcox Solar Observatory measurements.

post-maximum phase of the previous sunspot cycle (see Fig. 1).

Although the model includes the effect of higher-order multipoles, the basic properties of the IMF intensity variation can be understood in terms of the dominant solar dipole component (Wang and Sheeley, 1988; Wang, 1993). The combined effect of the heliospheric volume currents and of the current sheet at the dipole equator is to make the asymptotic field about 3 times stronger along the dipole axis than near the current sheet. The magnitude of B_r at any given latitude is then determined by the amplitude and relative orientation of the dipole moment. The dipole attains its greatest strength at sunspot minimum, when its axis is aligned with the solar poles, whereas it is weakest at sunspot maximum, when its axis lies near the ecliptic. As a result, the greatest solar-cycle variation in $|B_r|$ should occur above the poles (where the dipole is seen "head-on" when it is strongest and "edge-on" when it is weakest), and the least variation should occur in the ecliptic (where the dipole is seen "head-on" when it is weakest and "edge-on" when it is strongest). Ulysses will be visiting the solar poles near sunspot minimum, when the difference between the polar and ecliptic IMF intensities should be greatest.

References

King, J. H.: 1979, *J. Geophys. Res.* **84**, 5938
Pneuman, G. W., and Kopp, R. A.: 1971, *Solar Phys.* **18**, 258
Schatten, K. H.: 1971, *Cosmic Electrodyn.* **2**, 232
Schatten, K. H., Wilcox, J. M., and Ness, N. F.: 1969, *Solar Phys.* **6**, 442
Slavin, J. A., Jungman, G., and Smith, E. J.: 1986, *Geophys. Res. Lett.* **13**, 513
Wang, Y.-M.: 1993, *J. Geophys. Res.* **98**, 3529
Wang, Y.-M., and Sheeley, N. R., Jr.: 1988, *J. Geophys. Res.* **93**, 11227

ALFVENIC FLUCTUATIONS IN THE SOLAR WIND: A CASE STUDY USING ULYSSES MEASUREMENTS

PETE RILEY, C. P. SONETT*
Lunar and Planetary Laboratory University of Arizona, Tucson. U.S.A.
**Also at the Department of Planetary Sciences, University of Arizona. U.S.A.*

A. BALOGH, R. J. FORSYTH
The Blackett Laboratory, Imperial College, London. U.K.

E. E. SCIME, W. C. FELDMAN
Los Alamos National Laboratory, Los Alamos, New Mexico, U.S.A.

Abstract. We present preliminary results of a study of solar wind Alfvénic fluctuations using magnetometer and plasma measurements from the spacecraft Ulysses. We focus on a particular case study, and use 'wave language' to discuss the fluctuations. The observations are consistent with a mixture of convected structures and Alfvén wave trains. Furthermore, our results tend to suggest that the Alfvénic-fluctuations propagate in the radial direction.

1. Introduction

The exact nature of the structure and fluctuations of the Interplanetary Magnetic Field (IMF) is important for many areas of heliospheric research. For example, in numerical simulations of the modulation of galactic cosmic rays, the diffusion of these particles is very sensitive to the assumed fluctuations in the magnetic field on essentially all length-scales.

The study of fluctuations in the solar wind (SW) has fallen into two main areas: *Coleman* (1968) suggested the fluctuations were of a turbulent nature, while *Belcher and Davis* (1971) proposed that they were due to propagating Alfvén waves, particularly at the trailing edges of high speed streams. Recently, *Tu and Marsch* (1992) proposed a model of solar wind fluctuations aimed at reconciling these two viewpoints. Their idea supposes that the fluctuations are composed of two main elements: (*i*) Convective structures, which they associate with the quasi-2-dimensional turbulent component suggested by *Matthaeus et al.*(1990); and (*ii*) Alfvén waves propagating both inwardly and outwardly along magnetic field lines.

There is a plethora of phenomena that cause the interplanetary magnetic field to deviate from the pure Parker spiral. For the type of fluctuations we are investigating we can reduce these to three: (*i*) Alfvén waves; (*ii*) rotational discontinuities (RDs); and (*iii*) tangential discontinuities (TDs). Although we have made a distinction between Alfvén waves and RDs, in reality, the latter is simply a sharply crested Alfvén wave. The important differences then lie between TDs and Alfvénic fluctuations. And an important distinction between the two is that the former are non-propagating, i.e. they are static structures that are convected past the spacecraft by the IMF.

A fluctuation is said to be Alfvénic in nature if the following relationship between the velocity fluctuations ($\Delta\mathbf{v}$) and magnetic field fluctuations ($\Delta\mathbf{B}$) is satisfied:

Space Science Reviews **72**: 197–200.
© 1995 *Kluwer Academic Publishers.*

$$\Delta \mathbf{v} = \pm \frac{\Delta \mathbf{B}}{(\mu_o \rho)^{1/2}} \tag{1}$$

Equation (1) is satisfied by Alfvén waves and RDs. In general, there is no such requirement for tangential discontinuities. For a set of parallel tangential discontinuities we would expect the direction of minimum variance to be aligned with the normal to the discontinuities. Taking the model of Tu and Marsch as *a priori* correct, we would expect: (*i*) The direction of minimum variance to be perpendicular to <B>; and (*ii*) The normal component to be zero. For Alfvénic fluctuations, on the other hand, we would expect (*i*) A non-zero component of **B** along the direction of minimum variance; and (*ii*) the direction of minimum variance to be aligned with <B>.

In order to investigate the Tu and Marsch idea, we focus on a case study, and use minimum-variance analysis as a means of separating the two components.

2. The Case Study

The event we chose to present in this report occurred on the 8th January, 1991, centred at 08:30 UT with a half-window of 400 mins. At this time Ulysses was located at 1.67 AU from the Sun, at a heliospheric latitude of −1.2°.

In figure 1 we compare Δv and ΔB. Although there are some significant deviations; the overall conclusion is that equation (1) is at least qualitatively satisfied.

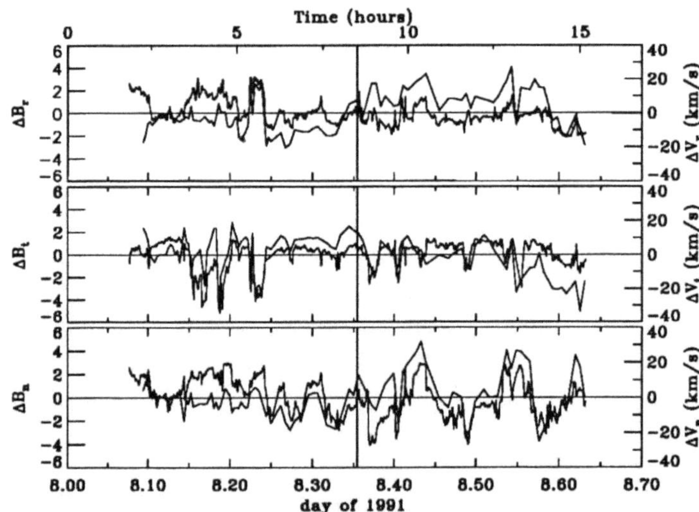

Figure 1. Velocity and magnetic field perturbations in rtn-coordinates for 8th January, 1991. In each component, the lowest resolution data corresponds to the velocity perturbations.

The ordered structure in the magnetic field is best viewed in minimum variance coordinates (e.g. *Sonnerup and Cahill*, 1967, *Lichtenstein and Sonett*, 1980). In figure 2 we present the magnetic field components in the minimum, intermediate, and maximum variance directions. We note the following points: (*i*) For almost the entire day, the magnetic field magnitude remained essentially constant; (*ii*) The magnetic field in a plane perpendicular to the direction of minimum variance remained approximately constant in magnitude although rotated through significant angles; (*iii*) Although the component along the direction of minimum variance varied significantly, for the entire period the perpendicular component was significantly larger; (*iv*) The direction of <B> is qualitatively consistent with the direction of a Parker spiral at this heliospheric position; and (*v*) The direction of minimum variance is approximately perpendicular to the direction of <B>.

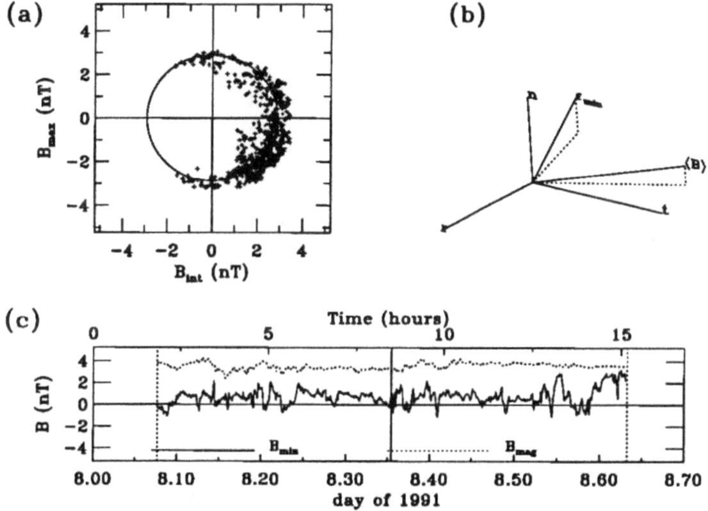

Figure 2. Magnetic field components in the minimum variance frame. (a) shows the relationship between the intermediate and maximum variance components; (b) shows the relationship between the minimum variance directions and <B> with the rtn-coordinate system; and (c) shows the absolute magnitude of the magnetic field as well as the field strength in the minimum variance direction.

Figure 3 is a 3-dimensional plot of the data contained in figure 2 (a), with time plotted as the z-coordinate. It is readily apparent that the field rotations are not smooth, and in some cases field rotations of >90° can occur on a timescale of the resolution of the magnetometer data (1 min).

The data shown in figure 1 were analysed on shorter timescales. Specifically, we reduced the half-window to 60 minutes and moved the minimum variance window along in steps of 15 minutes. For almost every window there was a well-defined field rotation (as defined by a hodograph of B_{int} versus B_{max}). Furthermore, the results suggested two distinct classes of field rotation:

(1) Events where B_{min} is essentially zero, and the angles between ε_{min} and <B> are clustered around 90°.

(2) Events where B_{min} is significantly different from zero (i.e. of the same order as B_{max}). Furthermore, it appears that ε_{min} is loosely clustered along (or anti-parallel to) the radial direction.

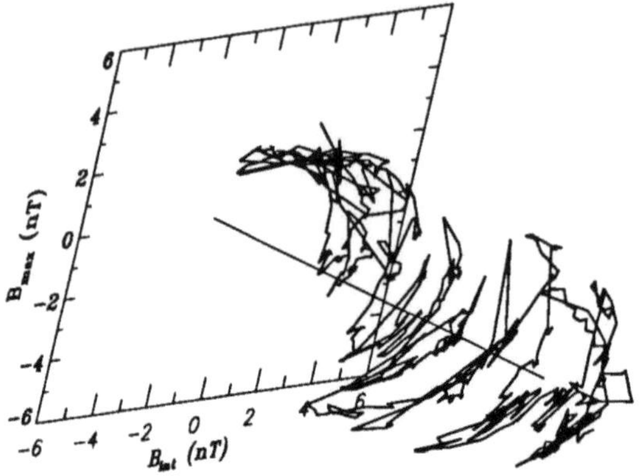

Figure 3. 3-D plot of the intermediate- and maximum-variance magnetic field components as a function of time. The time-axis labeling is removed for clarity, but extends for the same period as figure 2(c).

3. Conclusions

The above results tend to support the Tu and Marsch model. In fact it is difficult to imagine a different configuration of the magnetic field that could reproduce all the observations. The case study we have presented appears to be a set of tangential discontinuities, with propagating structures embedded within them. However, these propagating structures seem to align themselves with the radial direction, and not along the magnetic field. Such a conclusion is at best tentative, however, and requires further study.

References

Belcher, J. W. and Davis, L.: 1971, *J. Geophys. Res.*, **76**, 3534

Coleman, P. J. Jr.: 1968, *Astrophys. J.*, **153**, 371.

Lichtenstein, B. R., and Sonett, C. P.: 1980, *Geophys. Res.Lett.*, **7**, 189

Matthaeus, W. H., Goldstein, M. L., and Roberts, D. A.: 1990, *J. Geophys. Res.*, **95**, 20673

Sonnerup, B. U. Ö., and Cahill, L. J.: 1967, *J. Geophys. Res.*, **72**, 171.

Tu, C. Y. and Marsch, E.: 1992, in E. Marsch and R. Schwenn (eds.), *Solar Wind Seven*, Cospar Colloquia Series V3, Pergamon Press, pp. 549-554.

Acknowledgments

This research was supported by JPL contract 959377 with the Jet Propulsion Laboratory.

MAGNETIC HOLES IN THE SOLAR WIND AND THEIR RELATION TO MIRROR-MODE STRUCTURES

D. WINTERHALTER[1], M. NEUGEBAUER[1], B. E. GOLDSTEIN[1], E. J. SMITH[1], B. T. TSURUTANI[1], S. J. BAME[2], AND A. BALOGH[3]

[1] *Jet Propulsion Laboratory, California Institute of Technology, Pasadena, CA 91109*
[2] *Los Alamos National Laboratory, Los Alamos, NM 87545*
[3] *Blackett Laboratory, Imperial College, London SW7 2BZ, UK*

Abstract. Data obtained by the Ulysses magnetometer and solar wind analyzer have been combined to study the properties of magnetic holes in the solar wind between 1 and 5.4 AU and to 23° south latitude. Although the plasma surrounding the holes was generally stable against the mirror instability, there are indications that the holes may have been remnants of mirror mode structures created upstream of the points of observation. Those indications include: (1) For the few holes for which proton or alpha-particle pressure could be measured inside the hole, the ion thermal pressure was always greater than in the plasma adjacent to the holes. (2) The plasma surrounding many of the holes was marginally stable for the mirror mode, while the plasma environment of all the holes was significantly closer to mirror instability than was the average solar wind. (3) The plasma containing trains of closely spaced holes was closer to mirror instability than was the plasma containing isolated holes. (4) The near-hole plasma had much higher ion β (ratio of thermal to magnetic pressure) than did the average solar wind.

1. Introduction

Localized depressions in the magnitude of the interplanetary magnetic field (IMF) were first observed in 1971 by Explorer 43 (Turner *et al.*; 1977). These "magnetic holes" were distinct entities in otherwise average IMF conditions; i.e., they were not random depressions in a region of noisy or weak fields. Similar features observed in the Earth's magnetosphere, in the magnetosheaths of Earth, Jupiter and Saturn, and in the cometosheath of comet Halley were thought to be caused by the mirror instability. Against this background of magnetosheath observations, Tsurutani *et al.* (1992) interpreted a series of magnetic dips (which they did not call "magnetic holes") observed by Ulysses in the plasma behind an interplanetary shock as being caused by the mirror instability. The ion β during that period was high (~4), but the plasma anisotropy was not examined.

In this paper we pursue the question of the possible relation of magnetic holes in the solar wind to the mirror instability. The methodology was: (1) to search 26 months of solar-wind data acquired by the magnetometer on Ulysses for magnetic holes, (2) to compile statistics concerning their nature and occurrence rate as a function of distance

Space Science Reviews **72**: 201–204.
© 1995 *Kluwer Academic Publishers.*

from the Sun and relative to the solar wind stream structure, and (3) to select a subset of linear holes that were reasonably simple and isolated from other field variations, calculate the mirror instability criterion for each event in the subset, and compare the results to the average wind observed by Ulysses. An expanded account of this study is given in Winterhalter *et al.* (1994).

The data were obtained by the Ulysses magnetometer experiment (Balogh *et al.*, 1992) and by the Ulysses solar wind plasma experiment Solar Wind Observations Over the Poles of the Sun (SWOOPS) (Bame *et al.*, 1992).

2. Observations and Analysis

Examples of two linear magnetic holes are given in Figure 1. In order to gather statistical information about the magnetic holes, Ulysses magnetic field data (one second averages) covering the period from launch to the end of 1992 were electronically scanned. A hole was defined to be a dip in the field strength such that $B_{min}/B_0 \leq 0.5$ where B_{min} and B_0 are the minimum and average field magnitudes within a sliding window 300 seconds in length. The program also determined the width δt of each hole, in seconds, as well as the field rotation $\delta\theta$ across the hole.

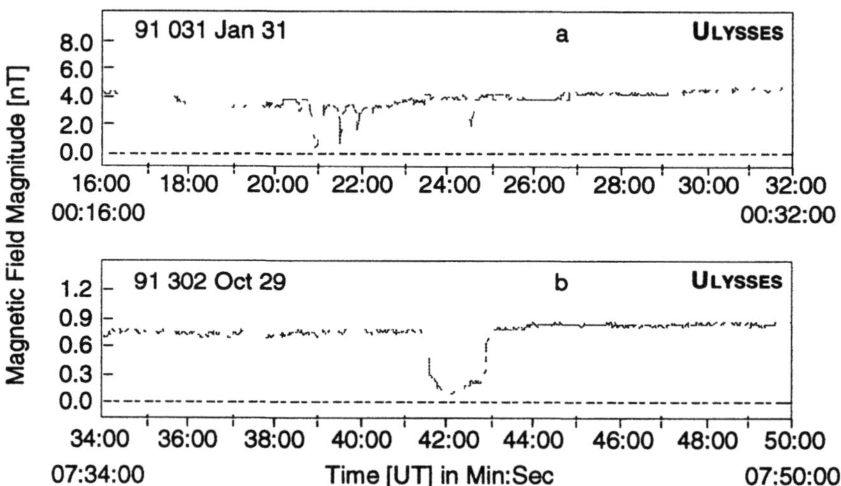

Figure 1. Field magnitude variations in two selected magnetic holes. The events have in common that $B_{min}/B_0 < 0.5$, and $\delta\theta < 5°$ ("linear holes").

The search program found 4127 magnetic holes during the 26 month period Oct. 1990 to Dec. 1992, including 428 events downstream of the Jovian bow shock. Magnetic holes were found over the full ranges of heliocentric distance and latitude sampled by Ulysses through 1992. There is no obvious indication that the occurrence of holes was related to heliocentric distance. Typically, B_{min} was about 0.01 nT, but many holes had minimum field strengths that were larger. There were 41 holes with $B_{min} \leq 0.005$ nT. The "width" of the holes was between 5 seconds and 25 seconds, with a most probable value lying in the 10-15 seconds bin and a median value of 22 s. These numbers compare well with those found by Turner *et al.* (1977).

If the magnetic holes are mirror-mode structures, the dip in magnetic pressure must be compensated by an increase in the perpendicular plasma pressure. However, the plasma instrument can only obtain a spectra every 4 minutes, so it was usually not possible to measure the ion density and temperature within a hole. There were a few lucky exceptions when either the proton or the alpha-particle distribution (seldom both) were measured within a magnetic hole. The data shows that on average, $nT/n_0T_0 = 1.38.(\pm 0.17)$, where the subscript 0 denotes the quantities measured outside the hole. It is not possible, however, to carry out a detailed calculation of pressure balance because we do not have simultaneous observations of protons, alphas, and electrons inside the hole.

The stability criterion for mirror mode waves can be stated as the ratio

$$R = \frac{\beta_\perp/\beta_{||}}{1 + 1/\beta_\perp}$$

where instability corresponds to $R > 1$. This criterion is valid for the kinetic treatment of the instability (Winterhalter *et al.*, 1994). Figure 2a shows the distribution of R computed from hourly averages of Ulysses field and plasma data between Dec. 18, 1990 (when the spacecraft nutation stopped) and the end of 1992.

Figure 2b shows the values of R computed from the plasma spectra in the undisturbed plasma adjacent to each of the magnetic holes in our smaller data set. The values of β, $T_{||}$, and $T_\perp$ used to calculate R correspond to total ion values, including protons and alphas, but not electrons. From Figure 2a it is seen that the solar wind is almost always very stable against the mirror instability, with a median value of $R = 0.32$. R exceeded 1.0 for only 4% of the hours in the interval sampled. For the near-hole observations, however, R was usually between 0.5 and 1.0 (Figure 2b), with a cut-off near R=1. These features are consistent with a state of marginal stability. On the other hand, there are a number of cases which are closer to R=0.5 than they are to R=1, and it is difficult to argue that the plasma conditions are only marginally stable for those cases. We have found (Winterhalter *et al.*, 1994) that it is primarily single isolated holes (like in Figure 1b) that appear to have evolved away from the R=1 line, as opposed to multiple holes (like in Figure 1a) which cluster more closely around R=1. Our interpretation is that Ulysses observed structures generated by mirror mode instabilities which remained after the distribution relaxed to a marginally stable state. The single holes may be older remnants of wave trains whose other members have decayed away, and whose distribution nearby is much more isotropized.

3. Acknowledgments

This research was supported by the Jet Propulsion Laboratory, California Institute of Technology, under contract with the National Aeronautics and Space Administration. Work at Los Alamos was performed under the auspices of the U.S. Department of Energy.

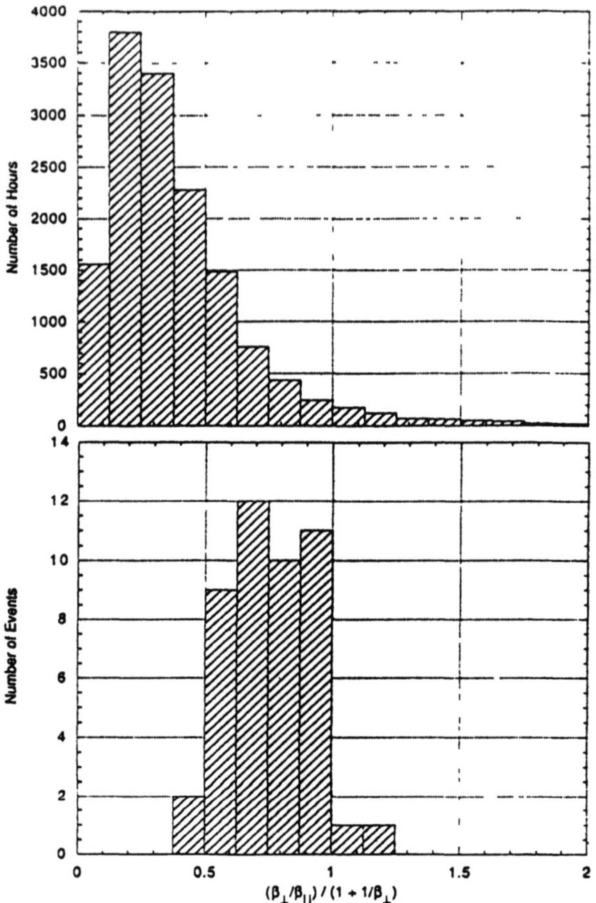

Figure 2. Distribution of the stability criterion for mirror mode waves R = $(\beta_\perp/\beta_{||})/(1 + 1/\beta_\perp)$, which is > 1 for instability to occur, for (a) hourly averages of all the solar wind data acquired between Dec. 18, 1990 and Dec. 31, 1992, and (b) in the plasma adjacent to the subset of magnetic holes selected for detailed study.

References

Balogh, A., T. J. Beek, R. J. Forsyth, P. C. Hedgecock, R. J. Marquedant, E. J. Smith, D. J. Southwood and B. T. Tsurutani: 1992, *Astron. Astrophys. Suppl.*, **92**, 221.

Bame, S. J., D. J. McComas, B. L. Barraclough, J. L. Phillips, K. J. Sofaly, J. C. Chavez, B. E. Goldstein and R. K. Sakurai: 1992, *Astronomy and Astrophysics Supplement*, **92**, 237.

Tsurutani, B. T., E. J. Smith, D. J. Southwood and A. Balogh: 1992, *Geophy. Res. Lett.*, **19**, 1267.

Turner, J. M., L. F. Burlaga, N. F. Ness and J. F. Lemaire: 1977, *J. Geophys. Res.*, **82**, 1921.

Winterhalter, D., M. Neugebauer, B. E. Goldstein, E. J. Smith, S. J. Bame and A. Balogh: 1994, *J. Geophys. Res., (submitted).*

INTERPLANETARY DISCONTINUITIES AND ALFVÉN WAVES

BRUCE T. TSURUTANI, EDWARD J. SMITH, CHRISTIAN M. HO, MARCIA NEUGEBAUER, BRUCE E. GOLDSTEIN AND JOHN S. MOK
Jet Propulsion Laboratory, California Institue of Technology, Pasadena, CA 91109-8099, U.S.A.

ANDRE BALOGH AND DAVID SOUTHWOOD
Imperial College of Science and Technology
The Blackett Laboratory, Prince Consort Road, London SW7 2BZ, *England*

and

WILLIAM C. FELDMAN
Los Alamos National Laboratory, Los Alamos, New Mexico 87545, U.S.A.

(Accepted 3 May , 1994)

Abstract. The rate of occurrence of interplanetary discontinuities (ROID) is examined using Ulysses magnetic field and plasma data from 1 to 5 AU radial distance from the Sun and at high heliospheric latitudes. It is found that there are two regions in interplanetary space where the ROID is high: in stream-stream interaction regions and in Alfvén wave trains. This latter feature is particularly obvious at high heliographic latitudes when Ulysses enters a high speed stream associated with a polar coronal hole. These streams are characterized by the presence of continuous, large-amplitude ($\Delta\vec{B}/|B| \sim 1-2,$) Alfvén waves and an extraordinarily high ROID value (~ 150 discontinuities/day). In a number of intervals examined, it is found that (rotational) discontinuities are an integral part of the Alfvén wave: they represent ~ 90° phase rotation of the wave out of the full 360° rotation of the wave. These large amplitude nonlinear Alfvén waves thus appear to be phase steepened. The nonlinear Alfvén waves are spherically polarized, i.e., the tip of the perturbation vector resides on the surface of a sphere (a consequence of constant |B|). The best description of this wave plus discontinuity is a "spherical arc polarization".

1. Introduction

The rate of occurrence of interplanetary discontinuities (ROID) is highly variable (Burlaga, 1971). Variations in the number of discontinuities per day can increase or decrease by factors of 10 or more from one day to the next. We hope to explain some of this variablility in this paper.

Both the temporal and radial distance variations of discontinuity occurrence rates were previously studied using a two-spacecraft (Pioneers 10 and 11) technique (Tsurutani and Smith, 1979, hereafter referred to as TS). The two (radially-aligned) spacecraft measurements were used to remove the temporal variations. It was found that there was an apparent decrease in the ROID with radial distance from the sun, a decrease that was interpreted as a simple thickening of the discontinuities with decreasing ambient magnetic field strength. Because TS (1979) used identification criteria that depended on

Space Science Reviews **72**: 205–210.
© 1995 *Kluwer Academic Publishers.*

the thickness of discontinuities, at larger distances where the field was weaker, some of the thicker discontinuities fell outside of detection. An empirical relationship to normalize the rate of occurrence to 1 AU was derived, $e^{-(r-1)/4}$, where r is in units of AU from the sun.

The purpose of this note is to examine the ROID as a function of solar wind stream structures, as Ulysses first travels from 1 to 5 AU and then as the spacecraft travels to high heliospheric latitudes. A second objective of the note is to examine and discuss the detailed relationship between discontinuities and Alfvén waves, particularly those occurring in the trailing portions of high-speed streams.

2. Discontinuity Selection Criteria

Discontinuities are identified by sharp changes in the direction and/or magnitude of the interplanetary magnetic field. The criteria for a "sharp" change and how much of a

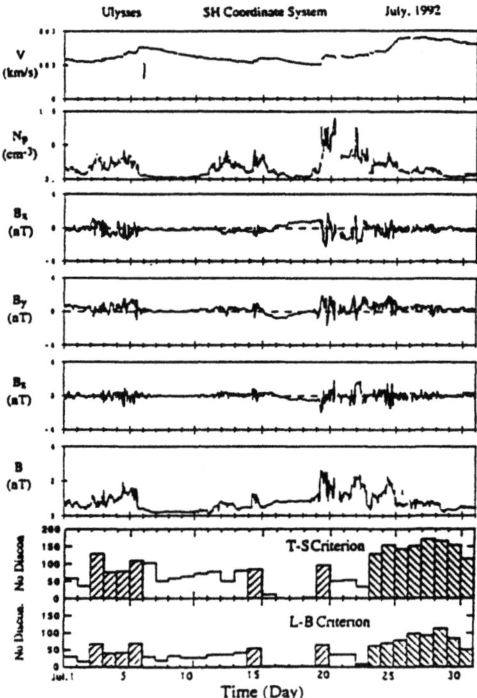

Fig. 1. The relationship between the interplanetary stream structures and the rate of occurrence of discontinuities at 5.2 AU. Discontinuities occur most frequently where large amplitude transverse fluctuations (Alfvén waves) are present (see B_Z panel). These waves are present primarily in two regions: stream-stream interaction regions (possibly a mixture of freshly created waves/turbulence and fossil waves) and in the trailing portions of high speed streams (fossil waves).

change, is, of course, arbitrary. TS (1979) have used $|\Delta\vec{B}|/|B| > 0.5$ and $|\Delta\vec{B}| \geq 2\delta$ where the value δ is the field variance on either side of the discontinuity. $\Delta\vec{B}$ is the change in one minute average vectors separated by 3 minutes. Lepping and Behannon (1986, hereafter referred to as LB) have used a criteria that the field change direction by at least 30° within one minute. The expression for the angular change is $\theta = \cos^{-1}(\vec{B}_1 \cdot \vec{B}_2)/|B_1| \cdot |B_2| \geq 30°$, where $\vec{B}_1$ and $\vec{B}_2$ are the upstream and downstream magnetic field vectors, respectively. There are slight differences in the above two discontinuity selection criteria, and there are thus some differences in the rates of occurrence. The TS criteria is less stringent and therefore it gives a greater rate of occurrence. Values deduced from both criteria will be shown. Both will be normalized to 1 AU by the $e^{-(r-1)/4}$ factor.

Figure 1 illustrates the day-to-day ROID variation for this month-long interval, in July of 1992. The top two panels are the hourly averages of the solar wind velocity and the density. Below are the 3 components and magnitude of the interplanetary magnetic field in Solar Heliospheric (SH) coordinates. In this system, $\hat{X}$ is directed radially outward from the sun, $\hat{Y}$ is $\hat{\Omega} \times \hat{X} / |\hat{\Omega} \times \hat{X}|$, where $\hat{\Omega}$ is the solar rotation axis, and $\hat{Z}$ completes the right-hand system. The bottom two panels are the number of discontinuities per day using the TS and the LB criteria. Data gaps were taken into account in determining these normalized values.

There are several noteworthy features found in the Figure. First, high ROID values occur at stream-stream interaction regions. Examples can be noted in July 3-6 and 20, and to a lesser extent in July 15. However, the highest ROID values for this interval occur from July 24-31, in the trailing portion of a high speed stream. This region is characterized by large directional changes in the field with time scales of 10 min. to days. These fluctuations have been examined and are found to be Alfvén waves (discussed later). The ROID value reaches ~ 150 discontinuities/day (normalized) and remains at this value even as the amplitude of the Alfvén waves decreases.

From 16-19 July, the ROID value is exceptionally low, essentially zero/day. This interval contains a magnetic cloud (Klein and Burlaga, 1982), a region where there are few Alfvén waves or discontinuities (Zwickl et al., 1983; Tsurutani et al., 1988, Neugebauer and Alexander, 1991; Tsurutani and Gonzalez, 1993). This region has been identified by its magnetic field properties.

Figure 2 has the same format as Figure 1, but at -37° latitude in September of 1993. Here Ulysses is below the heliospheric current sheet (Smith et al., 1993), and is in a continuous high speed stream associated with a polar coronal hole (Phillips et al., 1994). The velocity of ~ 800 km/s is relatively constant. The region is characterized by continuous, large amplitude, $|\Delta\vec{B}/|B| \sim 1-2$, transverse waves. Comparison of $\Delta\vec{B}$ to $\Delta\vec{V}\sqrt{4\pi\rho}$ indicates that these are Alfvén waves propagating radially outward from the sun (not shown to conserve space). The normalized discontinuity occurrence rate is almost continuously high, ~ 100-200 for the TS criteria and ~ 75-150 for the LB criteria. These values are 4-5 times the ROID value typically detected in the ecliptic plane. It is clear that there is a latitude dependence which presumably is due to the fraction of the time that Ulysses spends within coronal hole high speed streams.

To determine the relationship between Alfvén waves and discontinuities, we examine the interval 0519 to 0614 UT, day 17, 1992. Ulysses is at a radial distances of 5.2 AU and at a latitude of 1°. By examining the cross-correlation between V_x and B_x and V_y and B_y, we find that this and neighboring intervals consist of Alfvén waves propagating away from the sun (~ 0.6 correlation coefficient at zero lag).

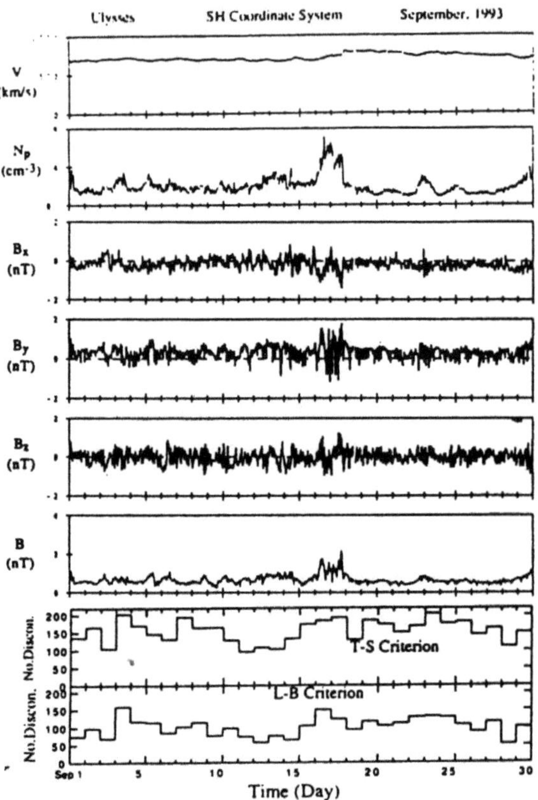

Fig. 2. Same as for the above Figure, but at high (-37°) heliographic latitudes (4.4 AU). Note that large amplitude $|\Delta\vec{B}/|B|| \sim 1-2$, waves are almost continuously present. The (normalized) rate of discontinuity occurrence is approximately 4-5 times higher than in the ecliptic plane.

We have divided the above interval into two parts: from 0519 to 0610 UT, where the field changes slowly due to the wave and from 0610 to 0614 UT where there is a discontinuity. The entire interval is also examined as well.

The B_1 - B_2 hodograms for the interval 0519 - 0614 , day 17, 1992, and the two sub-intervals are shown in Figure 3. B_1 corresponds to the field component in the direction of maximum variance, B_2 the field in the direction of intermediate variance and B_3 the minimum variance direction (Sonnerup and Cahill, 1967; Smith and Tsurutani, 1976). The hodogram for the slow rotation is shown on the left, the discontinuity in the center and the whole interval on the right. The field magnitude is relatively constant throughout the interval (not shown). The slowly rotating part of the Alfvén wave is therefore a spherical wave (the perturbation vector rotates on the surface of a sphere, a consequence of a constant B magnitude). The hodogram indicates that this portion of the wave has an arc-like profile. It consists of a phase rotation of ~ 270°. The perturbation vector rotates across the arc and the partially back. The discontinuity (found to be rotational) also has an arc-like polarization (because the rotation is much faster, higher time resolution data are used). The ~ 90° phase rotation of the discontinuity completes the 360° phase rotation. The (total) perturbation vector sequence does not complete a symmetric ellipse, but is more like the tip of a windshield wiper. It rotates across and back. This can be seen in the hodogram of the whole interval.

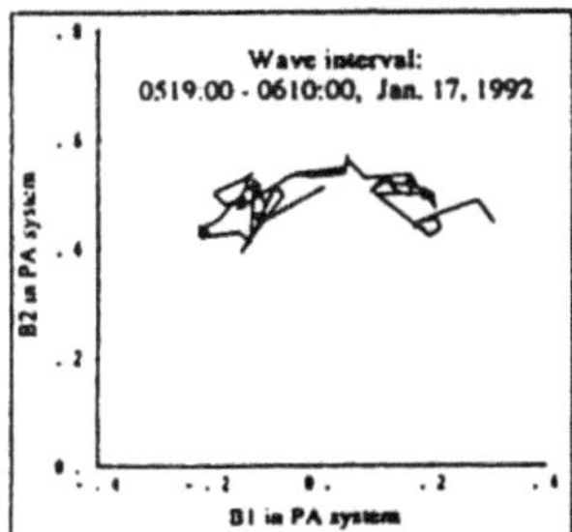

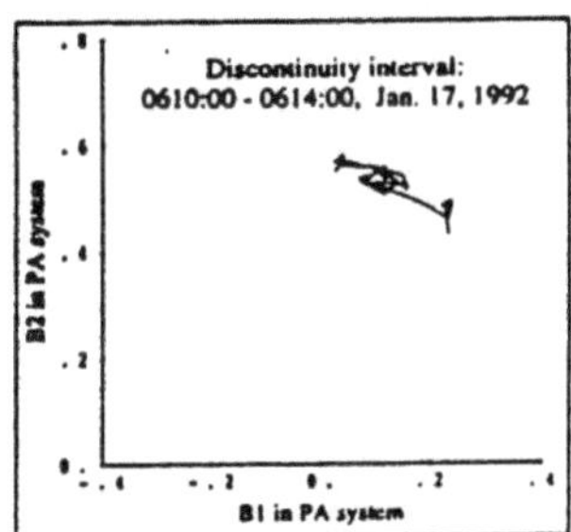

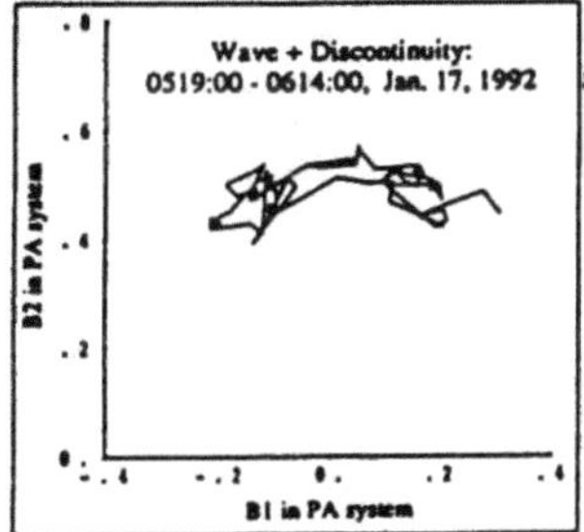

Fig. 3. The B_1 - B_2 hodogram for the Alfvén wave (without the trailing discontinuity) at 0519 - 0610 UT January 17, 1992 (left panel) and for the trailing discontinuity (middle panel). The wave portion consists of a ~ 270° phase rotation, linearly-polarized (on a spherical surface). The discontinuity portion has a ~ 90° phase rotation. The Alfvén wave plus discontinuity comprise a 360° wave (right panel). The RD is the steepened part of the nonlinear Alfvén wave.

We have examined the relationship between discontinuities and Alfvén waves for other intervals. We find that discontinuities occur at the edges of Alfvén waves between ~ 35% and ~ 65% of the time, depending on the level of interplanetary activity. At this time a couple dozen events have been examined in detail using principal axis techniques. We find that the above Alfvén wave/discontinuity arc polarizations are typical.

4. Conclusions and Discussion

We have shown the ROID value is highest where there are Alfvén waves present. This is most obvious at high heliographic latitudes where high speed streams emanate from coronal holes. These streams are dominated by Alfvén waves. We verify that there is, in general, a lack of correlation between ROID values and V_{SW} and $|B|$. This is because there are interplanetary intervals with high V_{SW} and $|B|$, which are devoid of both Alfvén waves and discontinuities (magnetic clouds) and also intervals of moderate V_{SW} and low $|B|$ where discontinuities are abundant (trailing portions of high speed streams). There are, of course, also intervals where the ROID, V_{SW} and $|B|$ values are all high, e.g., stream-stream interaction regions and the beginning parts of high speed streams. Taking all four regions together, one can now understand why a general correlation between the ROIDs, V_{SW} and $|B|$ does not exist.

We have examined a number of cases of the detailed relationship between Alfvén waves and discontinuities. In this study we have limited our examination to Alfvén wave train intervals only. In Figure 3, we have shown one such example. In this particular case, we find that the discontinuity is actually part of the Alfvén wave, comprising approximately ~ 90° or less of the phase rotation of the total wave (360°). Thus, this large amplitude, $|\Delta\vec{B}/|B| \sim 1-2$, nonlinear Alfvén wave appears to be phase-steepened, with the rotational discontinuity being the steepened edge (at this time we do not know whether the discontinuities occur at the leading or trailing edges or both). From our initial examination of the data, we cannot tell if this is a stable configuration or not. Computer simulation results and theoretical arguments are somewhat mixed on the topic as well. This will remain as a topic of future research.

Acknowledgements

Portions of this research effort was done at the Jet Propulsion Laboratory, California Institute of Technology under contract with the National Aeronautical and Space Administration.

References

Burlaga, L. F.: 1971, *J. Geophys. Res.*, **76**, 4360.

Klein, L. W. and L. F. Burlaga: 1982, *J. Geophys. Res.*, **87**, 613.

Lepping, R. P. and K. W. Behannon: 1986, *J. Geophys. Res.*, **91**, 8725.

Neugebauer, M. and C. J. Alexander: 1991, J. *Geophys. Res.*, **96**, 9409.

Phillips, J. L., A. Balogh, S. J. Bame, et al.: 1994, submitted to *Geophys. Res. Lett.*.

Smith, E. J. and B. T. Tsurutani: 1976, *J. Geophys. Res.*, **81**, 2261.

Smith, E. J., M. Neugebauer, A. Balogh, S. J. Bame, G. Erdos, R. J. Forsythe, B. E. Goldstein, J. L. Phillips and B. T. Tsurutani: 1993, *Geophys. Res. Lett.*, **20**, 2327.

Sonnerup, B. U. and L. J. Cahill, Jr.: 1967, *J. Geophys. Res.*, **72**, 171.

Tsurutani, B. T. and W. D. Gonzalez: 1993, *Phys. Fluids B*, **5**, 2623.

Tsurutani, B. T., W. D. Gonzalez, F. Tang, S.-I. Akasofu, and E. J. Smith: 1988, *J. Geophys. Res.*, **93**, 8519.

Tsurutani, B. T. and E. J. Smith: 1979, *J. Geophys. Res.*, **84**, 2773.

Zwickl, R. D., J. R. Asbridge, S. J. Bame, W. C. Feldman, J. T. Gosling, and E. J. Smith: 1983, *Solar Wind Five*, NASA Conf. Publ., CP-2280, **711**.

INTERPLANETARY SCINTILLATION OBSERVATIONS OF THE HIGH- LATITUDE SOLAR WIND

WM. A. COLES
E.C.E. Dept., University of California, La Jolla, CA 92093-0407
phone 619-534-2703; fax 619-534-2486; email bcoles@ucsd.edu

27 April 1994

Abstract.
Until the ULYSSES spacecraft reached the polar regions of the solar wind, the only high-latitude measurements available were from indirect techniques. The most productive observations in regions of the solar wind between 5 $R_{\odot}$ and 200 $R_{\odot}$ have been the family of radio scattering techniques loosely referred to as Interplanetary Scintillation (IPS) (Coles, 1978). Useful observations can be obtained using a variety of radio sources, for example spacecraft beacons, planetary radar echoes and compact cosmic sources (quasars, active galactic nuclei, pulsars, galactic masers, etc.). However for measurement of the high-latitude solar wind cosmic sources provide the widest coverage and this review will be confined to such observations. IPS observations played a very important role in establishing that polar coronal holes (first observed in soft x-ray emission) were sources of fast solar wind streams which occasionally extend down to the equatorial region and are observed by spacecraft. Here I will review the IPS technique and show the variation of both the velocity and the turbulence level with latitude over the last solar cycle. I will also outline recent work and discuss comparisons that we hope to make between IPS and ULYSSES observations.

Key words: solar wind – plasma – interplanetary scintillation

1. Introduction

Interplanetary Scintillation (IPS) refers to fluctuations in the apparent intensity of a distant radio source which occur when the source is observed near the sun (Hewish, Scott and Wills, 1964). The phenomenon is caused by small-scale density fluctuations in the solar wind which phase-modulate the wave from the distant source as it passes through the scattering region where the line of sight is closest to the sun. This phase modulation is converted to intensity modulation by diffraction as the wave propagates from the scattering region to the observer. As this diffraction pattern blows past the observer at the solar wind velocity, it appears as a temporal variation of intensity.

The temporal spectrum of intensity is directly related to the spatial spectrum of electron density (scaled by the velocity). IPS observations can be inverted to provide estimates of the electron density spectrum in the scale range to which IPS is sensitive, which is from about 3 km to 300 km. If the IPS pattern is observed with several suitably spaced antennas one can determine the time delay from one antenna to another and thus estimate the average flow velocity in the scattering region.

Space Science Reviews **72**: 211–222.
© 1995 *Kluwer Academic Publishers.*

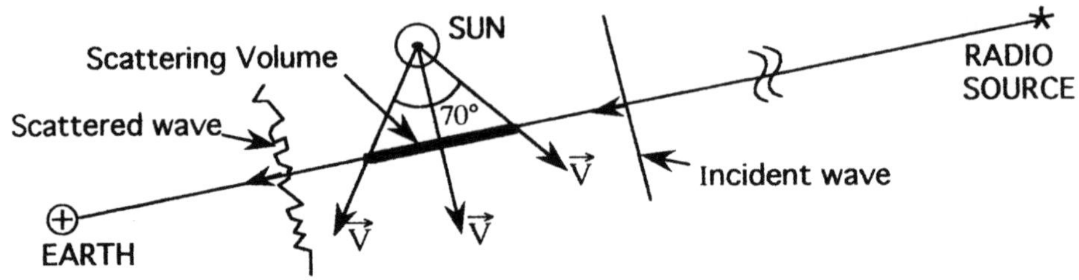

Fig. 1. The geometry of a radio scattering observation.

A sketch of the geometry of an IPS observation is shown in Fig. 1. The scattering is proportional to the spatial spectrum of electron density, which goes approximately as R^{-4}. Thus the scattering volume, defined as the region in which the scattering is at least half of the peak value, subtends about 70° at the sun. This scattering volume is long but very thin, since typical scattering angles are μrad. When the scattering volume is mapped back to the solar surface it is a fragment of a great circle as shown in Fig. 2.

The refractive index n of the solar wind plasma deviates from unity by a fraction δn which is proportional to the electron density variation δN_e and the square of the radio wavelength λ. When the scintillations are weak, i.e. $\delta N_e \rightarrow 0$, and $\delta I/I < 1$, they increase with λ and with decreasing solar distance as sketched in Fig. 3.

However the IPS saturates as $\delta I/I \rightarrow 1$ and then decreases rapidly with further increase in δN_e. Since radio telescopes are sensitivity limited for IPS observations, the range of solar distance over which one can obtain useful information is strongly wavelength dependent.

The solar wind velocity was first modeled with a magneto-hydrodynamic theory by Parker (1965). Spacecraft observations confirmed Parker's prediction that the wind would be supersonic, but found that the speed was somewhat higher than the maximum that could be expected using Parker's theory (Neugebauer and Snyder, 1966). Subsequent MHD theoretical models have shown that some additional momentum must be added to the solar wind after the supersonic point in order to match observations at 1 AU. Such models are essentially independent of distance outside of about 50 $R_\odot$. One must also bear in mind that the models are spherically symmetric and the solar wind is often far from spherically symmetric.

Clearly, to study the latitude-longitude structure of the solar wind speed one must be careful to isolate the effects of acceleration. This suggests that one should observe outside of 50 $R_\odot$ where the speed is independent of

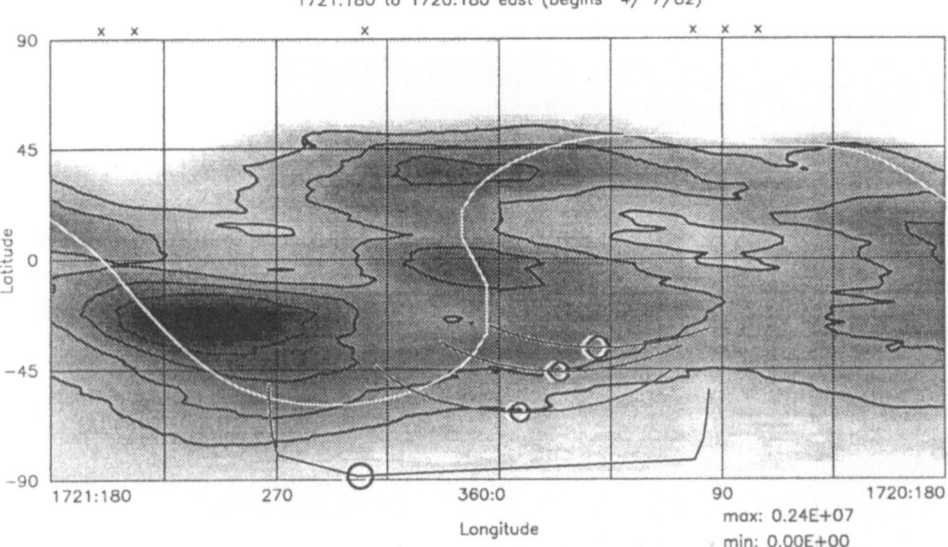

Fig. 2. Four examples of the IPS scattering volume projected back to the source surface. The observations are CTA21 on four successive days in 1982, ending on the day of closest approach. The center of the scattering volume is marked with a circle whose area is proportional to the strength of turbulence. A line exactly at the equator would be straight and extend over 70° longitude. A line exactly at the pole would consist of two sections extending over 35° of latitude, separated by 180° of longitude. The IPS data are overlaid on a plot of the electron density derived from white light corona measurements at 1.3 $R_\odot$ from Mauna Loa. The white line is the magnetic neutral line as estimated from the Wilcox Solar Observatory data (Hoeksema et al., 1983; Hoeksema and Scherrer, 1985).

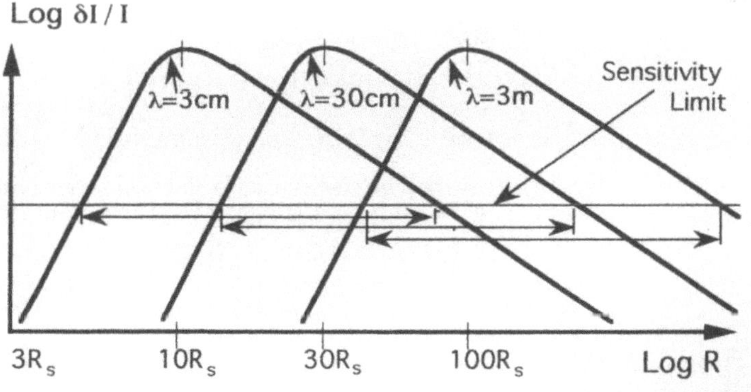

Fig. 3. The normalized rms intensity fluctuation as a function of distance and radio wavelength. The peak occurs for $\delta I/I$ less than unity, but typically close to unity depending on the structure of the radio source and the receiver bandwidth. In practice the sensitivity limit will also be wavelength dependent.

distance. Such IPS observations must be made at wavelengths of 1 m or longer. On the other hand, to study the distance profile of speed V(R) one must be careful to isolate the effects of latitude-longitude structure. Of course one must also observe where there is substantial acceleration i.e. inside of 50 $R_{\odot}$. Such IPS observations must be made at wavelengths of 30 cm or less and selected such that other spatial and temporal variations are negligible. Clearly the polar coronal hole which forms at solar minimum is a good place to test acceleration models because the flow is locally spherically symmetric and transients are infrequent.

To study the latitude-longitude structure of the solar wind density fluctuations one must compensate for the variation with distance which is very strong. It is also necessary to separate the turbulence which is convected out from the sun, from that which is locally generated. Observations of the distance variation have been summarized by Manoharan (1993). All reported observations are consistent with $\delta N_e^2 \propto R^{-4.1\pm0.1}$ for $20 < R < 100\ R_{\odot}$. There has been an unconfirmed report of anomalous variation for $10 < R < 20\ R_{\odot}$ (Lotova et al., 1985), but other observations show that the power law range extends inwards at least to 5 $R_{\odot}$ (Bourgois and Coles, 1992). At distances greater than about 100 $R_{\odot}$ there is substantial locally generated turbulence in stream interaction regions where a fast stream follows a slow stream (Ananthakrishnan et al., 1980; Marsch and Tu, 1990). However the power law range may also be extended out to 200 $R_{\odot}$ if corotating enhancements are excluded (Armstrong and Coles, 1978). These measurements are made in the equatorial regions since one must follow the same radio source for months. It is not known if the same power law applies in the polar regions and it will be interesting to see what light ULYSSES can throw on this question.

2. Latitude Structure of Velocity

The greatest contribution of IPS to our knowledge of the solar wind is undoubtedly in the understanding of the solar cycle variation of the velocity at high latitudes. Between the 74 MHz observations from the University of California at San Diego and the 327 MHz observations from the Solar Terrestrial Environment Laboratory of Nagoya University, we have continuous observations over two solar cycles from 1972 to the present. These observatories are able to follow 10 or 20 sources per day, of which a few will be at high latitudes. Because high latitude sources are sparse it takes 2 to 6 months to assemble a reasonably complete velocity map with a resolution of about 15° in heliographic coordinates. Such maps can only display large scale quasi-static structures but they have been very useful.

The regular IPS measurements from UCSD and STELab are processed to yield the average velocity across the line of sight. If the scattering region

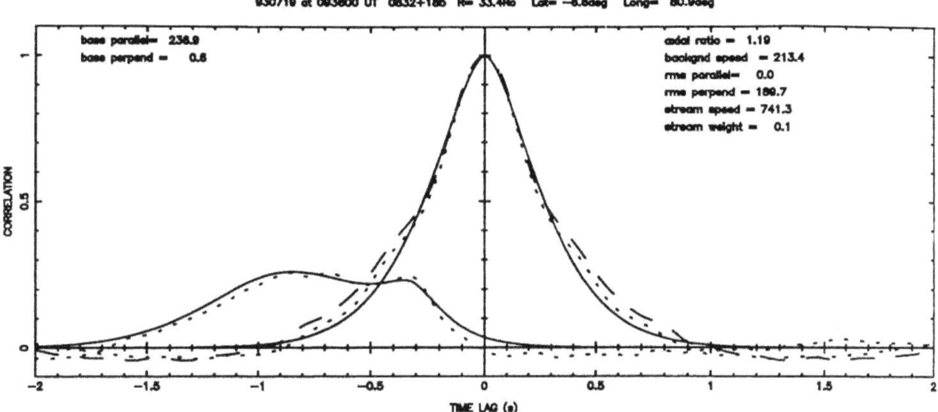

Fig. 4. The cross correlation of intensity for a case in which there are two distinct streams in the scattering volume. The dotted and dashed lines are the observations, the solid lines a simple model. The narrow peak at a lag of about 0.3 s is from the fast stream, the broader peak at a lag of 0.9 s is from the remainder of the scattering region. The mean velocity is increased only 50 km/s by this stream.

is in a uniform spherically symmetric flow, this will be about 18% below the mean flow speed. The IPS velocity measurements have been calibrated against spacecraft measurements during periods when the solar wind was dominated by corotating streams. Under these conditions the spacecraft data can be used to predict the IPS measurement and generally good agreement was found (Coles et al., 1978). The IPS measurements identified all streams more than one or two days wide, but missed occasional narrow streams such as those emanating from isolated equatorial coronal holes. More sensitive observations can identify fast streams when they contribute as little as 10% of the total scattering power, as shown in Fig. 4. On this day a narrow equatorial coronal hole was evident on the soft x-ray image from the Yohkoh spacecraft.

Early IPS measurements showed that from 1972 through 1976 the speed over the poles was nearly twice that in the equatorial region (Coles and Maagoe, 1972; Coles and Rickett, 1976) and it was later found that this latitude gradient completely disappeared at solar maximum (Coles et al., 1980). Latitude-longitude maps were first prepared by Sime and Rickett (1978) who realized that the fast polar streams were exactly the polar coronal holes which had been discovered on ATM soft x-ray images and were later identified in the white light corona (and various other coronal observations). This interpretation agreed with the identification of the equatorial extensions of these polar holes as the source of fast streams observed by spacecraft (Hundhausen, 1977) and established this explanation conclusively.

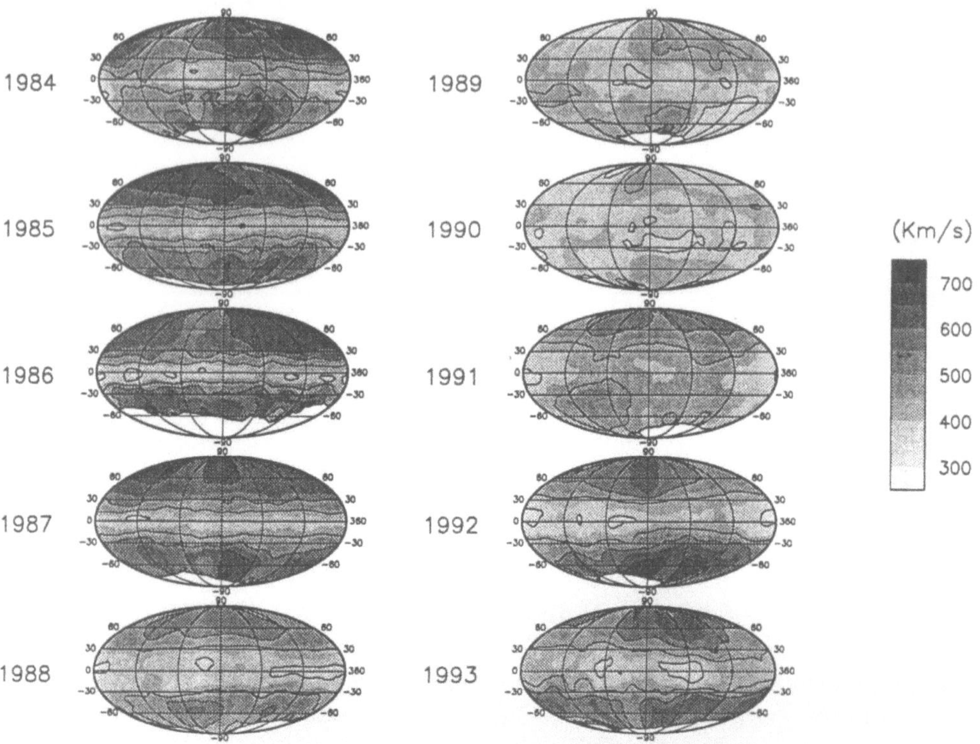

Fig. 5. Latitude-longitude maps of the velocity in an equal area projection, from the Solar-Terrestrial Environment Laboratory of Nagoya University.

A series of velocity maps covering the last 10 years is shown in Fig. 5. These maps were prepared by H. Misawa and M. Kojima at the STE Lab. They were formed using a back projection algorithm to reduce the effect of the extended scattering volume shown in Fig. 2. Such maps have been analyzed by both the San Diego and Nagoya groups, most recently by Kakinuma and Kojima (1987) and Rickett and Coles (1991). Quantitative analyses require model fitting to compensate for the line of sight integration. However, these maps display very clearly the evolution of the large scale structure. Years of maximum activity are characterized by a uniform mean speed of just under 400 km/s over the entire sun (recognizing that the averaging process will eliminate transients). Years of minimum activity are characterized by high speed flow everywhere except in a low velocity trough over the equatorial streamer belt that occupies a latitude range of ±15 to 20°. The transition from minimum to maximum configuration is very rapid, occuring in less than 6 months (Rickett and Coles, 1991). The transition from maximum to minimum is slower, going through a phase of several years in which the polar streams are relatively narrow and not centered on the rotational pole. This transition appears to have been different in 1971-74 than in 1982-85. The present cycle appears to be declining early and it will be interesting to see if it follows the 1971-74 model. A thorough study of the declining phase requires comparison with magnetic field and white light corona observations and is inappropriate for this review. It should be noted that the published IPS observations would be insensitive to a polar hole of width less than about 40° because the scattering from the polar stream is smaller than from the equatorial regions.

It is clear from the 1993 map in Fig. 5 that ULYSSES will have been be in high speed flow south of $-40°$ latitude and indeed this has been observed. In fact the gradient of the ULYSSES observations is much steeper than the apparent gradient in the IPS velocity because of the line of sight integration. It is not clear that the solar wind in 1994 will be in a full minimum condition typical of 1985-87. We hope that this will be so, because that would provide an excellent opportunity to test acceleration models.

3. Latitude Structure of Turbulence Level

The turbulence level can be measured with IPS in several ways. The most obvious is simply to measure δI^2. However this is only proportional to the turbulence level in weak scintillation, a constraint that becomes more difficult to satisfy near the sun. It is also difficult to compare δI^2 measurements from different sources because it depends on the structure of the source, which is not always known. A more robust method is to measure the coherence bandwidth of the IPS in strong scintillation (Bourgois and Coles, 1992). This statistic has the advantage of being useful near the sun and being relatively unaffected by source structure. A long series of IPS bandwidth measurements has been made at Nancay observatory by our late colleague Gabriel Bourgois. The bandwidth versus distance has been measured for two sources each year since 1982. One source 3C279 is equatorial, it shows very little change from one year to the next. The other source CTA21 passes about 6 $R_\odot$ south of the sun at closest approach. It shows much less turbulence at closest approach during solar minimum. We have used the equatorial source to normalize the other source, so the observations provide a yearly measurement of the variation of turbulence level (at scales of 5 to 50 km) versus latitude. Of course there is also longitude structure involved and this becomes very apparent in the declining phase of solar activity.

We have found that the best normalization of the equatorial data is to a model of $\delta N_e^2 \propto N_e^2$ where we have used the model of Muhleman and Anderson (1981) as the best estimate of $N_e(R)$ at distances between 5 and 20 $R_\odot$. The resulting estimate of turbulence level versus latitude is shown

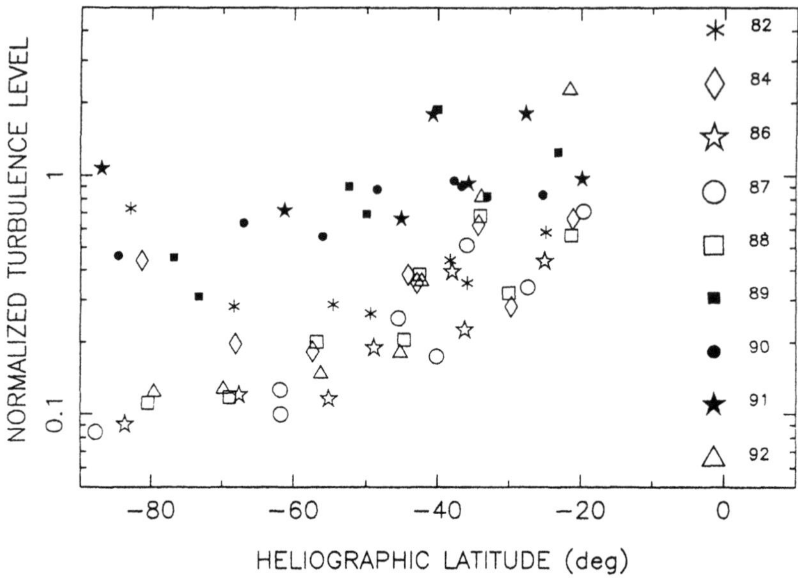

Fig. 6. An estimate of the turbulence level δN_e^2 vs latitude at various times in the solar activity cycle. The observations of CTA21, a high latitude source, have been normalized by observations of the equatorial source 3C279. Years of maximum solar activity are marked by filled symbols.

in Fig. 6. This latitude dependence is somewhat steeper than that shown earlier by Bourgois and Coles (1992), because the normalization function is steeper.

Here it is evident that the turbulence level in the polar hole is about a factor of 10 below that in the equatorial regions. However one must be careful to compensate for the line of sight integration. We have modeled this with a simple circular polar hole and a linear transition region of 20°. This model indicates that the actual turbulence level is about a factor of 15 lower in the polar hole. The model is drawn over the actual observations for 1987 in the lower panel of Fig. 7. The upper panel of Fig. 7 shows a comparison of the same model with the velocity observations further from the sun made by the STE Lab of Nagoya University.

Such comparisons agree equally well at other years of minimum activity. At solar maximum we do not see any latitude variation of turbulence level, so the model is trivial. However during the declining phase such model comparisons have not worked very well. We believe that the reason is that the IPS observations do not sample the polar stream uniformly when it is substantially offset from the rotational pole of the sun.

The turbulence level δN_e^2 is not a good measure of the development of the turbulence. What is needed is $\delta N_e^2/N_e^2$. Although we do not have estimates

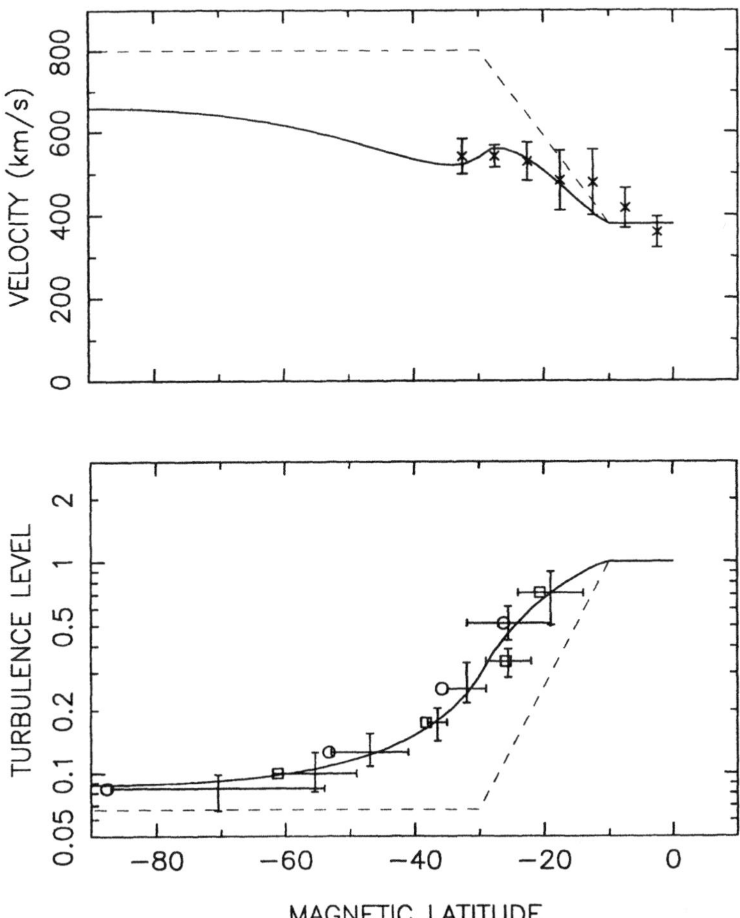

Fig. 7. A model fit to the observations of velocity and turbulence level vs latitude for 1987, a period of minimal solar activity. The model is shown in dashed lines, the IPS velocity and turbulence level calculated from this model are shown in solid lines, and the actual IPS observations are shown as symbols with vertical error bars. The horizontal bars indicate the latitude extent of the scattering region. The symbol is at the peak of the weighting function. During this year the fit to magnetic latitude was significantly better than to heliographic latitude. This was not true in the very similar year of 1986, which was better fit in heliographic latitude.

of N_e coincident with our IPS observations the white light coronagraph estimates of N_e are quite close to our observation region. They indicate that N_e is at least a factor of 10 lower than normal in the polar holes (Withbroe, 1988). Therefore it seems likely that $\delta N_e^2/N_e^2$ is actually at least a factor of 6 higher in the polar streams than over the equatorial regions at distances near 10 $R_\odot$. Related observations by Manoharan (1993) suggest that $\delta N_e^2/N_e^2$

remains substantially enhanced in the polar stream out to distances of 30-45 $R_{\odot}$. We will be very interested in the ULYSSES observations of density and the fluctuation spectra to see if the enhancement in $\delta N_e^2/N_e^2$ persists to greater distances.

4. The Distance Profile of Velocity

Much of the recent work of the UCSD group has been to measure V(R) and to compare the measurements with theoretical acceleration models. As discussed earlier, this must be done at cm-wavelengths because the greatest acceleration occurs inside of 20 $R_{\odot}$. An extensive set of observations has been made at the European Incoherent Scatter Facility (EISCAT) since 1990 (Coles et al., 1991). Perhaps the most important feature of these observations is that we have been able to secure sufficient observing time to select cases in which several sources are radially aligned with respect to the sun so the observations sample a single flow tube.

The theoretical model we have used for comparison is one of a class of models in which momentum is added to the flow by the ponderomotive force of Alfven waves. Although there are significant variations in such MHD models (e.g. Marsch and Tu, 1990) the published models that we have seen have similar V(R) profiles. Typical comparisons are shown in Fig. 8.

We find that more than half of the well-determined measurements agree with the MHD model as well as shown in the top panel. However, at least a quarter of the well-determined measurements show disagreement similar to that shown in the bottom panel. We refer to these latter cases as *delayed acceleration* streams. We have not been able to adjust the model parameters to match the data for these streams and we do not believe that it can be done with this class of MHD model.

In an attempt to understand the origin of delayed acceleration streams we have mapped them back to the sun for comparison with white light density measurements and magnetic field measurements. This was done for nearly 100 streams observed between 1990 and 1993 and no clear pattern was found with respect to the magnetic neutral line or the equatorial streamer belt. However we have not yet observed a stream under true solar minimum conditions when the IPS line of sight lies entirely within the polar hole. Furthermore we have not been able to use Earth-orbiting spacecraft observations very well because we seldom get a latitude-longitude match with our streams. Also the velocity profile can be significantly modified by stream-stream interactions before 1 AU.

We hope to obtain good comparisons with ULYSSES during this summer and next, because the flow structure is much more uniform in the polar regions and exact latitude-longitude alignment is not as important as it is in the equatorial regions. The most important ULYSSES observations

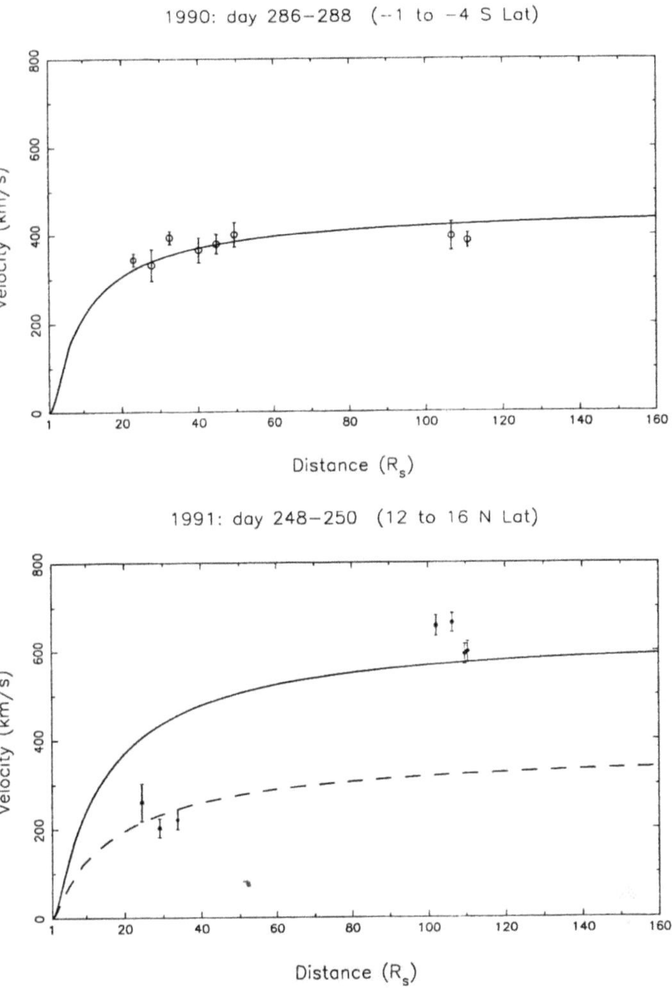

Fig. 8. A comparison of an MHD acceleration model with measured velocity profiles. The observations were taken in the latitude range noted on the panel. The longitude range, about ±30°, was small compared with the longitude extent of the scattering volume.

for this purpose are the velocity, density, magnetic field and the spectra of their fluctuations. To test acceleration models effectively we also need good estimates of the density, pressure and magnetic field at the coronal base. We are working with ground-based optical and radio observations, Yohkoh soft x-ray observations, and SPARTAN optical observations to place bounds on these parameters.

Acknowledgements

I would like to acknowledge the assistance of Hiroaki Misawa and Masayoshi Kojima, who provided their latitude-longitude maps of the solar wind speed in advance of publication. Mike Klinglesmith provided plots of Gabriel Bourgois' observations of turbulence level versus latitude and did the model fits thereto. Russ Grall provided the plots of the acceleration profile from EISCAT data. I am also indebted to Mike Bird for his careful review of the manuscript.

References

Ananathakrishnan, S., Coles, W. A. and Kaufman, J. J.: 1980, *J. Geophys. Res.* **85**, 6025.
Armstrong,J. W. and Coles, W. A.: 1978, *Astrophys. J.* **220**, 346.
Bourgois, G. and Coles, W. A.:1992, in Marsch and Schwenn, eds., *Solar Wind Seven*, Pergamon:Oxford, 158.
Coles, W. A. and Maagoe, S.: 1972, *J. Geophys. Res.* **77**, 5622.
Coles, W. A. and Rickett, B. J.: 1976, *J. Geophys. Res.* **81**, 4797.
Coles, W. A.: 1978, *Space Science Rev.* **21**, 411.
Coles, W. A., Harmon, J. K., Lazarus, A. J. and Sullivan, J. D.: 1978, *J. Geophys. Res.* **83**, 3337.
Coles, W. A., Rickett, B. J., Rumsey, V. H., Kaufman, J. J., Turley, D. G., Ananthakrishnan, S., Armstrong, J. W., Harmon, J. K., Scott, S. L. and Sime, D. G.: 1980, *Nature* **286**, 239.
Coles, W. A., Esser, Ruth, Loevhaug, Unni-Pia and Markkanen, Jussi: 1991, *J. Geophys. Res.* **96**, 13849.
Kojima, M. and Kakinuma, T.: 1987, *J. Geophys. Res.* **92**, 7269.
Hewish, A., Scott, P. F. and Wills, D.: 1964, *Mon. Not. Roy. Astron. Soc.* **126**, 469.
Hoeksema, J. T., Wilcox, J. M. and Scherrer, P. H.: 1983, *J. Geophys. Res.* **88**, 9910.
Hoeksema and Scherrer, P. H.: 1985, *Rep. CSSA-ASTRO* **8511**, Stanford University, Stanford California.
Hundhausen, A. J.:1977, in Zirker ed., *Coronal Holes and High Speed Streams*, University of Colorado Press:Boulder, 225.
Lotova, N. A., Blums, D. F. and Vladimirskii, K. V.: 1985, *Astron. Astrophys.* **150**, 266.
Marsch, E. and Tu, C.-Y.: 1990, *J. Geophys. Res.* **95**, 11945.
Manoharan, P. K.: 1993, *Solar Phys.* **148**, 153.
Muhleman, D. O. and Anderson, J. D.: 1981, *Astrophys. J.* **247**, 1093.
Neugebauer, M. and Snyder, C. W.: 1966, *J. Geophys. Res.* **71**, 4469.
Parker, E. N.: 1965, *Space Science Rev.* **4**, 666.
Rickett, B. J. and Coles, W. A.: 1991, *J. Geophys. Res.* **96**, 1717.
Sime, D. G. and Rickett, B. J.: 1981, *J. Geophys. Res.* **86**, 8869.
Withbroe, G. L.: 1988, *Astrophys. J.* **325**, 442.

DOPPLER SCINTILLATION MEASUREMENTS OF THE HELIOSPHERIC CURRENT SHEET AND CORONAL STREAMERS CLOSE TO THE SUN

RICHARD WOO[1], JOHN W. ARMSTRONG[1] and PAUL R. GAZIS[2]
[1]Jet Propulsion Laboratory, California Institute of Technology, Pasadena, CA 91109, USA;
[2]SJSU Foundation, NASA Ames Research Center, Moffett Field, CA 94035, USA

Abstract. Prominent enhancements in Doppler scintillation lasting a fraction of a day (solar source several degrees wide) and overlying the neutral line represent the signature of the heliospheric current sheet and the apparent interplanetary manifestation of coronal streamers near the Sun. This first detection of coronal streamers in radio scintillation measurements provides the link between *in situ* measurements of the spatial wavenumber spectrum of electron density fluctuations beyond 0.3 AU and earlier measurements deduced from radio scintillation and scattering observations inside 0.3 AU. Significant differences between the density spectra of fast streams and slow solar wind associated with the heliospheric current sheet near the Sun reinforce the emerging picture that high- and low-speed flows are organized by the large-scale solar magnetic field, and that while the contrast between solar wind properties of the two flows is highest near the Sun, it undergoes substantial erosion in the ecliptic plane as the solar wind expands.

1. Introduction

Compressive structures and fluctuations spanning an extensive range of scale sizes are ubiquitous in the solar wind. Investigations of these fluctuations have been based on *in situ* plasma measurements (Intriligator and Wolfe, 1970; Goldstein and Siscoe, 1972; Neugebauer et al., 1978; Marsch and Tu, 1990) as well as remote sensing radio scintillation measurements using both natural and spacecraft radio sources (Hewish, 1971; Coles, 1978; Woo and Armstrong, 1979). The latter represent essentially our only means for studying electron density fluctuations inside 0.3 AU. Amongst the wide range of scintillation phenomena (Coles, 1978; Woo, 1993), Doppler or equivalently phase scintillation measurements have been especially useful, because they not only probe the fullest range of spatial wavenumbers, but also a heliocentric distance range that starts near the Sun and extends to near Earth orbit. In spite of some progress (Montgomery et al., 1987; Marsch and Tu, 1990), the nature of compressive fluctuations is still not yet fully understood.

Variations in compressive fluctuations, reflecting solar wind changes, produce enhancements [also referred to as transients, see Woo (1993)] in Doppler scintillation. While many of these enhancements are caused by propagating interplanetary disturbances, some of which drive interplanetary shocks (Woo and Armstrong, 1981), others are due to quasi-stationary structure such as coronal streamers, and still others arise from dynamic interaction in the solar wind with increasing heliocentric distance, such as the formation of compressed plasma in interaction regions ahead of fast streams (Houminer and Hewish, 1972). One of the advantages of observing the solar wind near

Space Science Reviews **72**: 223–228.
© 1995 *Kluwer Academic Publishers.*

the Sun is that variations of solar origin can readily be distinguished from those that evolve as a result of dynamic interaction.

With the availability of high time resolution Doppler scintillation observations for comparisons with both solar and *in situ* plasma measurements, there is growing evidence that the morphology of Doppler scintillation is organized by the large-scale solar magnetic field (Woo and Gazis, 1993). In this paper, we provide further evidence for this morphology, including the first apparent signature of the heliospheric current sheet, and differences between the spatial wavenumber spectrum of electron density fluctuations for fast and slow solar wind. Finally, we summarize the emerging global picture of large scale solar wind structure in the vicinity of the Sun based on recent radio scintillation and scattering results.

2. Doppler Scintillation Measurements

Doppler scintillation is a path-integrated measurement that responds to electron density fluctuations and solar wind speed transverse to the radio path (Woo and Schwenn, 1991). Correlation of Doppler scintillation measurements with solar source surface magnetic field maps (Hoeksema and Scherrer, 1986) reveal that scintillation enhancements are generally observed in the vicinity of the neutral line near the Sun. In the case of the 1984 Pioneer Venus measurements investigated in Woo and Gazis (1993), these enhancements extended over a longitude range of 90-140°. However, enhancements lasting only a fraction of a day (corresponding to an extent of several degrees), probably reflecting those occasions on which coronal mass ejection (CME) activity is either absent or low, also appear frequently. Two examples of these transients observed by Pioneer Venus in 1987 are shown in Fig. 1, along with the contour map of the source surface magnetic field strength produced by the Wilcox Solar Observatory (see e.g., Hoeksema and Scherrer, 1986) and corresponding to the relevant Carrington Rotation CR 1772. The corresponding closest approach distances of the Pioneer Venus radio path in solar radii and AU, shown at the top of the time series panel, indicate that the S-band (13 cm wavelength) measurements probed the solar wind around 35 R_{O}.

The Doppler scintillation time series in Fig. 1 has been scaled to 1 AU by multiplying the observed rms Doppler scintillation (3-min values based on 1 per 10 sec Doppler measurements) by $R^{1.5}$, representing a R^{-2} fall-off in electron density fluctuation with heliocentric distance R. The dots on the magnetic field map represent the points of closest approach of the Pioneer Venus radio path on the indicated days of year (DOY) at 0000 UT mapped back to the surface of the Sun assuming a constant radial solar wind speed of 450 km/s. For convenience of comparison, the time axis of the normalized Doppler scintillation has been reversed and displayed in such a manner that DOY lines up approximately with the corresponding dots on the magnetic field map. The scintillation enhancements occur during crossings of (or when the measurements are very close to) the neutral line, and in agreement with previous results (Woo and Gazis, 1993, 1994), Doppler scintillation levels away from neutral line

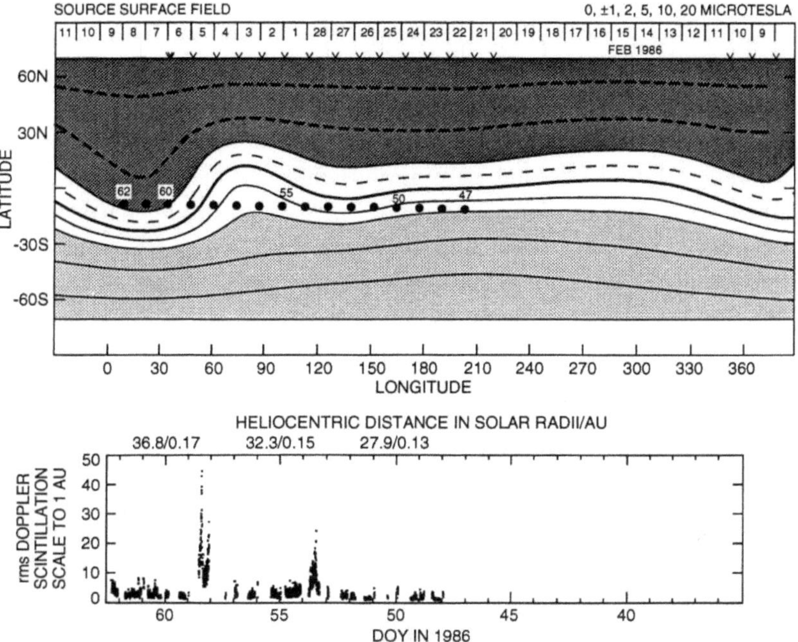

Fig. 1 Contour map of the source surface magnetic field strength produced by Wilcox Solar Observatory corresponding to Carrington Rotation 1772, and time series of Doppler scintillation scaled to 1 AU.

(presumably associated with the fast wind) are depressed and exhibit low variability. Recent Ulysses ranging measurements (Bird et al., 1994), show that peaks in mean density occur during enhancements like those in Fig. 1 (Woo et al., in preparation). Coincidence of Doppler scintillation enhancements like those in Fig. 1 with peak densities and the neutral line leads to the first identification in radio scintillation measurements of the signature of the heliospheric current sheet and the apparent interplanetary manifestation of coronal streamers near the Sun. *In situ* fields and particles measurements beyond 0.3 AU have shown that minimums in helium abundance, solar wind flow speed and proton temperature, coincide with the peak density observed at the sector boundary (Gosling et al., 1981).

3. Spatial Wavenumber Spectrum of Electron Density Fluctuations

In situ measurements of proton density fluctuations (limited to frequencies lower than 6 x 10^{-3} Hz) in the heliocentric distance range of 0.3-1.0 AU by Helios have shown that they are significantly different between high and low-speed solar wind flows especially near 0.3 AU (Marsch and Tu, 1990). In the slow wind near the sector boundary, the compressive fluctuations are more fully developed and intense, and exhibit a spatial wavenumber spectrum that is radially invariant and approximately Kolmogorov. In contrast, fast stream turbulence is significantly less compressive in terms of relative density fluctuations, but becomes increasingly compressive as the solar wind expands, with its density spectrum showing a flatter high-frequency part that evolves with heliocentric distance.

Radio scintillation and scattering observations complement these *in situ* plasma measurements, because they not only provide the only measurements of the spatial wavenumber spectrum of electron density fluctuations inside 0.3 AU, but over an extensive range of spatial scales as well, including scales smaller than those observed by *in situ* measurements. Yet, the lack of simultaneous measurements of solar wind speed has so far precluded discriminating high and low-speed flows (Woo and Armstrong, 1979; Coles et al., 1991).

That enhancements like those in Fig. 1 coincide with sector boundary crossings, where *in situ* measurements beyond 0.3 AU have shown that the solar wind is slow and highly compressive, strongly suggests that at least some of the enhancements identified as 'transients' in Coles et al. (1991) represent similar slow solar wind, while the 'transient'-free solar wind more likely represents the fast streams. Evidently, although many Doppler scintillation enhancements, especially those during the high activity phase of the solar cycle, represent interplanetary disturbances characterized by fast moving solar wind (speeds exceeding about 500 km/s) (Woo and Schwenn, 1991), the enhancements in Fig. 1 are associated with slow solar wind. While correlations between Doppler scintillation near the Sun and *in situ* solar wind speed measurements near 1 AU (but 90° apart in longitude) conducted in 1984 show a close association between scintillation enhancements in the vicinity of the neutral line and the slow solar wind (Woo and Gazis, 1993), more direct evidence has recently been obtained (Woo and Martin, in preparation) with solar wind speeds deduced from Voyager 1 and 2 intensity scintillation measurements of the near-Sun solar wind during 1979-1982 (Martin, 1985).

Although the range of overlapping spatial scales is narrow, further support comes from the similarity in behavior of density spectra deduced from phase scintillation and spectral broadening measurements inside 0.3 AU (Woo and Armstrong, 1979; Coles et al., 1991) to that obtained from *in situ* Helios plasma measurements beyond 0.3 AU (Marsch and Tu, 1991). Voyager phase scintillation and spectral broadening measurements conducted in 1979-1980 (Coles et al., 1991) reveal that electron density spectra of 'transient'-free [fast wind as shown in some of cases by velocity estimates deduced from simultaneous intensity scintillation measurements (Woo and Martin, in preparation)] tend to be steep (approximately Kolmogorov) at large spatial scales (10^3-10^6 km), but which show flattening at smaller scales (10-100 km). The inflection between the steep and flatter regions is abrupt and occurs in the vicinity of 100-300 km. In the 'transient' wind [corresponding to slow wind as shown in some cases by wind speeds deduced from simultaneous intensity scintillation measurements (Woo and Martin, in preparation)], there is an overall increase of power in the density spectra at large scales and a steepening in spectra at small scales, resulting in spectra that are not only significantly higher than those of the 'transient'-free solar wind, but which are (as in the case of *in situ* measurements for slow wind) approximately Kolmogorov.

Inflections similar to those observed in the Voyager measurements were not as readily apparent (if at all) in a comprehensive study of the electron density spectrum based on 1976 Viking phase scintillation and spectral broadening measurements (Woo and Armstrong, 1979). The Viking radio measurements took place essentially in the ecliptic plane at a time when the neutral line was confined to the vicinity of the ecliptic plane (flat heliospheric current sheet) (Hoeksema and Scherrer, 1986), so that the 1976 Viking measurements essentially observed slow solar wind much of the time. On the other hand, the Voyager measurements took place in 1979-1982 when the neutral line experienced large latitudinal excursions (warped heliospheric current sheet) resulting in the probing of both slow and fast solar wind flows.

4. Global Large-Scale Solar Wind Structure

Detailed studies are still in progress, but a broad picture of global large-scale solar wind structure near the Sun can be described. Beyond 0.5 AU, we know from multiple station intensity scintillation (IPS) observations that solar wind speed is slow over the streamer belt and fast away from it (Kakinuma and Kojima, 1990; Rickett and Coles, 1991). Inside 0.3 AU, scintillation measurements conducted with coherent spacecraft radio signals have provided more details of the association of solar wind properties with large-scale solar magnetic fields. Over the streamer belt where coronal streamers and CMEs prevail, density fluctuations are enhanced and varied (Woo and Gazis, 1993), and their spatial wavenumber spectrum is close to Kolmogorov. The fast stream wind emanating from regions away from the streamer belt exhibits compressive fluctuations characterized by depressed levels and low variability, and spectra that tend to show flattening at small spatial scales. Variations in solar wind speed and mass flux are low (Woo and Gazis, 1994). *In situ* measurements have shown that the solar wind speed dividing high and low speed flows is around 450 km/s. During solar minimum conditions, when polar coronal holes serve as sources of polar fast streams, and slow wind is confined mainly to the equatorial region, the differences between fast and slow winds are manifested as latitudinal variation showing pole-to-equator increases in mass flux and pole-to-equator decreases in solar wind speed (Woo and Goldstein, 1994).

Because their sources are physically separate at the Sun, contrast in properties of the high- and low-speed solar winds is highest near the Sun and tends to be abrupt. With solar wind expansion, dynamic interaction, which is greatest in the ecliptic plane, leads to erosion of this contrast, and is manifested by: (1) an increase in compressive fluctuations with heliocentric distance within fast streams (Marsch and Tu, 1991), (2) a trend showing decreasing differences with heliocentric distance between the intensity (magnitude) and spatial wavenumber spectrum of density fluctuations of both flows (Marsch and Tu, 1991; Woo and Gazis, 1993, 1994), (3) a slight increase in mass flux with heliocentric distance in fast streams (Schwenn, 1990; Woo and Gazis, 1994), and (4) a corresponding decrease in mass flux with heliocentric distance in the slow solar wind (Schwenn, 1990). Due to reduced interaction with slow wind, density fluctuations in the polar fast wind currently being encountered by Ulysses at high latitude but beyond 1 AU might be expected to possess a spatial wavenumber spectrum that more closely resembles that of the fast wind observed by radio scintillation measurements in the vicinity of the Sun (showing flattening at small scales).

Although contrasting high- and low-speed solar wind in the vicinity of the Sun is most evident near solar minimum conditions when the large-scale solar magnetic field configuration is more stable and coronal mass ejections less frequent (Howard et al., 1986; Hundhausen, 1993), results obtained to date indicate that it is discernible at other times of the solar cycle, both in and out of the ecliptic plane.

Acknowledgments

It is a pleasure to acknowledge the Pioneer Venus Project and the NASA DSN. We thank C. Chang for computer efforts, G. Goltz for data retrieval, J.T. Hoeksema for discussions and W.A. Coles for useful comments. This paper describes research carried out at JPL, Caltech, under a contract with NASA, and at NASA Ames.

References

Bird, M.K. et al., *Astrophys. J.* **426**, 373.
Coles, W.A.: 1978, *Space Sci. Rev.* **21**, 411.
Coles, W.A., Liu, W., Harmon, J.K. and Martin, C.L.: 1991, *J. Geophys. Res.* **96**, 1745.
Goldstein, B. and Siscoe, G.L.: 1972, in *Solar Wind*, NASA Spec. Publ., SP-308, 506.
Gosling, J.T. et al.: 1981, *J. Geophys. Res.* **86**, 5438.
Hewish, A.: 1971, *Astrophys. J.* **163**, 645.
Hoeksema, J.T. and Scherrer, P.J.:1986, WDC-A for Sol-Terr. Phys., *Rept. UAG-94*.
Houminer, Z. and Hewish, A.: 1972, *Planet. Space Sci.* **20**, 1703.
Howard, R.A., Sheeley, Jr., N.R., Michels, D.L. and Koomen, M.J.: 1986, in *The* Sun *and the Heliosphere in Three Dimensions* (ed. R.G. Marsden), D.Reidel, Norwell, Mass, 107-111.
Hundhausen, A.J.: 1993, *J. Geophys. Res.* **13**, 177.
Intriligator, D.S. and Wolfe, J.H.: 1970, *Astrophys. J.* **196**, 879.
Kojima, M. and Kakinuma, T.: 1990, *Space Sci. Rev.* **53**, 173.
Martin, J.M.: 1985, Ph.D. Thesis, Stanford University, Stanford, CA.
Marsch, E. and Tu, C.-Y.: 1990, *J. Geophys. Res.* **95**, 11945.
Montgomery, D., Brown, M.T. and Matthaeus, W.H.: 1987, *J. Geophys. Res.* **92**, 282.
Neugebauer, M., Wu, C.S. and Huba, J.D.: 1978, *J. Geophys. Res.* **83**, 1027.
Rickett, B.J. and Coles, W.A.: 1991, *J. Geophys. Res.* **96**, 1717.
Schwenn, R.: 1990, in *Physics of the Inner Heliosphere*, eds. R. Schwenn and E. Marsch, Springer-Verlag, Berlin, pp. 99-181.
Woo, R.: 1993, in *Wave Propagation in Random Media* (Scintillation), eds. V. Tatarskii, A. Ishimaru and V. Zavorotny, SPIE, Bellingham, Wash., pp.50-83.
Woo, R.: 1993, *J. Geophys. Res.* **98**, 18999.
Woo, R. and Armstrong, J.W.: 1979, *J. Geophys. Res.* **84**, 7288.
Woo, R. and Armstrong, J.W.: 1981, *Nature* **292**, 608.
Woo, R. and Gazis, P.R.: 1993, *Nature* **366**, 543.
Woo, R. and Gazis, P.R.: 1994, *Geophys. Res. Letters* (in press).
Woo, R. and Goldstein, R.M.: 1994, *Geophys. Res. Letters* **21**, 85.
Woo, R. and Schwenn, R.: 1991, *J. Geophys. Res.* **96**, 21227.

LATITUDINAL VARIATION OF SOLAR WIND VELOCITY

S. ANANTHAKRISHNAN
National Centre for Radio Astrophysics, TIFR,
PO Box 3, Ganeshkhind, Pune 411 007, India

V. BALASUBRAMANIAN
Radio Astronomy Centre, TIFR,
PO Box 8, Udhagamandalam 643 001, India

and

P. JANARDHAN
Physical Research Laboratory,
Navrangpura, Ahmedabad 380 009, India

Abstract. Single station solar wind velocity measurements using the Ooty Radio Telescope (ORT) in India (operating at 327 MHz) are reported for the period August 1992 to August 1993. Interplanetary scintillation (IPS) observations on a large number of compact radio sources covering a latitudinal range of $\pm 80°$ were used to derive solar wind velocities using the method of fitting a power law model to the observed IPS spectra. The data shows a velocity versus heliographic latitude pattern which is similar to that reported by Rickett and Coles (1991) for the 1981-1982 period. However, the average of the measured equatorial velocities are higher, being about 470 km s^{-1} compared to their value of 400 km s^{-1}. The distribution of electron density variations (ΔN_e) between $50R_\odot$ and $90R_\odot$ was also determined and it was found that ΔN_e was about 30% less at the poles as compared to the equator.

1. Introduction

Interplanetary scintillation (IPS) observations of compact extragalactic radio sources have been used extensively for investigations on several properties of the solar wind plasma. Multi station IPS observations for determining the velocity of the solar wind have been described by Rickett & Coles (1991) and references therein. These three station IPS studies show that the annual average velocity of solar wind increases markedly with heliospheric latitude (both in the northern and southern hemispheres), during times of low and declining solar activity. Although these observations were made on 20 to 50 radio sources only, they were distributed widely in space. Hence, velocities of solar wind originating from a large range in heliographic latitude (γ) could be determined ($-60° \leq \gamma \leq 60°$). This is in contrast to *in situ* spacecraft measurements which are restricted mostly to the ecliptic plane. The notable exception will be the measurements by the spacecraft Ulysses during its solar polar passage in 1994-95.

Manoharan & Ananthakrishnan (1990) used model fitting techniques to estimate successfully the solar wind velocity from a single radio telescope on a routine basis. This was applied to IPS observations of a large number of sources using the Ooty Radio Telescope (ORT) during the period August

Space Science Reviews **72**: 229–232.
© 1995 *Kluwer Academic Publishers.*

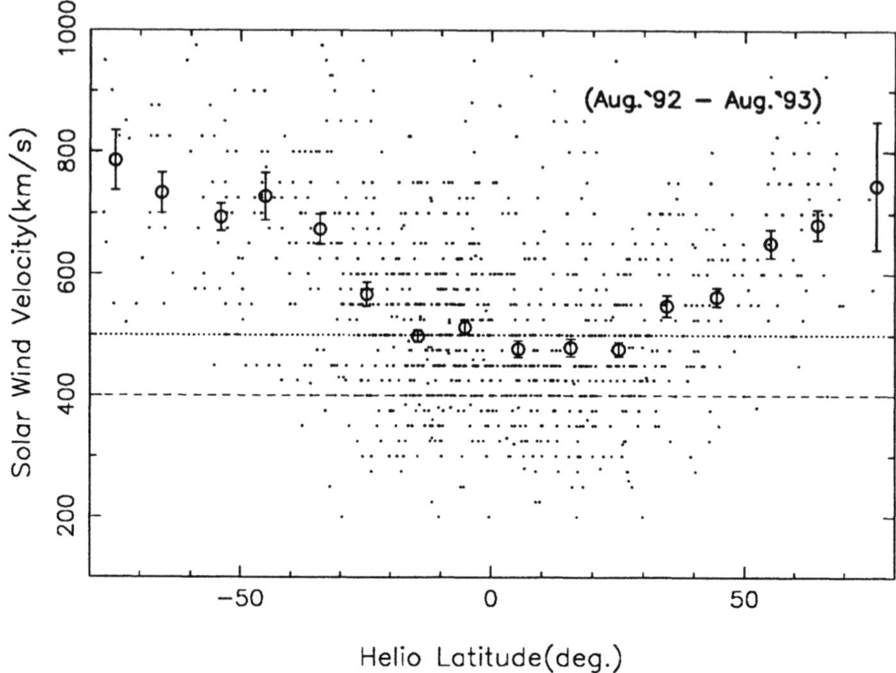

Fig. 1. Scatter plot of the single station solar wind velocities as a function of heliographic latitude. Each velocity was determined using spectra with signal to noise ratio greater than 20 db and sources in with $15^o \leq \epsilon \leq 50^o$.

1992 to August 1993 (Balasubramanian *et.al.*, 1993). About 1000 strongly scintillating sources were observed, at three different solar elongations (ϵ) between 10° and 50°. The heliographic latitudes of these sources were in the range $\pm 80^o$. In this paper we present the results obtained from the above data on variation of solar wind velocities with heliographic latitude. Also described are the latitudinal distribution of electron density variations in the solar wind.

2. Analysis and Results

2.1. Latitudinal Variation of Observed Velocities

The velocities derived from the spectral fitting method were first sorted to remove spectra derived from observations at solar elongation $\epsilon < 15^o$. This was done so as to remove the effects of strong scattering, which could extend a few degrees beyond $\epsilon = 12^o$ during disturbed conditions, and cause one to overestimate velocities. Next, the spectra are sorted to remove those having signal to noise ratio (S/N) $<$ 20 db. The velocities, determined from spectra with S/N greater than 20 db and $\epsilon \geq 15^o$ are plotted in figure 1, as a function

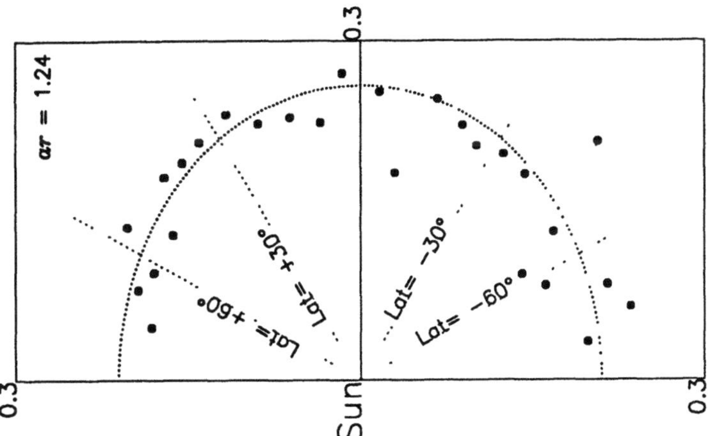

Fig. 2. Polar plot of scintillation index as a function of helio latitudes for sources in the distance range $50R_\odot$ to $92R_\odot$. The radial dotted lines are drawn at intervals of 30° in helio latitude and the finely dotted semi-circle is a reference circle. The dotted semi-ellipse is a best fit ellipse to the data and has an axial ratio of 1.24.

of heliographic latitude, for each observation. The individual velocity measurements are shown by small dots, while the large open circles represent the mean velocities in each 10° bin of latitude. The error bars are $\pm 1\ \sigma$ and are weighted by the number of points in each bin. The individual very low or very high values for velocities could have larger errors in the model fitting. Only the average trend is considered here. It can be seen that the velocities begin to increase beyond $\sim$ 10° on either side of the equator, from $\sim$ 450 km s^{-1} to reach a value of $\sim$ 800 km s^{-1}. The average equatorial velocity V_o and the gradient of velocity increase can be determined (Manoharan, 1991) by using the simple empirical relation $V = V_o(1+\eta\sin^2\gamma)$ where, γ is the heliographic latitude and η is called the activity factor and varies with solar cycle. The straight line fit gives V_o = 469 km s^{-1}. The factor η is equal to 0.68 and the velocity gradient over 90° of latitude is 3.6 km s^{-1} deg^{-1}. These values are in agreement with the determinations of annual averages for the same parameters for the declining period of the previous cycle; however, the average equatorial velocity V_o seems greater than that reported by Rickett and Coles (1991) by $\sim$ 70 km s^{-1}.

2.2. Latitudinal Variation of Electron Density Fluctuations

In the weak scattering region m is directly proportional to the rms electron density fluctuations (ΔN_e) and is a function of elongation (ϵ), angular source size (θ) and heliographic latitude (γ). This fact can be used to study the variations in ΔN_e with latitude, by comparing measurements of m at the same ϵ for sources of the same angular size. From the spectral fitting method (Manoharan & Ananthakrishnan, 1990) source sizes can also be reliably determined for spectra having S/N in excess of 20 db. For this purpose we selected all sources with 100 mas $\leq \theta \leq$ 150 mas (1 mas = 1 milli arcsecond) observed in the range $13^o \leq \epsilon \leq 25^o$, corresponding to distances of 50 to 92 solar radii from the sun. The results are presented in figure 2 which is a polar plot of scintillation index m vs γ with the sun at the origin. As all observations were made east of the sun, we have an one-sided plot. Each of the plotted points, shown by filled circles, represents the average value of m over a 6° bin of γ. It is clear from figure 2 that the ΔN_e distribution deviates from spherical symmetry as one approaches the poles. The axial ratio of the best fit ellipse was found to be 1.24, while the reduction in ΔN_e at the poles was about 19% which implies reduction in the electron density towards the poles.

3. Discussion and Conclusion

A recent study (Manoharan, 1993) used the turn over distances (which occurs at $\sim 40 R_\odot$) of compact scintillating sources at 327 MHz and examined the variation of ΔN_e as a function of γ. It was shown that the ΔN_e distribution changed from being spherically symmetric at solar maximum to a distribution that was flattened towards the polar regions at solar minimum with a reduction of ΔN_e in the poles by a factor of $\sim$2.5. Our observations show a much smaller departure from spherical symmetry in the ΔN_e distribution which is to be expected in the early part of the declining phase of the current solar cycle. The velocity gradient, too, compares well with the previous measurements (Rickett & Coles, 1991). The observed ΔN_e distribution and the velocity gradient suggests that the polar coronal hole regions have begun to expand equatorward. It will be very interesting to compare these results with those from the Ulysses satellite.

4. Acknowledgements

The authors thank Dr. P.K. Manoharan, for suggestions and help.

References

Balasubramanian, V. et al.: 1993, 'IPS Survey at 327 MHz for detection of compact radio sources', *Bul.Astr.Soc.ofIndia*, **21**, 469.

Manoharan, P.K. and Ananthakrishnan, S.: 1990, 'Determination of solar-wind velocities using single station measurements of interplanetary scintillation', *Mon.Not.Roy.Astr.Soc.*, **244**, 691.

Manoharan, P.K.: 1991, 'Study of Solar wind using single station interplanetary scintillation', *Ph.D Thesis*, Bombay University, India.

Manoharan, P.K.: 1993, 'Three-dimensional structure of the solar wind: variation of density with solar cycle', *Sol.Phys.*, **148**, 153.

Rickett, B.J. and Coles, W.A.: 1991, 'Evolution of the solar wind structure over a solar cycle: interplanetary scintillation velocity measurements compared with coronal observations', *J.Geophys.Res.*, **96**, 1717.

MOTION OF THE HELIOSPHERIC TERMINATION SHOCK AT HIGH HELIOGRAPHIC LATITUDE

Aaron Barnes

Theoretical Studies Branch, NASA Ames Research Center, Moffett Field, California, USA

Abstract. We expect the mean distance of the heliospheric termination shock to be greater (smaller) at polar latitudes than at equatorial latitudes, depending on whether the mean dynamic pressure of the solar wind is greater or smaller at high latitudes The heliospheric termination shock is expected to move in response to variation in upstream solar wind conditions, so that at any particular instant the termination shock will resemble a distorted asymmetric balloon with some parts moving inward and others moving outward If the shock is a gasdvnamic or magnetohydrodynamic shock the results of the analysis depend only very weakly on the nature of the upstream disturbance, typical speeds of the disturbed shock are ~100 to 200 km/s In the absence of a significant latitude gradient of the typical magnitude of solar wind disturbances typical motions of the disturbed shock at polar latitudes would be about twice as fast, due to the higher speed of the high-latitude wind If the dynamics of the termination shock are dominated by acceleration of the anomalous component of the cosmic rays, the motion of the shock in response to a given disturbance is substantially slower than in the gasdynamic case Conceivably, particle acceleration might be a less important effect at higher latitudes, and we envision the possibility of a termination shock that is dominated by particle acceleration at lower latitudes and is an MHD shock at high latitudes In this event high latitude solar wind disturbances would produce substantially larger inward and outward motions of the shock in the polar regions

1. Morphology of the Heliospheric Termination Shock

It is widely accepted that the solar wind flow changes from supersonic to subsonic through a termination shock whose location, roughly speaking, is where the solar wind dynamic pressure is balanced by pressure due to the very local interstellar medium (ISM) A number of physical processes can produce asymmetries in the termination shock For example, motion of the local interstellar medium toward the heliosphere will tend to compress the heliosphere, including the termination shock, so that the distance to the termination shock will show a distinct fore-aft asymmetry with respect to the direction of the interstellar flow, the magnitude of this asymmetry will be substantial if the flow of the local interstellar medium is supersonic relative to the heliosphere Asymmetry in the termination shock can originate internally in the solar wind as well as externally in the interstellar medium It may well be the case that in the dynamic pressure at a given distance is substantially higher (or lower) at polar latitudes than at equatorial latitudes, especially in the years around the solar activity minimum, resulting in a prolate (oblate) figure for the steady-state shock

Space Science Reviews **72** 233–236
© 1995 *Kluwer Academic Publishers*

2. Motion of the Termination Shock

In reality the configuration of the termination shock will be much more complicated than the above picture suggests, because the dynamic pressure of the solar wind undergoes large variations on the time scale of days and weeks. Hence the termination shock is unlikely ever to be in equilibrium, and will move inward and outward in response to solar wind variations (e.g., Barnes, 1993; Belcher et al., 1993; Grzedzielski and Lazarus, 1993; Suess, 1993; Whang and Burlaga, 1993). For the most part the solar wind disturbances will be confined to limited ranges of heliographic latitude, so that at a given moment some portions of the termination shock will move outward while others are simultaneously moving inward.

The local response of the termination shock to various upstream disturbances has been analyzed for a pure gasdynamic shock (Barnes, 1993, Naidu and Barnes, 1994a) and a magnetohydrodynamic shock (Naidu and Barnes, 1994b). The results are qualitatively similar in all these analyses, so it is sufficient to review the simplest case, for which the initial upstream disturbance is a gasdynamic contact discontinuity. This disturbance may be parameterized by the density ratio ρ_1/ρ_u, where the subscript u refers to the solar wind initially just upstream of the termination shock, and the subscript 1 refers to the solar wind upstream of the disturbance. Representative results are given in the first three columns of Table I (adapted from Barnes, 1994). Here v_1 is the speed of the upstream solar wind, V_1 is the speed of the termination shock after interaction with the disturbance, and V_2 is the speed of the downstream propagating signal (a sound wave or weak shock) that carries information of changed upstream conditions into the far downstream medium. Note that V_1 is positive (outward propagation) for a density enhancement ($\rho_1/\rho_u>0$), and negative for a density decrease, as one would expect intuitively.

Density disturbances of a factor of order 10 are not unusual in the outer solar wind, so that for $v_1 \sim 400$ km/s, as is typical of the outer solar wind at equatorial latitudes, one would expect the termination shock to move inward and outward at ~100 to ~200 km/s. Thus inward and outward excursions of several AU could be typical. Clearly the first encounter of an outer heliospheric spacecraft with the termination shock would be due to the shock's inward motion past the spacecraft, as opposed to the spacecraft's passing a stationary shock.

Although the trajectory of ULYSSES is limited to the inner heliosphere, high-latitude data from this spacecraft will be relevant to the motion of the termination shock at polar latitudes. Intermediate-latitude ULYSSES observations strongly suggest that at the present epoch the high-latitude solar wind is roughly twice as fast as the equatorial wind (Gosling *et al.*, 1993). Therefore a given enhancement (or decrease) of dynamic pressure would produce an excursion in the termination shock moving roughly twice as fast as the excursion produced by the same variation at equatorial latitudes. Hence it is possible that the excursions of the polar termination shock are systematically larger than the equatorial shock. This conclusion should hold if typical

TABLE I.
Response to Upstream Disturbance (γ=5/3)

	λ=0		λ=0.5		λ=0.9	
ρ_1/ρ_u	V_1/v_1	V_{sig}/v_1	V_1/v_1	V_{sig}/v_1	V_1/v_1	V_{sig}/v_1
0	−2.24	0.809	-1.36	0.513	-0.558	0.203
0.1	-0.553	0.809	-0.383	0.513	-0.182	0.203
.25	-0.334	0.809	-0.237	0.513	-0.116	0.203
0.5	-0.166	0.809	-0.121	0.513	-0.061	0.203
1	0	0.809	0	0.513	0	0.203
2	0.160	0.895	0.123	0.592	0.066	0.251
4	0.309	0.985	0.245	0.684	0.137	0.313
10	0.485	1.10	0.401	0.816	0.242	0.419
∞	1	1.5	1	1.42	1	1.35

variations in dynamic pressure at a given heliocentric distance are comparable at polar and equatorial latitudes. On the other hand, it is quite plausible that the flow is substantially smoother at high latitudes, implying smaller variations of dynamic pressure, and mitigating the high-latitude excursions of the termination shock. This issue should be settled by high-latitude ULYSSES plasma observations, at least for epochs not near the solar activity maximum.

3. Possible Role of Anomalous Cosmic Rays

So far we have assumed that the termination shock may be modeled as a gasdynamic or magnetohydrodynamic shock. However, it is possible that the heliospheric termination shock is quite different. It is thought that the anomalous component cosmic rays originates as interstellar neutral atoms that penetrate the heliosphere, become ionized and then are accelerated to energies on the order of 100 MeV/nucleon (Fisk *et al.*, 1974), probably at the termination shock itself (e.g., see the reviews of Jokipii (1990) and McKibben (1990)). Extrapolations of outer heliospheric cosmic-ray observations suggest that the energy density of the anomalous component could be comparable to the solar wind energy density within less than 100 astronomical units (AU) from the Sun (Jokipii and Kota, 1990; McKibben, 1990). If so, acceleration of the anomalous component is an essential element in the dynamics of the shock itself.

Barnes (1994) has modeled such a termination shock as a gasdynamic shock modified by inclusion of a sink (representing particle acceleration) in the energy jump conditions. The equilibrium position of this shock is slightly farther from the Sun than in the pure gasdynamic case. The most striking modification, however, is in the response of the shock to upstream disturbances (Table I). The shock is parameterized by the energy-sink strength λ (=rate of energy loss by particle acceleration/rate of flow of energy into the shock); $\lambda=0$ is the pure gasdynamic case. It is apparent from the table that for a given upstream disturbance, diversion of energy into particle acceleration tends to reduce the velocity of the disturbed termination shock.

So the termination shock may have its dynamics dominated by acceleration of the anomalous component, or it may be very like a pure gasdynamic shock; for convenience let us label the two kinds of shock as AC and GD, respectively. An interesting speculation is that some portions of the shock may be of the AC variety, others of the GD variety. In fact, according to the models of Donohue and Zank (1993), a shock that is initially AC may convert to GD and vice versa.

If the distance of the termination shock is of order 100 AU, the spiral angle of the heliospheric magnetic field is fairly near 90 degrees even at fairly high latitudes, so that the termination shock on average would be expected to be quasi-perpendicular at essentially all latitudes. On the other hand, if the solar wind is much faster at polar latitudes, the shock transition involves more energy per particle at high latitudes. In addition, heliospheric magnetic fields may deviate considerably from the Archimedean spiral model (Jokipii and Kota, 1989). Such variations might or might not be sufficient to effect the transition between AC and GD, but if the transition can occur it is conceivable that there is a systematic equator-pole transition from one kind of shock to the other. Then, for example, if polar shock tends to be GD and the equatorial shock AC, the polar shock would move back and forth with much larger amplitude than the equatorial shock.

References

Barnes: A.: 1993, *J. Geophys. Res.* **98**, 15,137.

Barnes, A.: 1994, *J. Geophys. Res.* **99**, 6553.

Belcher, J.W., A.J. Lazarus, R.L. McNutt, Jr., and G.S. Gordon, Jr.: 1993, *J. Geophys. Res.* **98**,15,177.

Donohue, D.J., and G.P. Zank: 1993, *J. Geophys. Res.* **98**, 19,005.

Fisk, L.F., B. Koslovsky, and R. Ramaty: 1974, *Astrophys. J., Pt. 2* **190**, L35.

Grzedzielski and Lazarus: *J. Geophys. Res.* **98**, 5551.

Gosling, J.T.: 1994, *EOS Trans. Am. Geophys. Un.* **74** Supp. , 479.

Jokipii, J.R.: 1990, 'The Anomalous Component of Cosmic Rays', in S. Grzedzielski and D.E. Page (eds.), *Physics of the Outer Heliosphere* , Pergamon, New York, pp. 169-178.

Jokipii, J.R., and J. Kota: 1989, *Geophys. Res. Lett.* **16**, 1.

Jokipii, J.R., and J. Kota: 1990, *Proceedings of the 21st International Cosmic Ray Conference* **6**, 198.

McKibben, R.B.: 1990, 'Cosmic rays in the local interstellar medium', in S. Grzedzielski and D.E. Page (eds.), *Physics of the Outer Heliosphere* , Pergamon, New York, pp. 107-118.

Naidu, K., and A. Barnes: 1994a, '*J. Geophys. Res.*, in press .

Naidu, K., and A. Barnes: 1994b, submitted to *J. Geophys. Res.*

Suess, S.T.: 1993, *J. Geophys. Res.* **98**, 15,147.

Whang, Y.C., and L.F. Burlaga: 1993, *J. Geophys. Res.* **98**, 15,221.

THE MULTIFUID SOLAR WIND TERMINATION SHOCK AND ITS INFLUENCE ON THE THREEDIMENSIONAL PLASMA STRUCTURE UPSTREAM AND DOWNSTREAM

S.V. CHALOV
Institute for Problems in Mechanics of the Russian Academy of Sciences, Prospect Vernadskogo,101, 117526 Moscow, Russia

and

H.J. FAHR
Institut für Astrophysik and Extraterrestrische Forschung der Universität Bonn, Auf dem Hügel 71, D-53121 Bonn, Germany

Abstract. The solar wind termination shock is described as a multi-fluid phenomenon taking into account the magnetohydrodynamic self-interaction of a multispecies plasma consisting of solar wind ions, pick-up ions and shock-generated anomalous cosmic ray particles. The spatial diffusion of these high energy particles relative to the resulting, pressure- modified solar wind flow structure is described by a coupled system of differential equations describing mass-, momentum-, and energy-flow continuities for all plasma components. The energy loss due to escape of energetic particles (MeV) from the precursor into the inner heliosphere is taken into account. We determine the integrated properties of the anomalous cosmic ray gas and the low-energy solar wind. Also the variation of the compression ratio of the shock structure is quantitatively determined and is related to the pick-up ion energization efficiency and to the mean energy of the downstream anomalous cosmic ray particles. The variation of the resulting shock structure and of the solar wind sheath plasma extent beyond the shock is discussed with respect to its consequences for the LISM neutral gas filtration and the threedimensional shape of the heliosphere.

Key words: Anomalous Cosmic Rays, Heliospheric Shock, Multifluid Plasma

1. Introduction

The diffusive acceleration of ionized interstellar particles convected as pick-up ions towards the termination shock is considered as source of so-called anomalous cosmic rays (ACR) (Fisk, 1976, Fisk et al.,1974, Pesses et al.,1981, Jokipii, 1992). According to most recent observations, these anomalous particles represent sufficiently high energy densities to act by means of their pressure gradients on the solar wind flow at large solar distances ($\geq 30AU$) (Jokipii, 1990; Fahr et al., 1992; Grzedzielski et al., 1992; Fichtner et al., 1993/1994). The diffusive termination shock wave, thus, has to be treated as a multifluid shock including thermal plasma and cosmic rays leading to the formation of both an entropy-generating subshock and an extended non-dissipative precursor. Here we present a two-fluid model of such a planar diffusive shock wave, assuming that a critical energy level E_{max} exists above which energetic particles will not be retained in the precursor by the turbulent plasma waves, instead escape upstream into the inner heliosphere.

Space Science Reviews **72**: 237–242.
© 1995 *Kluwer Academic Publishers.*

The energy-averaged diffusion coefficient of energetic particles enters our calculations as a scalar quantity.

2. Brief sketch of the theoretical approach

The cosmic-ray-modulated shock treated here consists of a shock precursor and an entropy-generating subshock. At the subshock we assume an injection of MeV-energetic particles originating from KeV-pick-up ions. These anomalous cosmic ray particles diffuse opposite to the solar wind flow and lead to a scalar anomalous pressure p_c. If the planar subshock is located at $x = 0$ on the x-axis, then the stationary state is expressed through the conservation of mass-, momentum- and energy-flows by the following differential equations:

$$\frac{d(\rho u)}{dx} = \frac{d\mu}{dx} = -q\delta(x), \tag{1}$$

$$\mu\, u + p_g + p_c = const, \tag{2}$$

$$\mu \left[\frac{u^2}{2} + \frac{\gamma_g}{\gamma_g - 1} \frac{p_g}{\rho} \right] + F_c = const, \tag{3}$$

where ρ, u, μ and p_g are the density, velocity, mass flow, and pressure of the solar wind plasma, and p_c is the anomalous plasma pressure. γ_g is the polytropic index of the solar plasma, q is the mass loss rate from the solar plasma to the anomalous plasma, connected with the particle injection via losses of pick-up ions at the subshock. $\delta(x)$ is the Dirac delta function. The cosmic ray energy flux is given by

$$F_c = \frac{\gamma_c}{\gamma_c - 1} u p_c - \frac{\bar{\kappa}}{\gamma_c - 1} \frac{dp_c}{dx}, \tag{4}$$

where γ_c is the polytropic index of the anomalous plasma, and $\bar{\kappa}$ is the energy-averaged diffusion coefficient. The energy flux F_c is connected with the pressure p_c by the diffusion equation for high energetic particles

$$\frac{dF_c}{dx} = u \frac{dp_c}{dx} + Q\delta(x), \tag{5}$$

where Q denotes the energy source for the shock-generated cosmic rays which is related to the mass loss rate q and the mean energy gain E_{inj} in the energization process by $Q = qE_{inj}/m_p$. The anomalous pressure is assumed to behave continuous at the subshock. Hence one obtains from equations (1) to (5) the following jump relations at the subshock

$$[\mu] + q = 0, \quad (6) \qquad [\mu\, u + p_g] = 0, \quad (7)$$

$$\left[\mu \left(\frac{u^2}{2} + \frac{\gamma_g}{\gamma_g - 1} \frac{p_g}{\rho} \right) \right] + Q = 0, \quad (8) \qquad [p_c] = 0, \quad (9) \qquad [F_c] = Q. \quad (10)$$

We take into account a loss of energetic particles upstream from the shock and suppose that the most energetic anomalous particles leak out from the shock system if their diffusion velocity dominates their convection velocity. Thus, they can move upstream to large distances $|x| \geq |x_1|$ from the subshock. In this planar approximation the functions ρ, u, p_g can be taken to be constants ρ_1, u_1, and p_{g1} at $|x| \geq |x_1|$. The variation of these quantities in the precursor is described by equations (1) through (5) and can be determined by introduction of a new function $f(x) = u/u_1 = \rho_1/\rho$. From equations (2) through (5) one can then obtain all dynamical quantities like ρ, u, p_g, p_c as functions of $f(x)$. As shown in detail in Chalov and Fahr (1994) for the function $f(x)$ one can derive the following differential equation:

$$\frac{\bar{\kappa}}{u_1(\gamma_c - 1)}\frac{df}{dx} = \frac{G(f)}{1 - 1/M_{g1}^2 f^{\gamma_g+1}} \tag{11}$$

with the function $G(f)$ given by:

$$\begin{aligned} G(f) &= \frac{\gamma_c}{\gamma_g M_{g1}^2}\frac{(1-f)}{(\gamma_c - 1)} \times \\ &\left[\gamma_g M_{g1}^2 \frac{\gamma_c + 1}{2\gamma_c}(f - \beta_c) - 1 + \frac{\gamma_g - \gamma_c}{\gamma_c(\gamma_g - 1)}\frac{1 - f^{1-\gamma_g}}{1 - f}\right] - \frac{1}{4}\frac{p_{c1}}{\mu_1 u_1} \end{aligned} \tag{12}$$

Here the denotation $\beta_c = (\gamma_c - 1)/(\gamma_c + 1)$ was used. Equation (11) has to be solved with the boundary conditions $f(x = x_1) = 1$ and $f(x = 0) = f_0$. The value f_0 at the subshock has to be determined from the jump relations (6) through (10). Quantities at the upstream side of the subshock may be denoted with the index "0", yielding $f_0 = u_0/u_1$. With η denoting the fraction of pick-up ions converted per unit of time to anomalous particles relative to the number of pick-up ions convected to the subshock the mass transfer rate is given by: $q = \eta\rho_0 u_0 = \eta\mu_1$. From equation (6) we thus derive:

$$\rho_2 = (1 - \eta)\rho_1/z f_0, \tag{13}$$

where z is defined by: $z = u_2/u_0$. From relation (10) one then obtains:

$$\frac{\gamma_c}{\gamma_c - 1}(z - 1)f_0\left[1 - f_0 + \frac{1 - f_0^{-\gamma_g}}{\gamma_g M_{g1}^2}\right] + G(f_0) = \frac{Q}{mu_1^2} \tag{14}$$

In addition from relation (8) one derives:

$$(1 - \eta)z^2 - 1 + \frac{2\gamma_g}{\gamma_g - 1}\left[(1 - z)z - \frac{1 - z}{\gamma_g M_{g1}^2 f_0^{\gamma_g+1}} + \eta z^2\right] + \frac{2Q}{mu_1^2 f_0^2} = 0 \tag{15}$$

where Q is related with η by: $Q = \eta E_{inj}\mu_1/m_p$. As is shown in detail by Chalov and Fahr (1994) the energy input Q into the anomalous regime can

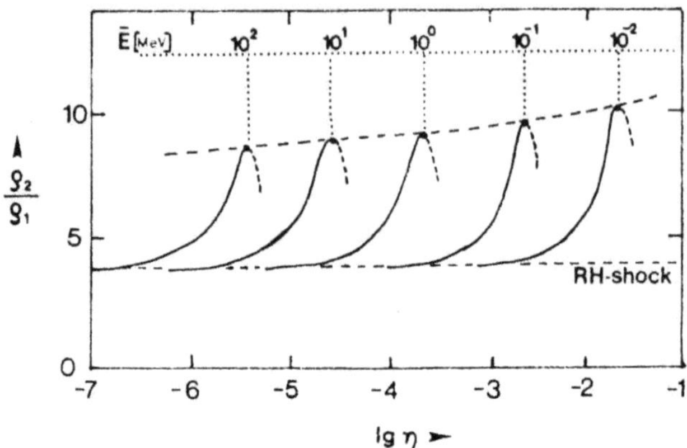

Fig. 1. The total compression ratio ρ_2/ρ_1 across the shock as function of the injection rate η. The parameters used are: $E_{max} = 300MeV$, $E_{inj} = 1keV$, $(u_1/c) = (4/3) \cdot 10^{-3}$, $M_{g1} = 10$.

be easily related to the energy flux F_{c1} leaking out from the shock structure in upstream direction by:

$$2\eta \frac{E_{max}}{m_p c^2} \left(\frac{c}{u_1}\right)^2 = \frac{2zf_0}{\gamma_c - 1} \frac{E_{max}}{\bar{E}} \left[1 - f_0 + \frac{1}{\gamma_g M_{g1}^2} \left(1 - f_0^{-\gamma_g}\right)\right] + \epsilon \quad (16)$$

where the normalized energy loss $\epsilon = 2\gamma_c p_{c1}/m_p u_1(\gamma_c - 1)$ has been introduced, and where E denotes the mean energy of the downstream anomalous particle population. With equations (14), (15), and (16) we have obtained three equations for the three unknown quantities ϵ, z and f_0. The above equations also contain boundary parameters: M_{g1}, (u_1/c), E_{max}, E_{inj}, $\bar{E}$ and η, which parametrize the solutions derived from the above equations.

3. Numerical results and conclusions

Equations (14), (15) and (16) have been solved numerically for the most reasonable set of boundary parameters, i.e $E_{max} = 300MeV$;$\bar{E} = 100MeV$; $E_{inj} = 1KeV$; $(u_1/c) = (4/3) \cdot 10^{-3}$. In Fig.1 we show results for the total compression ratio $\rho_2/\rho_1 = u_1/u_2 = 1/(zf_0)$ as function of the injection rate η for various fractions of $\bar{E} = 100MeV$. As can be seen, the compression ratio may substantially exceed the classical Rankine-Hugoniot value of 4 for a strong ($M_{g1} \leq 10$)-shock at higher values of the injection rate η. The postshock solar wind plasma passing through an ACR-producing shock hence will be much denser and cooler than expected from classical shock relations. This result is supported in similar investigations by Barnes (1994) who included an energy-sink in the usual shock relations in order to describe the ACR production. In view of papers by Osterbart and Fahr (1992), Baranov and Malama (1993), and Fahr et al. (1993) this must have important consequences for the filtration of interstellar neutral atoms at their passage into the inner heliosphere due to a cooler and denser solar sheath plasma region. Especially the downwind/upwind density ratio of strongly charge-exchanging atoms like hydrogen and oxygen will be strongly reduced by this fact. In addition the linear extent of the precursor region may be of interest here. Even in the presence of ACR productions, the upwind subshock may still be located at $R_S = 80AU$ since its location is nearly unchanged by the increase of entropy generation in the shock (see Barnes, 1994), but one may ask for the associated upstream extent $|x_1|$ of the precursor region within which particles up to an energy E_{max} are diffusively bound to the shock yielding the following formula:

$$|x_1| = \frac{\kappa(E_{max})}{\gamma_c(E_{max})u_1}. \tag{17}$$

Near the ecliptic one can adopt a perpendicular MHD-shock, i.e an interplanetary magnetic field perpendicular to the shock normal. Thus the diffusion upstream is described by a coefficient $\kappa_\perp$ which according to most recent results by Reinecke et al. (1993) is given by:

$$\kappa_\perp(E) = \kappa_{\perp 0}(v/c)(E/E_0)(B_E/B) \tag{18}$$

where v and E are the velocity and the energy of the cosmic ray particle. B is the local magnetic field, and B_E a reference value at $r_E = 1AU$. For a reference energy of $E_0 = 1GeV$ a reference diffusion coefficient with the value $\kappa_{\perp 0} = 6 \cdot 10^{20} cm^2/s$ is recommended by Reinecke et al.(1993). For particles with energies of $E_{max} = 300MeV \approx 0.3E_0$ and a subshock at $R_S = 80AU$ in the ecliptic we thus obtain, on the basis of an Archimedian spiral field yielding $B(R_S) \approx B_E/80$, as the relevant diffusion coefficient:

$$\kappa_\perp(E_{max}) \approx \sqrt{1-0.3^2}(0.3)(80)\kappa_{\perp 0} \approx 5 \cdot 10^{21} cm^2/s. \tag{19}$$

With these values for $\kappa_\perp$ and $\gamma_c = 1.5$ one obtains from Eq. (19): $|x| \approx 6AU$. One should notice that the shock structure shall look very different at polar regions. While the ecliptic shock is a perpendicular MHD shock, the polar shock will be of a parallel type. This has several consequences in the context discussed here. At polar regions the cosmic ray diffusion is controlled by the coefficient $\kappa_{||}$ rather than by $\kappa_\perp$ which there may be larger by a factor $\kappa_{||0}/\kappa_{\perp 0} \approx 10$ (see e.g. Palmer 1982). Keeping in mind in addition that here the magnetic field drops off by $(1/r^2)$, rather than by $(1/r)$ as in the ecliptic, this causes an additional increase in the effective diffusion coefficient and thus in the extent of the polar precursor region. Furthermore the pick-up ion conversion rate η may be much smaller at the poles in view of results from shock acceleration simulations by Scholer and Burgess (1992) and Scholer (1993). Thus the polar shock may be close to a classical Rankine-Hugoniot

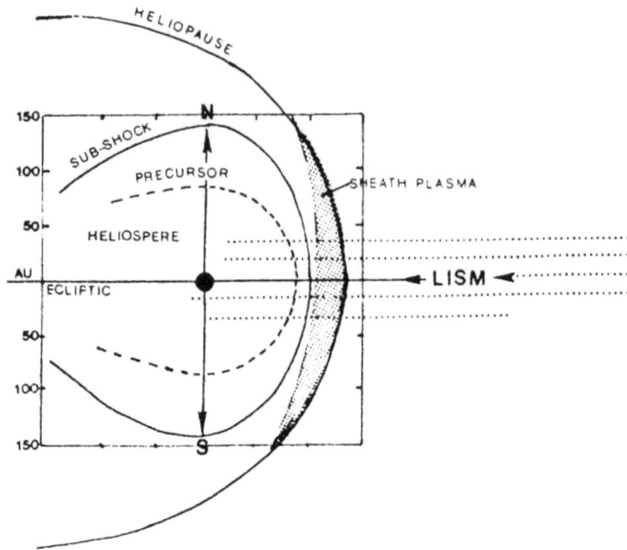

Fig. 2. Schematic illustration of the heliospheric configuration including the precursor region, the subshock, the condensed post-shock sheath plasma and the heliopause.

shock indicating that the postshock plasma density, due to the different compression ratios, will be higher in the ecliptic compared to polar regions. This is illustrated in Figure 2.

References

Barnes, A.: 1994, *J. Geophys. Res.* **99**, 6553
Baranov V.B. and Malama, Y.G.: 1993, *J. Geophys. Res.* **98**, 157
Chalov, S.V. and Fahr, H.J.: 1994, *Astron. Astrophys.*, in press
Fahr H.J., Fichtner H. and Grzedzielski S.: 1992, *Solar Phys.* **137**, 355
Fahr, H.J., Osterbart, R. and Rucinski, D.: 1993, *Astron. Astrophys.* **274**, 612
Fichtner, H., Fahr H.J. and Sreenivasan S.R.: 1993a, Proc. 23. Int. Cosm. Ray Conf., Calgary, Vol. 3, 423
Fichtner, H. et al.: 1994, *Astron. Astrophys.* **284**, 599
Fisk, L.A.: 1976, *J. Geophys. Res.* **81**, 4633
Fisk, L.A., Kozlovsky. B and Ramaty, R.: 1974, *Astrophys. J.* **190**, L35
Grzedzielski S., Fahr, H.J. and Fichtner, H.: 1992, '' in E. Marsch and R. Schwenn, ed(s)., *Solar Wind VII*, Pergamon Press: Oxford, 173
Jokipii, J.R.: 1990, '' in S. Grzedzielski and D.E. Page, ed(s)., *Physics of the outer heliosphere*, Pergamon Press: Oxford, 169
Jokipii J.R.: 1992, *Astrophys. J. Lett.* **393**, 41
Reinecke, J.P.L., Moraal, H. and McDonald, F.B.: 1993, *J. Geophys. Res.* **98**, 9417
Osterbart, R. and Fahr, H.J.: 1992, *Astron. Astrophys.* **264**, 260
Palmer, I.D.: 1982, *Space Sci. Rev.* **20**, 335
Pesses, M.E., Jokipii, J.R. and Eichler, D.: 1981, *Astrophys. J.* **246**, L85
Scholer, M.: 1993, *J. Geophys. Res.* **98**, 47
Scholer, M. and Burgess: 1992, *J. Geophys. Res.* **97**, 8319

HELIOSPHERIC RADIO EMISSIONS

DONALD A. GURNETT

Department of Physics and Astronomy, The University of Iowa, Iowa City, IA 52242

Abstract. A review is given of heliospheric radio emissions. Only radio emissions generated well away from the Sun (beyond a few solar radii) and well away from the magnetized planets are considered. These consist of (1) type III radio bursts, (2) type II radio bursts, and (3) heliospheric 2-3 kHz radio emissions. The physical processes involved in the generation of each of these radio emissions are described with an emphasis on recent developments. A prognosis is given of advances that can be expected from the forthcoming flight of Ulysses over the poles of the Sun and the flights of Voyagers 1 and 2 to the outer limits of the heliosphere.

I. Introduction

The heliosphere is the region around the Sun where the solar wind has a controlling influence. Five spacecraft, Pioneers 10 and 11, Voyagers 1 and 2, and Ulysses are now exploring the heliosphere. These missions are highly complementary. Pioneers 10 and 11, and Voyagers 1 and 2 are providing measurements at increasingly large radial distances from the Sun (58.6, 39.8, 54.5, and 41.9 AU, respectively, as of January 1, 1994), and Ulysses is providing measurements at high latitudes, over the poles of the Sun, at distances on the order of 2 AU. Three of these spacecraft, Voyagers 1 and 2, and Ulysses, carry radio and plasma wave instruments [Scarf and Gurnett, 1977; Stone et al., 1992]. In this paper we review the various types of radio emissions generated in the heliosphere. To restrict the scope of this review, we will only consider radio emissions generated well away from the influence of strong solar magnetic fields (beyond a few solar radii), and well away from the magnetized planets. Only three types of radio emissions fit these conditions. They are: (1) interplanetary type III radio bursts, (2) interplanetary type II radio bursts, and (3) heliospheric 2-3 kHz radio emissions. To carry out this review, it is convenient to start with a discussion of type III radio bursts, which have been studied in the greatest detail, and to end with a discussion of the heliospheric 2-3 kHz radio emissions, which have been discovered most recently.

II. Interplanetary Type III Radio Bursts

Type III radio bursts are produced by energetic electrons from solar flares and are characterized by an emission frequency that decreases rapidly with increasing time. These bursts were first studied by Wild [1950] who showed that they are associated with solar flares. Wild suggested that the radio emission is produced at or near the local electron plasma frequency, ($f_p = 9\sqrt{n_e}$ kHz, where n_e is the electron density in cm^{-3}),

Space Science Reviews **72**: 243–254.
© 1995 *Kluwer Academic Publishers.*

as the particles ejected by the flare move outward through the solar corona. Lin [1970] was the first to show that the particles involved are electrons. The electron energies are typically in the range from a few keV to several hundred keV. Radio direction-finding measurements [Fainberg et al., 1972] show that the electrons stream outward along the interplanetary magnetic field lines as shown in Figure 1. Since most solar flares occur at relatively low solar latitudes, less than 30° [Hundhausen, 1993], type III radio bursts are basically a low latitude phenomena [Dulk et al., 1986].

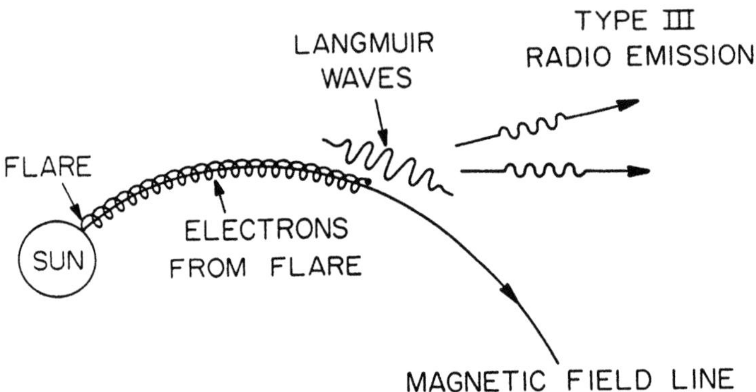

Fig. 1. Type III radio bursts are produced by energetic electrons from solar flares. The electrons stream outward from the Sun along the solar wind magnetic field lines and produce electrostatic oscillations called Langmuir waves. The Langmuir waves then mode convert to electromagnetic radiation via nonlinear wave-wave interactions.

The downward frequency drift of a type III solar radio burst is caused by the decreasing electron plasma frequency encountered by the energetic electrons as they stream outward from the Sun. According to presently accepted ideas, the radio emission is produced by a two-step process in which (1) the energetic electron beam first excites electrostatic oscillations called Langmuir waves at the plasma frequency, f_p, and (2) the Langmuir waves are then converted to electromagnetic radiation via non-linear wave-wave interactions (see Figure 2). This mechanism was first proposed by Ginzburg and Zheleznyakov [1958], and later confirmed by direct in situ observations with the Helios spacecraft [Gurnett and Anderson, 1976]. It is well known that Langmuir waves are generated whenever a beam of sufficient intensity streams through an otherwise quiescent plasma (see Krall and Trivelpiece [1973]). From very general considerations, it can be shown that the Langmuir wave growth rate is proportional to $\partial f/\partial v$, where f is the electron distribution function. An example of a type III burst showing the arrival of electrons from the solar flare, the positive slope in the distribution function ($\partial f/\partial v > 0$), and the simultaneous generation of Langmuir waves is shown by Lin et al. [1986].

Dynamic spectrum measurements [Wild et al., 1954a] show that type III radio emissions are sometimes observed at both the fundamental (f_p) and the harmonic ($2f_p$).

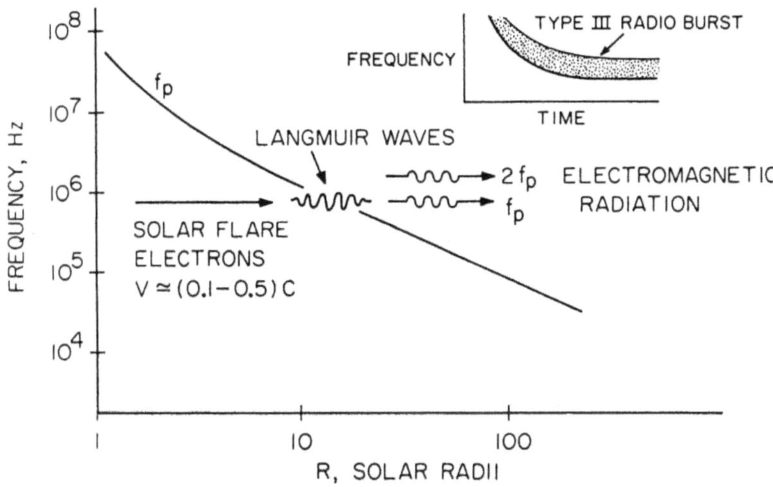

Fig. 2. The downward frequency drift of a type III radio burst is caused by the decreasing electron plasma frequency, f_p, encountered by the solar flare electrons as they stream outward from the Sun. Electromagnetic radiation is produced at f_p and $2f_p$ by mode conversion from Langmuir waves.

According to current ideas the fundamental emission is believed to be produced by a wave-wave interaction between the beam-generated Langmuir wave (L) and a low-frequency ion-sound wave (S). This interaction is often represented by the process L + S → T, where T is the transverse electromagnetic wave. From very general principles (conservation of energy and momentum) it can be shown that the frequencies and wave vectors of the interacting waves must obey the relations $\omega_L + \omega_S = \omega_T$, and $\vec{k}_L + \vec{k}_S = \vec{k}_T$. Since $\omega_S \ll \omega_L \simeq \omega_p$, the energy conservation condition shows that the frequency of the emitted electromagnetic wave must be slightly above the electron plasma frequency. Since the wave number of the Langmuir wave is much larger than the wave number of the electromagnetic wave, $k_L \gg k_T$, the momentum conservation condition shows that $k_S \simeq - k_L$. This relationship shows that the ion-sound wave has a wavelength comparable to the Langmuir wave, and is propagating in the opposite direction. The emission at the harmonic is believed to be produced by a wave-wave interaction between the beam-driven Langmuir wave (L) and a second Langmuir wave (L′) such that L + L′ → T. The frequencies and wave vectors of the interacting waves must satisfy the relations $\omega_L + \omega_L' = \omega_T$, and $\vec{k}_L + \vec{k}_L' = \vec{k}_T$. Since $\omega_L \simeq \omega_L' \simeq \omega_p$, the energy conservation condition shows that $\omega_T \simeq 2\omega_p$. Since again $k_L \gg k_T$, the momentum conservation condition shows that $k_L' \simeq -k_L$. This relationship shows that the second Langmuir wave has a wavelength comparable to the beam-driven Langmuir wave, and is propagating in the opposite direction.

Considerable theoretical attention has been given to the origin of the oppositely propagating waves (S and L′). These waves are thought to arise spontaneously from the beam-driven Langmuir wave via nonlinear decay processes. Although the general principles involved are known, the detailed mechanisms involved are poorly understood and still subject to experimental verification. At least two types of low-frequency waves (one electrostatic, and the other electromagnetic) have been observed in association with beam-driven Langmuir waves [Lin et al., 1986; Kellogg et al., 1992]. Since harmonic emission is usually dominant at the radial distance where direct in situ satellite observations are available [Kaiser, 1975], it is not clear which, if any, of these waves plays the role of S in the fundamental emission process. Recently high-time resolution electric field waveform measurements from the Galileo spacecraft have shown that Langmuir waves associated with a type III radio burst have a characteristic beat pattern [Gurnett et al., 1993a]. This beat pattern is produced by two closely spaced narrowband components. One of these components is believed to be the beam-driven Langmuir wave (L), and the other is believed to be the oppositely propagating Langmuir wave (L′) involved in the harmonic emission process. Cairns and Robinson [1992] have suggested that the oppositely propagating wave is produced by a nonlinear parametric instability.

If the Langmuir wave electric field becomes sufficiently intense a highly nonlinear process called soliton collapse can occur [Zakharov, 1972]. In this process the electric field pressure of the Langmuir wave becomes so large that a depression is produced in the plasma density. The density depression then acts to focus the Langmuir waves into the depression, further increasing the electric field pressure. This feedback process ultimately leads to the collapse of the wave field into small, extremely intense soliton-like packets. Because of the intense fields generated in the collapsed wave packets, these regions are expected to be strong sources of electromagnetic radiation. The change in the wavelength as the collapse proceeds also has the effect of shifting the waves out of resonance with the beam. This shift away from resonance ($\omega/k = V_b$) is believed to answer a long-standing question, which is why the beam is able to propagate such long distances without being disrupted by the resonant Langmuir waves. For a discussion of these and other effects see Papadopoulos and Freund [1978].

Because of the importance of the soliton collapse process, there has been considerable interest in determining whether the Langmuir waves observed in association with type III radio bursts have the expected soliton-like structure. For many years it has been known that the Langmuir waves associated with type III bursts are very spiky, with intense bursts occurring on time scales of a fraction of a second or less [Gurnett and Anderson, 1976]. However, until recently, the available measurements did not have sufficient temporal resolution to resolve the extremely small spatial scales (~ten Debye lengths) at which Langmuir solitons are expected to occur. The first clear observation of Langmuir wave solitons was recently reported by Kellogg et al. [1992], using a Fast Envelope Sampler on the Ulysses spacecraft. These data show the existence of very intense spikes on time scales of about one millisecond, with peak amplitudes of a few mV/m. These spikes are usually imbedded in a background of Langmuir wave emissions with intensities of a few hundred μV/m. At a nominal solar wind velocity of 400 km/s the duration of the spikes corresponds to spatial scales of roughly ten Debye lengths, which is the expected size of Langmuir solitons.

III. Interplanetary Type II Solar Radio Bursts

Interplanetary type II radio bursts are produced by shock waves propagating outward through the solar corona and are characterized by an emission frequency that decreases slowly with increasing time. These radio bursts were first studied in detail by Wild et al. [1954b] who suggested that they were produced by shock waves. It is now known that the shock waves responsible for type II radio bursts are produced by coronal mass ejections (CMEs), which are transient ejections of material from the Sun [Hundhausen, 1993]. CMEs are often, but not always, associated with solar flares and other forms of solar activity. Since CMEs can occur at any solar latitude, the distribution of type II radio bursts is expected to be essentially independent of latitude. For a discussion of the relationship between these various phenomena, see Gosling [1993].

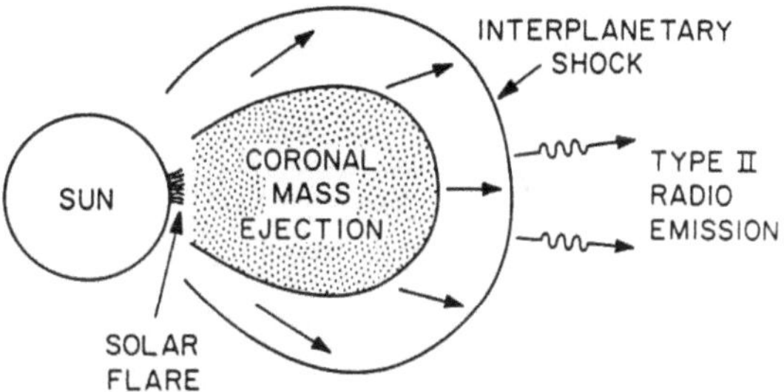

Fig. 3. Type II solar radio bursts are produced by interplanetary shocks driven by coronal mass ejections.

When a shock is produced by a CME, it expands outward as illustrated in Figure 3. As with type III radio bursts, type II radio bursts are believed to be produced by mode conversion from Langmuir waves, very similar to the two-step process involved in the generation of type III radio emissions. However, the electron beam originates from the shock rather than from the Sun. Radiation is observed at both the electron plasma frequency, f_p, and the harmonic $2f_p$. Since the plasma frequency decreases with increasing radial distance from the Sun, the emission frequency decreases with increasing time (see Figure 4), very similar to a type III solar radio burst. However, since the shock propagation speed is much less than the speed of the electrons responsible for type III radio bursts, the frequency drift rates of type II bursts are much slower. Typical shock propagation speeds in the solar wind are in the range from five hundred to one thousand km/s.

Unfortunately, because the source of the type II radio emission seldom extends out to the orbit of the Earth, where in situ spacecraft measurements are available, it has not been possible to verify the details of the type II radio emission process to the same extent that has been possible for type III radio bursts. From various interplanetary spacecraft it is known that Langmuir waves often occur ahead of interplanetary shocks [Gurnett et al., 1979; Kennel et al., 1982]. These Langmuir waves are believed to be produced by electrons accelerated at the shock, very similar to the electron acceleration and generation of Langmuir waves at the planetary bow shocks [Scarf et al., 1971; Filbert and Kellogg, 1979]. At the Earth's bow shock, Langmuir waves are known to produce electromagnetic radiation at f_p and $2f_p$ via nonlinear wave-wave interactions

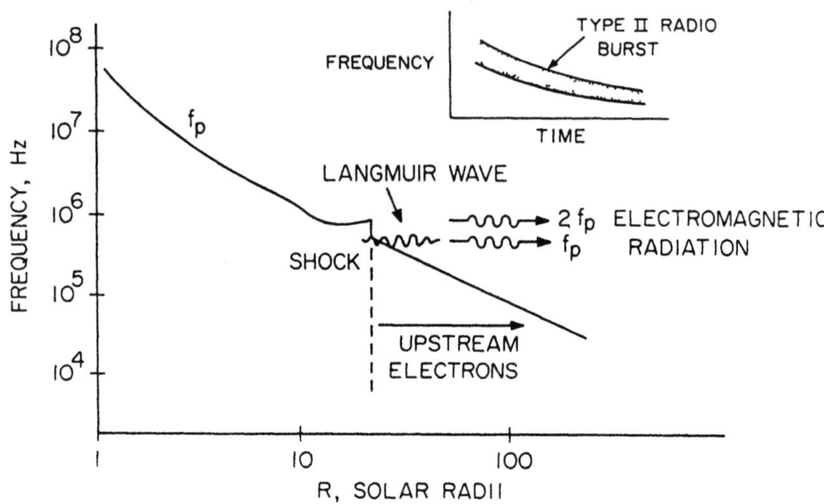

Fig. 4. The downward frequency drift of a type III radio burst is caused by the decreasing electron plasma frequency encountered by the interplanetary shock as it propagates outward from the Sun. Electrons accelerated at the shock are believed to produce Langmuir waves which mode convert to electromagnetic radiation at f_p and $2f_p$.

similar to those believed to occur in type II radio bursts [Dunckel, 1974; Gurnett and Frank, 1975; Hoang et al., 1981].

Langmuir waves are not always observed ahead of interplanetary shocks [Kennel et al., 1982]. The controlling factor is believed to be the orientation of the solar wind magnetic field. From studies of the Earth's bow shock, it is known that the electron acceleration at the shock is most intense and energetic for quasi-perpendicular shocks, where the magnetic field is nearly perpendicular to the shock normal [Anderson et al., 1979]. This magnetic control may explain why the type II radio emission is not generated uniformly over the surface of the shock front, but instead originates from small isolated regions [Nelson and Robinson, 1975].

Although radiation from an electron beam streaming out ahead of an interplanetary shock provides an appealing picture of the generation of type II radio emissions, it is by no means clear that this picture provides a complete description of the type II radio emission process. Comparisons of radio direction-finding observations with spacecraft coronagraph images of coronal mass ejection events [Wagner, 1983] and other techniques [Lengyel-Frey, 1992] show that the radio emission source is often located behind the shock front. Such observations suggest that other processes may be operative to excite Langmuir waves and radio emission in the turbulent region behind the shock front. For a further discussion of type II radio emission mechanisms, see Nelson and Melrose [1985].

IV. Heliospheric 2-3 kHz Radio Emissions

For over ten years the Voyager 1 and 2 spacecraft have been detecting an unusual radio emission in the outer heliosphere in the frequency range from about 2 to 3 kHz. Two particularly strong events have occurred, the first in 1983-84 [Kurth et al., 1984] and the second in 1992-93 [Gurnett et al., 1993b]. Five other much weaker events have also been observed, one in late 1985, one in 1989, and three in 1990-91 [Kurth et al., 1987; Kurth and Gurnett, 1991]. Several possible sources have been considered, including (1) planetary, (2) heliospheric, and (3) stellar. Since the solar wind electron plasma frequency does not drop below 2 kHz until roughly the orbit of Saturn, the source of these radio emissions must be located relatively far from the Sun, at least R $\gtrsim$ 10 AU. The total radiated power has been estimated by Gurnett et al. [1993b] to be at least 10^{13} W, which effectively rules out planetary sources (also see Gurnett and Kurth [1994]). Stellar sources, such as pulsars, are difficult to rigorously rule out, but tend to involve implausibly large radiated powers when the propagation must extend over interstellar distances.

The best current explanation of the 2-3 kHz radio bursts is that the emissions are produced in the outer regions of the heliosphere by a disturbance propagating outward from the Sun. The size of the heliosphere is controlled by pressure balance between the solar wind and the interstellar medium [Parker, 1963]. As currently envisaged, the interface between the solar wind and the interstellar medium consists of a shock, called the terminal shock, where the solar wind flow becomes subsonic, and a contact discontinuity, called the heliopause, where the solar wind pressure is balanced by the pressure of the interstellar medium. Because the Sun is moving through the local interstellar medium at a speed of about 26 km/s [Lallement et al., 1993], the heliopause is expected to form a bullet-shaped boundary around the Sun. The distance from the Sun to the nose of the heliosphere is estimated to be on the order of 100 AU. If the interstellar plasma flow is supersonic, a second shock, called the bow shock, is expected to form in the interstellar plasma flow around the heliosphere. For a further discussion of the structure of these boundaries, including the effect of neutral gas interactions, see the reviews by Axford [1990] and Suess [1990]. No spacecraft has yet reached either the terminal shock or the heliopause.

McNutt [1988] first suggested that a solar wind disturbance could "trigger" a burst of 2-3 kHz radio emission. He proposed that the radio emission was produced by an interaction of a high-speed solar wind stream with the terminal shock. This idea was further explored by Grzedzielski and Lazarus [1993], who identified a series of dynamic pressure increases in the solar wind that they believed were responsible for the 1983-84, 1985, and 1989 events, again assuming an interaction with the terminal shock. Despite the possible merit of this idea, the cause-effect relationship was not convincingly established. Later, in the process of studying the 1992-93 radio emission event, Gurnett et al. [1993b] discovered that the large radio emission events in 1983-84 and 1992-93 both occurred about 400 days after large Forbush decreases in the cosmic ray intensity. This relationship is illustrated in Figure 5, which shows the cosmic ray intensity from the Deep River neutron monitor in the top panel, and the 3.1-kHz radio emission intensity from Voyager 1 in the bottom panel. The two Forbush decreases, marked A and B in Figure 5, were produced by periods of intense solar activity in mid-July, 1982, and late-May/early-June, 1991. These were the two largest Forbush decreases (21% and 30%, respectively) observed in over 30 years for which such data are available. As can be seen, the 1983-84 radio emission event started about 412 days after event A, and the 1992-93 radio emission event started about 419 days after event B. It is well known that Forbush decreases are caused by a global interplanetary shock and associated magnetic disturbances propagating outward from the Sun. In both cases the shocks and associated disturbances were detected by a variety of interplanetary spacecraft. For a discussion of these events, including estimates of the propagation speeds, see Van Allen and Randall [1985], Webber et al. [1986], Van Allen and Fillius [1992], and Webber and Lockwood [1993].

Based on the above observations, a strong case can be made that the 2-3 kHz radio emissions are produced by the interaction of an interplanetary shock with the outer

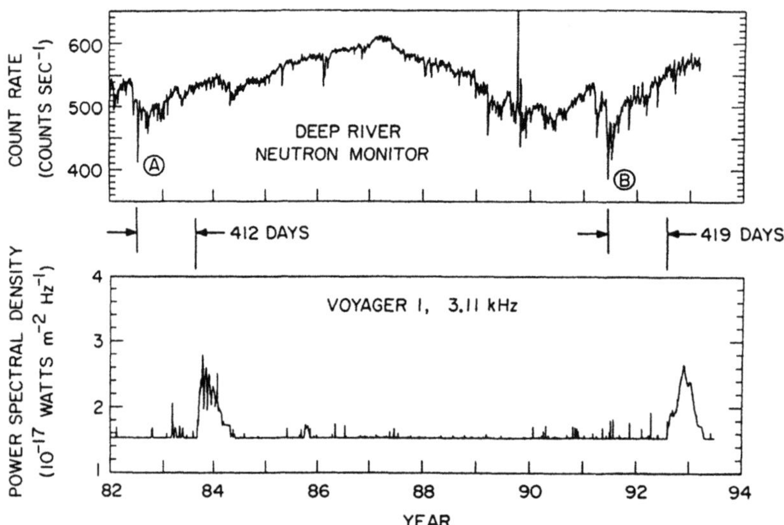

Fig. 5. The top panel shows the cosmic ray intensity from the Deep River neutron monitor over a 12-year period from 1982 to 1994. The bottom panel shows the corresponding 3.1-kHz heliospheric radio emission intensity detected by Voyager 1. The strong radio emission events in 1983-84 and 1992-93 occurred about 400 days after two large Forbush decreases, marked A and B. These Forbush decreases were caused by interplanetary shocks and associated disturbances propagating outward from the Sun after periods of intense solar activity in 1982 and 1991.

boundaries of the heliosphere. Two obvious boundaries could be involved in this interaction, the terminal shock and the heliopause. For the typical shock propagation speeds involved, which are in the range from 550 to 800 km/s, the long transit time, ~400 days, would indicate that the interaction must take place at a distance from the Sun of at least 100 AU. At the present time, there is only one known mechanism that could produce these radio emissions, namely mode conversion at f_p and $2f_p$ from Langmuir waves excited by a shock [see Cairns et al., 1992]. To decide between a source at the terminal shock and a source at the heliopause, it is necessary to consider the radial variation of f_p. For a constant solar wind velocity, it is easy to show that the electron plasma frequency must vary inversely with distance from the Sun ($f_p \sim 1/R$). A representative radial variation of f_p is shown in Figure 6. Since an average value for f_p at 1 AU is 20 kHz, the electron plasma frequency at 100 AU should be about 200 Hz. According to conventional MHD theory, the jump in the plasma frequency at the terminal shock can be no more than a factor of two. Thus, the maximum plasma frequency at the terminal shock should be about 400 Hz. This plasma frequency is too low to account for radio emissions at 2-3 kHz. Thus, it is believed that an interaction at the terminal shock cannot account for the heliospheric 2-3 kHz radio emission. The only known way in which the terminal shock could be the source would be if the density jump substantially exceeds the MHD limit due to the effect of cosmic rays, as has been suggested by Donohue and Zank [1993], or if the terminal shock were to be located much closer than 100 AU, which would invalidate the cause-effect relationship suggested by Figure 5.

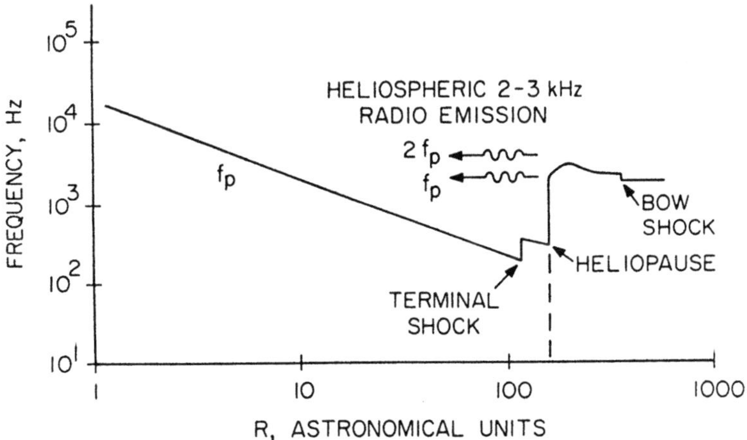

Fig. 6. A representative electron plasma frequency profile through the terminal shock, the heliopause, and the bow shock. Using nominal parameters the heliopause is believed to be the most likely source of the 2-3 kHz radiation, since the plasma frequency is too low in the vicinity of the terminal shock.

At the heliopause, the situation is much more favorable for generating radio emissions in the required frequency range. Since the heliopause is a contact discontinuity, the density can increase by whatever factor is required to maintain pressure balance. Present estimates are that the electron density in the local interstellar medium is in the range from 0.06 to 0.1 cm^{-3} [Lallement et al., 1993]. Since the bow shock is expected to be a weak shock, the interstellar electron density in the vicinity of the heliopause should be approximately 2.2 to 2.8 kHz. This plasma frequency is consistent with the frequency range of the 2-3 kHz radio emissions. For this reason, Gurnett et al. [1993b] proposed that the 1983-84 and 1992-93 heliospheric radio emission events were caused by an interaction involving the heliopause and the interplanetary shocks responsible for the giant Forbush decreases observed in 1982 and 1991. For propagation speeds in the range from 550 to 800 km/s, which are consistent with the available observations, and a travel time of ~400 days, the radial distance to the heliopause is estimated to be in the range from 106 to 177 AU. The corresponding distances to the terminal shock are about 80 to 133 AU.

V. Prospects for Future Progress

The forthcoming passage of the Ulysses spacecraft over the poles of the Sun and the continued progress of Voyagers 1 and 2 to greater distances from the Sun are both likely to lead to significant advances in our understanding of heliospheric radio emissions. The ability of the Ulysses plasma wave instrument to capture and store Langmuir wave amplitudes on time scales of 1 ms, or less, is an important capability that may lead to a better understanding of the fine structure of Langmuir waves, and the generation of type II and type III radio emissions. The two-dimensional radio direction-finding capability of Ulysses [Stone et al., 1992] will provide a unique capability for studying the latitudinal dependence of solar radio bursts as the spacecraft passes over the poles of the Sun. This direction-finding capability would also be of great interest for studying the heliospheric 2-3 kHz radio emissions. Although the average electron plasma frequency at the Ulysses orbit appears to be too high to allow the detection of the 2-3 kHz radiation, if the spacecraft passes through a region of sufficiently low plasma density (f_p < 3.6 kHz), every effort should be made to detect these emissions.

References

Anderson, K.A., Lin, R.P., Martel, F., Lin, C.S., Park, G.K., and Reme, H.: 1979, 'Thin sheets of energetic electrons upstream from the Earth's bow shock', *Geophys. Res. Lett.* **6**, 401-404.

Axford, W.I., 1990: 'Introductory Lecture - The Heliosphere', in S. Grzedzielski and D.E. Page (eds.), *Physics of the Outer Heliosphere*, COSPAR Colloquia Series, **1**, Pergamon Press, Oxford, p. 7-15.

Cairns, I.H., and Robinson, P.A.: 1992, 'Theory for low-frequency modulated Langmuir wave packets', *Geophys. Res. Lett.* **19**, 2187-2190.

Cairns, I.H., Kurth, W.S., and Gurnett, D.A.: 1992, 'Outer heliospheric radio emissions, 2. Foreshock source models', *J. Geophys. Res.* **97**, 6245-6259.

Donohue, D.J., and Zank, G.P.: 1993, 'Steady state and dynamical structure of a cosmic-ray-modified termination shock', *J. Geophys. Res.* **98**, 19,005-19,025.

Dulk, G.A., Steinberg, J.-L., Hoang, S., and Lecacheux, A.: 1986, 'Latitude distribution of interplanetary magnetic field lines rooted in active regions', in R.G. Marsden (ed.), ***The Sun and the Heliosphere in Three Dimensions***, D. Reidel Publising Company, 229-233.

Dunckel, N.: 1974, 'Low-frequency radio emissions from the Earth and Sun', *Tech. Rep. 3469-2*, Radioscience Laboratory, 114-131.

Fainberg, J., Evans, L.G., and Stone, R.G.: 1972, 'Radio tracking of solar energetic particles through interplanetary space', *Science* **178**, 743-745.

Filbert, P.C., and Kellogg, P.J.: 1979, 'Electrostatic noise at the plasma frequency beyond the Earth's bow shock', *J. Geophys. Res.* **84**, 1369-1381.

Ginzburg, V.L., and Zheleznyakov, V.V.: 1958, 'On the possible mechanism of sporadic solar radio emission (radiation in an isotropic plasma)', *Sov. Astron.* **AJ2**, 653.

Gosling, J.T.: 1993, 'The Solar Flare Myth', *J. Geophys. Res.* **98**, 18,937-18,949.

Grzedzielski, S., and Lazarus, A.J.: 1993, '2- to 3-kHz continuum emissions as possible indications of global heliospheric "breathing"', *J. Geophys. Res.* **98**, 5551-5558.

Gurnett, D.A., and Anderson, R.R.: 1976, 'Electron plasma oscillations associated with type III radio bursts', *Science* **194**, 1159-1162.

Gurnett, D.A., and Frank, L.A.: 1975, 'Electron plasma oscillations associated with Type III radio emissions and solar electrons', *Solar Physics* **45**, 477-493.

Gurnett, D.A., Hospodarsky, G.B., Kurth, W.S., Williams, D.J., and Bolton, S.J.: 1993a, 'Fine structure of Langmuir waves produced by a solar electron event', *J. Geophys. Res.* **98**, 5631-5637.

Gurnett, D.A., and Kurth, W.S.: 1994, 'Evidence that Jupiter is not the source of the 2-3 kHz heliospheric radio emission', *Geophys. Res. Lett.*, in press.

Gurnett, D.A., Neubauer, F.M., and Schwenn, R.: 1979, 'Plasma wave turbulence associated with an interplanetary shock', *J. Geophys. Res.* **84**, 541-552.

Gurnett, D.A., Kurth, W.S., Allendorf, S.C., and Poynter, R.L.: 1993b, 'Radio emission from the heliopause triggered by an interplanetary shock', *Science* **262**, 199-203.

Hoang, S., Fainberg, J., Steinberg, J.L., Stone, R.G., and Zwickl, R.D.: 1981, 'The $2f_p$ circumterrestrial radio radiation as seen from ISEE-3', *J. Geophys. Res.*, **86**, 4531-4536.

Hundhausen, A.J.: 1993, 'The size and location of coronal mass ejections: SMM observations from 1980 and 1984-89', *J. Geophys. Res.* **98**, 13,177.

Kaiser, M.L.: 1975, 'The solar elongation distribution of low-frequency radio bursts', *Solar Phys.* **45**, 181-187.

Kellogg, P.J., Goetz, K., Lin, N., Monson, S.J., Balogh, A., Forsyth, R.J., and Stone, R.G.: 1992, 'Low-frequency magnetic signals associated with Languir wave packets', *Geophys. Res. Lett.* **19**, 1299.

Kennel, C.F., Scarf, F.L., Coroniti, F.V., Smith, E.J., and Gurnett, D.A.: 1982, 'Nonlocal plasma turbulence associated with interplanetary shocks', *J. Geophys. Res.* **87**, 17-34.

Krall, N.A., and Trivelpiece, A.W.: 1973, *Principles of Plasma Physics*, McGraw-Hill, New York.

Kurth, W.S., and Gurnett, D.A.: 1991, 'New observations of the low frequency interplanetary radio emissions', *Geophys. Res. Lett.* **18**, 1801-1804.

Kurth, W.S., Gurnett, D.A., Scarf, F.L., and Poynter, R.L.: 1984, 'Detection of a radio emission at 3 kHz in the outer heliosphere', *Nature* **312**, 27-31.

Kurth, W.S., Gurnett, D.A., Scarf, F.L., and Poynter, R.L.: 1987, 'Long-period dynamic spectrograms of low-frequency interplanetary radio emissions', *Geophys. Res. Lett.* **14**, 49-52.

Lallement, R., Bertaux, J.-L., and Clark, J.T.: 1993, 'Deceleration of interstellar hydrogen at the heliospheric interface', *Science* **260**, 1095-1098.

Lengyel-Frey, D.: 1992, 'Location of the radio emitting regions of interplanetary shocks', *J. Geophys. Res.* **97**, 1609-1617.

Lin, R.P.: 1970, 'The emission and propagation of 40 keV solar flare electrons', *Solar Phys.* **12**, 266.

Lin, R.P., Levedahl, W.K., Lotko, W., Gurnett, D.A., and Scarf, F.L.: 1986, 'Evidence for nonlinear wave-wave interactions in solar type III radio burst', *Astrophys. J.* **308**, 954-965.

McNutt, R.L., Jr.: 1988, 'A solar-wind "trigger" for the outer heliospheric radio emissions and the distance to the terminal shock', *Geophys. Res. Lett.* **15**, 1307-1310.

Nelson, G.J., and Melrose, D.: 1985, 'Type II bursts', in D.J. McLean and N.R. Labrum (eds.), *Solar Radiophysics*, Cambridge University Press, Cambridge, 331-359.

Nelson, G.J., and Robinson, R.D.: 1975, 'Multi-frequency heliograph observations of type II bursts', *Proc. Astron. Soc. Aust.* **2**, 370.

Papadopoulos, K., and Freund, H.P.: 1978, 'Solitons and second harmonic radiation in type III bursts', *Geophys. Res. Lett.* **5**, 881-884.

Parker, E.N.: 1963, *Interplanetary Dynamical Process*, Interscience, Rochester, NY, 137.

Scarf, F.L., Fredricks, R.W., Frank, L.A., and Neugebauer, M.: 1971, 'Nonthermal electrons and high-frequency waves in the upstream solar wind: 1. Observations', *J. Geophys. Res.* **76**, 5162-5171.

Scarf, F.L., and Gurnett, D.A.: 1977, 'A plasma wave investigation for the Voyager mission', *Space Sci. Rev.* **21**, 289-308.

Stone, R.G., et al.: 1992, 'The unified radio and plasma wave investigation', *Astron. Astrophys. Suppl. Ser. 92*, 291-316.

Suess, S.T.: 1990, 'The heliopause', *Rev. Geophys.*, **28**, 97-115.

Van Allen, J.A., and Fillius, R.W.: 1992, 'Propagation of a large Forbush decrease in cosmic-ray intensity past the Earth, Pioneer 11 at 34 AU, and Pioneer 10 at 53 AU', *Geophys. Res. Lett.* **19**, 1423-1426.

Van Allen, J.A., and Randall, B.A.: 1985, 'Interplanetary cosmic ray intensity: 1972-1984 and Out to 32 AU', *J. Geophys. Res.* **90**, 1399-1412.

Wagner, W.J.: 1983, 'SERF studies of mass motions arising in flares', *Adv. Space Res.* **2**, 203.

Webber, W.R., and Lockwood, J.A.: 1993, 'Giant transient decreases of cosmic rays in the outer heliosphere in September 1991', *J. Geophys. Res.* **98**, 7821-7825.

Webber, W.R., Lockwood, J.A., and Jokipii, J.R.: 1986, 'Characteristics of large Forbush-type decreases in the cosmic radiation, 2. Observations at different radial distances', *J. Geophys. Res.* **91**, 4103-4110.

Wild, J.P.: 1950, 'Observations of the spectrum of high-intensity solar radiation at metre wavelengths, III, Isolated bursts', *Aust. J. Sci. Res. Ser. A*, **3**, 541.

Wild, J.P., Murray, J.D., and Rowe, W.C.: 1954b, 'Harmonics in the spectra of solar radio disturbances', *Aust. J. Phys.* **7**, 439.

Wild, J.P., Roberts, J.A., and Murray, J.D.: 1954a, 'Radio evidence of the ejection of very fast particles from the Sun', *Nature* **173**, 532.

Zakharov, V.E.: 1972, 'Collapse of Langmuir Waves', *Zh. Exp. Theoret. Fiz.* **62**, 1745, (English transl. in *Soviet Phys. - JETP* **35**, 908).

INTERPLANETARY TYPE III RADIO BURSTS OBSERVED FROM HIGH SOUTHERN HELIOGRAPHIC LATITUDES

M. J. REINER
Hughes STX, Lanham, MD, 20706, USA

and

J. FAINBERG and R. G. STONE
NASA/GSFC, Greenbelt, MD, 20771, USA

Abstract. Type III radio bursts observed from high southern latitudes are analyzed for the first time. The continual observation of these radio bursts by Ulysses from after the Jupiter swing-by to 50°S latitude argues for a wide latitudinal directivity of type III radiation. From this high latitude perspective, type III radio sources that lie in the far hemisphere of the Sun with respect to Ulysses are unambiguously resolved for the first time. Using the Ulysses direction-finding capabilities, the radio source locations in the 3-D heliosphere are derived for a radio event on 31 January 1994 when Ulysses was ~45°S latitude. The source locations describe a spiral-like trajectory originating from the far side of the Sun. The angular radii of these radio sources are compared to angular radii that were previously derived from in-ecliptic observations.

1. Introduction

Interplanetary (kilometric) type III radio bursts are produced by suprathermal electrons ejected from the Sun that propagate along open magnetic field lines in the interplanetary medium (IPM). It has been directly established that this radiation is generated at both the fundamental and harmonic of the local plasma frequency (Reiner *et al.*, 1992). Type III radio emissions are characterized by increasingly later onset times with decreasing observing frequencies, which results from the propagation of electrons through regions of decreasing plasma density. Since the interplanetary plasma density falls off roughly uniformly as r^{-2} (r = heliocentric distance), a radio source at a given frequency must lie somewhere on a sphere of known radius--the so-called plasma sphere. Several interplanetary density scales have been derived from observations (Fainberg and Stone, 1971; Bougeret *et al.*, 1984a, 1984b). Thus type III radio sources, observed at decreasing frequency, are located at increasing heliocentric distances.

Type III radio emission beams are characterized by a wide longitudinal directivity (~60°), which results from the finite width of the electron stream as well as from the intrinsic beamwidth at the source (Steinberg *et al.*, 1984, 1985; Reiner and Stone 1988, 1989). Scattering and refraction of the radiation by density inhomogeneities in the IPM have been shown to play an important role, for observations made within the ecliptic plane (Steinberg *et al.*, 1984, 1985). Indeed, many observed type III radio bursts actually originate behind the solar limb (Dulk *et al.*, 1985; Lecacheux *et al.*, 1989) and are only visible to an observer on the near side of the Sun due to scattering of the radiation in the

Space Science Reviews **72**: 255–260.
© 1995 *Kluwer Academic Publishers.*

IPM. Reiner and Stone (1990) have presented indirect evidence that these behind-the-limb or far hemisphere events may sometimes appear, to spacecraft located in the ecliptic plane, as halo-like sources surrounding the Sun.

The use of dipole antennae on spinning spacecraft to locate radio sources in the IPM was first developed by Fainberg (1979). This technique was applied to interplanetary type III radio bursts, permitting the first remote tracking of suprathermal electron streams and consequently mapping of global interplanetary magnetic field topologies (Fainberg *et al.,* 1972). The derived radio source locations revealed the spatial trajectory of the electron stream through the IPM and confirmed the spiral shape (Parker, 1963) of interplanetary magnetic fields.

Prior to Ulysses, type III radio source characteristics and locations were determined exclusively from spacecraft located within the ecliptic plane. Since these radio sources generally remain close to the ecliptic plane (Dulk *et al.,* 1986), as illustrated schematically in Fig. 1a, only a foreshortened view of a spiral-like type III radio trajectory is observed from spacecraft located near the ecliptic plane. Trajectories for far hemisphere radio bursts, however, are obscured by the intervening plasma sphere so that only a complex (perhaps halo-like) radio source is visible, which is difficult to distinguish from near hemisphere sources using dipole antennae.

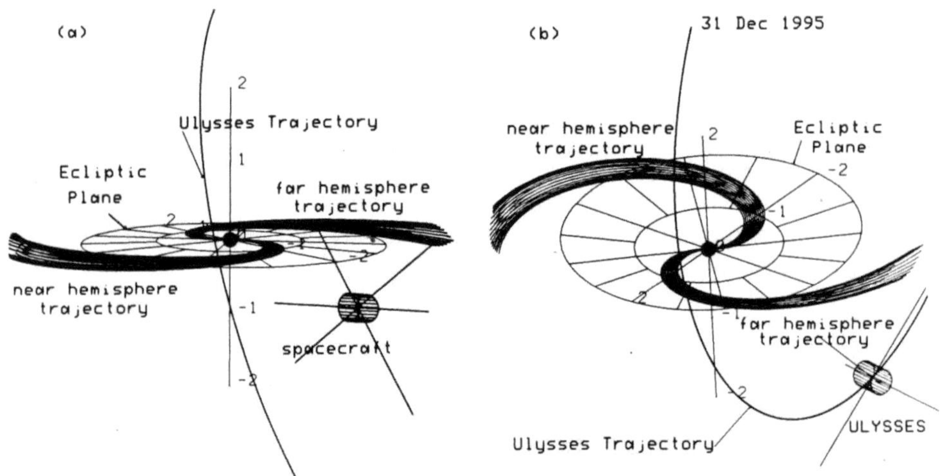

Fig. 1. (a) Foreshortened view of near and far hemisphere spiral-like type III radio trajectories viewed from 8°N ecliptic latitude. (b) View of the same radio trajectories from 30°S latitude.

Ulysses provides the first opportunity for observing interplanetary type III radio bursts from high heliographic latitudes. Such observations provide new information on the latitudinal dependence of the radio beamwidth and on scattering of the radiation in the source region. In addition, high-latitude observations provide, for the first time, a clearer view of far hemisphere radio sources and their radio trajectories, as illustrated in Fig. 1b. As Fig. 1b indicates, near and far hemisphere type III radio bursts may be distinguished by the fact that their spiral-like radio trajectories drift in opposite directions with decreasing observing frequency: the near hemisphere trajectory drifting from west to east and the far hemisphere trajectory from east to west. This of course assumes that the beamwidth for these radio sources extends to sufficiently high latitudes that the both remain visible to Ulysses.

2. Ulysses High-Latitude Observations

Type III radio bursts have been observed continuously by Ulysses since it advanced to high (southern) ecliptic latitudes after the Jupiter swing-by in February 1992. These observations demonstrate that the latitudinal directivity for type III radio bursts extends at least 50° in latitude. There is no evidence to suggest that the radio intensities of these bursts have diminished, even at high frequencies (~1MHz) where the beamwidth might be expected to be narrower. Furthermore, as we will illustrate below, far hemisphere events are also readily observed from high latitudes.

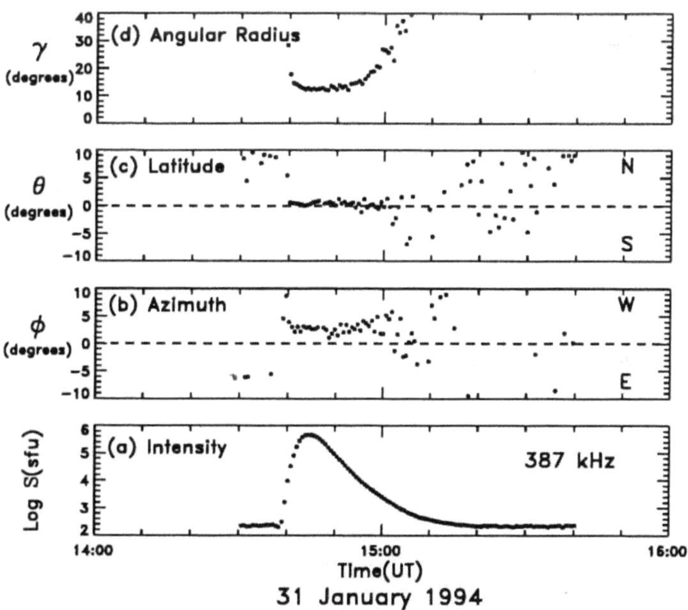

Fig. 2. a.) Radio intensity profile at 387 kHz from 14:00 to 16:00 UT on 31 January 1994 that was observed by Ulysses at 44°S ecliptic latitude. b.) Azimuthal angle of the radio source relative to the azimuth of the Sun (dashed line) as observed from Ulysses. c.) Latitudinal angle of the radio source relative to the latitude of the Sun (dashed line). d). Angular radius of the radio source assuming a uniform circular disk.

The intensity-time profile at 387 kHz of a type III radio burst that was observed by Ulysses at ~14:30 UT on 31 January 1994 is shown in Fig. 2a. At the time of these observations, the Earth was 56° east of the Ulysses-Sun line and Ulysses was ~44° south of the ecliptic plane, 3.7 AU from the Sun. No solar flare was observed from Earth at the time of the type III burst. However, a flare may have occurred behind the solar limb as viewed from Earth.

The direction-finding capabilities on Ulysses are achieved by analyzing the modulation in the RF signals from the two spacecraft antennae as the spacecraft spins, with a 12s period. Figs. 2b and 2c show the azimuth and latitude of the radio emission at 387 kHz, relative to the Sun direction, as viewed from Ulysses. These angles are expressed in a coordinate system centered at Ulysses with x axis passing through the Sun and z axis in the plane containing the ecliptic pole. The observed source azimuth indicates that, at 387 kHz, the radio source was west of the Ulysses-Sun line.

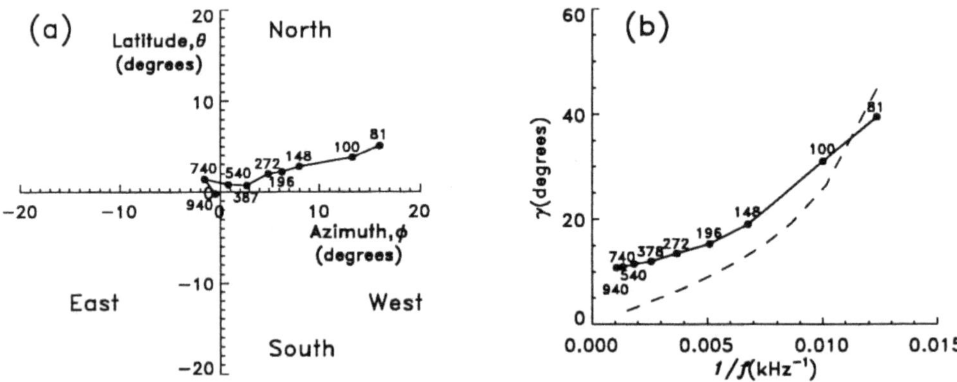

Fig. 3. a.) Derived radio azimuth ϕ and latitude θ, relative to the Sun direction, as observed from Ulysses. The numbers near the data points indicate the frequency in kHz. b.) Source angular radius γ as a function of inverse frequency. The dashed line represents the expectation based on previous analyses by ISEE-3, scaled to the distance of Ulysses.

Fig. 3a shows that at higher frequencies (940, 740 kHz) the type III radio sources were east of the Ulysses-Sun line, but as the observing frequency decreased the radio source directions steadily drifted to the west. This azimuthal angular drift is just opposite to that expected for suprathermal electrons propagating along spiral field lines originating in the near hemisphere of the Sun, as viewed from Ulysses. This radio burst is therefore believed to be a far hemisphere event, seen for the first time, from high southern latitudes. The latitudinal angle, also indicated in Fig. 3a, was slightly north of the Sun direction for the entire range of observing frequencies.

The direction-finding capabilities on Ulysses also yields the effective angular radius of a radio source. Fig. 2d shows that the angular radius of the radio source at 387 kHz was ~12° (assuming that this source can be represented by a circular disk of uniform intensity). It has been established from in-ecliptic observations that the angular radius of a radio source scales as the inverse of the frequency (Steinberg *et al.*, 1985). It is of interest to determine if a similar dependence exists for the high-latitude Ulysses observations. As Fig. 3b shows, the source angular radius for the 31 January burst does not appear to scale as *1/f*. For comparison, the dashed curve shows the angular radius that would be expected for a "typical" type III radio burst derived from in-ecliptic observations, corrected geometrically for the larger heliocentric distance to Ulysses (3.7 AU). The derived angular radius for the 31 January burst is generally larger than expected. This may be a consequence of its far hemisphere origin, which renders it visible to Ulysses primarily due to interplanetary scattering in the source region.

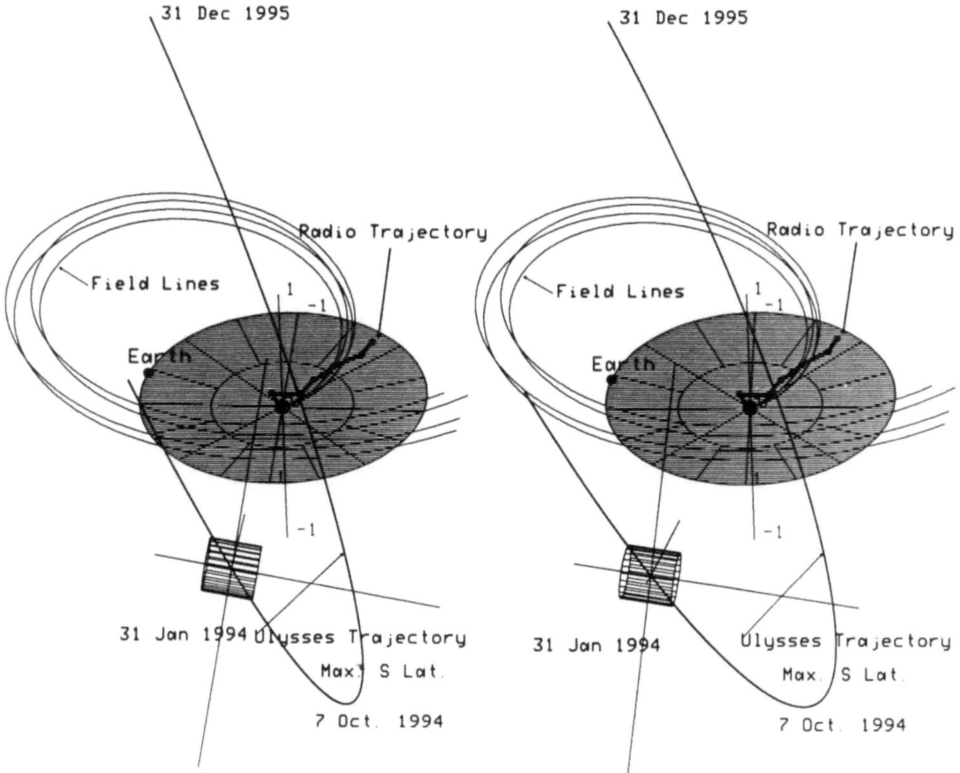

Fig. 4. Stereo view of the trajectory derived from the Ulysses direction-finding for the type III event on 31 January 1994. To view the 3-D image, relax eyes (by focusing at a distant object through the page) and allow the two images to merge into a third image, which will be seen in 3-D.

Type III bursts have also been observed whose angular radii at low frequencies were significantly smaller than those indicated in the dashed curve. This may be an indication of reduced interplanetary scattering of the radiation near Ulysses due to the low plasma density and absence of interplanetary structures. In general, interplanetary scattering depends on the square of the plasma density at the spacecraft so that at Ulysses the angle broadening of a radio source is expected to be significantly less than for in-ecliptic spacecraft (Steinberg *et al.*, 1985).

The angles in Fig. 3a give the 3-D direction to the radio source at each observing frequency, but they do not indicate where along the line-of-sight from Ulysses the source is located. To determine the radio source location, additional modelling is required. The simplest way to derive the source location is to assume that the interplanetary density falls off uniformly in all directions in the 3-D heliosphere and that the radio source, at a given frequency, lies on the corresponding plasma sphere. The 3-D radio source location is then given as the intersection of the line-of-sight with this plasma sphere. Using the RAE density scale (Fainberg and Stone, 1971), we find that for the direction angles in Fig. 3a the source locations yield a type III radio trajectory in the far hemisphere of the Sun. A stereo view of this trajectory, as viewed from Ulysses, is shown in Fig. 4. The derived radio locations at decreasing frequency follow a spiral-like path from the Sun that lies north of the ecliptic plane. This is an example of a far hemisphere type III trajectory, which because of the high heliographic latitude of Ulysses, is seen in full view.

3. Conclusion

Type III radio bursts have been continuously observed by Ulysses at high southern ecliptic latitudes, indicating a wide latitudinal directivity of type III radio emission beams, in addition to the well-established wide longitudinal beamwidth. From high latitudes, far hemisphere type III radio events are observed, presumably due to interplanetary scattering in the source region, and their radio trajectories are unambiguously resolved for the first time. The angular radii of type III radio sources observed at high latitudes appears to differ significantly from the in-ecliptic results; they does not appear to scale as *1/f*.

References

Bougeret, J.-L. , King, J. H., and Schwenn, R.: 1984a, *Solar Phys.* **90**, 401.
Bougeret, J.-L., Fainberg, J. and Stone, R. G.: 1984b, *Astron. Astrophys.* **141**, 17.
Dulk, G.A., et al.: 1985, *Astron. Astrophys.* **150**, L28.
Dulk, G.A., et al.: 1986, in R. G. Marsden (ed.), *The Sun and the Heliosphere in Three Dimensions,* D. Reidel Publ. Co., Dordrecht, pp. 229-233.
Fainberg, J., Evans, L.G., and Stone, R.G.: 1972, *Science* **178**, 743.
Fainberg, J. and Stone, R.G.: 1971, *Solar Phys.* **17**, 392.
Fainberg, J.: 1979, NASA TM 80598.
Lecacheux, A, et al.: 1989, *Astron. Astrophys.* **217**, 237.
Parker, E. N.: 1963, *Interplanetary Dynamical Processes*, Interscience Publishers, New York.
Reiner, M.J. and Stone, R.G.: 1988, *Astron. Astrophys.* **206**, 316.
Reiner, M.J. and Stone, R.G.: 1989, *Astron. Astrophys.* **217**, 251.
Reiner, M.J. and Stone, R.G.: 1990, *Solar Physics.* **125**, 371.
Reiner, M.J., Stone, R.G., Fainberg, J.: 1992, *Astrophys. J.* **394**, 340.
Steinberg, J.-L., Hoang, S. and Dulk, G.A.: 1985, *Astron. Astrophys.* **150**, 205.
Steinberg, J.-L. et al.: 1984, *Astron Astrophys.* **140**, 39.

ULYSSES/GALILEO OBSERVATIONS OF TYPE III RADIO BURSTS AND ASSOCIATED IN-SITU ELECTRONS AND LANGMUIR WAVES

M. J. REINER[1], K. A. ANDERSON[2], E. ROELOF[3], T. ARMSTRONG[4],
G. B. HOSPODARSKY[5], J. FAINBERG[6], R. G. STONE[6], L. LANZEROTTI[7],
D.A. GURNETT[5], J. L. PHILLIPS[8], R. FORSYTH[9]

[1]*Hughes STX, Lanham, MD, 20706, USA*
[2]*Space Science Lab., Univ. of Cal., Berkeley, CA, 94720, USA*
[3]*Johns Hopkins Univ./APL, Laurel, MD, 20723, USA*
[4]*Dept. of Phys. & Astron., Univ. of Kansas, Lawrence, KS, 66045, USA*
[5]*Dept. of Phys. & Astron., Univ. of Iowa, Iowa City, IA, 52242, USA*
[6]*NASA/GSFC, Greenbelt, MD, 20771, USA*
[7]*ATT, Bell Labs., NJ, 07974, USA*
[8]*Los Alamos National Lab., Los Alamos, NM, 87545, USA*
[9]*The Blackett Lab., Imperial College, London, SW7 2BZ, UK*

Abstract. Both the Ulysses and Galileo spacecraft detected energetic electrons and Langmuir waves that were associated with a type III radio burst on 10 December 1990. At the time of these observations, these spacecraft were in the ecliptic plane and separated by 0.4 AU, with Galileo near the Earth at 1 AU and Ulysses at 1.36 AU. From the measured electron arrival times, the propagation path lengths of the electrons to both Ulysses and Galileo were estimated to be significantly longer than the length of the Parker spiral. These long path lengths are interpreted as due to draping of the interplanetary magnetic field lines around a CME. The onset times of the Langmuir waves at Ulysses and Galileo coincided with the estimated arrival time of the 9 keV and 14 keV electrons, respectively.

1. Introduction

Type III solar radio bursts are produced by streams of suprathermal electrons that propagate along open magnetic field lines into the interplanetary medium. Most analyses of these electrons have been restricted to measurements made by a single spacecraft located within the electron stream (Lin, 1985). Observations of series of flare associated electron events, originating from a common solar active region, indicate that the longitudinal width of electron streams is $\geq 60^{\circ}$ (Lin, 1970a). Propagation path lengths and solar injection times of electrons were deduced from measured electron velocity dispersive onsets (Potter *et al.*, 1980). The derived path lengths were often consistent with the length of the Parker spiral, suggesting scatter-free propagation.

The onset of type III associated Langmuir waves is directly correlated with the formation of a "bump-on-tail" in the measured electron distribution function (Lin *et al.*, 1981; 1986). The Langmuir waves generate radio emissions at the fundamental and harmonic of the local plasma frequency (Reiner *et al.*, 1992). The remotely observed type III radio emission is characterized by increasing onset times with decreasing frequency due to the fall off of the interplanetary density with heliocentric distance.

Space Science Reviews **72**: 261–266.
© 1995 *Kluwer Academic Publishers.*

During November-December 1990 the relative configuration of Ulysses and Galileo was such that both spacecraft were often simultaneously located within electron streams associated with western solar flares. This unique configuration permits the characteristics of the electrons and associated Langmuir waves to be simultaneously measured and compared at two spatially separated points.

2. Ulysses/Galileo Observations of a Type III Radio Burst

We examine a type III radio burst on 10 December 1990 that was previously analyzed by Gurnett *et al.* (1993) using the Galileo data alone. Both Ulysses and Galileo observed the frequency-drifting radio emission from this type III burst, which was associated with a western 2F flare (N15°W47°). Figure 1 shows a dynamic spectrum of the radio emission observed by Ulysses. Many type III radio bursts were observed during this time. The one of interest is the intense burst beginning at ~07:45 UT at 940 kHz. The onset of the ~5 MHz radio emission observed at 07:42 UT at Galileo suggests that the type III associated electrons were injected into the interplanetary medium at ~07:34 UT. The frequency-drifting radiation was observed to very low frequencies (~20 kHz).

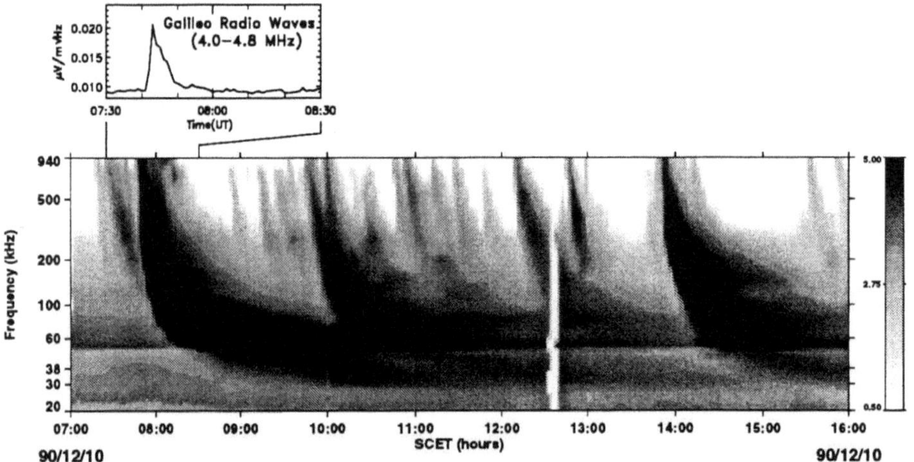

Fig. 1. Dynamic spectrum of the radio emission from 20 to 940 kHz observed at Ulysses from 07:00 to 16:00 UT on 10 December 1990. The intensity-time profile of the radio emission observed at Galileo at ~5 MHz is shown in the insert.

At this time, Galileo was passing near Earth at 1 AU and Ulysses was about 7° west of the Earth-Sun line and 1.36 AU from the Sun. The distance between Ulysses and Galileo was 0.4 AU and both spacecraft were located in the ecliptic plane. Their relative configuration is illustrated in Figure 2.

Figure 3 shows the energetic electrons and Langmuir waves observed at Ulysses and Galileo that were associated with this type III radio burst. An impulsive electron event was detected between ~08:10 and 8:30 UT at Galileo (Figure 3a) and between 08:30 and

08:40 UT at Ulysses (Figure 3c). In both cases, the pitch angle distributions (not shown) indicate field-aligned beaming of the electrons in the antisunward direction. This is characteristic of a prompt electron event (Lin, 1970b; Anderson *et al.*, 1981) and implies essentially scatter-free propagation. Due to the relative separation of Ulysses and Galileo, these electrons must have propagated along interplanetary field lines that originated at different heliographic longitudes (separated by ~20°) near the Sun. The electron fluxes observed at Galileo in the ~42-60 keV energy range were significantly (~30 times) higher than those observed at Ulysses. This may be due to the fact that, for electrons ejected from a W47° flare site, Galileo should lie nearer the centroid of the electron stream.

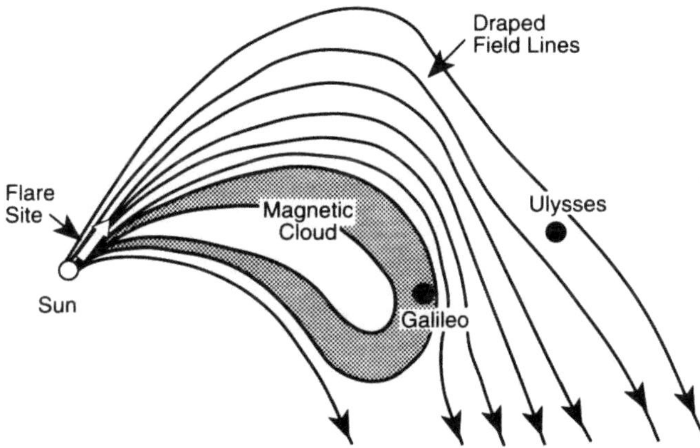

Fig. 2. Relative location of Ulysses and Galileo on 10 December 1990. This cartoon illustrates the possibility, discussed in the text, that Ulysses was located in the region of magnetic field lines draped around a coronal mass ejection.

Although a velocity dispersion in the electron onset times at both spacecraft is suggested by the data, it is not particularly well-defined. This makes a precise quantitative analysis of the beam dynamics somewhat ambiguous. Spatial structures in the interplanetary medium may be partly responsible for these ill-defined electron onsets. Our "best" estimate for the arrival times for the electrons in the five energy channels shown in Figure 3 are plotted in Figure 4 versus $1/\beta$, where β=v/c. If it is further assumed that the electrons detected at both spacecraft were injected into the interplanetary medium at 07:34 UT (as determined from the radio data), then the "best" straight line fits to these data give a propagation path length of ~1.7 AU to Galileo and of ~3.4 AU to Ulysses. These lengths are significantly longer (~1.4 times for Galileo; ~2 times for Ulysses) than the length of the Parker spirals (~1.15 AU for Galileo; ~1.7 AU for Ulysses) that correspond to the measured 400 km/s solar wind speed at Ulysses.

The onset time of the Langmuir waves at Galileo (Figure 3b) at 08:36 UT corresponds to the time that it would take ~14 keV electrons (assuming no pitch angle scattering), propagating along the same path length, to arrive at Galileo. The maximum intensity of these Langmuir waves was estimated by Gurnett *et al.*, (1993) to be ~1.7mV/m. The 14 keV electrons were expected to arrive at Ulysses along the inferred 3.4 AU path at ~09:38 UT, but no Langmuir waves were observed at this time. Langmuir waves at Ulysses (Figure 3d) were not observed until ~10:10 UT (~30 minutes later), corresponding to the estimated arrival time of ~9 keV electrons. The maximum intensity of the Langmuir waves at Ulysses was ~7mV/m, i.e., 5 times higher than at Galileo.

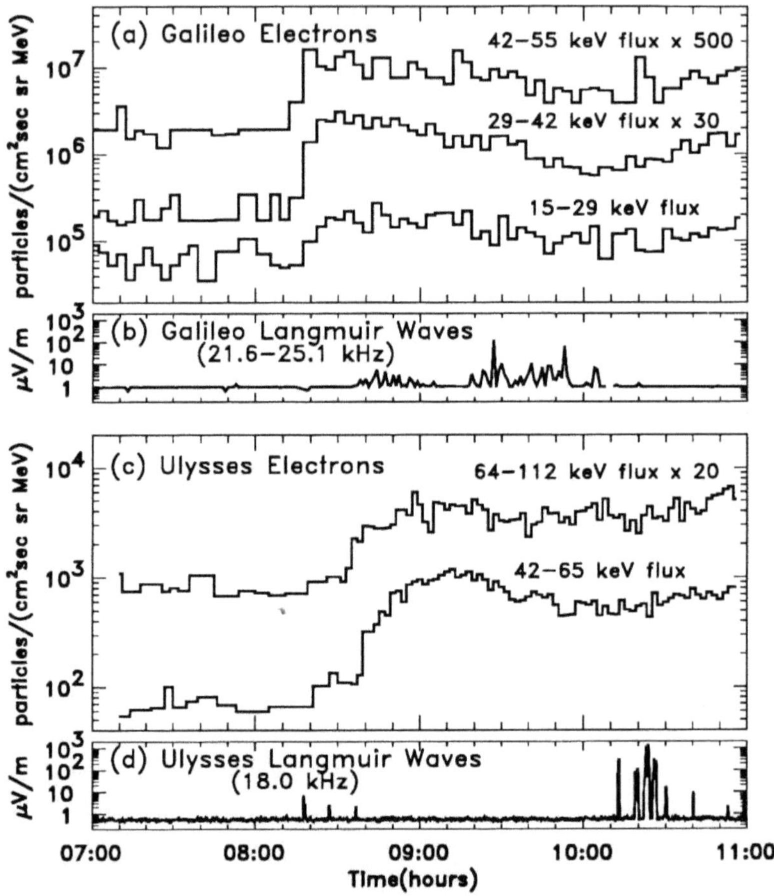

Fig. 3. (a) 15 - 55 keV peak electron fluxes at Galileo from 07:00 to 11:00 UT. (b) Langmuir waves detected near the local plasma frequency at Galileo. (c) Two minute averaged 42 - 112 keV background subtracted electron fluxes at Ulysses. (d) Langmuir waves observed near the local plasma frequency at Ulysses.

3. Discussion

Simultaneous observations by longitudinally separated spacecraft provide a direct measure of the longitudinal extent of a type III burst associated electron stream. The observations for the 10 December event imply that this electron stream extended over at least a ~20° heliographic longitudinal range of field lines above the solar active region.

There is *a priori* no reason that solar electrons at different heliographic longitudes should be injected simultaneously onto open interplanetary magnetic field lines or that the length of the field lines to longitudinally separated spacecraft should be the identical. Simultaneous electron observations by two spacecraft offers the unique opportunity for independently measuring both the injection times of the suprathermal electrons at the Sun and their respective propagation path lengths to each spacecraft by the analysis suggested in Figure 4. Because of the imprecise onset times for the electron events considered here, however, we were not able to utilize the full potential of this technique.

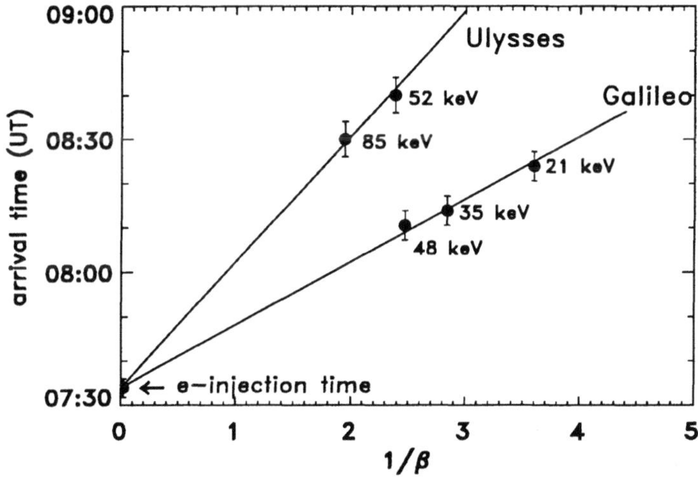

Fig. 4. Plot of arrival time versus 1/β for the energetic electrons detected at Galileo and at Ulysses. The values for β were obtained using the geometric mean energy in each channel. The straight lines assume a common injection time at the Sun. The slopes of the lines are proportional to the length of the propagation path.

The chronology of the *in-situ* observations at Ulysses and Galileo for the December 10 type III radio burst differed significantly from expected. The analysis suggests that the effective electron path lengths to both Ulysses and Galileo were significantly longer than the nominal Parker spiral. The electron anisotropies observed at both Ulysses and Galileo suggest that pitch angle scattering alone cannot account for such large path lengths. Although interplanetary spatial structures may have influenced the dispersive onsets for some of these electrons, our analysis does not depend critically on this. It is unlikely that spatial structures would have significantly delayed the electron arrival times at both spacecraft simultaneously. It is therefore reasonable to conclude that the derived path lengths represent the approximate length of the field lines from the Sun to each spacecraft. On the other hand, the Ulysses plasma and field data indicate that a CME, with field characteristics identifying it as a magnetic cloud, passed Ulysses on 11 December 1990. Thus, at the time of these observations, Ulysses was antisunward from the CME in the region of the draped magnetic field lines (Gosling and McComas, 1987), as illustrated schematically in Figure 2. The significantly longer path length to Ulysses (and possibly also to Galileo) may therefore correspond to "lengthening" of the field lines that were draped around the CME (McComas *et al.*, 1989).

There was also a difference in the expected Langmuir wave onset times at the two spacecraft. The Langmuir waves observed at Ulysses involved lower energy electrons (9 keV) than those at Galileo (14 keV). One possible explanation for this delay at Ulysses is that the Langmuir waves were not produced uniformly across the electron stream. Thus, although 14 keV electrons may have been present at Ulysses at 09:38 UT, the bump-on-tail in the velocity distribution function may not have formed in this region of the interplanetary medium at this time. This may be consistent with the observations of Reiner *et al.* (1992), that local radio emissions and associated Langmuir waves are produced only in localized regions of the interplanetary medium. Another possibility is that the effective electron exciter speed to Ulysses (near the stream edge) was lower than for Galileo (near the stream centroid) (Reiner and Stone, 1989). Unfortunately, since the low-energy electrons (2- 40 keV) were not detectable at either spacecraft, the parallel velocity distribution function, which would directly reveal the onset time of the beam instability, could not be determined.

4. Conclusion

Simultaneous *in-situ* observations of electrons at Ulysses and Galileo, at longitudinally separated points within a suprathermal electron beam, provide direct information on the longitudinal extent of the electron stream associated with a type III radio burst on 10 December 1990. The measured velocity dispersion in the electron arrival times suggest that the propagation path lengths to Ulysses and Galileo were significantly larger than the length of the nominal Parker spiral. These observations may represent the first direct evidence for lengthening of field lines due to the presence of a CME. The observed Langmuir waves were also significantly delayed at Ulysses compared to those observed at Galileo.

References

Anderson, K.A., McFadden, J.P. and Lin, R.P.: 1981, *Geophys. Res. Letters* **8**, 831.
Gosling, J.T and McComas, D.J.: 1987, *Geophys. Res. Letters* **14**, 355.
Gurnett, D. A. et al.: 1993, *J. Geophys. Res.* **98**, 5631.
Lin, R.P.: 1970a, *Solar Phys.* **15**, 453.
Lin, R.P.: 1970b, *J. Geophys. Res.* **75**, 2583.
Lin, R.P., Potter, D.W. Gurnett, D.A. and Scarf, F.L.: 1981, *Astrophys. J.* **251**, 364.
Lin, R.P.: 1985, *Solar Phys.* **100**, 537.
Lin, R.P., *et al.*:1986, *Astrophys. J.* **308**, 954.
McComas, D.J. *et al.*: 1989, *J. Geophys. Res.* **94**, 1465.
Potter, D.W., Lin, R.P. and Anderson, K.A.: 1980, *Astrophys. J.* **236**, L97.
Reiner, M.J., Stone, R.G.: 1989, *Astron. Astrophys.* **217**, 251.
Reiner, M.J., Stone, R.G. and Fainberg, J.: 1992, *Astrophys. J.* **394**, 340.

INTERPLANETARY TYPE III RADIO BURSTS OBSERVED SIMULTANEOUSLY BY ULYSSES AND ICE

S. HOANG and J.-L. STEINBERG
Observatoire de Paris, DESPA, URA 264 CNRS, 92195 Meudon, France

and

M.J. REINER and R.G. STONE
NASA Goddard Space Flight Center, Greenbelt, MD 20771, USA

Abstract. We analyze two solar type III radio bursts that were observed simultaneously by the ICE and Ulysses spacecraft. Both bursts originated behind the solar limb as viewed from either spacecraft. At the time of these events, ICE was in the ecliptic plane at ~1 AU and Ulysses was ~ 35° south of the ecliptic plane at ~4 AU. For one event on 931117, the ratios of the peak flux densities measured at each spacecraft, at each observing frequency, were consistent with the most probable source locations relative to ICE and Ulysses. The second event on 931004 was a complex burst consisting of two distinct components at high frequencies. At low frequencies, the intensity of the first component decreased rapidly at each spacecraft. The second component, however, dominated the low frequency emission observed at Ulysses but not at ICE. These differences in the observed radiation must be related to the different viewing geometries of the two spacecraft. The measured onset times as a function of observing frequency were consistent with a constant exciter speed through the interplanetary medium and suggest that there are significant propagation delays, especially for the radiation propagating within the ecliptic plane.

1. Introduction

Type III radio bursts are generated in the interplanetary medium (IPM) when fast electrons ejected from solar active regions travel outward along spiral magnetic field lines and trigger the radio emission at the local plasma frequency f_p and/or $2f_p$. Since f_p decreases with increasing heliocentric distance R as $\simeq R^{-1}$, type III bursts are observed to drift from high to low frequencies. At a given frequency, the type III intensity-time profile rises rapidly at the "onset time" to a maximum flux, followed by a slower exponential decay. However, observed burst profiles often consist of a superposition of many individual profiles that may correspond to multiple electron injections. Previous analyses of IP type III bursts observed simultaneously by longitudinally separated spacecraft (s/c) in the ecliptic plane (Steinberg et al., 1984; Dulk et al., 1985; Lecacheux et al., 1989) provided information on their directivity and on propagation effects in the IPM. The comparison of the radiation characteristics of IP type III bursts observed simultaneously by the radio receivers on ICE (Knoll et al., 1978) and Ulysses (Stone et al., 1992) at high heliographic latitudes provides a unique opportunity to study the latitudinal directivity and the influence of the IPM on radio propagation outside the ecliptic plane. However, during 1993, only a dozen type III

Space Science Reviews **72**: 267–270.
© 1995 *Kluwer Academic Publishers.*

bursts were simultaneously recorded at both ICE and Ulysses. This is due primarily to the limited telemetry coverage of ICE (2 to 4 hours a day). Here we consider two events that occurred on 931004 and 931117.

2. Observations

The 931117 event was associated with a B2.2 X-ray flare at ~22:07 UT. It is reasonable to assume that this flare was located in the only sunspot group on the Boulder map at ~ 8°N and 18°E; so this flare was located 130°E of Ulysses and 123°E of ICE. The ecliptic latitude of Ulysses was -37.1°S. The second event on 931004 was associated with a flare at 7:25 UT in sunspot group 7590 at 7°N and 14°W, ~140°E from Ulysses and 106°E from ICE. At this time, the ecliptic latitude of Ulysses was -33.5°S. The relative configuration of the s/c, projected onto the ecliptic plane, for these events is illustrated in the insert in Fig. 1a and 1d. The type III radio sources at decreasing frequency are assumed to be located at increasing heliocentric distance along the spiral field line originating from the flare site. Since for both events the flare longitudes were more than 106°E from either s/c, the radio sources were located entirely in the far hemisphere of the Sun as viewed from the s/c. The radiation was visible at the two s/c only as a scattered image through the IPM. The primary source for such "behind the limb" events is in general difficult to locate (Reiner and Stone, 1990).

Fig. 1a and 1d show, for each event, the peak flux density S observed at each s/c as a function of f^{-1}, which is essentially a distance scale, since 1/f ~ R. For both bursts the flux density decreases with increasing 1/f and, as expected, the flux is generally higher at ICE which is significantly nearer the radio source. In Fig. 1b and 1e, we plot the observed flux ratio r = S_I(ICE)/S_U(Ulysses). For comparison, the dotted lines are the ratio $d = (R_U/R_I)^2$, where R_U and R_I are the distances from the source to Ulysses and ICE, respectively. The ratio d was computed assuming (1) the source was located at the Sun at all frequencies; the source was located along the spiral rooted at the flare site at a heliocentric distance determined by assuming the IPM plasma density varies as R^{-2}, and that at 1 AU (2) f_p is 20 kHz or (3) 40 kHz. If the source radiated isotropically and there were no strong propagation effects along the path to the s/c, we should find $r \simeq d$. Indeed, for the 931117 event at low frequency, $r \sim 10 - 15$, in general agreement with the range of acceptable d ratios. At high frequency, however, r decreases to ~ 5. This is very likely due to occultation of the radio emission by the plasma sphere, which is expected to be less effective for Ulysses located outside the ecliptic plane. The situation is quite different for the 931004 event (Fig. 1e). At high frequency, $r \sim 10$, twice as high as on 931117. This difference may be explained from the different viewing geometry. For this case, the radio sources along the spiral at high frequency are located far to

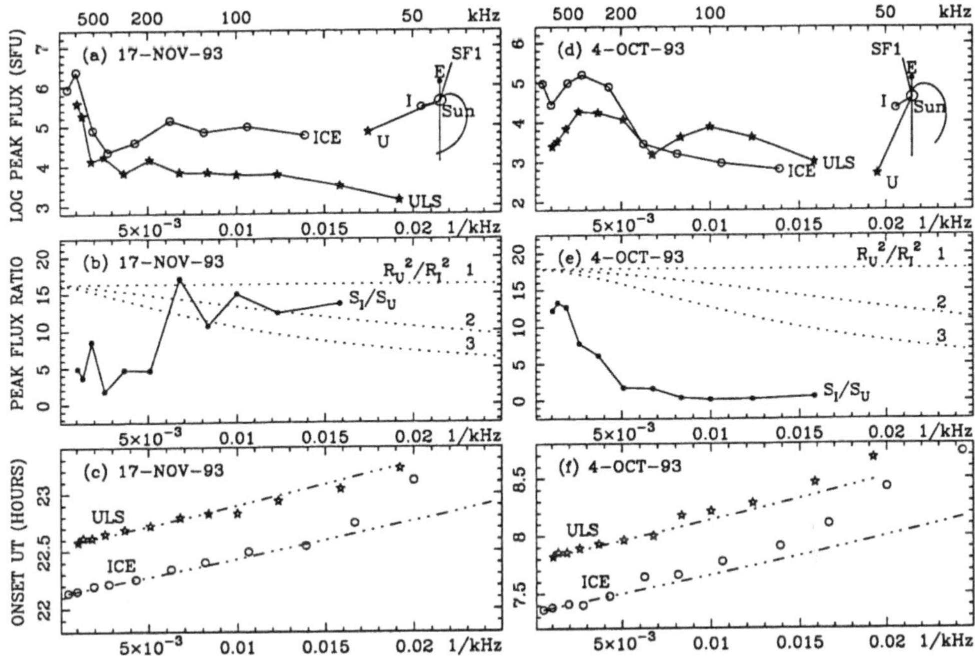

Fig. 1. Observations of two type III events, 17 Nov 1993 (left panels, a-c) and 4 Oct 1993 (right panels, d-f). (a,d) Peak flux densities vs. inverse frequency. (Insert) Geometry projected onto the ecliptic plane of the observations and associated solar flares (SF); on 931117 (931004), Ulysses was at 4(4.2) AU and 37.1(33.5) deg below the ecliptic plane. (b,e) The ratio S_I/S_U of intensities at ICE and Ulysses vs. 1/f (full line). The ratio $(R_U/R_I)^2$ of the distances from the radio source to the s/c is plotted as dotted lines (see text). (c,f) Onset times vs. 1/f. The dotted lines are the predicted times (see text).

the east of ICE. The radiation is now more visible at ICE and r is closer to the expected values. At low frequency, however, r differs dramatically from the expected values. Below $\sim$ 160 kHz, S_I is now much lower than S_U so $r < 1$. The reason for this is suggested from a more detailed examination of the burst profiles. The burst profile for the 931004 event consists of at least two components that are readily distinguishable and separated by several minutes at the high frequencies. The first of these components is observed to rapidly decrease in intensity with decreasing frequency at both s/c. The second component also decreases rapidly in intensity at ICE, but at Ulysses it increases in intensity until by 272 khz it dominates the radio emission. The reason that this second component is significantly more visible at Ulysses than at ICE must be related to the difference in viewing geometries from ICE and Ulysses. It is unlikely that this could result from a greater occultation due to the sources being farther behind the limb at low frequency since this should diminish both components by the same factor.

In Fig. 1c and 1f, we plot the onset times t_o as a function of f^{-1}, for both s/c. The dotted lines are derived values of t_o, assuming that $f_p \propto R^{-1}$ in the IPM and the exciter electrons travel along spiral field lines (with curvature set by $V_{SW} = 400$ km/s) at constant speed determined from t_o for the first and fifth frequency data points. For the 931117 event (Fig. 1c), these computed times fit the Ulysses data quite well over a very wide frequency range. The line through the ICE data is nearly parallel to that through the Ulysses data, which is expected since the exciter speeds should be identical. The ICE data follow the computed t_o quite well at the high frequencies but differ significantly at the low frequencies. This can probably be explained in a qualitative way : at high frequencies, the radiation is not significantly modified by propagation effects, so the same frequency drift, which is related to the speed of the type III electrons moving along spiral field lines, is observed at both s/c. But at low frequencies, the radiation is expected to be sigificantly modified by propagation effects, especially within the ecliptic plane. This may explain the relative delay in the onset of the radiation at ICE at low frequencies. For the 931004 event (Fig. 1f), the computed times fit the data reasonably well at the high frequencies as expected. The deviation from the computed t_o for $f < 148$ kHz at Ulysses is interpreted as due to the dominance of the second burst component. Again the lines are nearly parallel for Ulysses and ICE suggesting identical exciter speeds. Also the lines for the 931004 and the 931117 events are nearly parallel suggesting that the exciter speeds were essentially the same for both events.

3. Concluding remarks

From this preliminary study, we find that the observed ratios of intensities at the two spacecraft often differed from the expected ratios. We interpret these differences as due to occultation by the plasma sphere and by the different visibility of multiple burst components. We find evidence for significant propagation delays, especially within the ecliptic plane. These simultaneous observations may provide a better understanding of behind-the-limb events, propagation effects, and multiple component type III emissions.

References

Dulk, G.A., Steinberg, J.-L., Lecacheux, A., Hoang, S. and MacDowall, R.J.: 1985, *Astron. Astrophys.*,**150**, L28

Fainberg, J. and Stone, R.G.: 1974,*Space Sci. Rev.*,**16**, 145-188

Knoll, R. et al.: 1978, *IEEE Trans. geosci. Electronics*, **GE-16**, 199-204

Lecacheux, A., Steinberg, J.-L., Hoang, S. and Dulk, G.A.: 1989, *Astron. Astrophys.*,**217**, 237-250

Reiner, M.J. and Stone, R.G.: 1990, *Solar Phys.*,**125**, 371-388

Steinberg, J.-L., Dulk, G.A., Hoang, S., Lecacheux, A. and Aubier, M.: 1984,*Astron. Astrophys.*,**140**, 39-48

Stone, R.G. et al.: 1992, *Astron. Astrophys. Suppl. Series*, **92**, 291-316

SESSION 3

ENERGETIC PARTICLES IN THE HELIOSPHERE

ENERGETIC PARTICLES AT HIGH LATITUDES

R. MÜLLER-MELLIN and G. WIBBERENZ
Institut für Kernphysik, Universität Kiel, 24118 Kiel, Germany

Abstract. We review work on diffusion coefficients of energetic particles with an attempt to extract implications on their behaviour at high latitudes. In the ecliptic plane results from solar energetic particle propagation between the Sun and about 5 AU can be described by an effective radial mean free path λ_r which is approximately constant as a function of distance r. When particle propagation in three dimensions in the heliosphere is considered it is not sufficient to consider λ_r only. Jovian electrons can be used as probes to determine the parameters of three-dimensional diffusion. In the polar regions diffusion is dominated by its parallel component. Some predictions how $\lambda_{||}$ should vary with latitude are discussed. For different choices of this variation we present expectations for intensity-time profiles of solar particle events during the Ulysses polar passages.

Key words: heliosphere – galactic cosmic rays – solar cosmic rays – Jovian electrons – propagation parameters – Ulysses

1. Introduction

Some time after the successful Jupiter flyby of Ulysses expectations were expressed that "poles of the Sun will be a region of simplicity"and that "there may be what amounts to a funnel for cosmic rays over the polar regions" (quotations of E.C. Smith after Appenzeller, 1992). Results presented at this symposium (see e.g. Heber et al., 1994; McKibben, 1994) show that at least up to latitudes of about 60°S there is no indication of an easy access of cosmic rays. The latitudinal gradients of protons are small or even consistent with zero. This is a rather unexpected result for the present $A > 0$ cycle. Here we summarize some of the ideas related to energetic particle transport in the polar regions. In section 2 we will start with a few remarks on transport models and the behaviour of perpendicular and parallel diffusion coefficients (DC). In section 3 we summarize observations related to cosmic ray DC in the ecliptic plane at distances up to about 5 AU, in an attempt to combine results from solar and Jovian electrons. These observational results will be used in combination with theoretical arguments to predict cosmic ray DC along the expected Ulysses orbit in section 4. Solutions of the focused transport model will be used to predict intensity profiles of solar particle events observed during polar passage.

2. Cosmic Ray Transport

Cosmic ray particles in the heliosphere undergo outward convection with the solar wind, inward diffusion driven by the radial gradient, and adiabatic deceleration resulting from the expansion of the solar wind. In the diffusion

Space Science Reviews **72**: 273–284.
© 1995 *Kluwer Academic Publishers.*

approximation the cosmic ray number density U per unit interval of kinetic energy T satisfies a Fokker-Planck equation (Parker,1965):

$$\frac{\partial U}{\partial t} = \frac{\partial}{\partial X_i}\left(K_{ij}\frac{\partial U}{\partial X_j} - V_i U\right) + \frac{1}{3}\left(\frac{\partial V_i}{\partial X_i}\right)\frac{\partial}{\partial T}(\alpha T U) \qquad (1)$$

where V_i is the solar wind velocity, K_{ij} the generalised diffusion tensor, and $\alpha = (T + 2m_0c^2)/(T + m_0c^2)$. The drift in the large scale magnetic field is implicitely given in the off-diagonal elements of the diffusion tensor.

Since the nature of particle transport is anisotropic, the transport can be visualized as being split into motions parallel and perpendicular to the magnetic field. Therefore, we need to know the large scale structure of $\mathbf{B}(\mathbf{r}, t)$. This can be the Parker spiral field or the modifications introduced by Jokipii and Kota (1989) or Smith and Bieber (1991). The parameters which control the *systematic* and *stochastic* forces are listed in the following table.

systematic		stochastic	
Convection + Adiab. Decel.	Drift	Perpendicular Diffusion	Parallel Diffusion
⇓	⇓	⇓	⇓
$V_w(\mathbf{r}, t)$	K_T	$K_\perp$	$K_\parallel$

This means that we need to know the solar wind speed - with a transition from about 400 km/s to 800 km/s when we move from equatorial to polar latitudes - and the diagonal as well as antisymmetric terms of the diffusion tensor.

The major cosmic ray transport processes for galactic cosmic rays are depicted in Fig. 1, see the discussion by Moraal (1993). The outward convection SW establishes a density gradient which prompts an inward diffusion DF. This flux is somewhat off-radial due to higher cosmic ray densities expected above the solar poles. The drift velocity pattern is visualised for the example of positive particles during positive polarity ($qA > 0$): in a narrow cylinder (1 AU radius) above the poles, the polar drift PD is sunward. Outside the polar cylinder the particles follow almost circular drift paths DR from the polar regions toward the equatorial plane. The neutral sheet drift ND transports the particles outward along the equatorial plane.

It should be noted that equation (1) describes cosmic ray transport in the diffusive approximation which is only valid for small anisotropies. It can break down for solar cosmic rays when the mean free path is sufficiently large and when focusing effects in the diverging interplanetary magnetic field also play a role. We will discuss the necessary modification in section 3.

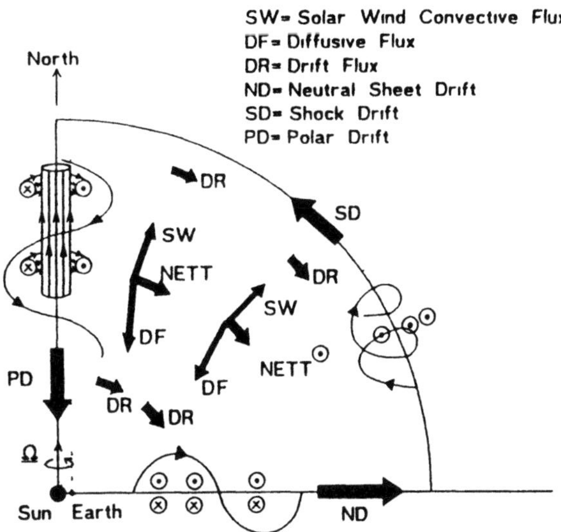

Fig. 1. Meridional projection of a quarter heliosphere showing the major cosmic ray transport processes. From Moraal (1993).

In many applications it is important to consider how particles move from one radial shell to the other. This is characterized by the effective radial DC.

$$K_{rr} = cos^2\phi\, K_{\|} + sin^2\phi\, K_{\perp} \tag{2}$$

where ϕ is the spiral angle given by $\tan\phi = (\Omega r \sin\theta)/V_w$. Here the latitude is $\theta = 0°$ at the poles, $\theta = 90°$ at the equator. In the ecliptic plane and for a solar wind speed of the order of 430 km/s we can rewrite equation (2) in the simple form

$$K_{rr} = \frac{1}{1+r^2} K_{\|} + \frac{r^2}{1+r^2} K_{\perp} \tag{3}$$

where r is in AU. Depending on the ratio

$$a = \frac{K_{\perp}}{K_{\|}} \tag{4}$$

between the two DC we can determine at which locations the radial transport is essentially controlled by parallel or perpendicular diffusion, respectively, and where both effects have to be considered.

We note here that there is no a priori reason why the two perpendicular components should be the same. It may be too simple to work with one value for $K_{\perp}$ only. Discussion of the Jovian electrons will show that we may have to distinguish two values, $(K_{\perp})_1$ and $(K_{\perp})_2$, within and out of the ecliptic plane.

Even without this complication, there is a great variety of possibilities to select proper values for K_{ij} as a function of location in the heliosphere and of particle properties. In many modulation models $K_{\|}$ is assumed to vary inversely with the magnitude B of the magnetic field (e.g. Jokipii and Davila, 1981; Potgieter et al., 1993, see also Reinecke et al., 1993, for slight modifications). This gives

$$K_{\|} = K_o \beta K_p(P) \frac{B_o}{B} \tag{5}$$

with suitable choices for a rigidity dependence $K_p(P)$ (see section 3 below). The value for $K_{\perp}$ can then by obtained by keeping the ratio a (equation (4)) constant, independent of the location in the heliosphere and of particle properties, leading to

$$K_{\perp} = a K_{\|} \propto \frac{1}{B}. \tag{6}$$

It is argued that this set of assumptions makes the diffusion tensor simple to handle, gives reasonable agreement with observed energy spectra at different positions in the outer heliosphere, and can be theoretically justified. Jokipii and Davila (1981), in the absence of a generally accepted spatial dependence of the diffusion coefficient, had chosen it to be proportional to the gyroradius, which leads to the scaling with 1/B (equation (5)). They also had shown that this choice gives similar results as when one uses more detailed calculations of the spatial dependence presented by Morfill and Völk (1979). It should be noted that the functional form of equations (5) and (6) makes it very easy to consider the influence of modified magnetic field structures on cosmic ray transport (see Haasbroek and Potgieter, 1994).

3. Observations

There are various sources of energetic particles, and their propagation in the heliosphere ought to be described by the same set of transport parameters.

1. **Solar energetic particles in the ecliptic plane.** Many fits of solar particle events within about 1 AU have been obtained by assuming an effective radial DC which does not vary with radial distance (see Palmer, 1982, Kallenrode, 1993, Bieber et al., 1994). In these fits, it has turned out to be important, a) to take finite injections close to the Sun into account, b) to work with the proper model of focused transport (Roelof, 1969) because in many cases the mean free path is sufficiently large to make the simple diffusion model invalid. Kunow et al. (1991) show solutions to the equation of focused transport for 9 MeV protons and a fairly large mean free path of 0.5 AU where a strong coherent peak is followed by a diffusive wake.

However, a coherent peak is only observed for sufficiently short solar injections. For longer solar injections, interplanetary intensity-time profiles are indistinguishable from those in diffusive transport. In order to separate this effect from long lasting events caused by a small mean free path, we have to fit intensity *and* anisotropy data simultaneously. Fits to actual data are shown e.g. by Kallenrode (1993).

In Fig. 2 we show as summary the parallel mean free paths for solar particle events over a large range of rigidities (modified from Bieber et al., 1994). Circles denote actual determinations of the mean free path, while upward pointing triangles denote lower limit values. Palmer (1982) had concluded that the empirical mean free paths could be characterized by a "consensus" band shown shaded in Fig. 2. The dotted line shows the prediction of conventional quasilinear theory (Jokipii and Coleman, 1968), using slab geometry and representative magnetic spectra derived from survey observations acquired over multi-year periods. We have added to Fig. 2 what we believe represents an average variation of $\lambda_{\|}$ with rigidity (full line). This curve corresponds to a slope of 0.09 below 100 MV, of 0.3 between 100 and 1000 MV, and increases $\propto P$ beyond. We have also added as a dashed curve the variation of the mean free path which could be considered as characteristic for choices in describing galactic cosmic ray modulation. This curve represents $\lambda = 3K_{\|}/v$ at 1 AU taken from Haasbroek and Potgieter (1994). It is interesting to note that the results averaged over a number of solar particle events - which is characteristic for the scattering conditions within 1 AU - agree rather well with near 1 AU results obtained from modulation studies.

A crucial question is how the radial DC varies beyond 1 AU. If we insert $K_{\perp} = aK_{\|}$ into equation (3) and keep a constant, we can relate $K_{\|}$ and K_{rr} as follows

$$K_{\|} = \frac{1+r^2}{1+ar^2} K_{rr}, \tag{7}$$

Similarly, we obtain for the perpendicular DC

$$K_{\perp} = a\frac{1+r^2}{1+ar^2} K_{rr}. \tag{8}$$

When we keep $K_{rr} = const$ as before we get a transition from $K_{rr} = K_{\|}$ at small distances to $K_{rr} = K_{\perp}$ at large distances from the Sun. This also means that $K_{\|}$ increases by a factor $1/a$ when going from small to large distances. In the ecliptic plane the radial diffusion coefficient K_{rr} does not seem to vary greatly with radial distance from the Sun between about 1 and 5 AU (Zwickl and Webber, 1977; Wibberenz et al., 1992; Müller-Mellin et al., 1993). The discussion by Hamilton (1977) shows that some increase with radial distance might be more appropriate. He finds for MeV electrons $\lambda_r = (0.03 - 0.06)\,(r/r_E)^b$ AU with $b = 0.4 \pm 0.2$. In what follows we shall however use $\lambda_r = const$ as a simple approximation in the ecliptic plane between 1 and 5 AU. Note that the absolute value determined by Hamilton near 1 AU is in excellent agreement with results obtained on Ulysses for MeV electrons from an SEP event at 3.2 AU on day 149, 1991 (see Wibberenz et al., 1992). Let us see now how this choice agrees with results from Jovian electrons.

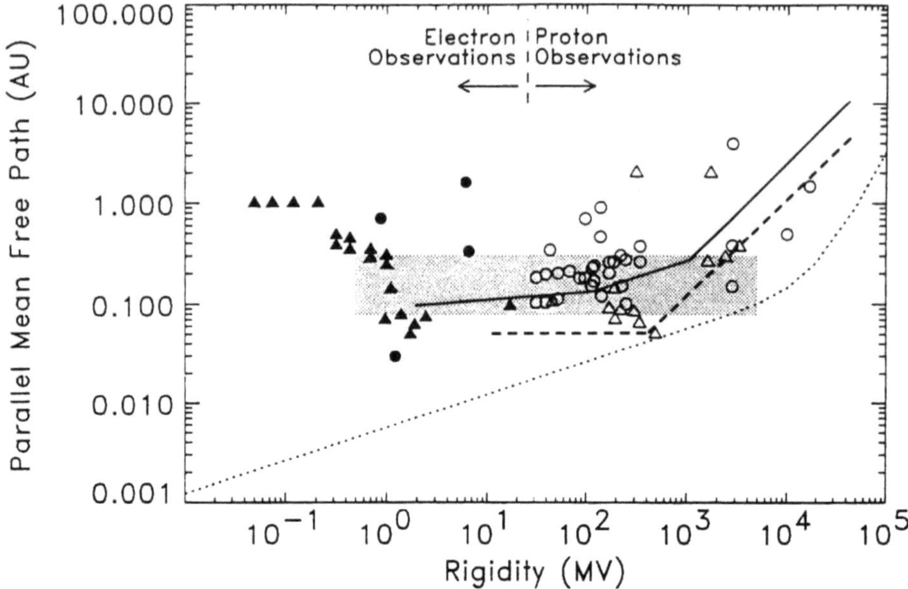

Fig. 2. Cosmic ray mean free path versus particle rigidity.

2. Jovian electrons. An approximation of the transport equation (1) for Jovian electrons was obtained by Conlon (1978) under the following assumptions: steady state solar wind, neglecting adiabatic energy loss, idealized magnetic field with parallel field lines, and constant and diagonal diffusion tensor K_{ij} in a coordinate system where X is measured from Jupiter across the average field direction, Y is directed along the field lines, and Z is perpendicular to the plane of the ecliptic (see Chenette, 1980). With an analytic solution to this approximation it was possible to fit the envelope of Jovian electron observations reasonably well.

The study of Jovian electrons is of particular importance in order to obtain an estimate for the ratio a. Ferrando et al. (1994) show that in- and off-ecliptic Jovian electron data can be described by a diffusion-convection model when the two perpendicular diffusion coefficients are identical, with $K_{xx} = K_{zz} = 8 \cdot 10^{20}$ cm^{2}s^{-1}. The parallel DC is consistent with a value found earlier by Hamilton and Simpson (1979), $K_{yy} = 5 \cdot 10^{22}$ cm^2 s^{-1}. This leads to $a = 1.6\%$.

Note that this ratio might be slightly smaller when one takes the earlier result by Hamilton and Simpson (1979) of $K_{zz} = 2 \cdot 10^{20}$ which is still marginally consistent with the Ulysses data (Ferrando et al., 1994).

Now it can be seen easily that it is impossible to fulfill the two conditions $K_{rr}(r) = const$ and $a = const$ simultaneously for a *fixed* value of $K_{\|}$. From equations (7) and (8) we see that when keeping K_{rr} and a constant, both

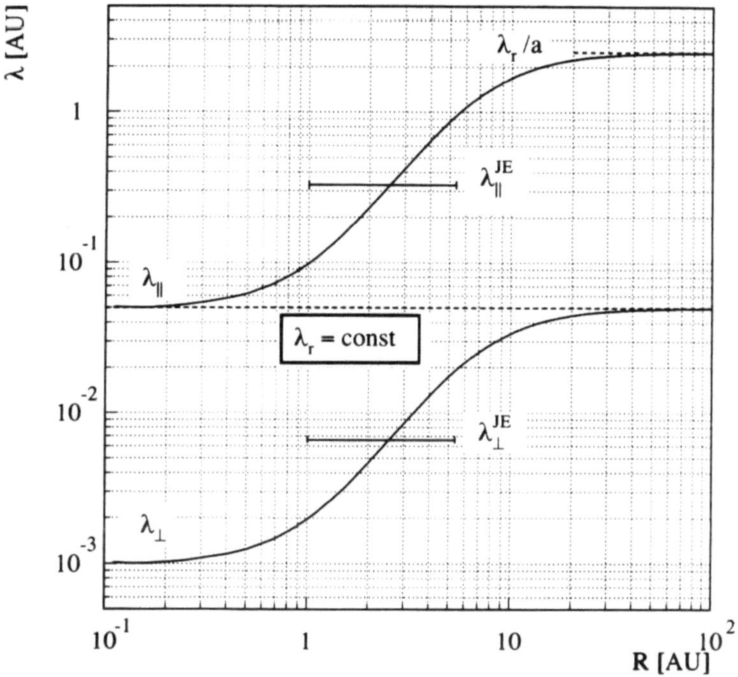

Fig. 3. Cosmic Ray mean free path versus radial distance

$K_{\|}$ and $K_{\perp}$ increase considerably with distance r. This is shown in Fig. 3 for the special cases of $\lambda_r = 0.05$ AU, $a = 0.02$. These are values taken to be representative for electrons in the MeV range. The upper curve shows the behaviour of $\lambda_{\|}$. It starts out with the value of λ_r near the Sun and approaches λ_r/a at infinity. The lower curve shows the behaviour of $\lambda_{\perp}$, starting out with λ_r/a near the Sun and approaching λ_r at infinity.

This variation of λ is not compatible with the previous analysis of Jovian electrons which is based on the assumption that both the parallel and perpendicular DC are spatially constant. However, this does not seem to be a serious objection. In Fig. 3 we have also inserted values of $\lambda_{\|}^{JE} = 0.33$ AU and $\lambda_{\perp}^{JE} = 6.6\text{x}10^{-3}$ AU. These values correspond to the DC for Jovian electrons quoted above.

The suggested increase of $K_{\|}$ with radial distance has one additional advantage. According to Fig. 3, $\lambda_{\|}$ is of the order of 1 AU close to the position of Jupiter. This would be consistent with the angular distributions of Jovian electrons detected in jet-like structures close to Jupiter and after Jupiter flyby. The strong anisotropy detected in these jets (Ferrando et al., 1994) suggests nearly scatter-free propagation in this region of space, consistent with the large parallel mean free path suggested by Fig. 3.

4. Prediction for the Polar Regions

Various physical parameters and processes change when we move from the equatorial region towards the poles.

1. The transition from spirals to straight lines. The transport of particles from one radial shell to the other is governed by equation (2) where ϕ varies with radial distance r and latitude θ. Note that this variation is different in the Parker field and in the modifications treated by Jokipii and Kota, 1989, or Smith and Bieber, 1991. In the limiting cases we obtain at the equator

$$\lim_{r\to\infty} K_{rr} = K_{\perp} \tag{9}$$

and at the poles

$$\lim_{r\to\infty} K_{rr} = K_{||}. \tag{10}$$

Let us first assume that for a fixed r the parallel diffusion coefficient does not vary with latitude. This could occur when the Sun generates the same degree of turbulence independent of latitude, e.g. by the emission of Alfvén waves. If these waves travel essentially parallel to the average magnetic field direction we end up with $\partial\lambda_{||}/\partial\theta = 0$. This leads then to the important result that for large distances the radial diffusion coefficient K_{rr} at the pole is by the factor $1/a$ larger than at the equator, corresponding to a factor of 50 with the value of $a = 0.02$ adopted in the previous section. This considerable increase of the effective diffusion coefficient for radial transport is only due to the different pathlengths along the average magnetic field. It was already pointed out by Völk (1976) that this shorter geometric path may not be the main effect for an easy access of cosmic rays over the poles (see next point).

2. Power spectrum and resonance condition. There is no a priori reason why the parallel mean free path should be invariant with respect to latitude. We have to consider how the spectrum of magnetic field fluctuations might vary with latitude, and how this influences the mean free path via the resonance condition. Völk (1976) considers particle scattering by radially propagating Alfvén waves and magnetic field power spectra given by

$$S(f) = \frac{C}{1 + (f/f_o)^{1.5}} \tag{11}$$

The frequency f_o characterizes the transition from $S(f) = const$ to $S(f) \propto f^{-1.5}$.

One important effect is the change in the resonance condition due to a different variation of B with radial distance r. This leads to a "transition distance" for the polar regions beyond which the DC increases very fast with r, $\propto r^3$ or r^4.

An additional effect was considered by Morfill and Völk (1979). They argue that at low frequencies close to the equator the power comes mainly from solar wind corotating interaction streams. Since these are missing in the polar regions, the power is assumed to be a factor of 10 less for $f < f_s$ where $f_s = 1.37 \cdot 10^{-6}$ corresponds to the average rate of solar wind streams near the equator. An interesting prediction from the Morfill and Völk (1979) computations is that the most dominant increase in λ_r is restricted to a range in latitudes of about 20° from the polar axis, conditions to be expected for the polar passages of Ulysses.

Discussion. We have shown that no simple variation of $\lambda_{||}$ with θ is to be expected. At large distances from the Sun the expections for an "easy access" of galactic cosmic rays are based on a combination of three effects, (a) a shorter geometrical path, (b) a decrease of the magnetic field magnitude which is more pronounced than close to the equator, leading to a different resonant condition, (c) the additional removal of power at low wave numbers, due to the lack of stream-stream interactions.

At small radial distances, close to 1 AU, the situation is quite different. Here a *decrease* of $\lambda_{||}$ towards the poles is expected, based on the assumption (Völk, 1976; Morfill and Völk, 1979) that cosmic ray scattering is partly due to scattering by *radially* propagating Alfvén waves. It remains to be seen whether the explanation of scattering near 1 AU in the ecliptic plane by oblique Alfvén waves is the correct explanation for the discrepancy between QLT predicitons and solar cosmic ray observations (see also Wanner et al., 1994, Jaekel et al., 1994, for discussion). Alternative explanations for this discrepancy are given by Bieber et al. (1994) based on 2D-turbulence perpendicular to the average magnetic field and by the behaviour of the wave spectrum at high frequencies (Schlickeiser and Achatz, 1993; Dröge, 1994).

For the near-polar transport of energetic particles from the Sun to distances of several AU we can distinguish therefore three different possibilities:

- The parallel mean free path does not vary greatly with latitude. This would result from a situation where in the wavenumber range relevant for energetic particle scattering the number of waves and the wave vector distribution with respect to the average magnetic field remain constant.
- The parallel mean free path decreases with latitude. This result is expected when the wave vector distribution of fluctuations with respect to the magnetic field direction changes considerably, e.g. leading to a smaller amount of oblique Alfvén waves. This particular situation is interesting because one would expect in this case that predictions from a combination of QLT and the slab model of field fluctuations ought to be in good agreement with solar particle results.
- The parallel mean free path increases with latitude. We have to face the fact that the nature of magnetic field fluctuations leading to particle scattering is not yet fully understood. This means that it could happen that part of these fluctuations are missing over the polar region with a much smoother expansion of the solar wind free of colliding streams. As a result, the mean free path might increase even further.

Model calculations for solar particle transport. The study of solar cosmic ray events at high latitudes and comparison of the resulting mean free paths with estimates based on measured magnetic field spectra is ideally suited to check the various predictions in the inner solar system given above. In what follows we present results for relativistic solar electrons (0.5 MeV) with pulse-like injection close to the Sun. Solutions for focused transport are given for a distance of 2.81 AU which will be reached by Ulysses upon entry into the polar passage phase at 70°. For the spiral structure we use an unmodified Parker field with solar wind velocities of v_w = 400 km/s at the equator and v_w = 750 km/s above the poles. At this distance, we neglect perpendicular diffusion and compare results at 0° and 70° (Fig. 4) for the *same* parallel mean free path. As a numerical value we choose $\lambda_r^{equ} = const = 0.05$ AU which was presented in section 3 as a reasonable average for MeV electrons in the equatorial plane. Since a is small we essentially have $\lambda_{||} \propto (1 + r^2)\lambda_r^{equ}$. Due to the shorter geometrical path the effective radial mean free path off the equator then has to increase. This explains the faster intensity profile for a polar observer (dashed) compared to an equatorial observer (full line), both marked with $\lambda = 0.05$ AU in Fig. 4.

We also show in Fig. 4 results when the parallel mean free path is not invariant with latitude. We discussed above the possibilites that $\lambda_{||}$ could systematically decrease or increase with latitude. This is indicated by the two remaining dashed curves marked with $\lambda = 0.01$ AU and 0.2 AU, respectively.

Though we are aware of the large fluctuations in the actual mean free paths from one solar event to the other, the study of a few events only and the comparison of the results with predictions based on QLT and measured magnetic field spectra should give an indication which one of the three possibilities for a variation of $\lambda_{||}$ with latitude actually exists.

5. Summary

For our understanding of diffusive transport the composition of the effective radial DC from parallel and perpendicular diffusion is of crucial importance. We started by discussions in the ecliptic plane. Beyond about 10 AU, the radial diffusion is almost exclusively determined by perpendicular diffusion. Depending on the ratio of $a = K_\perp / K_{||}$ the contribution from the two may be of equal importance around 5 AU. Here we have crucial observations from Jovian electrons leading to $a \approx 2\%$. Further inward radial diffusion is essentially controlled by $K_{||}$. For simplicity we have adopted K_{rr} to be independent of radial distance, consistent with solar particle observations within 1 AU. A slight increase of K_{rr} between 1 and 5 AU is also feasible. In any case, the *parallel* mean free path has to increase considerably between 1 and 5 AU. In order to make these conclusions based on solar particle events consistent with data from Jovian electrons, further clarification is necessary.

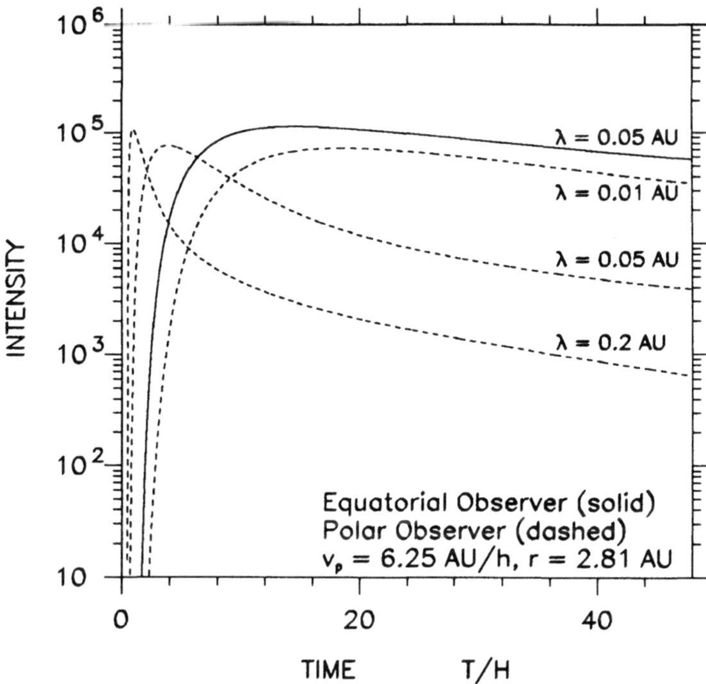

Fig. 4. Intensity-time profiles for an observer at 2.81 AU in the equatorial plane ($\theta = 0°$, solid line) and above the poles ($\theta = 70°$, dashed lines)

We have identified three problems,

- whether or not $(K_\perp)_1$ and $(K_\perp)_2$ in and out of the ecliptic plane are the same,
- whether or not the ratio $K_\perp/K_{||}$ remains independent of radial distance,
- whether or not both coefficients increase considerably with radial distance in order to end up with $K_{rr}(r)$ being approximately constant or increasing only slightly with r.

It is suggested to get rid of the coupling between the two diffusion coefficients, i.e. that $K_\perp$ is proportional to $K_{||}$. In particular, it does not seem justified to request that $K_{||} \propto 1/B$ for *all* locations in the heliosphere. From the result in the ecliptic plane we concluded $K_{||} \propto r^2$ somewhere beyond 1 AU which is not too far from $1/B$. But this condition certainly breaks down within 1 AU where $K_{||}$ must be roughly constant (see Wanner, 1993; Kallenrode, 1993).

In the polar regions the numerical values of $K_{||}$ are decisive for radial diffusion out to far larger distances than in the equatorial region. We have presented arguments why $K_{||}$ might not be invariant with latitude. Particles of solar origin are ideal probes to test this prediction within the Ulysses orbit. The small latitudinal gradients observed so far make it quite likely that the cosmic ray DC have to be reduced when moving towards the poles in contrast to some of the early expectations about the easy access.

Acknowledgements. We thank M.-B. Kallenrode for providing solutions of the focused transport equation presented in Fig. 4. One of the authors (G.W.) thankfully acknowledges an invitation to the University of Potchefstroom where he has profited from stimulating discussions with M.S. Potgieter, H. Moraal, and J.P.L. Reinecke. Part of this work was supported by the German DARA under grant 50 ON 9106.

References

Appenzeller, T 1992, *Science* **257**, 1478-1479
Bieber, J W et al 1994, *Astrophys J* **420**, 294-306
Chenette, D L 1980, *J Geophys Res* **85**, 2243
Conlon, T F 1978, *J Geophys Res* **83**, 541
Droge, W 1994, *Astrophys J Suppl* **90**, 567-576
Ferrando, P , Ducros, R , Rastoin, C , Raviart, A 1994, *Planet Space Sci*
Gleeson, L J 1969, *Planet Space Sci* **17**, 31-47
Haasbroek, L J , Potgieter, M S 1994, *this volume*
Hamilton, D C 1977, *J Geophys Res* **82**, 2157
Hamilton, D C , and Simpson, J A 1979, *Astrophys J* **228**, L123-L127
Heber B et al 1994, *this volume*
Jaekel, U , Wanner, W , Schlickeiser, R , and Wibberenz, G 1994, *Astron Astrophys* , submitted
Jokipii, J R 1971, *Rev Geophys Space Phys* **9**, 27-87
Jokipii, J R , and Coleman, P J 1968, *J Geophys Res* **73**, 5495
Jokipii, J R , and Kota, J 1989 *Geophys Res L* **16**, 1-4
Jokipii, J R , and Davila, J M 1981, *Astrophys J* **248**, 1156-1161
Kallenrode, M -B 1993, *J Geophys Res* **98**, 19 037-19 047
McKibben, R B 1994, *this volume*
Moraal, H 1993, *Nuclear Physics B (proc Suppl)* **33A,B**, 161-178
Morfill, G E , and Volk, H J 1979, *J Geophys Res* 84, 4446-4450 *Ap J*
Muller-Mellin et al 1993, *Adv Space Res* **13**, (6)85-(6)93
Palmer, I D 1982, *Rev Geophys* **20**, 335
Parker, E N 1965, *Planet Space Sci* **13**, 9-49
Potgieter, M S , Le Roux, J A , Burlaga, L F , and McDonald, F B 1993, *Astrophys J* **403**, 760-768
Reinecke, J P L , Moraal, H , and McDonald, F B 1993, *J Geophys Res* **98**, 9417 9431
Roelof E C 1969, in *Lectures in high energy astrophysics*, edited by H Ogelmann and J R Wayland, NASA SP-199, 111
Schlickeiser, R , and Achatz, U 1993 *J Plasma Phys* **49**, Part 1, 63
Smith C W , and Bieber, J W 1991, *J Geophys Res* **370**, 435-441
Volk, H J 1976, *J Geophys Res 81* 5570-5572
Wanner W 1993, *Dissertation*, Kiel University
Wanner W et al 1994, *Astron Astrophys* , submitted
Wibberenz, G et al 1992, *Geophys Res Lett* **19**, 1279
Wong K Y 1982, *Master's thesis*, University of Malaya
Zwickl R D , and Webber W R 1977 *Solar Phys* 54, 457

ENERGETIC PARTICLE OBSERVATIONS AT HIGH HELIOGRAPHIC LATITUDES

E. KEPPLER, M. FRÄNZ, A. KORTH, N. KRUPP, M.K. REUSS, B. WILKEN
Max-Planck-Institut für Aeronomie, Katlenburg-Lindau, Germany

A. BALOGH, R.J. FORSYTH, J.J. QUENBY
The Blackett Laboratory, Imperial College, London, U.K.

B. BLAKE
The Aerospace Corporation, SSL, Los Angeles, CA, USA

Abstract. Energetic particles, accelerated in shocks which were associated with recurrent fast solar wind streams, were observed in high heliographic latitudes; fifteen such streams were included in the present study. Intensity variations ranged up to four orders of magnitude. Energy spectra were typically steeper near forward shocks than near reverse shocks. Electrons were observed only related to the reverse shocks. Composition ratios in accelerated streams resembled those observed in fast CIR's. In periods between the recurrent acceleration regions elemental abundance ratios were similar to those of the anomalous cosmic rays (ACR). The intensity of the accelerated particles declined as the latitude of ULYSSES increased, probably due to the weakening of the shocks.

1. Introduction

The ULYSSES space probe was launched on October 6, 1990 and, after Jupiter flyby, left the ecliptic plane. Since then it has been moving towards higher southern heliographic latitudes, progressing in latitude by about 2 degrees per month; it will reach 80.2 degrees in September 1994 (Wenzel *et al.*, 1992). When the spacecraft had reached a heliographic latitude of 13.4 degrees South in June 1992, it started to encounter fast solar wind streams, rotating with the Sun's rotation period of about 27 days. Peak velocities of up to more than 900 km/s (Bame *et al.*, 1993) were observed in these streams. The leading edges of the stream interaction regions were bounded by forward/reverse shock pairs (Bame *et al.*, 1993) up to Days 115-120, 1993, afterwards only reverse shocks were seen. We report on the observation of a Southern Solar latitude zone of regular and efficient interplanetary shock acceleration. This zone coincides with the fast streams observed by Bame et al. (1993). Most of the energetic particle fluxes seen by the EPAC instrument on ULYSSES are likely to have been accelerated in diffusive shock acceleration processes.
Superimposed on this feature were several CME-related particle events as well as particles released from flares on the sun (Reuss *et al.*,1994). The pattern was seen over and over again, even beyond the period discussed in this paper to at least Day 200/1993.

Space Science Reviews **72**: 285–290.
© 1995 *Kluwer Academic Publishers.*

2. Instrumentation

The EPAC instrument on board of Ulysses consists of 4 identical three element semiconductor telescopes mounted at different angles relative to the S/C spin axis so that by virtue of the spin rotation of the S/C the sphere could be sampled in 32 bins. The energy range covered is 0.4 to 1.5 MeV for protons, and 0.4 to 6 MeV/N for heavier ions. Electrons are measured in two channels (L: 0.1<E<0.38 MeV and H: E>0.18 MeV). By means of the E-dE/dx technique elements up to iron can be separated (Keppler *et al.*, 1992).

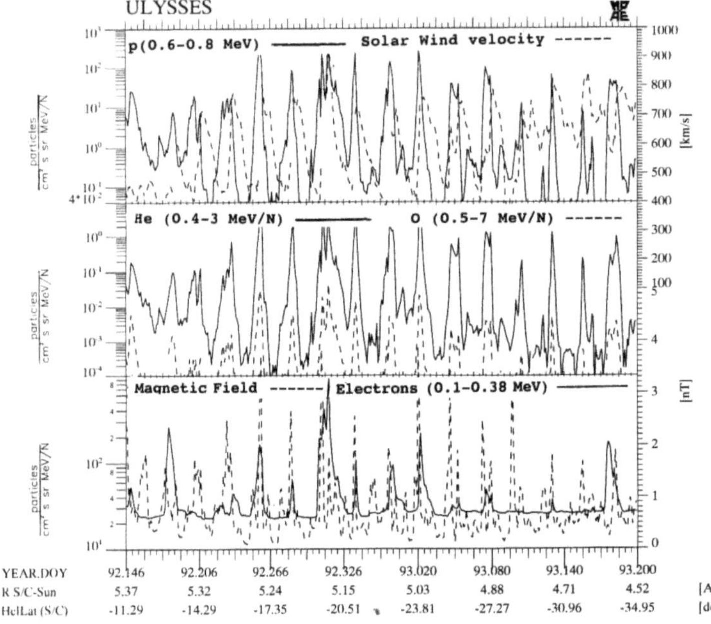

Fig.1: Fluxes of energetic protons, electrons, Helium and Oxygen plotted along with solar wind velocity (derived from SWICS data, courtesy G. Gloeckler) and magnetic field strength (from HED-Instrument).

3. Observations

Fig. 1 shows omnidirectional fluxes of protons, electrons, Helium, and Oxygen along with the solar wind velocity and the magnetic field intensity during the period from Day 146, 1992 to Day 200, 1993. The data show the first encounter with the fast stream which apparently emerged from the southern solar polar hole (Balogh *et.al.*, 1993, Smith *et al.*, 1993) following Day 176, 1992 (stream #1 in Fig. 1). Fig. 2 is an expanded view of the encounters number #10, 11 and 12 as indicated in Fig. 1. On this scale it can be more clearly seen that associated with the forward/reverse shock pairs (rise in solar wind speed and magnetic field intensity), the charged particle fluxes rise dramatically (up to four orders of magnitude) in intensity, similar for other species.

The forward shock was always associated with steeper ion spectra (γ~3 if expressed as a power law in energy) as shown for encounter #9 in Fig.3. Electrons with energies above 100 keV were observed above the RTG background only in the reverse shock. Higher intensities in the energetic particle fluxes were associated with the reverse shock. Typical energy spectra for all ion-species with γ~2 were observed. Spectra for all species have been reported as exponentials in particle velocity over a wide range in energy (Gloeckler *et al.*, 1979), which is understood theoretically quite well (Fisk & Lee, 1980). Converted to power laws in energy/N, our spectra are in reasonable agreement with their predictions. (for the forward shock the Fisk & Lee theory predicts a steeper slope of the energy spectrum -4.6 for a power law in E/N, whereas we observe -3. For the reverse shock they predict --2.25; we observe -2 for all species). Typical features known from observations of shock acceleration at low heliographic latitudes were also observed: A short ramp in front of the forward shock, a peak in ion-intensity at the shock surface; subsequent decay in the region between the shock pair; a longer ramp towards the reverse shock; a much larger peak at this shock followed by a much slower decay after it passes the S/C. Electron acceleration at the same rigidity as protons is slower because the diffusion coefficient is proportional to velocity and hence the more efficient R-shock is more likely to accelerate electrons. The problem of getting a sufficiently fast time constant to achieve acceleration at the observed energies may be overcome by appealing to fluctuations in the field-shock angle during the shock propagation (Savopulos *et al.*, 1994). These observations have been made within 4.5 to 5.4 AU from the sun, while the S/C climbed in heliographic latitude from 11 to 35 degrees South.

4. Results

(1) Particle intensities dropped very rapidly by the time the fast speed stream had reached its maximum velocity which almost coincided with the passage of the reverse shock and remained low until the next encounter with the corotating fast stream (Fig. 1). Fifteen

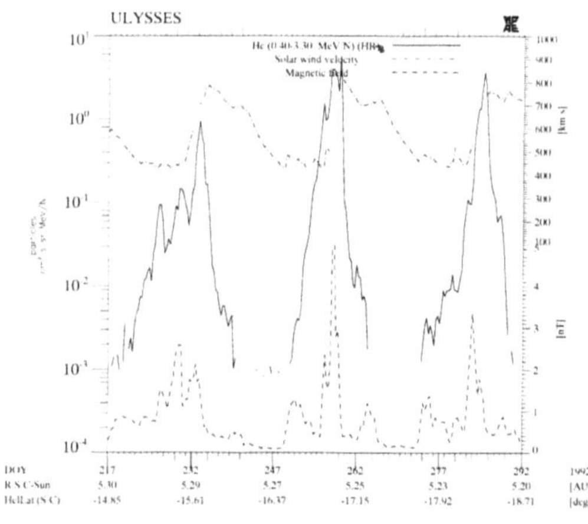

Fig.2: Expanded view of proton fluxes, solar wind speed and magnetic field intensity during encounters with two recurrent high speed solar wind streams (# 10, 11, 12 in Fig.1).

encounters have been observed so far. During the time periods between the recurrent observations at ULYSSES of shocks and their associated energetic particles, the counting rates drop very low. Plots of E-dE/dx show the clear presence of the anomalous cosmic rays; we observe a C/O ratio of the order of 5%. The other events are believed to be largely background. In our analysis we require that accepted events fall along the elemental tracks. An estimate for background in the tracks indicates a negligible contribution to the observations (see also figs.5).

(2) The maxima in intensity of all species observed diminished with time in an exponential fashion where successive maxima varied in a zig-zag manner, shown in Fig. 4.

(3) After Day 100, 1993, the S/C has left the streamer belt (as noted by Bame *et al.*, 1993). The forward shocks disappear while reverse shocks still were observed including their signature in particle measurements. After this time electrons were no longer visible to our instrument.

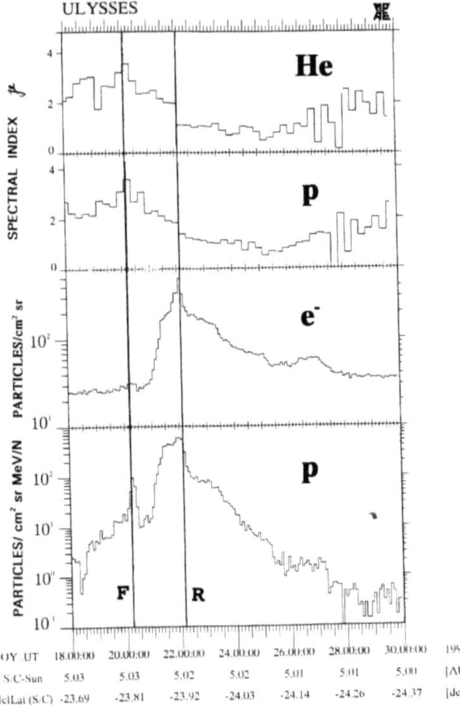

Fig.3: Proton and electron flux shown along with spectral index γ for p and He ($j = k*E^{-\gamma}$). Vertical lines indicate passage of forward (F) and reverse (R) shock for stream #9 in Fig.1.

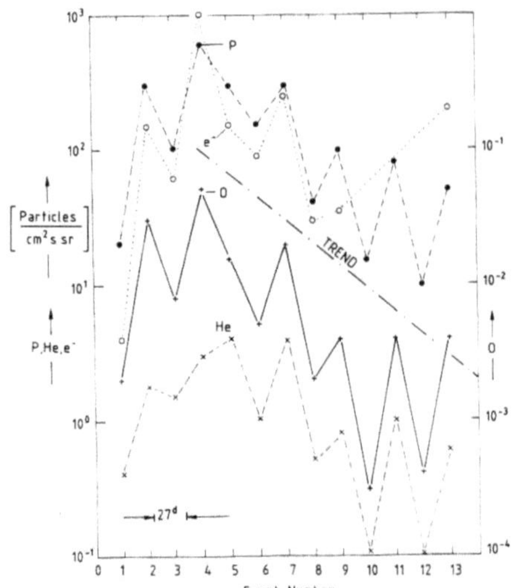

Fig.4: General decline of observed peak intensities of accelerated particles during 15 recurrent fast solar wind stream encounters for p, He, O and Electrons during the high latitude path of ULYSSES.

(4) There were quite remarkable variations in composition, the ranges of which are summarized in Table 1. Details are given in Fränz *et al.*,1994. Around the time of the peak fluxes of shock accelerated particles most of the observed ratios resemble the composition observed in Corotating Interaction Regions with v>600 km/s. The data also show an anisotropy in the solar direction, i.e. there is a net outward convected flow of the observed particles.

(5) It is interesting to note that before Day 200/1992, i.e. below about 14°S heliographic latitude, the variation of the p/He ratio varies from event to event between 80 and 25 (as observed in high speed streams). The Fe/He ratio is comparable to the ratio in fast solar wind streams, but is much higher in the "valleys". Also, we do not see any significant change in composition with the change in the solar wind speed variation after Day 100/1993.

(6). The composition of the accelerated particle streams is obviously similar to that observed in various types of Solar Wind streams or the driver gas behind CME´s. One can argue that we observe particles accelerated from a seed population supplied by small flares on the sun. Indeed, particles observed between Days 58 and 97/1993 could be related to solar releases, but there is no evidence for this always to be true. On the other hand, we observe particle fluxes to rise fairly rapidly by several orders of magnitude when the shock approaches the S/C (Figs.1,2). Also, in the valleys, i.e. before the F-shock arrives, we observe particles obviously belonging to the anomalous component and not to solar released seed particles. This could then suggest that acceleration of charged particles to MeV energies is likely to occur from solar wind energies on. In fact, Barnes and Simpson (1976) and Simpson *et al.*,(1986) came to very similar conclusions when discussing their observations with Pioneer 10 and 11.

(7) In the "valleys" ("V"), energetic particle fluxes were low. To determine relative abundances, integration of events along tracks of the various ions in the E-dE/dx matrix (Figs. 5a,b) has been performed as described by Fränz *et al.*,1994. The observed ratios of C/O=0.04±0.017 and N/O=0.119±0.030 determined in this manner suggest that we observe anomalous cosmic ray particles. The Oxygen flux at 6 MeV/N if averaged over 6 months increased by a factor of 3 when comparing the first and the second half year of 1993. This is consistent with what has been observed by e.g. SAMPEX instruments (D.Hovestadt, private communication). The measurements are similar for all four telescopes: in the "valleys" there is no obvious anisotropy to be seen.

Table 1: Composition of energetic particles in high latitudes (E >0.5 MeV/N). (1):This work; (2):from Klecker et al.,1977, (3): from Richardson et.al., 1993)

Ratio	Peak (1)	Valley (1)	ACR (2)	Solar part. (3)	Driver Gas(3) (CME)	CIR (3)
He/O	200-400	3	9.8	52-120	55	54-130
p/O	4000	< 100	-	$2\text{-}6*10^{3}$	1170	2300-2500
Fe/O	0.08-0.15	0.08	0.03	0.03-0.3	0.15	0.1-0.14
C/O	0.55-0.9	0.040 ±0.017	0.04	0.35-0.5	0.47	0.6-0.9
p/He	10-20	33	-	17-115	21	19-43
Fe/He	$3*10^{-4}$	0.03	$3*10^{-3}$	$5*10^{-4}$-$2.3*10^{-3}$	$2.7*10^{-3}$	$3*10^{-3}$-$8*10^{-4}$
C/He	$1.5*10^{-3}$	0.013	$4*10^{-3}$	$2.9\text{-}9*10^{-3}$	$8.5*10^{-3}$	$6.8\text{-}10*10^{-3}$

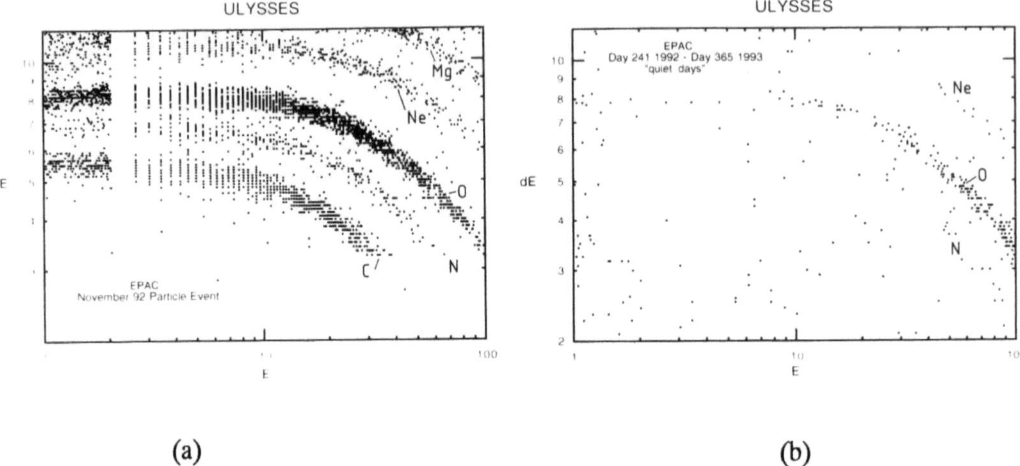

(a) (b)

Fig.5: E-dE/dx plots during a solar particle event (5a) and during the "valleys" accumulated over 1993 (5b).

References

Bame, S.J., B.E. Goldstein, J.T. Gosling, J.W. Harvey, D.J. McComas, M. Neugebauer, J.L. Phillips: 1993, Geophys. Res. Letters **20**, 2323-2326.

Balogh, A., G. Erdös, R.J. Forsyth, E.J. Smith: 1993, Geophys. Res. Letters **20**,2331-2334.

Barnes, C.W. & J.A.Simpson, 1976, Astrophys. J., **210**: L91-L96.

Fisk, L.A. & M.A. Lee: 1980, Astrophys. J., **237**, 620-626.

Fränz, M., E. Keppler, N. Krupp, M.K. Reuss, and B. Blake, 1994: 28th ESLAB Symposium on "The High Latitude Heliosphere", Friedrichshafen.

Gloeckler, G., D. Hovestadt, L.A. Fisk: 1979, Astrophys. J. (Letters) **230**, L191.

Keppler, E., J.B. Blake, D. Hovestadt, A. Korth, J. Quenby, G. Umlauft, J. Woch, 1992: Astronomy & Astrophysics, Suppl.Ser. **92**, 317-331.

Klecker, B., D. Hovestadt, G. Gloeckler, C. Fan, 1977: Astrophys. J., **212**, 290-299.

Reuss, M. K. et al.: 1994, 28th ESLAB Symposium on "The High Latitude Heliosphere", Friedrichshafen

Richardson, I., L. Barbier, D. Reames, T. v. Rosenvinge: 1993, J. Geophys.Res., **98**(A1), 13-32.

Savopulos, M., J.J. Quenby, A.R. Bell: 1994, sub. to Solar Phys.

Simpson, J.A., E.J. Smith, B. Tsurutani in: 1986, The Sun and the Heliosphere in Three Dimensions, R.G.Marsden (ed.), p. 319-324, D.Reidel Publ.Comp.

Smith, E.J., M. Neugebauer, A. Balogh, S.J. Bame, G. Erdös, R.J. Forsyth, B.E. Goldstein, J.L. Phillips, B.T. Tsurutani: 1993, Geophys. Res. Letters **20**, 2327-2330.

Wenzel, K.-P., R.G. Marsden, D.E. Page, E.J. Smith: 1989, Adv. Space Res. **9**,(4)25-(4)29.

HIGH-LATITUDE OBSERVATIONS OF ENERGETIC IONS DURING THE FIRST ULYSSES POLAR PASS

T. R. SANDERSON, R. G. MARSDEN, K.-P. WENZEL

Space Science Dept. of ESA, Estec, The Netherlands

A. BALOGH, R. J. FORSYTH

Blackett Laboratory, Imperial College, London, UK

and

B. E. GOLDSTEIN

Jet Propulsion Laboratory, California Institute of Technology, Pasadena, CA, USA

Abstract We present observations of energetic ions from the Ulysses COSPIN Low Energy Telescope in the mid and high-latitude regions of the heliosphere prior to and during the first polar pass of the Ulysses spacecraft. After the encounter with Jupiter, Ulysses started on its journey out-of-the-ecliptic. Between ~13°S and ~29°S the spacecraft sampled the solar wind from both the streamer belt and the polar coronal hole. Here, co-rotating magnetic structures with forward and reverse shocks and containing accelerated energetic ions were observed. At latitudes greater than ~29°S, Ulysses was completely immersed in the solar wind from the polar coronal hole. Here the co-rotating magnetic structures were weaker, and in general had only reverse shocks, but were still capable of accelerating the energetic ions, albeit with reduced intensity. The most recent results show that beyond ~50°S, very few if any, reverse shocks are observed. However, accelerated ions from magnetic interaction regions are still observed. We report also on an intensity enhancement at ~50°S due to the passage of a high-latitude CME.

1. Introduction

The Ulysses spacecraft was launched on 6 October 1990, and in February 1992 a Jupiter-gravity-assist was used to place the spacecraft into an orbit inclined at 80° to the ecliptic plane to enable it to reach the high-latitude regions of the heliosphere. By July 1992 the spacecraft had reached a southerly latitude of 13°S. The solar wind at Ulysses began to be dominated by a high-speed stream made up of three or four interaction regions per rotation (Smith et al., 1993), recurring once per solar rotation (Bame et al., 1993). Most of the interaction regions were bounded upstream by a forward shock and downstream by a reverse shock, and contained intense fluxes of energetic ions, accelerated up to energies a few MeV.

In this paper we discuss the behaviour of the energetic ions associated with these recurrent streams as the Ulysses spacecraft moved out of the ecliptic to increasing heliospheric latitudes. Preliminary results of observations up to a southerly latitude of ~49° can be found in Sanderson et al. (1994) and Simnett et al. (1994).

Space Science Reviews **72**: 291–296.
© 1995 *Kluwer Academic Publishers.*

2. Instrumentation

The particle observations presented here were made with the Low Energy Telescope (LET) of the COSPIN (Cosmic Ray and Solar Particle Instrumentation) consortium instrument (Simpson et al., 1992), which measures protons, alphas, and heavy ions over a range of energies from ~1 to ~50 MeV/n. Data from the 1.2 to 3.0 MeV proton channel and the 1.0 - 5.0 MeV/n alpha channel are presented here. The magnetic field observations presented here were made with the Ulysses Magnetometer (Balogh et al., 1992), whilst the solar wind measurements were made with the ion sensor of the Ulysses Solar Wind Plasma experiment (Bame et al., 1992).

3. Observations

Figure 1 shows a plot of the projection of the orbit of the spacecraft perpendicular to the plane of the heliographic equator, together with the regions of interest discussed in this paper.

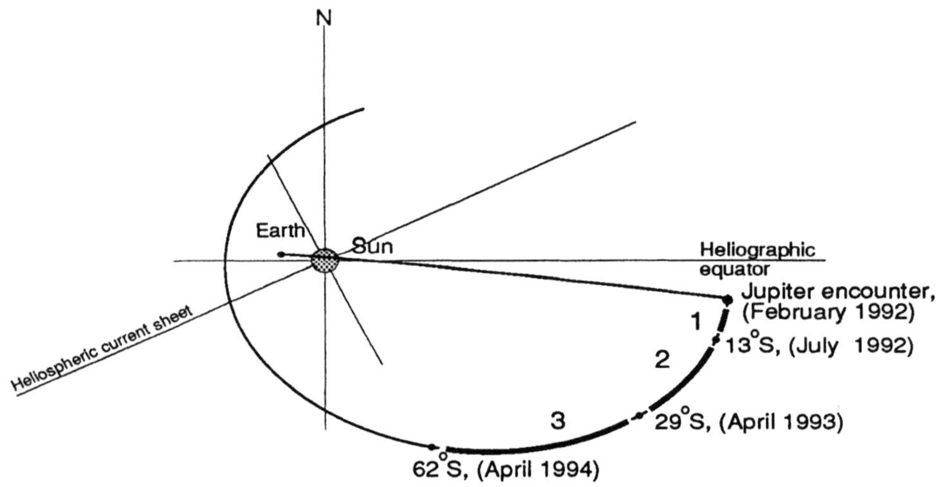

Fig. 1. The projection of the spacecraft orbit perpendicular to the plane of the heliographic equator, shown with a heliospheric current sheet with a tilt of ~26° as inferred by Gosling et al. (1993). The current sheet is shown tilted away from the spacecraft. The orbit is shown starting at the Earth, passing by Jupiter, over the southern polar coronal hole, and ending up over the north pole of the Sun. The three regions discussed in this paper are shown labelled 1, 2 and 3.

In Figure 2 we show observations of the regions shown in Figure 1, from day 1 of 1992 until day 120 of 1994 (1 January 1992 to 30 April 1994). In the top panel we show the intensity of 1.2 - 3.0 MeV protons (top trace) and 1.0 - 5.0 MeV/n alphas, whilst in the bottom panel we show the solar wind velocity. We have divided this period up into the three parts shown in Figure 1 as follows:

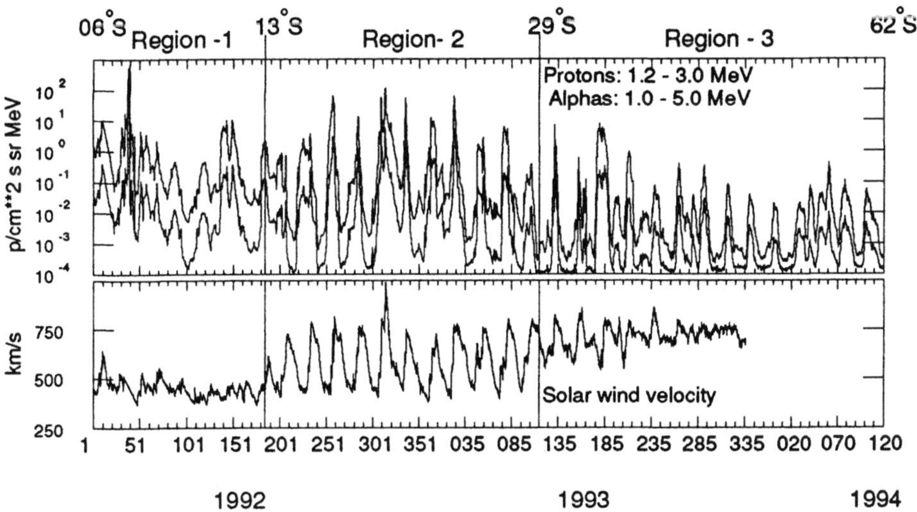

Fig. 2. Intensity of 1.2 - 3.0 MeV protons and 1.0 - 5.0 MeV/n alphas as observed by the LET during the period from day 1 of 1992 to day 120 of 1994 (1 January 1992 to 30 April 1994), together with the solar wind velocity. Three regions are shown, Region-1, where the spacecraft is completely immersed in the streamer belt, Region-2, where the spacecraft passes in and out of the polar coronal hole flow every solar rotation, and Region-3, where the spacecraft is completely immersed in the polar coronal hole flow.

3.1 REGION-1. HELIOGRAPHIC LATITUDES LESS THAN ~13°S.

Here the spacecraft was completely immersed in the streamer belt, and fluctuations of ~1 MeV protons were observed without any real pattern.

3.2 REGION-2. HELIOGRAPHIC LATITUDES BETWEEN ~13° AND ~29°S.

Here the spacecraft moves from the streamer belt to the polar coronal hole once per solar rotation, due to the combined effect of the rotation of the tilted heliospheric current sheet, and an equatorward extension of the polar coronal hole. A large recurrent corotating solar-wind stream was observed by Bame et al. (1993), made up of four magnetic structures (Smith et al., 1993). In Figure 3a we show an example of one of the interaction regions which contains both forward and reverse shocks. Here we show, from top to bottom, the 1.2 - 3.0 MeV proton intensity, the ratio of the 1.2 - 3.0 MeV proton and the 1.0 - 5.0 MeV/n alpha channels as a measure of the proton-to-alpha ratio, the three components of the solar wind velocity, and the magnetic field magnitude. The dashed lines show the times of the shocks, as identified by Balogh et al. (1994). In this example, as with most of the interaction regions, the reverse shock is dominant. We observe a peak in the particle intensity at both the forward and the reverse shocks, the largest peak being seen at the reverse shock.

There is a change in the flow direction across the interaction region (Gosling et al., 1993) and a drop in the proton to alpha ratio close to this change in the flow direction, similar to that observed at lower latitudes in co-rotating interaction regions (Barnes and Simpson, 1976, Marsden et al., 1993).

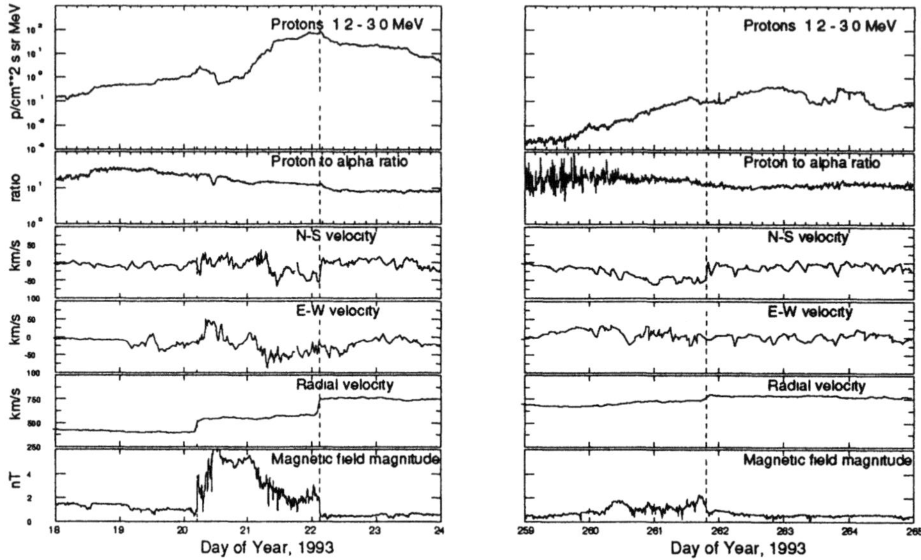

Fig. 3. (a) Data from day 18 to day 24 of 1993 (18 January to 24 January 1993), showing an interaction region containing a forward-reverse shock pair, and (b) data from day 259 to 265 of 1993 (16 to 22 September 1993). From top to bottom: the 1.2 - 3.0 MeV proton intensity, the proton to alpha ratio, the three components of the solar wind velocity, and the magnetic field magnitude.

3.3 REGION-3. HELIOGRAPHIC LATITUDES GREATER THAN ~29°S.

From ~29°S onwards the spacecraft was completely immersed in the flow from the polar coronal hole. In this region, there is a general decline in intensity at the peaks as Ulysses moves to higher latitudes and further into the flow from the polar coronal hole, the intensity at ~51°S (at the start of 1994) having dropped by nearly three orders of magnitude from the time Ulysses entered the polar coronal hole flow. An additional increase in intensity at the peaks, lasting for four rotations, was observed starting at around day 20 of 1994.

Interaction regions are still observed, but nearly all are bounded by a reverse shock at the trailing edge. These shocks are weaker than the shocks observed within Region-2, but are still able to accelerate ~1 MeV particles. An example of one of these regions is shown in Figure 3b. In contrast to before, the peak intensity is observed after the passage of the shock (i.e. in the upstream region of the reverse shock), and the intensity at the peak is lower.

In Figure 4 we show the proton and alpha particle data from Region-3 during the latter part of 1993 and the beginning of 1994, together with dashed vertical lines showing the shocks. The times of the shocks are taken from Balogh et al. (1994) and McComas et al. (1994). This shows the peaks which re-occur approximately every 26 days, the intensity at the peak slowly decaying with increasing latitude up to the end of 1993, and thereafter an increase for about four rotations starting at around day 20 of 1994. Additional peaks are also observed in between the main recurring peak.

Prior to day 215 of 1993, we observe proton to alpha ratios near the peaks of ~50, as expected from particles accelerated by the forward shocks. After day 215 (~36°S), the proton to alpha ratio at the peaks dropped to ~20. By around day 285 (~41°S) the ratio had dropped to ~10.

The last of these recurrent reverse shocks was observed on day 335 at ~45°S, but recurrent peaks in the particle intensity were still observed beyond this despite the lack of a locally observed reverse shock, implying acceleration by reverse shocks away from the location of Ulysses.

This pattern was disturbed early in 1994 by an increase in the intensity of the main peaks above the general decline, and the appearance of two new peaks in between the main peaks, the intensities at the peaks being higher than the main peaks, and having exceptionally high proton to alpha ratios of ~80 and ~50. These increases were associated with a high level of solar activity (e.g. a filament disappearance on day 36/37 and an M4/3B flare on day 51 (Solar Geophysical Data, 1994)), a CME at Ulysses together with a forward-reverse shock pair on day 57, and a reverse shock on day 69 (McComas et al., 1994).

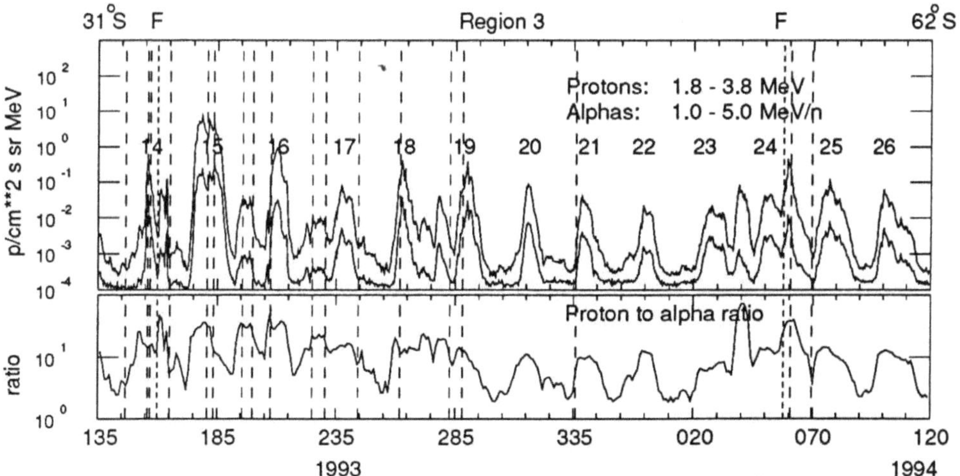

Fig 4. Proton and Alpha particle intensity, together with proton to alpha ratio for the period day 135 of 1993 to 120 of 1994 (15 May 1993 to 30 April 1994). The shocks are denoted by the vertical dashed lines, the forward shocks being labelled F. The shock times in 1993 are taken from Balogh et al. (1994), and in 1994 from McComas et al. (1994).

4. Summary

During the passage from Jupiter to the southern polar regions of the heliosphere, we have continued to observe accelerated energetic ions with energies ~1 MeV. We can summarise our observations as follows:-

1. From ~13°S to ~29°S we observe accelerated energetic ions recurring with one main peak in the energetic ion intensity every solar rotation. This main peak was observed at or close to the reverse shock of the dominant recurrent magnetic structure seen every solar rotation. Secondary peaks were observed at the forward shock of this structure. In between the main peaks, other minor peaks were observed together with the other magnetic structures.
2. At latitudes higher than ~29°S , where the spacecraft was completely immersed in the polar coronal hole flow, the peaks continued to be observed, the peak intensity decaying with increasing latitude. In between the peaks, we observe very low intensities, with the occasional secondary peak. Beyond ~35°S, where essentially no more forward shocks were seen, the peaks continued to be observed together with the reverse shock. Beyond ~45°S, increases were seen despite the lack of a locally observed reverse shock.
3. From ~50°S to ~60°S an increase in intensity at the recurrent peaks was observed lasting for around 4 rotations, coinciding with an increased level of solar activity. Two additional peaks, with intensities higher than the main recurrent peaks, were observed, one of which was associated with the passage of a CME.

References

Balogh, A. et al.: 1992, *Astron. and Astrophys.* **92,** 221.

Balogh, A., M. E. Burton, E. J. Smith, R. J. Forsyth, J.-A. Gonzales-Esparza, S. J. Bame, and B. E. Goldstein: 1994, *this issue.*

Bame, S. J. et al.: 1992, *Astron. and Astrophys.* **92,** 237.

Bame, S. J., B. E. Goldstein, J. T. Gosling, J. W. Harvey, D. J. McComas, M. Neugebauer, and J. L. Phillips: 1993, *Geophys. Res. Lett.* **20**, 2323.

Barnes, C. W., and J. A. Simpson: 1976, *Astrophys. J.* **210,** L91.

Marsden, R. G., T. R. Sanderson, K.-P. Wenzel, and S. J. Bame: 1993, *Adv. Space Res.* **13**, (6)95.

McComas, D.J., J. L. Phillips, S. J. Bame, J. T. Gosling, B. E. Goldstein, and M. Neugebauer: 1994, *this issue.*

Gosling, J. T., S. J. Bame, D. J. McComas, J. L. Phillips, V. J. Pizzo, B. E. Goldstein, and M. Neugebauer: 1993, *Geophys. Res. Lett.* **20**, 2789.

Sanderson, T. R., R. G. Marsden, K.-P. Wenzel, A. Balogh, R. J. Forsyth, and B. E. Goldstein: 1994, *Geophys. Res. Lett.* in press.

Simpson, J. A. et al.: *1992, Astron. and Astrophys.* **92**, 365.

Simnett, G. M., K. A. Sayle, E. C. Roelof, and S. J. Tappin: 1994, *this issue.*

Smith, E. J., M. Neugebauer, A. Balogh, S. J. Bame, G. Erdös, R. J. Forsyth, B. E. Goldstein, J. L. Phillips, and B. T. Tsurutani: 1993, *Geophys. Res. Lett.* **20**, 2327.

ELEMENTAL ABUNDANCES IN COROTATING INTERACTION REGIONS AT HIGH SOLAR LATITUDES

C. G. MACLENNAN and L. J. LANZEROTTI

AT&T Bell Laboratories, Murray Hill, NJ 07974 USA

Abstract. Throughout 1993, as the Ulysses spacecraft traveled from ~23° to ~45° south heliolatitude, the HI-SCALE instrument on the spacecraft measured a recurrent series of enhanced particle fluxes with a recurrence period of ~26.5 days. These particles are accelerated from a background seed population by the corotating interaction regions (CIRs) associated with a southern solar polar coronal hole. Using the Wart detector telescope of the HI-SCALE instrument, we have analyzed the elemental abundances of C, N, O, and Fe relative to He for 0.5 – 4.0 MeV/nucl ions and Ne, Mg, and Si for 1.0 – 4.0 MeV/nucl ions in the CIRs. We compare the relative abundances to some previous measurements reported from 1 A.U. as well as with solar photosphere abundances. We note that HI-SCALE measurements of the heliolatitude dependence of the oxygen abundance and spectrum as reported by Lanzerotti *et al.* (1994) suggest that a substantial fraction of the seed population for the CIR-accelerated oxygen is likely to be the anomalous oxygen component of the cosmic rays.

Key words: Heliosphere - Composition - CIRs - Ulysses

1. Introduction

From measurements made in the ecliptic plane, principally near 1 A.U., it has long been recognized that interplanetary particles can be accelerated in the interplanetary medium, often in association with corotating interaction regions (CIRs). This process was first identified for hydrogen and helium ions (e.g., McDonald *et al.*, 1976; Barnes and Simpson, 1976; Pesses *et al.*, 1978). The acceleration of higher mass nuclei by CIRs has been reported by a number of investigators from measurements made in the ecliptic plane near 1 A.U. (e.g., McGuire *et al.*, 1978; Gloeckler *et al.*, 1979; Scholer *et al.* 1979; Dietrich and Simpson, 1985) as well as at larger distances (e.g., Maclennan *et al.*, 1993, between 3 and 5 A.U.). The relative abundances of these higher mass nuclei tended to be different from the typical abundances of nuclei accelerated in solar events (discussion in Dietrich and Simpson, 1985).

We report in this paper determinations of the abundances of interplanetary ions measured in 1993 (day 356, 1992 to day 7, 1994) by the HI-SCALE instrument during the climb of the Ulysses spacecraft from 23° to 45° southern solar latitudes. During 1993, most of the interplanetary ions measured by HI-SCALE were those that occurred in association with CIRs sweeping across the spacecraft. The abundance determinations can be compared to those reported previously from the ecliptic measurements and from solar photospheric abundances in order to better understand the interplanetary acceleration process, the role of seed particles in the process, and the heliolatitude dependence of both of these.

The measurements presented herein were made with the Wart (W) aperture of the HI-SCALE instrument on Ulysses. The Wart consists of a three element Si solid state

Space Science Reviews **72**: 297–302.
© 1995 *Kluwer Academic Publishers.*

detector telescope with a 5μ first detector followed by two 200μ second and third detectors (Lanzerotti *et al.*, 1992). The detected particles are analyzed using two discrete energy channels as well as a four level priority-controlled pulse height analyzer (PHA). The results from the PHA analysis of the individual detected events (two events analyzed per spacecraft spin period of 12 s) are those reported here for the abundance determinations.

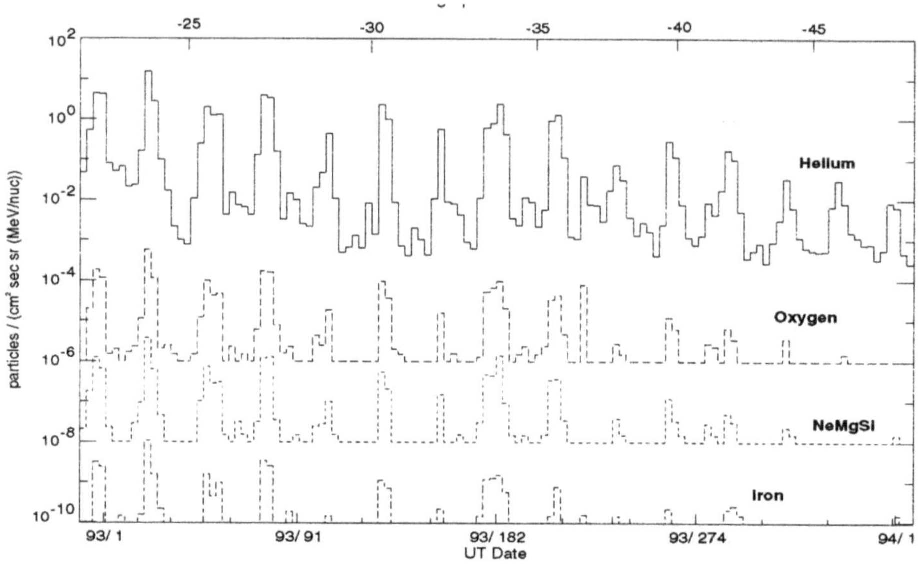

Fig. 1. Fluxes vs time of 0.5 to 1.0 MeV/nucl He, O ($\times 10^{-2}$), Ne-Mg-Si ($\times 10^{-4}$), and Fe ($\times 10^{-6}$) ions (spin-averaged, three-day averages) measured by the HI-SCALE instrument on the Ulysses spacecraft. For clarity, zero counts are set to a value of 10^{-4}

2. Observations

Plotted in Figure 1, top to bottom, are the spin-averaged, three-day average fluxes of 0.5 to 1.0 MeV/nucl helium, oxygen, NMS (Ne-Mg-Si), and iron ions measured during 1993. The dotted and dashed "baselines" connecting the enhancements in the O and NMS fluxes are used to guide the eye; no fluxes less than these lines were measured. The heliolatitude is given for selected times at the top of the figure. During this year, the helio-radius of Ulysses decreased from ~5 A.U. to ~3.8 A.U. The approximately regular large increases in the ion fluxes are produced by the CIR accelerations of existent interplanetary ions. There are approximately 15 CIRs evident during the year in the He fluxes, with a recurrence period of ~26.5 days (Thomson *et al.*, 1994). The decrease with time (and therefore with heliolatitude) of the ion intensities is particularly apparent. The amplitude change with latitude of the ion and electron populations during the year is discussed in Simnett *et al.* (1994).

Shown in Figure 2 are scatterplot and histogram distributions for the daily values of O/He and Fe/He abundances in the energy range 0.5 – 4.0 MeV/nucl. The number of points in the scatterplot corresponds approximately to the number of points in the Fe/He

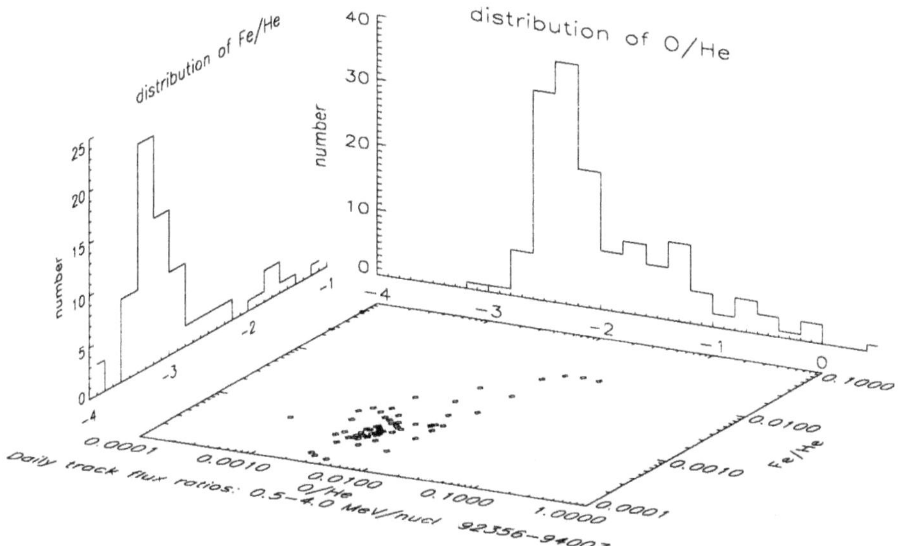

Fig. 2. Scatterplot of Fe/He to O/He abundances for 0.5–4.0 MeV/nucl particles for day 356, 1992 to day 7, 1994. The occurrence distributions of the two relative abundances are also shown.

distribution, as a point can only be plotted if both ratios are determined. The histogram distributions are plotted linearly in the logarithms of the relative abundances.

A scatterplot distribution of N/He vs O/He for 0.5 – 4.0 MeV/nucl ions is shown in Figure 3, together with the histograms of occurrence of the relative abundances. As in

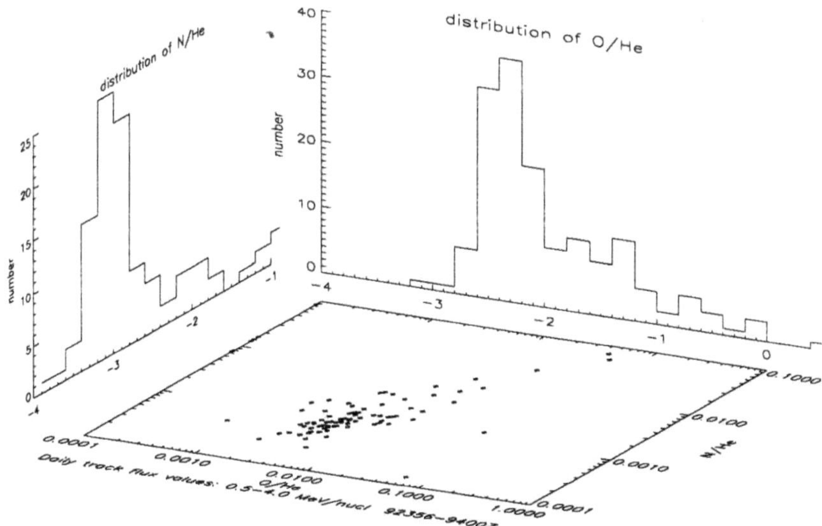

Fig. 3. Scatterplot of N/He vs O/He for 0.5–4.0 MeV/nucl particles.

Figure 2, there is a tendency for an increase in the relative abundances of one species to be accompanied by an increase in the relative abundances of another. Figure 4 shows a distribution scatterplot of the relative abundances of Ne/He vs Mg/He and Si/He for 1.0 – 4.0 MeV/nucl, energies where HI-SCALE can distinguish between these species.

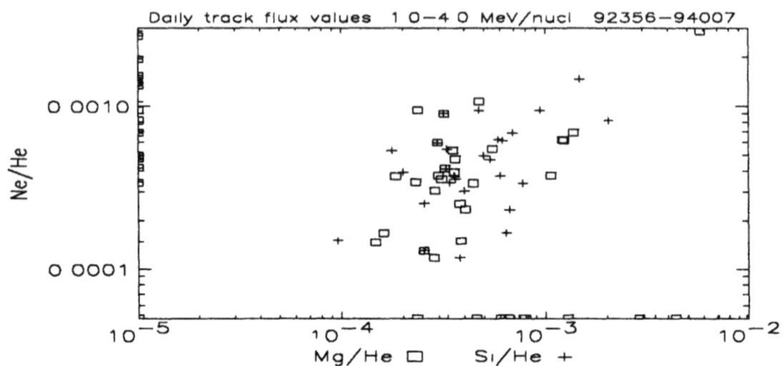

Fig. 4. Scatterplot of Ne/He vs Mg/He and Si/He.

3. Discussion

Listed in Table 1 are the abundances relative to He of C, N, O, and Fe for 0.5 – 4.0 MeV/nucl ions and of Ne, Mg, and Si for 1.0 – 4.0 MeV/nucl ions determined for the time interval studied here. The abundance values are those corresponding to the values of the center of the peaks in the abundance distributions (examples as shown in Figures 2 and 3) The uncertainty shown for each relative abundance value corresponds to plus or minus one-half of the full width at half height of the abundance histograms. Thus, these abundance values are not as large as if one took a simple average of the data.

TABLE 1. Abundances relative to Helium

	HI-SCALE '93		CIRs '82-'83[a]	Solar Photosphere[b]
	0.5–4.0 MeV/n	1.0–4.0 MeV/n	1.9–2.8 MeV/n	
C	.0025 ± .0010	-	.0070 ± .0020	.0041
N	.0009 ± .0004	-	.0014 ± .0004	.0010
O	.0045 ± .0015	-	.0081 ± .0023	.0087
Ne	-	.0004 ± .0002	.0013 ± .0004	.0014
Mg	-	.0003 ± .0001	.0012 ± .0004	.0004
Si	-	.0004 ± .0002	.0009 ± .0003	.0004
Fe	.0006 ± .0002	-	.0010 ± .0003	.0003

[a] from Table 2, Richardson *et al.*, 1993.
[b] from Table 1, Reames *et al.*, 1994.

Also listed in the Table are values of the relative abundances from CIR acceleration results reported from measurements made near 1 A.U. in the ecliptic plane by Richardson *et al.* (1993) at a similar phase of the last solar cycle as for the present Ulysses measurements. These Ulysses abundances relative to He at mid-heliolatitudes are systematically smaller than those reported for the ecliptic plane, 1 A.U. results. This could arise from several possibilities, including a weakening of the acceleration process(es) with increasing heliolatitude, a change in the acceleration process from forward and reverse shocks to predominantly reverse shocks (Simnett and Roelof, 1994), and changes in the seed population abundance with heliolatitude. Such considerations will be reported elsewhere.

The abundances relative to He obtained from HI-SCALE measurements are compared to solar photosphere abundances in Figure 5, plotted as a function of the first ionization potential (FIP) of the elements. Such comparisons are often done when relating interplanetary measurements of energetic solar-origin particles to solar abundances. It is used here because of the likelihood that at least some of the seed particles that are accelerated in the CIRs are originally of solar origin. The figure shows that there are no significant overabundances of the low FIP particles (e.g., Fe, Mg, Si) as has been reported for solar events at 1 A.U. (e.g., Reames *et al.*, 1981) and beyond 2 A.U. (Maclennan *et al.*, 1993). The results in Figure 5 indicate that, with the exception of Fe, the abundances are at, or below, those of the photosphere.

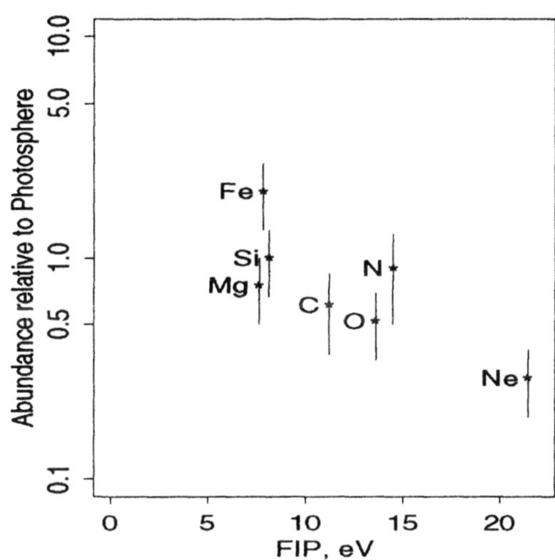

Fig. 5. Abundance vs first ionization potential.

A substantial fraction of the oxygen seed population may not be of solar origin, as reported by Lanzerotti *et al.* (1994) from HI-SCALE data. Plotted in Figure 6 as a function of Ulysses heliolatitude is the value of the spectral index of the best fit power law to 4-day average fluxes in the 0.47 –1.7 MeV/nucl and 1.7 – 19.1 MeV/nucl CNO W telescope channels. The quasi-periodic large negative values of the spectral indexes correspond to the occurrences of the CIRs. The increase in the hardness of the CIR-related spectra with increasing heliolatitude means that the higher energy CNO

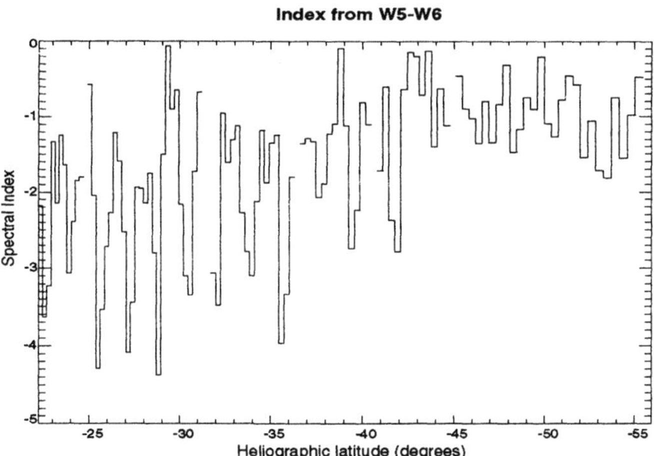

Fig. 6. Spectral index calculated from CNO fluxes.

ions are increasing in relative abundance as the spacecraft climbs to higher latitudes. In fact, the lower energy fluxes are found to be decreasing rapidly while the higher energy fluxes remain constant or increase slightly with time (latitude).

Lanzerotti *et al.* (1994) suggest that the absence of a decrease (and even an increase) in the fluxes of the higher energy O nuclei with increasing heliolatitude means that the higher energy fluxes are likely to be principally the anomalous component. The highest energy O fluxes (~16MeV/nucl) measured by HI-SCALE (Lanzerotti *et al.* 1994) correspond well to the lowest energy anomalous O fluxes measured on SAMPEX by Mewaldt *et al.* (1993). The decrease of the fluxes of the lower energy O nuclei results from a decreasing efficiency with increasing latitude of the CIR–acceleration process to accelerate the anomalous component seed population.

Acknowledgments

We thank G.M.Simnett for comments on this paper and our other HI-SCALE colleagues for informative discussions and beneficial collaborations on the instrument and the data.

References

Barnes, C.W., and Simpson, J.A.: 1976, 'Evidence for Interplanetary Acceleration of Nucleons in Corotating Interacting Regions', *Ap.J.Lett.*, **210**, L91.

Dietrich, W.F., and Simpson, J.A.: 1985, 'Elemental Abundance Differences between Nuclei Accelerated in CIR Shocks and Solar Flares', *19th ICRC,* **4**, 277.

Gloeckler, G., Hovestadt, D., Ipavich, F.V., and Mason, G.M.: 1979, 'Distribution Function Representation of Energy Spectra of H, He, C, O and Fe in Corotating Particle Streams', *16th ICRC,* **5**, 368.

Lanzerotti, L.J., *et al.*: 1992, 'Heliosphere Instrument for Spectra, Composition and Anisotropy at Low Energies', *Astron. Astrophys.,* **92**, 349.

Lanzerotti, L.J., *et al.*: 1994, 'Measurement of the Heliolatitude Dependence of Anomalous Cosmic Ray Oxygen', *Geophys. Res. Lett.,* , submitted.

Maclennan, C.G., Lanzerotti, L.J., Gold, R.E., Hawkins, S.E., III, and Tappin, S.J.: 1993, 'Composition of Interplanetary Ions in Recurrent Events: Measurements by HI-SCALE on Ulysses at 3–5 AU', *23rd ICRC,* **3**, 330.

McDonald, F.B., Teegarden, B.J., Trainor, J.H., von Rosenvinge, T.T., and Webber, W.R.: 1976, 'The Interplanetary Acceleration of Energetic Nucleons', *Ap.J.Lett.,* **203**, L149.

McGuire, R.E., Von Rosenvinge, T.T., and McDonald, F.B.: 1978, 'The Composition of Corotating Energetic Particle Streams', *Ap.J.Lett.,* **224**, L87.

Pesses, M.E., Van Allen, J.A., and Goertz, C.K.: 1978, 'Energetic Protons Associated with Interplanetary Active Regions 1-5 AU from the Sun', *J.Geophys.Res.,* **83**, 553.

Reames, D.V., Meyer, J.P., and von Rosenvinge, T.T.: 1994, 'Energetic Particle Abundances in Impulsive Solar-Flare Events', *Ap.J.Supp.Series,* **90**, 649.

Richardson, I.G., Barbier, L.M., Reames, D.V., and von Rosenvinge, T.T.: 1993, 'Corotating MeV/amu Ion Enhancements at ≤ 1 AU from 1978 to 1986', *J.Geophys.Res.,* **98**, 13.

Scholer, M., Hovestadt, D., Klecker, B., and Gloeckler, G.: 1979, 'The Composition of Energetic Particles in Corotating Events', *Ap.J.,* **227**, 323.

Simnett, G.M., Sayle, K.A., Roelof, E.C. and Tappin, S.J.: 1994, 'Corotating Particle Enhancements out of the Ecliptic Plane', *Geophys. Res. Lett.,* submitted.

Simnett, G.M., and Roelof, E.C.: 1994, 'Evidence for acceleration of electrons and protons observed by Ulysses at mid-heliolatitudes 5.3 AU –3.8 AU', *Space Sci. Rev.,* submitted.

Thomson, D.J., Lanzerotti, L.J., Maclennan, C.G., and Simnett, G.M.: 1994, 'Recurrence Rate of Interplanetary Interaction Regions at Southern Solar Latitudes', *Geophys.Res.Lett.*, submitted.

REVERSE SHOCK ACCELERATION OF ELECTRONS AND PROTONS AT MID-HELIOLATITUDES FROM 5.3-3.8 AU

G. M. SIMNETT
School of Physics and Space Research, University of Birmingham, B15 2TT, UK

and

E. C. ROELOF
Johns Hopkins University/Applied Physics Laboratory, Laurel, MD 20723, USA

May 23, 1994

Abstract. We examine the intensity, anisotropy and energy spectrum of 480-966 keV protons and 38-315 keV electrons observed by the HI-SCALE instrument on Ulysses associated with Corotating Interaction Regions (CIR) from mid-1992 to early 1994. The particle events are most clearly ordered by the reverse shocks bounding the CIRs. The bulk of the ion fluxes appear either straddling, or with their maximum intensity following, the reverse shock. The electron intensities rise sharply to their maximum upon the passage of the reverse shock, and are delayed with respect to the protons. We believe that following acceleration at the reverse shock the electrons re-enter the inner heliosphere and mirror, to return to the reverse shock for repeated acceleration. This process is more effective for electrons ($v \sim c/2$) than for ions, and also favours the higher velocity electrons, which accounts for the observed spectral hardening with latitude.

Key words: Heliosphere – Energetic particles – Shock acceleration – – Reverse shocks

1. Introduction

From the middle of 1992 through early 1994 Ulysses was moving from a heliolatitude, Λ, of 24°S to 54°S, and was sampling a recurrent high-speed solar wind stream (Bame *et al.* 1993). The Heliospheric Instrument for Spectra, Composition and Anisotropy at Low Energies (HI-SCALE; Lanzerotti *et al.* 1992) has measured the charged particle increases associated with CIRs formed by the high speed streams (Simnett *et al.* 1994). Sanderson *et al.* (1994) have discussed the relationship of the MeV ions observed by the COSPIN detector on Ulysses to the solar wind velocity and magnetic field strength. Here we focus on the relative timing of the 480-966 keV proton and the 38-53 keV electron increases and the electron spectral evolution.

When a high-speed solar wind stream overtakes the slower solar wind forward and reverse waves are formed; if they are strong enough a shock pair is formed, with the forward shock (FS) propagating outwards and the reverse shock (RS) propagating back towards the Sun. This is shown schematically for two solar rotations in Figure 1. The FS, RS and CS (contact surface) are shown as dashed lines; the solid lines represent magnetic field lines which intersect either the FS or RS, while the dotted lines reflect magnetic field

Space Science Reviews **72**: 303–308.
© 1995 *Kluwer Academic Publishers.*

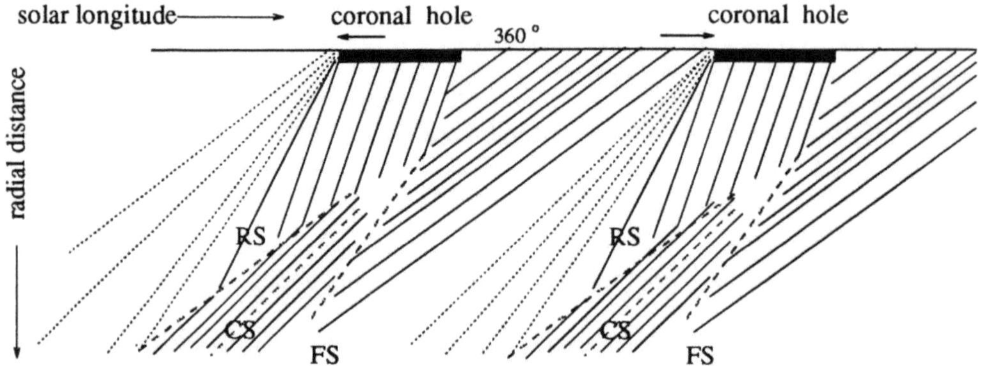

Fig. 1. The formation of the forward (FS) and reverse (RS) shocks by high-speed solar wind streams; the contact surface is CS. Oblique lines represent the interplanetary magnetic field, and are shown on an R-ϕ plot on the conical surface for constant solar latitude.

lines in the solar wind dwell - the region where the solar wind maps back to the same Carrington longitude.

Ions are accelerated at both the FS and the RS. Pizzo has modelled the evolution of the shock fronts both in the ecliptic plane (Pizzo, 1989) and in three dimensions (Pizzo, 1994). As the latitude increases the FS should disappear before the RS; therefore phenomena associated with the FS should also disappear before those associated with the RS. Previous studies of the latitude dependence of shock acceleration of ions <2 MeV associated with CIRs (Gold *et al.* 1988 ; and references therein) were from Voyager 1 (at 30°N) and Voyager 2 (in the ecliptic) at 15-25 AU. An intensity gradient of -3%/deg (decreasing northward) was found during 1984-5 (near solar minimum), << -23%/deg (decreasing southward) found on Ulysses by Simnett et al. (1994) at ~5 AU from 20°-47°S during 1992-3 (post solar maximum). The much stronger Ulysses gradient implies that either the latitude gradient of shock-accelerated ions is much larger in the inner heliosphere, or that it is larger closer to solar maximum (or that both of these statements are true).

2. The Observations

Above latitude 29°S the minimum solar wind velocity suddenly jumped to ~560 km s^{-1}(Bame *et al.* 1993), coinciding with the disappearance of the heliospheric current sheet (Smith *et al.* 1993). The last FS-RS pair was seen in association with CIR 15 (Gosling *et al.* 1993, Sanderson *et al.* 1994). The CIR numbering system is that used by Bame *et al.* (1993). At this time the electron intensity increases, which had been occurring approximately simultaneously with the protons, started to be delayed. This became more pronounced as the FS disappeared after CIR 15, although reverse shocks

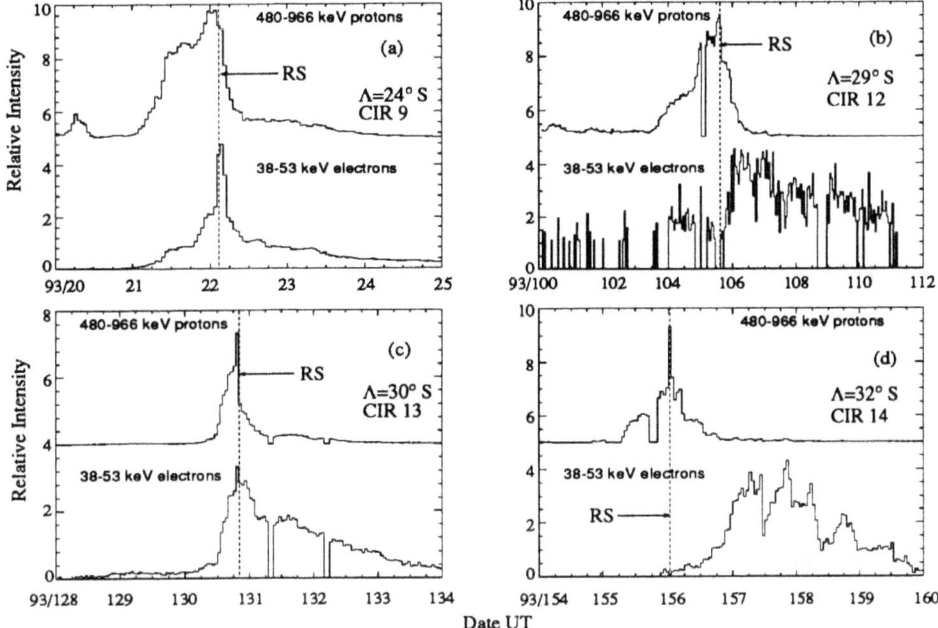

Fig. 2. The relative timing between 480-966 keV proton and 38-53 keV electron increases associated with CIR 9, 12, 13 and 14. In each panel the relative intensities are normalised.

continued to be observed for several more rotations (Gosling *et al.* 1993). The evolution of the delay near latitude 29°S is illustrated in Figure 2; in each panel we have indicated the approximate position of the RS. The times of the RS are given in paper 2.09 by Balogh *et al.* (this Symposium). Panel (a) shows the event associated with CIR 9, at Λ=24°S. The proton intensity peaks before the RS, whereas the electrons are more sharply peaked 2-3 hours later; this behaviour was typical of that seen for CIR 4-7.

Panel (b) is a much weaker electron event, which is delayed ∼ 6 hours, and lasts ∼ 5 days. The next rotation (c) has the intensity maxima coincident with the RS, but the long decay for the electrons is still present. By the next rotation (d) the electron event is delayed by ∼ 1 day, and is sporadic over ∼3 days. The delay for CIR 16 was four days; by this time Λ=35°S, and the minimum solar wind velocity had now jumped to ∼670 km $s^{-1}$and the magnetic field fluctuations diminished significantly (Sanderson *et al.* 1994). The next 7 rotations had delays in the relative maximum intensities from 1-5 days. Simnett *et al.* (1994) showed that the intensity of the CIR-associated events increased up to Λ=20°, and decreased thereafter. The decrease is more dramatic for the 480-966 keV protons. The electron increases diminish initially together with the protons, but after CIR 12 remain more nearly constant, and by early 1994 even increase.

3. Interpretation

The overall trends with heliolatitude to be explained are:

- (1) Lower latitudes (Λ <20°) Ion and electron events are associated with both the FS and RS, and both increase in intensity as Λ increases;
- (2) Middle latitudes (20° < Λ <35°) Ion and electron events are associated more with the RS. Ions tend to increase prior to the RS and then decline after its passage, while electrons tend to increase near the RS and decline more slowly than the ions. Maximum intensities diminish with increasing Λ. Forward shocks become rare (Gosling et al., 1993);
- (3) Higher latitudes (35° < Λ <60°) Ion events still rise and decay near R shock, although their maximum intensities diminish rapidly with Λ. Electron events rise days after the RS and then decay slowly. Unlike the ions, the electron events decay slowly after their delayed maximum.

Trend (1) is consistent with shock acceleration observed near the helioequator by Pioneer 10/11 and Voyager 1/2. The particle intensity increases because the CIR strength increases (as a function of latitude and/or time). To explain Trend (2) we argue that the ions, with their slower velocities, are efficiently accelerated, but do not easily outrun the shock on upstream field lines. Hence, they tend to appear in close association to both the FS and the RS. The electrons, with their higher velocities ($\gtrsim$ c/3), are accelerated less efficiently (in individual shock crossings), but easily outrun both the FS and RS. Hence the electrons are, in general, a population that is non-local with respect to the shocks. This ability of the electrons to outrun the shock may be a critical ingredient in their acceleration mechanism. For CIRs, the upstream field lines run back into the inner heliosphere (for both the FS and RS). Thus, a particle that can outrun either shock will travel into increasing field strength and eventually mirror. Upon returning to the CIR, the particle will encounter the increase in field strength (downstream of the shocks) within the CIR itself. Thus there is a natural mechanism for producing multiple encounters with the shocks for particles that can outrun the shocks. For electrons, the time between encounters is less than the transit time back to the Sun. At R=5 AU, this is ~ 4 hours for v=c/3 at 0° pitch angle, but much less for the bulk of the electrons which have larger pitch angles. This basic mechanism enhances the low single-encounter electron shock acceleration efficiency. Possibly the weakening of the FS simplifies the signatures at the RS (because of the lack of particles accelerated at the FS) and thus this pattern is clearer at Ulysses than at Pioneer 10/11 and Voyager 1/2 which were at low latitudes at a radial distance of 5 AU.

We propose that the dramatic ion/electron event separations in Trend (3) result from the foregoing velocity effects plus an additional consequence of the 3-D topology of CIRs. Figure 3 is an adaptation of the CIR configuration modelled by Pizzo (1994), which was applied by Gosling *et al.* (1993)

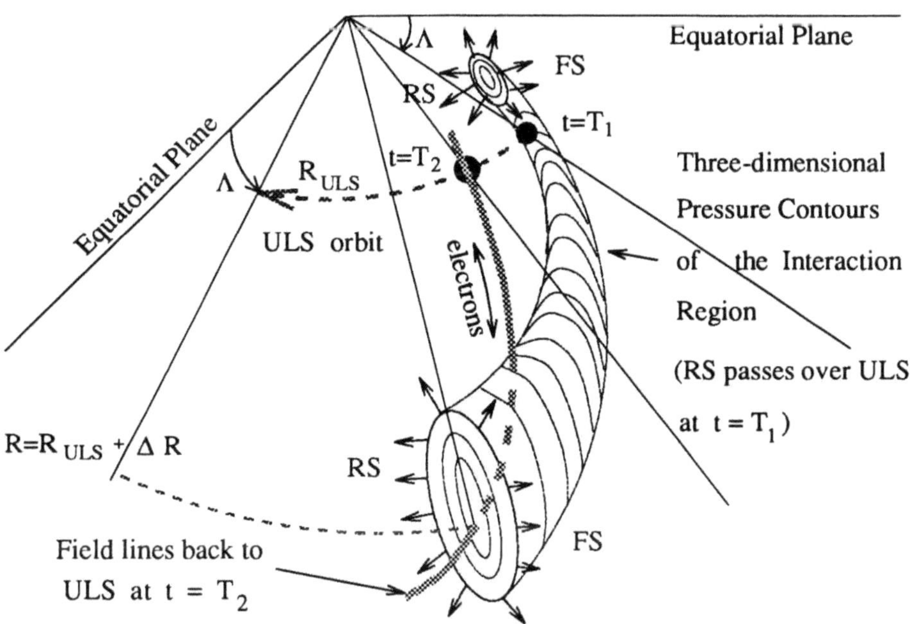

Fig. 3. Three-dimensional schematic of a magnetic CIR, (after Pizzo, 1994) that is consistent with the delay of the recurrent electron events with respect to the ion events at high latitudes. As the CIR (delimited by pressure contours shown in the cross-sections at two helio-radii) is expanding, the RS and the accompanying shock-accelerated ions are barely observed at the latitude of Ulysses ($t=T_1$), but the magnetic connection several days later ($t=T_2$) allows electrons accelerated by the expanding RS to propagate inward to Ulysses. The duration and intensity of the electron events should increase with latitude.

to explain the diminution of the FS with increasing latitude. If the CIR (concentric pressure contours in the two cross-sectional cuts) is viewed in a frame corotating with the Sun, then the orbit of Ulysses is approximately an arc at $R=R_{ULS}$ and $\Lambda = \Lambda_{ULS}$, chosen here to be $\approx 45°$. When the CIR passes over Ulysses at time $t=T_1$, the heliolatitude of Ulysses is such that only the southernmost extent of the CIR is sampled. The pressure gradients are weakest in this region, so the RS (which is expanding southward) is weak (and therefore inefficient in accelerating particles), while the FS is propagating northward and may not be observed at all. Thus the particle signatures are those of weak acceleration (local ions, low intensity electrons).

On the other hand, several days later at $t=T_2$, Ulysses is now in the highest speed solar wind (upstream of the RS) and the magnetic flux tube through Ulysses will connect to the CIR at a much larger radius. If V is the solar wind speed, and V_R is the radial velocity of the R shock in the spacecraft frame, it is easy to show that the radial difference ΔR between Ulysses and the RS is given by $\Delta R = (VV_R\Delta\phi/\Omega)/(V-V_R)$, where $\Delta\phi = \Omega(T_2 - T_1)$. Taking a time difference of 4 days, V=800 km $s^{-1}$and V_R=700 km s^{-1}, we obtain $\Delta R = 14AU$. Suppose the RS is expanding southward at $V_n = 50$ km s^{-1}. Then the distance that its southernmost extent at $R_{ULS} + \Delta R$ falls below the surface $\Lambda = \Lambda_{ULS}$ is $V_n(T_2 - T_1) = (V_n/V)\Delta R \approx$ 1 AU. Thus at t=T_2 Ulysses is magnetically connected to a very different region of the shock than is was at t=T_1 when the shock passed over it, *i.e.*, it is connected to a stronger RS at t=T_2. This explains not only the delayed electron events (after the RS) and the ion decay (they remain in the vicinity of the shock), but also their slow decays. Fast electrons can continue for days to reach Ulysses along field lines connected to the distant RS.

As the faster electrons make more shock encounters, this will feed back to the energy spectrum of the electron population. We have computed the spectral index γ (dJ/dE $\propto E^{-\gamma}$) derived from the 38-53 keV and 100-172 keV electron intensities. In general, following the maximum intensity the spectral index hardens. For the events associated with CIR 16-22 this hardening occurs at the time of the maxima of the 38-53 keV electron increases. For example, the derived spectral index γ for event 21 is -2.0, compared to -3.0 for event 17, which had a slightly lower intensity. The electron increases are isotropic, which is to be expected. They make multiple crossings of the reverse shock, which is beyond the spacecraft when the maximum fluxes are seen, so as many electrons move away from the shock as move towards it.

Acknowledgements

We thank the HI-SCALE team for their support and encouragement.

References

Bame, S.J., Goldstein, B.E., Gosling,J.T., Harvey,J.W., McComas, D.J. Neugebauer, M. and Phillips, J.L.: 1993, 'Ulysses observations of a recurrent high speed solar wind stream and the heliomagnetic streamer belt', *Geophys. Res. Lett.*,**20**, 2323.

Gold, R.E., Decker, R.B., Krimigis, S.M., Lanzerotti, L.J. and Maclennan, C.G.: 1988, 'The Latitude and Radial Dependence of Shock Acceleration in the Heliosphere', *J. Geophys. Res.***93**, 991.

Gosling, J.T., Bame, S., McComas, D.J., Phillips, J.L., Pizzo, V., Goldstein, B.E. and Neugebauer, M.: 1993, 'Latitudinal variation of solar wind corotating stream interaction regions: Ulysses', *Geophys. Res. Lett.*, **20**, 2789.

Lanzerotti, L.J. *et al.*, 'Heliospheric Instrument for Spectra, Composition and Anisotropy at Low Energies', *Astron. Astrophys.* , **92**, 1992.

Pizzo, V.J.: 1994, 'Global, Quasi-steady Dynamics of the Distant Solar Wind. 1. Origin of the North-South Flows in the Outer Heliosphere', *J. Geophys. Res.*, **99**, 4173.

Sanderson, T.R., Marsden, R.G., Wenzel, K.-P., Balogh, A., Forsyth, R.J. and Goldstein, B.E.: 1994, 'Ulysses High-latitude Observations of Ions Accelerated by Corotating Interaction Regions', *Geophys. Res. Lett.*, in press.

Simnett, G.M., Sayle, K.A., Roelof, E.C. and Tappin, S.J.: 1994, 'Corotating Particle Enhancements out of the Ecliptic Plane', *Geophys. Res. Lett.*, in press.

Smith, E. J., Neugebauer, M., Balogh, A., Bame, S.J. Erdös, G., Forsyth, R.J., Goldstein, B.E., Phillips, J.L. and Tsuratani, B.T.: 1993, 'Disappearance of the heliospheric sector structure at Ulysses', *Geophys. Res. Lett.*, **20**, 2327.

IMF CONNECTION FOR ENERGETIC PROTONS OBSERVED AT ULYSSES VIA MID-LATITUDE SOLAR WIND RAREFACTION REGIONS

E. C. ROELOF
Johns Hopkins University/Applied Physics Laboratory, Laurel, MD 20723, USA

G. M. SIMNETT
School of Physics and Space Research, University of Birmingham, B15 2TT, UK

and

T. P. ARMSTRONG
Department of Physics and Astronomy, University of Kansas, Lawrence, KS 66045, USA

June 14, 1994

Abstract. As Ulysses moved inward and southward from mid-1992 to early 1994 we noticed the occasional occurrence of "inter-events", lasting about 10 days and falling between the recurrent events, observed at proton energies of 0.48-97 MeV, associated with Corotating Interaction Regions (CIR). These inter-events were present for several sequences of two or more solar rotations at intensity levels around 1% of those of the neighbouring main events. When we compared the Ulysses events with those measured on IMP-8 at 1 AU we saw that the inter-events appeared at Ulysses after the extended emission (>10 days) of large fluxes of solar protons of the same energy that lasted at least one solar rotation at 1 AU. The inter-events fell completely within the rarefaction regions ($dv/dt < 0$) of the recurrent solar wind streams. The interplanetary magnetic field (IMF) lines in the rarefactions map back to the narrow range of longitudes at the Sun which mark the eastern edge of the source region of the high speed stream. Thus the inter-events are propagating at mid-latitudes to Ulysses along field lines free from stream-stream interactions. They are seen in the 0.39-1.28 MeV/nucleon He, which exhibit a faster decay, but almost never in the 38-53 keV electrons. We show that the inter-events are unlikely to be accelerated by reverse shocks associated with the CIRs and that they are more likely to be accelerated by sequences of solar events and transported along the IMF in the rarefactions of the solar wind streams.

Key words: Heliosphere – Energetic particles – Corotating high-speed streams

1. Introduction

From the middle of 1992 to early 1994 the HI-SCALE (Heliospheric Instrument for Spectra, Composition and Anisotropy at Low Energies; Lanzerotti *et al.*1992) instrument on Ulysses experienced recurrent increases of $\gtrsim 50$ keV ions and electrons which were associated with recurrent high speed solar wind streams observed by Bame *et al.*, 1993. Some of the characteristics of the particle increases have been presented by Simnett *et al.*, 1994 and Simnett and Roelof (1994) and the reader is referred to these papers for the

Space Science Reviews **72**: 309–314.
© 1995 *Kluwer Academic Publishers.*

salient details of the events. We are reporting here on an interesting but subtle effect which probably relates to the ordered nature of the inner heliosphere at this phase in the solar cycle plus the way in which ions are stored in this region of space. With such a small effect it is important to understand and separate the various temporal and spatial variations of particles injected in the inner heliosphere. We use as a reference the intensity of 0.55 - 1 MeV protons measured by the CPME (Charged Particle Measurements Experiment, Sarris *et al.*, 1976) instrument on IMP-8 at 1 AU.

2. The Observations

Figure 1 shows the intensity - time history of the 0.48 - 0.97 MeV protons and 0.39 - 1.28 MeV/nucleon helium nuclei from Ulysses, and the 0.5 - 1 MeV ions from IMP-8. The data cover the period from June 30, 1992 to February 13 1994 and the most obvious features are the recurrent increases labelled 2 - 24. These numbers correspond to the numbers assigned to the CIRs by Bame *et al.*, 1993; however, whereas the high speed streams merge into a quasi-permanent stream after event 16, the particle increases continue to recur approximately in phase with solar rotation up to the end of the dataset. The intensities are shown on a logarithmic scale in Figure 1, and shown shaded in the *middle* panel are the inter-events that are the subject of this paper. The increase between CIR 14 and 15 is associated with a coronal mass ejection (CME) and is not an inter-event.

The inter-events do not start until December 1992, subsequent to the solar flare activity which occurred at the time of CIR 6. There is no event between CIR 9 and 10, but they resume in March 1993 between CIR 10 and 11, and continue for two or three rotations. After CIR 12 the inter-events are certainly not strong, and it is debatable if they are even present during the remainder of 1993. The events are also generally seen in the helium nuclei, as is evident from Figure 1, although there is no He increase between CIR 7 and 8. The He inter-events have a tendency to decay faster than the protons, *e.g.* between CIR 10-11, CIR 11-12, CIR 15-16, CIR 18-19.

Figure 2 shows the proton intensity for the period covering events 6-14 in more detail, and we have added the 38-53 keV electron intensity measured by HISCALE. Unlike the protons, these electrons do not show any clear pattern of inter-events, whereas there is a one-to-one positive association with the recurrent events (Simnett *et al.*, (1994); Simnett and Roelof, 1994). After CIR 8 the 38-53 keV electron intensity does not decay immediately, but remains at a plateau until the protons reach maximum intensity in the inter-event. This is the only occasion where there might be some significant electron component in an inter-event. We do not think that the electrons are a significant component of the inter-events, nor do we expect them to be, because the correlation is known to be weak between proton and electron

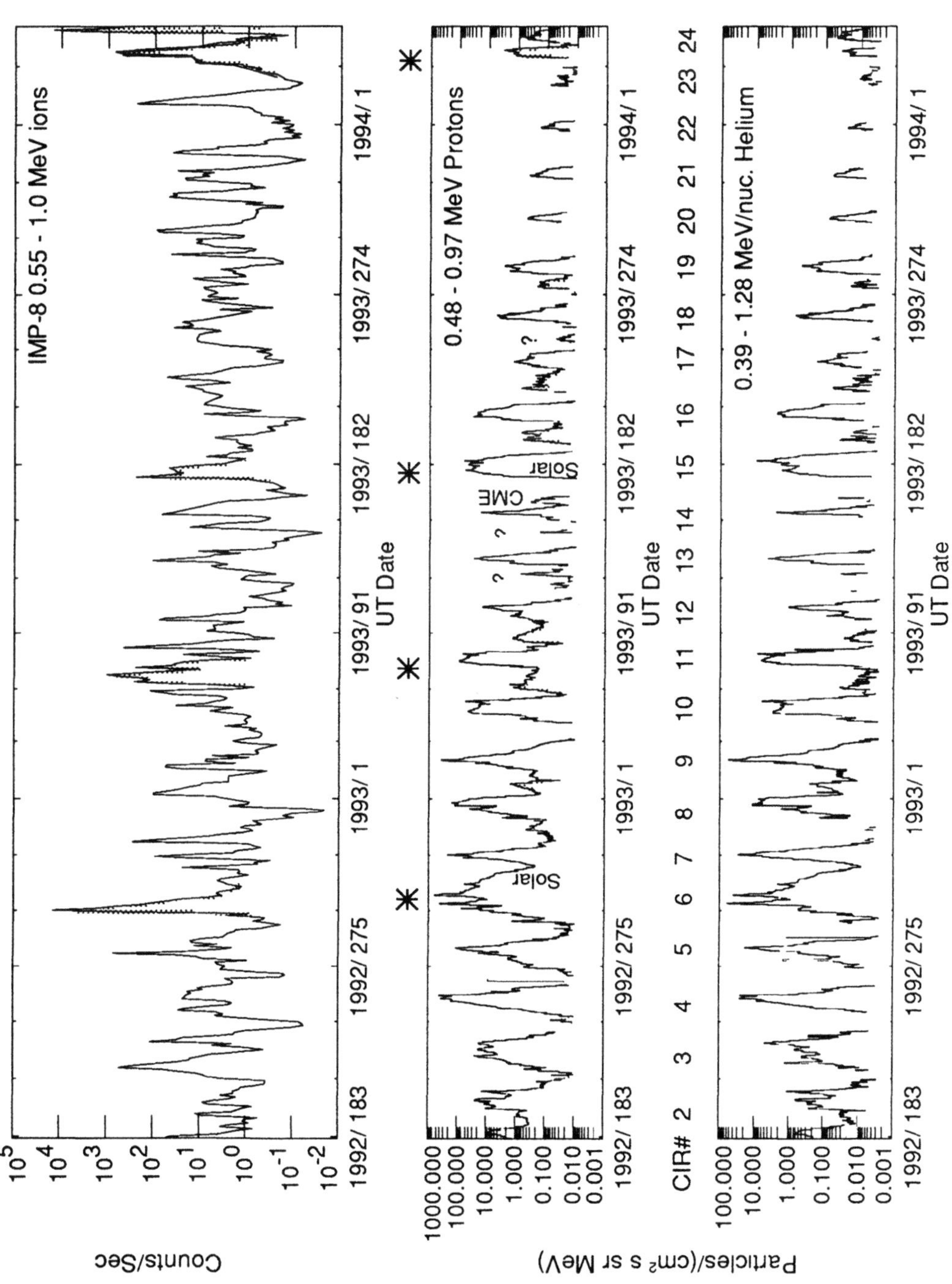

Fig. 1. The intensities of 0.48 - 0.97 MeV protons and 0.39 - 1.28 MeV/nucleon He from Ulysses; and 0.5 - 1.0 MeV ions from IMP-8. The numbers against the increases correspond to the associated CIR (see text). The shaded areas in the middle panel are the inter-events. Asterisks mark the pulses of solar proton activity at 1 AU, and the larger events are shown shaded in the upper panel.

events observed at 1 AU.

In the upper panel of Figure 1 we have shaded three periods of significant solar proton intensity increases seen by IMP-8 at 1 AU. The first of these in November, 1992 is clearly the biggest event during the period covering CIR 2-9. Centred around event 11 are several significant events at IMP-8, the largest of which is highlighted. Moreover, solar activity is increasing for at least one month before this event, and this is reflected in the resurgence of the inter-events after CIR 10. Figure 2 shows this series of three, or possibly 4, events in more detail. We note that as solar production of protons declines, so does the intensity of the inter-events. The next large solar event occurs around June 21, 1993, and it is clear that this is the last major event in 1993. It was followed by at least two, and probably 4, inter-events.

We have examined the solar wind characteristics at the times of the inter-events and we have noticed that they occur *within rarefaction regions* (solar wind velocity decays) of recurrent high speed streams. These are called solar wind "dwells" which were shown by Roelof and Krimigis (1973) to mark periods where the spacecraft is sampling plasma, magnetic field and energetic particles from a small range of solar longitudes. Some examples are given by Bame *et al.*, 1993 for solar rotations between August 1992 and June 1993. Simnett and Roelof (1994) (their Figure 1) showed schematically how the solar wind would map back to the Sun at the eastern edge of the coronal hole which was producing the high speed stream responsible for the CIR. It is clear that the inter-events are on magnetic field lines which also connect to this region.

The intensities of the inter-events in the 0.48-0.97 MeV energy range is from $\sim$1 - 10 (cm^2 s sr MeV)$^{-1}$. This is around 1% of the intensity of the CIR-associated events. They tend to appear subsequent to pulses of 0.5 - 1.0 MeV IMP-8 proton events observed at 1 AU which last >10 days at intensity levels that can exceed 10^3- 10^4 (cm^2 s sr MeV)$^{-1}$. From an integration of the proton fluence at IMP-8 we have determined that the four most significant inputs at 1 AU occur around days 305, 1992, 65, 1993, 173, 1993 and 39, 1994. These days are marked by asterisks in Figure 1. It can be seen from Figure 1 that the sporadic appearence of the inter-events at Ulysses follows these pulses of proton activity at IMP-8.

Note that there is a return of an inter-event in early 1994, once again at a flux level $\sim$ 1 (cm^2 s sr MeV)$^{-1}$. This flux is now larger than that seen in the recurrent events, suggesting an independent origin. The return follows the reappearence of large, long-lived 0.5 - 1 MeV proton events at IMP-8, which had been absent during the last half of 1993. The resumption of proton events at 1 AU and inter-events at Ulysses in 1994 indicates a clear causal relation between the two.

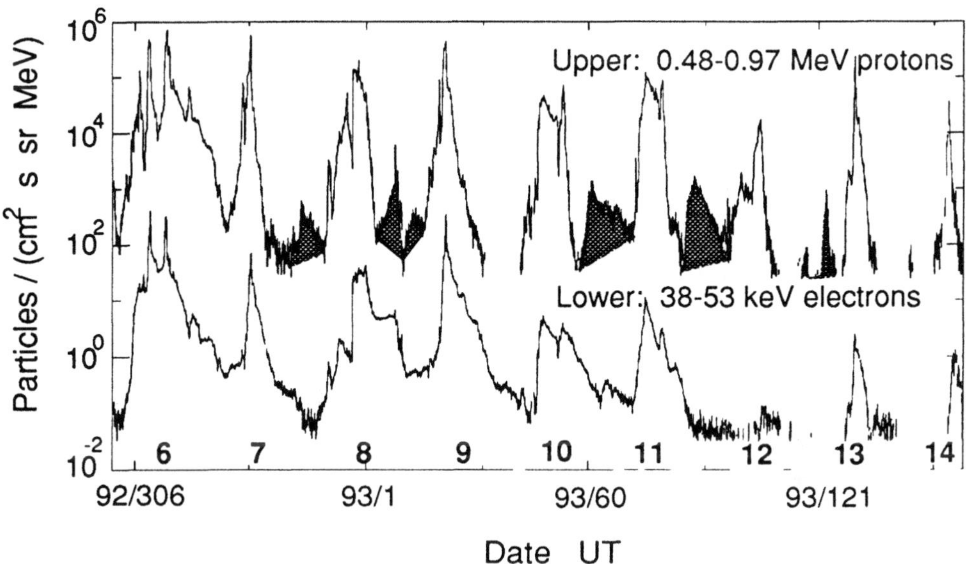

Fig. 2. The intensity of 0.48 - 0.97 MeV protons (× 1000) and 38 - 56 keV electrons (× 0.003) for the period between CIR 6 and 14. The numbers against the increases correspond to the associated CIR (see text). The shaded areas are the inter-events.

3. Discussion

We believe the most significant characteristics of the inter-events are:

1. They occupy IMF lines that connect *inward* to solar wind "dwells" near the eastern edge of the south polar coronal hole and *outward* to the edges of the CIRs.
2. They appear following pulses of solar proton activity at 1 AU.
3. The proton inter-events can occur without any clearly related electron increases. If the inter-events were accelerated via a shock connection beyond Ulysses we would expect a significant electron signature, as seen for the high latitude CIR-associated events (Simnett and Roelof, 1994). Therefore the origin is unlikely to be at radial distances beyond Ulysses.
4. The proton intensities at Ulysses of 1 - 10 $(\mathrm{cm}^2\ \mathrm{s\ sr\ MeV})^{-1}$ are comparable to the lower baseline limits of the IMP-8 intensities in the same energy range during pulses of activity. This implies relatively small non-impulsive baseline radial gradients for these small events. This is similar to the "reservoir" behaviour identified in the decays of large solar proton events during giant pulses of solar activity in February-June, 1991 (Roelof *et al.*, 1992).

5. The He increases in the inter-events tend to decay faster than those of the protons. This suggests that the storage is rigidity-dependent.

We have argued from (3) above that the origin of the inter-events is unlikely to be beyond Ulysses. We have further shown that the baseline intensity at IMP-8 is comparable to that at Ulysses. If the acceleration were taking place between Ulysses and IMP-8, such a coincidence would be most unlikely. Therefore we believe the origin must be well inside 1 AU. However, the inter-events occur in the recurrent rarefaction regions, where, by definition, there are no strong stream-stream interactions where efficient acceleration could take place. Therefore the origin of the events is most likely solar. This being the case, the fact that they appear to gradually die away in the periods between significant solar proton inputs to the interplanetary medium suggests that they undergo storage in the inner heliosphere.

Acknowledgements

We thank the HI-SCALE team for their support and encouragement. The efforts of E.C.R. and T.P.A. were supported in part by NASA.

References

Bame, S.J., Goldstein, B.E., Gosling, J.T., Harvey, J.W., McComas, D.J. Neugebauer, M. and Phillips, J.L.: 1993, 'Ulysses observations of a recurrent high speed solar wind stream and the heliomagnetic streamer belt', *Geophys. Res. Lett.*, **20**, 2323.

Gosling, J.T., Bame, S., McComas, D.J., Phillips, J.L., Pizzo, V., Goldstein, B.E. and Neugebauer, M.: 1993, 'Latitudinal variation of solar wind corotating stream interaction regions: Ulysses', *Geophys. Res. Lett.*, **20**, 2789.

Lanzerotti, L.J., Gold, R.E., Anderson, K.A., Armstrong, T.P., Lin, R.P., Krimigis, S.M., Pick, M., Roelof, E.C., Sarris, E.T., Simnett, G.M. and Frain, W.E.: 1992, 'Heliospheric Instrument for Spectra, Composition and Anisotropy at Low Energies', *Astron. Astrophys.*, **92**, 349.

Roelof, E.C. and Krimigis, S.M.: 1973 'Analysis and Synthesis of Coronal and Interplanetary Energetic Particle , Plasma and Magnetic Field Observations Over Three Solar Rotations,' *J. Geophys. Res.*, **78**, 5375.

Roelof, E.C., Gold, R.E., Simnett, G.M., Tappin, S.J., Armstrong, T.P. and Lanzerotti, L.J.: 'Low-energy Solar Electrons and Ions Observed at Ulysses February-April, 1991: The Inner Heliosphere as a Particle Reservoir,' *Geophys. Res. Lett.*, **19**, 1243.

Sarris, E.T., Krimigis, S.M. and Armstrong, T.P.: 1976, Observations of Magnetospheric Bursts of High Energy Protons and Electrons at 35 Re with IMP-7, *J. Geophys. Res.*, **81**, 2341.

Simnett, G.M., Sayle, K.A., Roelof, E.C. and Tappin, S.J.: 1994, 'Corotating Particle Enhancements out of the Ecliptic Plane', *Geophys. Res. Lett.*, **21**, in press.

Simnett, G.M.and Roelof, E.C.: 1994, 'Reverse Shock Acceleration of Electrons and Protons at Mid-heliolatitudes from 5.3 - 3.8 AU,' in *Proc. 28th ESLAB Symposium "The High Latitude Heliosphere", ed. R.Marsden, Space Science Reviews*, This Issue.

ULYSSES OBSERVATIONS OF A CORONAL ORIGIN PARTICLE EVENT AT 32° SOUTH HELIOGRAPHIC LATITUDE

M. PICK, A. BUTTIGHOFFER and A. KERDRAON
Observatoire de Paris, URA 1756 CNRS, France

T.P. ARMSTRONG
University of Kansas, Lawrence, USA

E.C. ROELOF
Applied Physics Laboratory, Laurel, USA

S. HOANG
Observatoire de Paris, URA 264 CNRS, France

L.J. LANZEROTTI
AT&T Bell Laboratories, Murray Hill, USA

G.M. SIMNETT
University of Birmingham, UK

and

J. LEMEN
Lockeed Palo Alto Research Laboratory, USA

Abstract. A remarkable streaming beam-like particle event of 60 keV-5 MeV ions and of 38-315 keV electrons has been reported previously. This event has been associated with the passage of a Coronal Mass Ejection (CME) over the Ulysses spacecraft on June 9-13, 1993. At this time, the spacecraft was located at 4.6 AU from the sun and at an heliolatitude of 32° south. It was proposed (Armstrong et al., 1994) that the particle injection source could have been of coronal origin. In this study, we analyse the solar activity during this period. We identify a region of solar radio noise storms in the corona and in particular, a flare on June 7 that presents all the required characteristics to produce the hot plasma beam observed in the interplanetary medium.

1. Introduction

A streaming beam-like event of 60 keV-5 MeV ions (energy spectrum peaked at ~270 keV) and of 38-315 keV electrons was identified on June 12, 1993 by the HISCALE (Heliosphere Instrument for Spectra, Composition and Anisotropy at Low Energies) instrument on Ulysses (Lanzerotti et al., 1992), then at 32°S heliolatitude and radial distance ~4.6 AU (Armstrong et al., 1994). These beams of ions and electrons occurred near the end of a coronal mass ejection (CME) detected by the solar wind plasma and magnetic field instrumentation on Ulysses and had the internal field structure of a flux rope (Gosling et al., 1993). The ion and electron pitch angle distributions measured by HISCALE were bidirectional in the outer regions of the flux

Space Science Reviews **72**: 315–320.
© 1995 *Kluwer Academic Publishers.*

rope, while there were strong unidirectional (antisunward) beams in the inner core of the structure. As this event presented a well-defined spatial structure and no signature of local interplanetary shock acceleration, it was proposed that the injection source could have been of coronal origin (Armstrong et al., 1994). To justify this hypothesis we have analysed here solar activity from May 31 to June 11, 1993. We have used the Yohkoh Soft X ray Telescope (SXT) (Tsuneta et al., 1991), the Nançay Radioheliograph (NRH) operating at 5 frequencies (The radioheliograph group, 1993), the Hα and K1 spectroheliograms of Meudon observatory and the Unified Radio and Plasma experiment (URAP) on Ulysses (Stone et al., 1992).

2. The Particle Event

Figure (1) shows the hourly-spin-average intensity in different energy channels, for electrons, protons and helium during the June 8-14 period. The figure shows successively: a decrease in the fluxes of 0.48-0.97 MeV protons and 53-103 keV electrons on June 8, an increase in the protons and electrons respectively starting on June 9 and 10; and a more impulsive event seen in all channels on June 12. The decrease is at the end of a CIR-associated event; the increase on June 9 is associated with the onset of the CME while the last increase corresponds to the strongly unidirectional beams described by Armstrong et al., (1994).

3. Solar Analysis

The June 1993 CME was associated with new magnetic loops on the southeast limb on May 31, as seen from the Earth (Gosling et al., 1993). At that time, Ulysses was located above this limb. The Yohkoh soft X-ray images show that these loops were not detected at 13:57 UT, but were visible on the next image at 17:30 UT; they were extended in the north-south direction and formed an arcade system elongated in longitude. They were barely discernable on the solar disk. In the southern hemisphere, the only candidate active region (AR 7518 in Solar Geophysical Data), located $\sim 10°$ south near the northern and western edge of the loops, developed rapidly and became very bright in Hα and soft X-rays on June 1, while its magnetic structure became more complex and reached maximum development around June 6. The active region was magnetically connected to Ulysses at the end of June 7 for a solar wind speed of ~ 700 km s^{-1}.

From June 1 to June 11, the NRH detected every day (except on June 8 and 9) sporadic noise storms as summarized in figure (2). Metric noise storm onsets or enhancements occur systematically in association with white light coronal dynamical changes near the radio source, sometimes corresponding to CMEs (Kerdraon et al., 1983; see also Lantos et al., 1981). Comparison

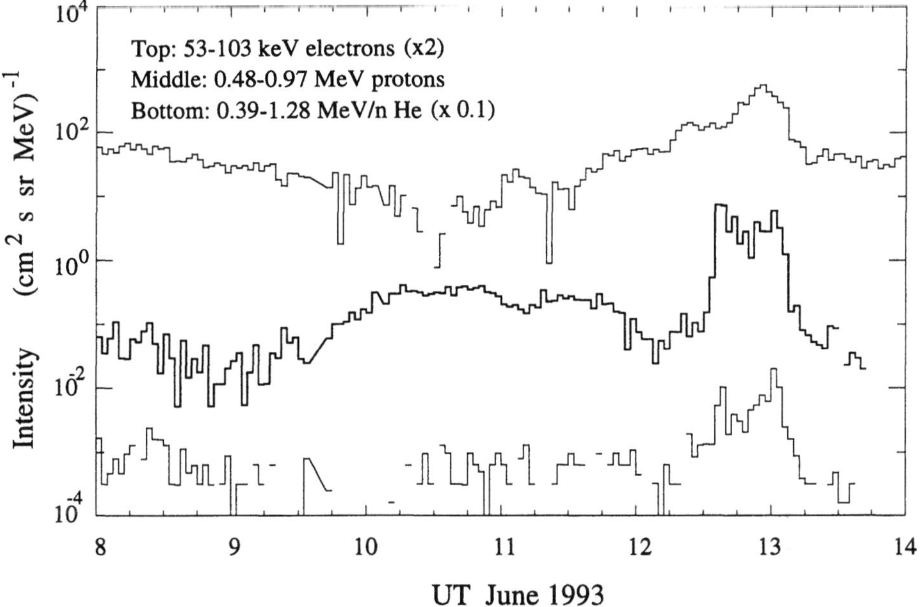

Fig. 1. Hourly-spin-average intensities of electrons, protons and He nuclei

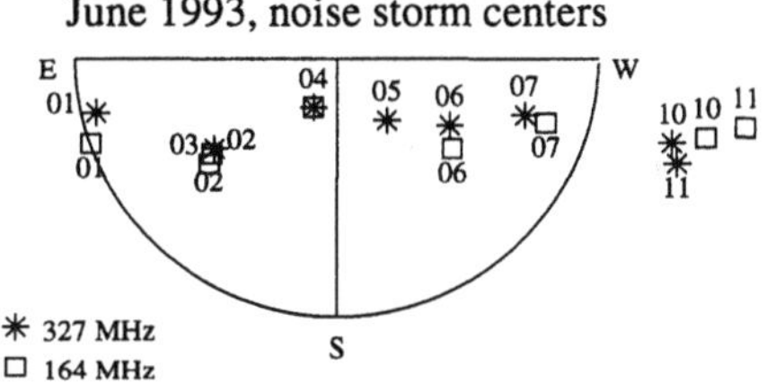

Fig. 2. Noise storm centers associated with AR 7518 during its passage on solar disk

with SXT images shows that between June 4 and 7, all the noise storm sources were located above the same X-ray bright feature (Figure (3)).

The largest hard X ray event associated with this active region was observed on June 7 by the BATSE(GRO) and GASOL(GRANAT) experiments. It occurred in conjunction with a type II/IV burst event observed by the NRH followed twenty minutes later by broad band continuum emission. Figure (4) presents at distinct single frequencies the time evolution of these radiobursts and also the contour plots, at 327 and 164 MHz, of east-west and south-north angles versus time of the brightness distribution. The NRH showed a double source in the corona with one above the same X-ray bright feature that contributed to the X-ray flare source (see figure (3)). This event was detected in the heliosphere by URAP from 10 kHz up to at least 940

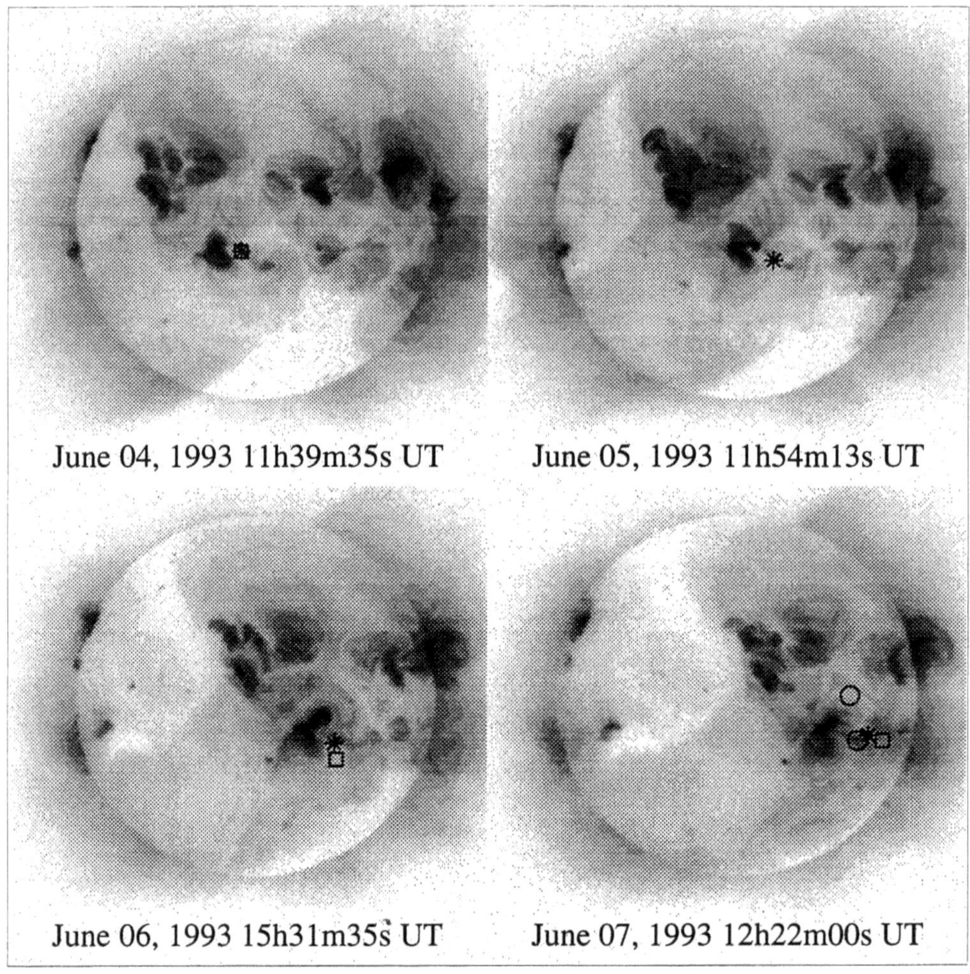

Fig. 3. Soft X-ray images. The positions of noise storms centers are reported : * 327 MHz; □ 164 MHz. For June 7, the type IV burst source is composed of two components; their positions have been also reported:○.

kHz (the highest URAP frequency).

The June 7 flare is the only candidate which could have accelerated ions up to energies seen in the Ulysses particle event (~2 MeV) on June 12. This is consistent with the onset of the lower energy ions; 60 keV protons will take approximatively 100 hours (~4 days), for a pitch angle=0°, to reach Ulysses. Two more observations appear to corroborate this conclusion:

- the June 7 event was associated with a particle event detected by GOES-7 and reported in SGD;
- following the interplanetary radio event, the only one during this period, the URAP experiment detected on several successive days ion acoustic and Langmuir waves with an onset on June 10 that coincided with the

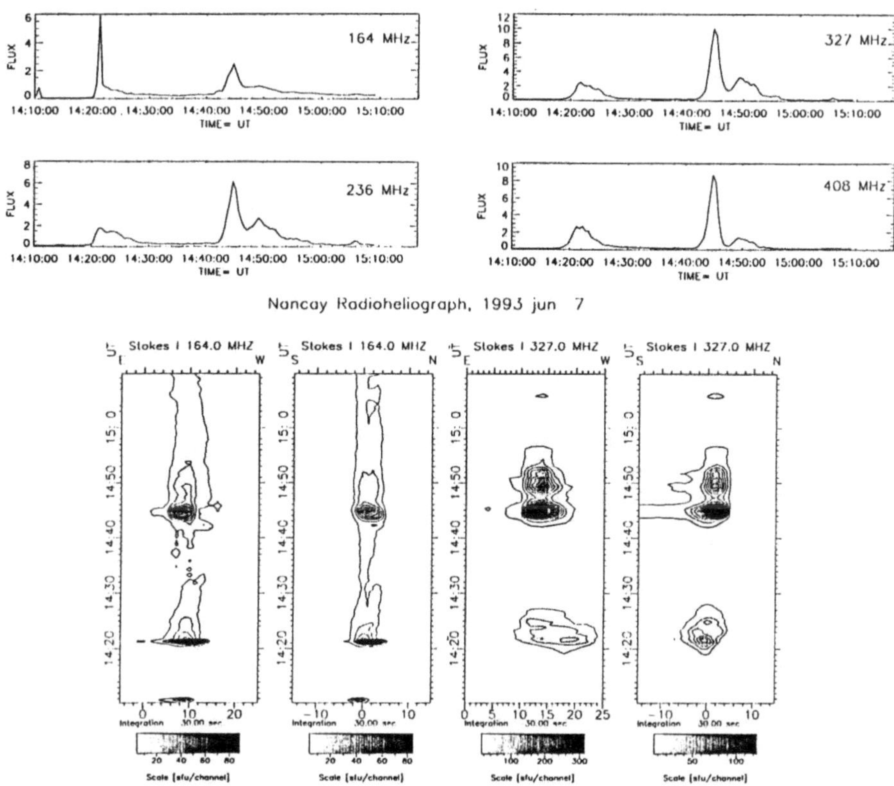

Fig. 4. June 7 event. Upper part: time evolution at different single frequencies; lower part: contour plots versus time of the brightness distribution for the east-west and north-south directions.

onset passage of the interplanetary CME. The Langmuir waves had to have been excited by electrons accelerated no earlier than the June 7 event and, consequently, the particle injection site had to be magnetically connected to the CME. The electron distribution was bidirectional on June 10 and evolved gradually into the beam on June 12, even though the Sun was rather quiet on June 8 and 9. This suggests some storage, either in the CME or in the outbound CME. On the other hand, because the electron beam on June 12 did not exhibit a bidirectional anisotropy and occurred during the passage of the CME, it must have originated at the Sun and been injected continuously on that day.

4. Conclusions

We believe the June 12 hot plasma beaming event had a coronal origin:

- The CME observed by Ulysses originated near active region 7518, responsible for contemporary activity in the southern hemisphere.
- One coronal site associated with energy release and efficient particle acceleration was identified over several successive days overlying the same bright X-ray structure; this site was very likely magnetically connected with the edge of the CME.
- The flare on June 7 exhibited all those characteristics to have it considered as an excellent candidate for producing the acceleration of protons of several MeV observed on June 12. However, the ion beam was measured to be peaked, with a "temperature" of order 270 keV. Therefore it is likely that the interplanetary mesurements provide some evidence of the coronal particle acceleration/release processes, as discussed in Armstrong et al. (1994). It is not certain, however, that the electron beam found its origin during the same flare, since noise storm activity started again on June 10 and 11.

This study extends to high heliocentric latitudes some earlier results obtained in the ecliptic plane: solar flare particles can propagate almost scatter free (*i.e.*, in strongly collimated beams) in selected regions of the solar wind (called "channels") (Anderson et al., 1991); Mangeney et al., 1994), and to large distances from the Sun (Buttighoffer et al., 1994).

5. Acknowledgements

We acknowledge Mrs I. Escaut for her help in the optical data analysis, Mrs J. Gapihan and Mrs M.P. Issartel for their help in the data reduction. The HISCALE investigation at JHU/APL is supported by NASA through Task I of contract N00039-94-C-0001 between the JHUAPL and the Department of Navy. The HISCALE investigation at Meudon is supported by CNRS and CNES.

References

Anderson, K.A. et al.: 1991, *Geophys. Res. Lett.***19**, 1283.
Armstrong, T.P. et al.: 1994, submitted to *Geophys. Res. Lett.*
Buttighoffer, A. et al.: 1994, submitted to *J. Geophys. Res.*
Gosling, J. T. et al.: 1993, *Geophys. Res. Lett.***vol.21 no.3**, 237-240.
Kerdraon, A. et al.: 1983, *Astrophys. J. Lett.***265**, L19-L21.
Lantos, P. et al.: 1981, *Astron. Astrophys.***101**, 33-38.
Lanzerotti, L. J. et al.: 1992, *Astron. Astrophys. Suppl. Ser.***92**, 207.
Mangeney, A. et al.: 1994, *Astron. Astrophys.***283**, 982.
The Radioheliograph group: 1993, *Adv. Space Res.***vol.13 no.9**, 411-414.
Stone, R. G. et al.: 1992, *Astron. Astrophys. Suppl. Ser.***92**, 207.
Tsuneta, S. et al.: 1991, *Sol. Phys.***136**, 37-67.

PROTON PHASE SPACE DENSITIES (0.5eV < E_p < 5MeV) AT MID-LATITUDES FROM ULYSSES SWICS/HI-SCALE MEASUREMENTS

G. GLOECKLER[1], E. C. ROELOF[2], K. W. OGILVIE[3], D. B. BERDICHEVSKY[4]

[1]*Dept. Physics and Astronomy, University of Maryland, College Park, MD 20742*
[2]*Johns Hopkins University/Applied Physics Laboratory, Laurel, MD 20723*
[3]*NASA/Goddard Space Flight Center, Greenbelt, MD 20771*
[4]*Hughes STX, Lanham, MD 20706*

Abstract. Proton phase space densities in the solar wind frame from suprathermal velocities 10 km s^{-1} to 30,000 km s^{-1} (0.5 eV–5 MeV) were derived from combined SWICS and HI-SCALE measurements when Ulysses was at 5 AU and –24° heliolatitude. The period (19–23 January 1993) encompasses a forward/reverse shock pair (20 January, 0500 UT and 22 January, 0300 UT). Strong evidence is found for shock acceleration of pickup protons from interstellar hydrogen at all energies measured.

1. Introduction

We are compiling a comprehensive atlas of proton phase-space densities (p.s.d.) using the intercalibrated SWICS (Gloeckler et al. 1992) and HI-SCALE (Lanzerotti et al. 1992) instruments on Ulysses. SWICS is a time-of-flight mass spectrometer for Ep < 60keV, whereas HI-SCALE has solid-state detectors in five collimated heads for 56keV < Ep < 4.75MeV. Thus p.s.d. can be constructed in the solar wind frame for proton velocities from thermal to 30,000 km s^{-1}. The analysis has been applied to data near the ecliptic plane while Ulysses was outbound from 2.5 and to 5 AU during 1991 and then during 1992–1993 as Ulysses was climbing southward to –30°. The p.s.d. in the vicinity of corotating interplanetary forward/reverse (F/R) shocks reveals a spectral break at $v = V_{sw}$ in the solar wind frame that becomes more and more well-defined as latitude increases. This is a clear and consistent signature of the presence of pickup protons from charge-exchanged interstellar hydrogen atoms. We present examples from the period 19–23 January 1993 at 5 AU and heliolatitude –24° of combined p.s.d. associated with one corotating interaction region (CIR) out of the long-lived sequence that lasted from mid-1992 until the velocity stream structure merged into the polar coronal hole stream in late 1993.

2. Measurement of Phase Space Densities

As is well known, the p.s.d. $F(\mathbf{p})$ for particles of momentum $\mathbf{p}$ may be constructed from the unidirectional flux $j(E,\mathbf{u})$ dE measured in the direction $\mathbf{u} = \mathbf{p}/p$ and in the differential energy band dE through the relationship $F(\mathbf{p}) = j(E,\mathbf{u})/p^2$. In this study, the ions all are nonrelativistic, so it is more convenient to use a p.s.d. based on velocity $\mathbf{v}$ rather than momentum $\mathbf{p} = m\mathbf{v}$, i.e., $f(\mathbf{v}) = m^3 F(\mathbf{p}) = mj(E,\mathbf{u})/v^2$. An important property of the

Space Science Reviews **72**: 321–326.
© 1995 *Kluwer Academic Publishers.*

p.s.d. is its invariance under Lorentz transformation, which nonrelativistically can be written as a Galilean transformation $\mathbf{v}' = \mathbf{v}-\mathbf{V}$ so that $f(\mathbf{v}) = f'(\mathbf{v}-\mathbf{V})$. If we take $\mathbf{V}$ as the solar wind bulk velocity so that f' is the p.s.d. in the local solar wind frame, then $f'(\mathbf{v}-\mathbf{V}) = mj(E,\mathbf{u})/v^2$. With Ulysses beyond 1 AU, the angle between the spacecraft spin axis (which always points toward the Earth) and the antisolar wind vector is usually <10°. To good approximation, we may take this angle as zero. Then any detector mounted at an angle θ to the spin axis will sample the solar wind p.s.d. at a speed $v' = (v^2-2vV \cos\theta + V^2)^{1/2}$, where $\mathbf{v}\cdot\mathbf{V} = vV\cos\theta$, and the clock angle will be the same in both frames.

The SWICS instrument has a collimator with a 57° width and a 4° aperture (Gloeckler *et al.*, 1992). One edge of the collimator coincides with the spacecraft spin axis, so it sweeps out a 57° swath of the sky with each rotation. The thermal solar wind beam is therefore always completely sampled by a small portion of the collimator near the spin axis, so we may take $v' \cong v-V$. However, for the hot nonthermal protons at the upper range of SWICS energy response ($E < 60$ keV) and for all of the HI-SCALE ion channels, the p.s.d. distribution is broader than the collimator of either instrument, and hence it is the total geometric function of the entire collimator that determines the response at the higher energies.

We utilize spin-averaged (i.e., averaged over the spin clock angle with θ fixed) fluxes $\langle j \rangle$ in our analysis, so if $f'(\mathbf{v}')$ is isotropic, then $\langle f'(\mathbf{v}')\rangle = f_0'(v') \cong m\langle j(E,\mathbf{u})\rangle/v^2$. Even if f' contains non-zero spherical harmonic coefficients (f_1', f_2', etc.) due to streaming along the magnetic field, the contribution of the first-order anisotropy to $\langle f'(\mathbf{v}')\rangle$ will be $O(f_1'\cos\theta\cos\theta_B)$ when the field is close to its nominal orientation $\theta_B \cong \pi/2$ at 5 AU (transverse to the radial direction), while the contribution of the f_2' term is $O(f_2'\sin^2\theta\sin^2\theta_B)$. Therefore when $\theta_B \cong \pi/2$ and $|f_2'| \ll f_0'$ we can estimate the isotropic component $f_0'(v')$ of the higher energy proton p.s.d. in the solar wind frame by plotting $m\langle j\rangle/v^2$ from SWICS at $v' = v-V$ and from HI-SCALE at $v' = (v^2-2vV\cos\theta+V^2)^{1/2}$, where the HI-SCALE ion telescopes are oriented at 30°, 60°, 120°, and 150°. Unlike SWICS, which can uniquely identify protons, 4 of the 5 HI-SCALE telescopes respond to all ions $Z \geq 1$. However, HI-SCALE also has a composition aperature (CA60) mounted at 60° to the spin axis which makes two-parameter separate identifications of H, He, CNO, and Ne-Fe, albeit at higher energies (>0.4 MeV/nuc). Under the assumption of similar velocity p.s.d.'s for all species, we can therefore estimate the contribution of the heavier ions ($Z \geq 2$) to the logarithmically-spaced ion rate channels up to 4.75 MeV, as well as compare the estimated ion p.s.d. to the directly-measured proton p.s.d. from the CA60 in the range $0.48 < E_p < 1.20$ MeV. For the event studied in this paper, we find that HI-SCALE is responding essentially to protons.

We are now prepared to combine the SWICS and HI-SCALE p.s.d.'s. This is the first time such p.s.d. have been constructed beyond 1AU. Gosling *et al.* (1981) had combined observations from instruments covering similar energy ranges to construct a p.s.d. of an energetic storm particle event at 1AU, although the instruments were on two different spacecraft (ISEE-2 and ISEE-3).

3. Observations of Phase Space Densities, 19–23 January 1993

The example of p.s.d. evolution that we present is associated with the forward/reverse shock pair (FS/RS) observed 20 January near 05:00 UT and 22 January near 03:00 UT, respectively. Times are taken from a Ulysses shock list compiled by Balogh et al. (1994). Figure 1(a–e) shows five multihour average p.s.d.'s during 19–23 January 1993; the averaging periods (a–e) are diagrammed in the top panel of Fig. 1. The abscissa is logarithmic in proton velocity v (km s^{-1}) relative to the solar wind (SW) frame, and the ordinate is also logarithmic, giving proton p.s.d. in units of s^3cm^{-6}. Note that the ordinate covers 16 orders of magnitude. The SWICS data are plotted as unconnected squares, and the error bars are indicated by points above and below the mean. The HI-SCALE ion data from all five heads are plotted by the symbols (labeled in Fig. 1 [a–e]) connected by straight lines. The error bars are smaller than the symbols. The response functions for both instruments have been calibrated independently; no ad hoc adjustment has been made to bring the SWICS and HI-SCALE p.s.d. into agree-

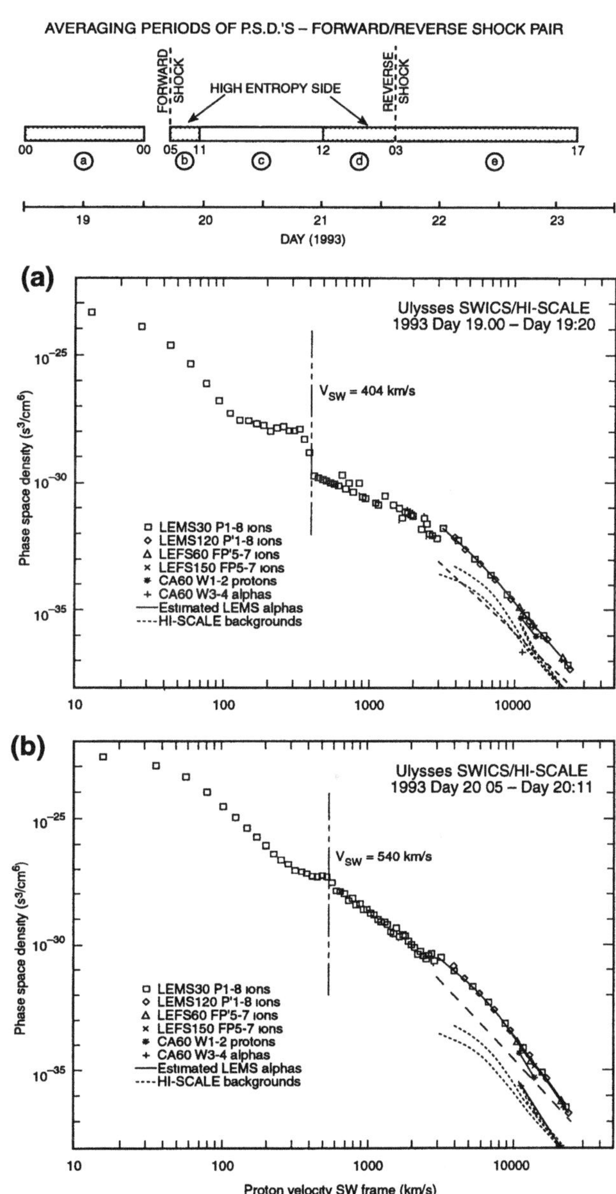

Figure 1. Proton p.s.d. measured by the combined SWICS and HI-SCALE experiments on Ulysses at 5AU and –24° heliolatitude. Panels (a)–(e) present multi-hour averages during the passage of a forward/reverse shock pair. See text for details.

ment (see the Appendix). Estimated backgrounds have been subtracted from the rates of each channel; the dotted curves give the backgrounds (converted to flux). The asterisks mark the two-point spectrum of the HI-SCALE two-parameter measurement of pure protons; its agreement with the p.s.d. derived from the ion channels indicates that the latter are responding mainly to protons. For the ion channels, an estimate of the helium contribution (dashed line) is made on the basis of the higher-energy two-parameter helium measurements.

The solar wind velocity V is indicated in Fig. 1 (a–e) by the vertical line labeled with its value. Since pickup ions all have an initial velocity **–V** in the solar wind frame, any spectral feature at this velocity in the solar wind frame is very likely associated with the pickup process. The strong "ledge" at v = V in Figs. 1a and 1e (where the p.s.d. has a discontinuity by a factor of 100), is a clear signature of pickup followed by adiabatic cooling in the expanding solar wind (Gloeckler *et al.* 1993, 1994). At velocities below the ledge (v < V), heating has broadened the p.s.d.; this is the "tail" noted by Ogilvie *et al.*, 1993. Note from Fig. 1a that at v = 30,000 cm s^{-1} the p.s.d.

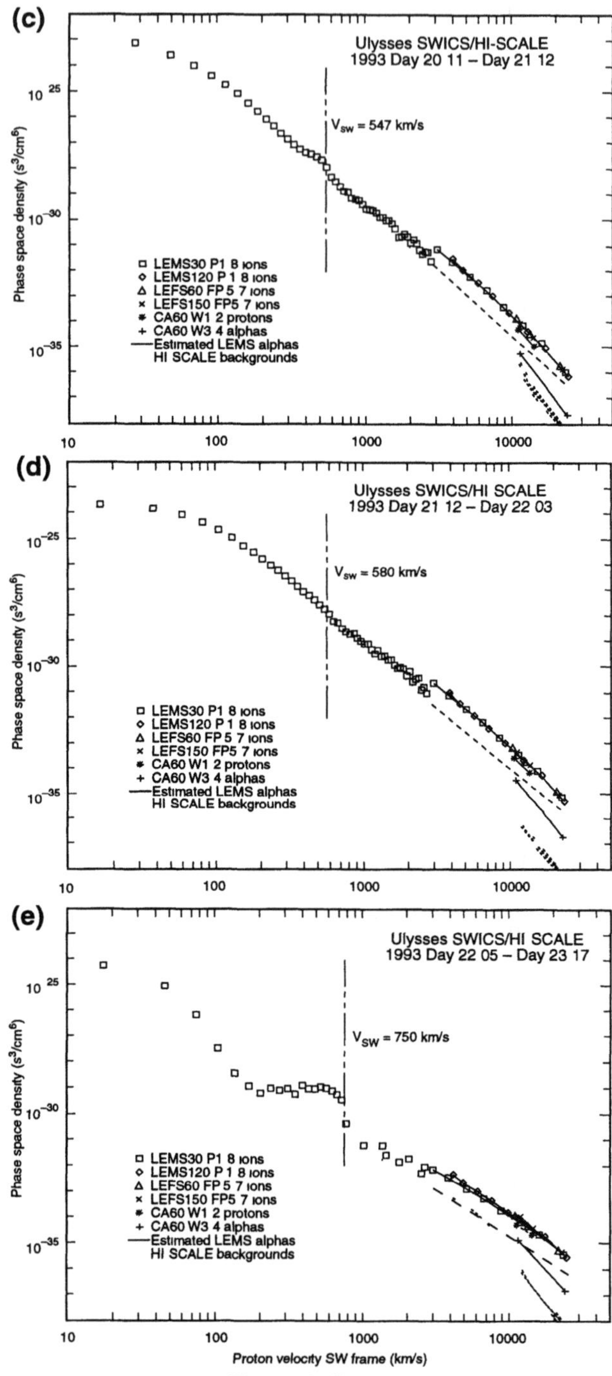

Figure 1 Continued

is about 10^{-37} s^3cm^{-6} upstream of the FS. Figure 1b shows the p.s.d. for the 6 hours immediately after the FS (20 January, 05:00–11:00). The ledge has increased by about a factor of 3, but the p.s.d. at immediately higher velocities ($v > V$) has increased by a factor of 100, although at v = 30,000 km s^{-1}, the increase is again only a factor of 3. In the average (Fig. 1c) over the next 25 hours (20 January, 11:00–21 January, 12:00) in the center of the CIR between the FS and the RS, the portion of the p.s.d. for $v > V$ (out to 30,000 km s^{-1}) has increased again by another factor of 3, although the p.s.d. for $v < V$ is roughly similar to those downstream of the shocks (Figs. 1b and 1d). In the average (Fig. 1d) for the 15 hours just prior (downstream) to the RS (21 January 12:00–22 January 03:00), the proton p.s.d. for $v < V$ has somewhat broadened, and is higher for $v > V$, to the degree that there no longer is a ledge. Nonetheless, a slight break in spectral slope remains at $v = V$, showing the pickup ion signature even though the trend is for the acceleration process to produce one continuous population. Finally, the 36-hour average in Figure 1e shows the p.s.d. after (upstream of) the RS (22 January, 05:00–23 January, 17:00). The ledge is as clear as it was at the FS in Fig. 1a, but at the RS it is a factor of 10 below the value prior to the FS. However, the p.s.d. at 30,000 km s^{-1} is about a factor of 30 higher than it was upstream of the FS, thus demonstrating the stronger effect of the RS structure on these high-velocity protons.

4. Discussion and Conclusions

The development of the p.s.d. in the vicinity of the FS/RS pair presented in Figs. 1(a–e) demonstrates the local acceleration of the pickup protons. The clear (factor of 100) proton "ledge" at $v = V$ upstream of both the FS and RS (Figs. 1a and 1e) is an unambiguous signature of the dominance of pickup protons at velocities just below $v = V$. That these pickup protons ($v < V$) are accelerated to velocities $v > V$ just downstream of the shocks is demonstrated by the dramatic reduction of the depth of the ledge downstream of the FS (Fig. 1b) or even more strongly by its complete absence (Fig. 1d) downstream of the RS, where the p.s.d. exhibits only a break in slope at $v = V$. These acceleration signatures are also seen in the midst of the CIR (Fig. 1c). This is intriguing, since magnetic field lines should map to the FS prior to the passage of the contact surface, or to the RS afterward. A more detailed analysis of the interior of the CIR is clearly called for.

In this particular CIR, the RS exhibits a stronger acceleration enhancement than the FS, not only just above the ledge at $v \geq V$, but also up to the highest energies measured (5 MeV). Since (1) the pickup protons dominate the p.s.d. just below the ledge ($v < V$), (2) the pickup protons are clearly accelerated in the vicinity just above the ledge ($v > V$), and (3) the spectrum for $v > V$ displays one continuous functional form out to 5 MeV (intermediate between an exponential and a power-law in energy), it is reasonable to conclude that a significant portion of the protons (up to at least 5 MeV) observed before, during, and after the CIR are accelerated pickup protons. This conclusion confirms and extends the results of Gloeckler *et al.* (1993, 1994).

Acknowledgments. These studies could not have been undertaken without the enthusiastic cooperation of the many members of both the SWICS and HI-SCALE teams. The efforts of the U.S. investigators were supported by NASA.

References

Balogh, A. *et al.*: 1994, *Space Sci. Rev.* this volume.

Gloeckler, G. *et al.*: 1992, *Astron. Astrophys. Supp.* **92**, 267.

Gloeckler, G. *et al.*: 1993, *Science* **261**, 70.

Gloeckler, G. *et al.*: 1994, *J. Geophys. Res.* **99**, *in press*.

Gosling, J. T. *et al.*: 1981, *J. Geophys.. Res.* **86**, 547.

Lanzerotti, L. J. *et al.*: 1992, *Astron. Astrophys. Supp.* **92**, 349.

Ogilvie, K. W. *et al.*: 1993, *J. Geophys. Res.* **98**, 3605.

Appendix: Agreement of SWICS and HI-SCALE Proton Measurements

Because the LEMS30 (Low Energy Magnetic Spectrometer) of HI-SCALE has a conical collimator of opening half-angle 22.5° centered at 30° off the spin axis, it sweeps out approximately the same swath of the sky as does SWICS. The lowest energy ion channel (p1) on LEMS30 is 56keV $< E_p <$ 78keV, so we can make a straightforward test of the dynamic response of the two instruments by plotting LEMS30 count rates against the summed SWICS upper velocity bins (34-60keV). We did so for a set of 1-to-8 day periods in 1991 selected for their nearly constant fluxes. The results are plotted below and demonstrate the colinearity of the SWICS and HI-SCALE response over a factor of 1000 in proton intensity.

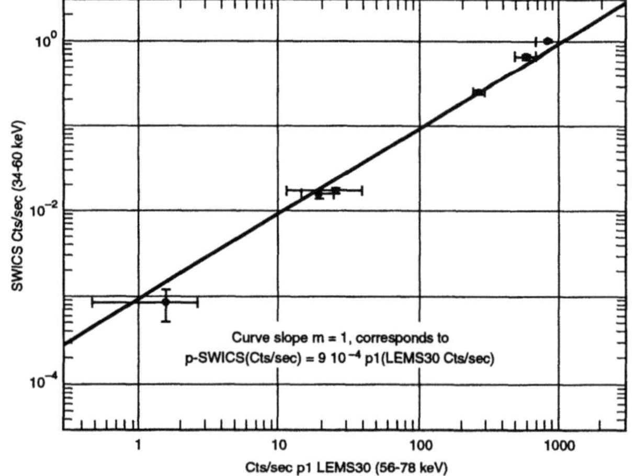

CO-ROTATING PARTICLE ENHANCEMENTS OUT OF THE ECLIPTIC PLANE

G.M. SIMNETT, K.A. SAYLE, S.J.TAPPIN
School of Physics and Space Research, University of Birmingham, B15 2TT, UK

and

E.C. ROELOF
Johns Hopkins University/Applied Physics Laboratory, Laurel, MD 20723, USA

May 18, 1994

Abstract. The characteristics of the recurrent electron (38-53 keV) and ion (>∼0.5 MeV) enhancements observed by Ulysses from mid-1992 to April 1994 are presented. The magnitude of the ion flux increases reached a maximum at a latitude of ∼20°S and decreased afterwards by ∼23%/degree until early 1994. The magnitude of the electron increases showed a similar trend until May, 1993, after which time it became approximately constant, until it started to increase again in early 1994. The electron enhancements have lagged the protons by up to 5 days once Ulysses left the heliospheric current sheet (mid-1993). The electron spectral index tended to harden (a) during the decay of the event and (b) as the latitude increased, up to 50°S. The events have recurred on a 26.0 day period, but with significant phase shifts over the 25 rotations studied. The H/He ratio decreases across the maximum intensity. The mean minimum value for H/He was 3.5 ±0.3, lower than that measured in previous studies in the ecliptic plane.

Key words: Heliosphere – Energetic particles – co-rotating interaction regions

1. Introduction

From the middle of 1992 through early 1994 Ulysses was moving from a heliolatitude of 24°S to 54°S, and was sampling a recurrent high-speed solar wind stream (Bame *et al.* 1993). The Heliospheric Instrument for Spectra, Composition and Anisotropy at Low Energies (HI-SCALE; Lanzerotti *et al.* 1992) has measured the charged particle events associated with co-rotating interaction regions (CIR) formed by the high speed streams. Preliminary results to the end of 1993 have been reported by Simnett *et al.* (1994).

2. The Observations

An overview of the electron and proton intensities from July 14, 1992 - April 19, 1994 is shown in Figure 1. The CIR numbering system is that used by Bame *et al.* (1993). Most events are due to particle acceleration at the CIR formed by the interaction of a high speed solar wind stream with slower solar wind. Some solar energetic particles are also present from time to time; (*e.g.* events 6 and 15), and at certain phases of events 24 and 25.

Space Science Reviews **72**: 327–330.
© 1995 *Kluwer Academic Publishers.*

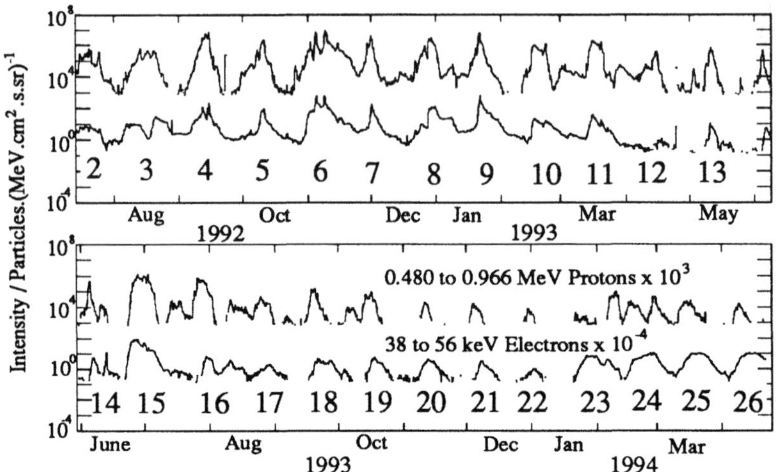

Fig. 1. Six-hour averages of 0.480-0.966 MeV protons (upper trace) and 38-53 keV electrons (lower trace) from July 14 1992 - April 19, 1994.

The latitude variation of the maximum intensity observed in each event is shown in Figure 2. From a maximum in November, 1992 the ion increases show a dramatic latitude decrease of ~23%/degree; the electron intensities decline initially, but after event 11 the decrease has largely disappeared.

A significant change to this pattern occurred in 1994. The maximum intensity of the electron events suddenly increased to levels not seen since event 11, and the duration of the events became broader. For events 23 and 24 this was due in part to solar energetic particles. The intensity of the proton events also increased, but only to levels experienced in event 20. The phase of the increases has been discussed by Simnett et al, (1994) and it exhibits a 26.0 day period, with several phase changes after events 9, 16, 20 and 23. The increases for events 21-23 are in phase with events 10-16, while events 24 - 26 appear to be in phase with the events <10.

The H/He ratio is shown in Figure 3, plotted on a 26.0 day period for all the events. The time of maximum proton intensity is indicated by a vertical dashed line; the ratio decreases through the period of maximum intensity. An average minimum value of 3.5 ± 0.3 is found for the events up to 20, less than that found by Barnes and Simpson (1976) from observations in the ecliptic plane. Note that we are observing a decreasing trend through the event; the ratio would be larger if the event average was used. The elemental composition with respect to O showed little variation with latitude.

The variation of the electron spectrum is shown in Figure 4 for the events after May 1993. The spectrum hardens both as the event progresses through the maximum intensity, and also with latitude, at least for events 16-22. The behaviour of events 25 and 26 is different. The spectrum is softest several

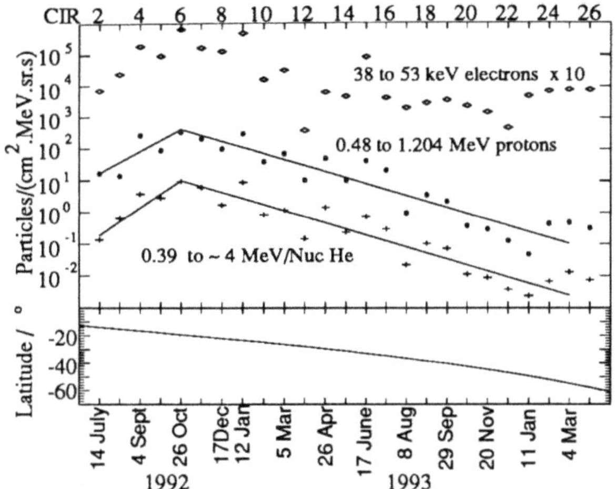

Fig. 2. The latitude variation of the maximum intensity (from a 24 hour running mean) seen associated with each CIR of the electrons, protons and He nuclei.

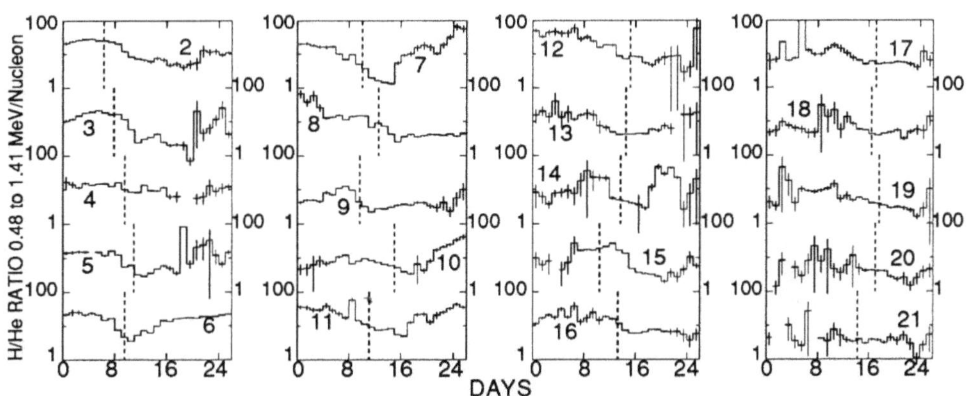

Fig. 3. The H/He ratio at 0.48-1.41 MeV/nucleon for events 2-24.

days before maximum intensity and hardens throughout the decline.

3. Discussion

The magnitude of the electron increase diminishes after event 6. However, once event 15 is removed, the declining trend vanishes after event 11. This is not true for the ions. The main difference between the early events and those at the end of 1993 is that Ulysses is never very close to the shocks produced by the CIRs. This suggests an ion acceleration mechanism which is a local phenomenon to the shocks, whereas the electron acceleration mechanism

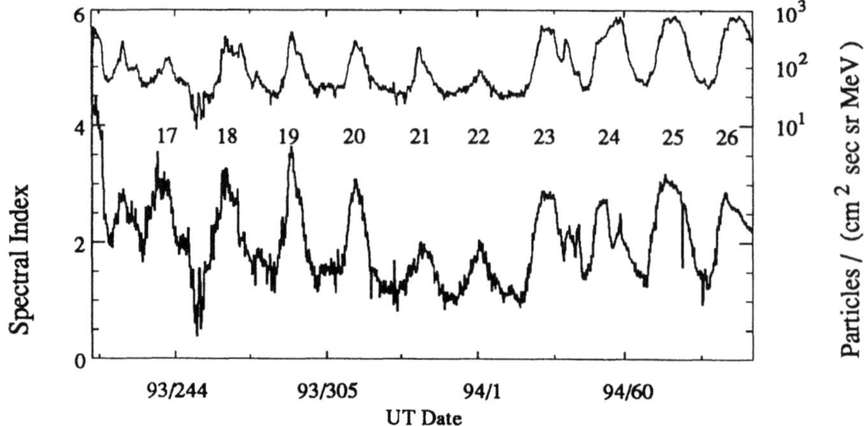

Fig. 4. The variation in the electron spectrum since the middle of 1993. The upper trace shows the 38-53 keV electron intensity and the lower trace shows the spectral index derived from this intensity and that from 100-172 keV.

is more remote. Simnett and Roelof (1994; this conference) have discussed their interpretation of the electron acceleration.

The change in the H/He ratio across the maximum intensity of the events is a persistent effect. This trend was observed by Barnes and Simpson (1976) from studies near the ecliptic plane; however, the minimum H/He ratios are lower than they found. It is not a systematic instrumental effect as the ratio for solar energetic particle events in the ecliptic plane is both consistent with other measurements and with those reported from Ulysses by Marsden *et al.* (1993) at higher energies.

The spectral evolution is a real effect. Whereas the computed values at low intensity levels is dependent on the background subtraction, this has little effect at high intensity levels. Confirmation of this may be seen by comparing the spectral index from event 17 with that from event 21, which had a significantly harder spectrum despite a lower maximum flux.

Acknowledgements

We thank the HI-SCALE team for their support and encouragement.

References

Bame, S.J. *et al.* , 1993, *Geophys. Res. Lett.*,**20**, 2323.
Barnes, C.W., and J.A.Simpson, 1976, *Astrophys. J., 210*, L91.
Lanzerotti, L.J. *et al.* , 1992, *Astron. Astrophys.* **92**.
Marsden, R.G. *et al.* , 1993, *Adv. Space Res.*, 13, No. 9.
Simnett, G.M. *et al.* , 1994, *Geophys. Res. Lett.*, in press.

SOLAR PARTICLES OBSERVED NEAR 33° S LATITUDE

S.J. TAPPIN and G.M. SIMNETT
School of Physics & Space Research, University of Birmingham, Birmingham, U.K.

Abstract. We have analysed the ions and electrons seen by the HI-SCALE instrument on the Ulysses spacecraft in the interval from 21-30 June 1993, ahead of an arrival of the CIR which has dominated the particle environment of Ulysses since July 1992. These particles show velocity dispersion effects usually associated with solar-flare particles. There was no flare activity of any consequence on the region of the Sun to which Ulysses was connected. We do however find a newly emerged active region (# 7525) close to the estimated site of connection, which we conclude could be the particle source.

1. Introduction

Since mid-July 1992 up to the end of 1993, the particle environment measured by the HI-SCALE instrument (Lanzerotti *et al.* 1992) on Ulysses has been dominated by ions and electrons accelerated by the corotating interaction region (CIR) resulting from a northward extension of the South-polar coronal hole. One major characteristic of these regions is that the ion fluxes are consistently seen to rise before the electron fluxes (Simnett *et al.* 1994). However the event which commenced on or around 21 June 1993 is an exception in that the electron fluxes rose first, it is this event which we shall consider in this paper. At the time of these observations Ulysses was located at a heliocentric distance of about 4.6 au and a heliographic latitude of -33°, the longitude of the spacecraft was about 110° east of Earth.

2. Observations

When we look in more detail at the rising phase of this event (Fig. 1) we can see that for all energies of electrons the start of the rise is simultaneous at 1200-1500 UT on June 21 to within the accuracy to which we can determine it. The higher-energy ions on the other hand show a clear velocity dispersion. This is apparently with all the particles having been accelerated simultaneously at a remote location. However the lower-energy ions show two distinct populations, firstly a component which begins abruptly at the same time as the electron onset, and secondly the velocity-dispersed population which was also seen in the higher energies.

For a solar wind speed of 650 km/s (Bame *et al.* 1993) the length of a spiral path from the Sun to Ulysses is approximately 7.6 au. This implies, from the electron onset time an initial injection time at the Sun near 1000 UT on June 21. The band on Fig. 1(b) gives the expected onset times (see figure caption); it is the earliest possible arrival of ions which started simultane-

Space Science Reviews **72**: 331–334.
© 1995 *Kluwer Academic Publishers.*

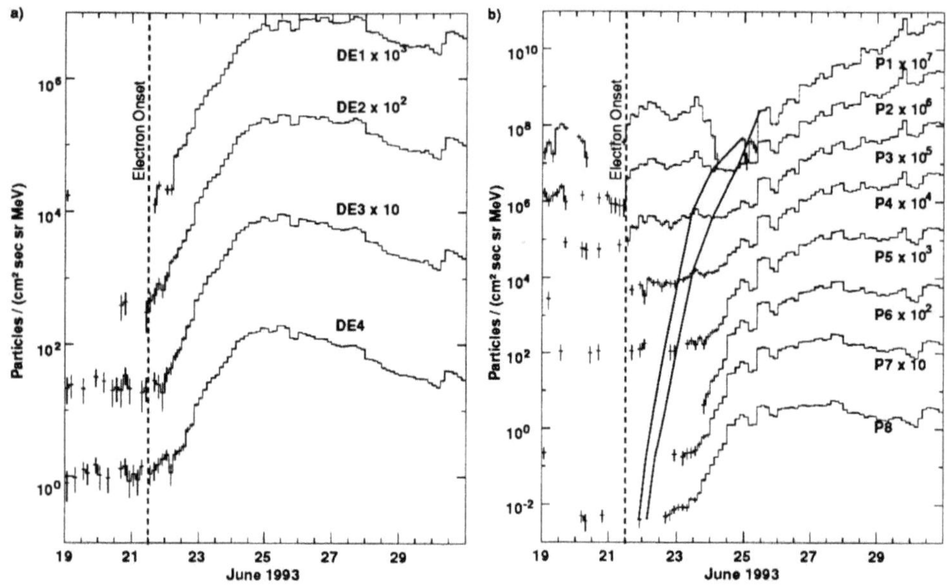

Fig. 1. (a) Plots of 38-53, 53-103, 103-175 and 175-315 keV electrons for the interval 19-28 June 1993. The vertical line gives the estimated onset time. (b) Similar plots for 56-78 78-130, 130-214, 214-337, 337-594, 594-1073, 1073-1802 and 1802-4752 keV ions. The sloping lines connecting the traces indicate the estimated onset times based on a 7.6 au path and injection at 1000 UT on 21 June. In both panels, channels are offset by successive factors of 10 from the channel above in energy. Both panels show 3-hour averages.

ously with the electrons. And indeed we see that the ions in the dispersed component are not seen until after these times.

For most of the time when there are sufficient counts to make meaningful determinations of the anisotropy, there is a unidirectional flow in both electrons and ions. Modelling (Tappin *et al.* 1994) shows that the distributions are consistent with a eam superposed on a isotropic population.

The ionic composition shows a falling proportion of heavy species up to June 29, the predicted time for the CIR, after which the ratio of heavier elements to hydrogen increases as observed by Simnett *et al.* (1994) for the other CIR events; this is also seen with solar particles (Scholer *et al.* 1979).

The spectral indices (γ) of all channels show considerable variations throughout this event, selected channels are shown in Fig. 2. The electron onset is indicated on all plots, and also the time of the maximum flux in the sum of the channels used for the index. For the electrons, initially $\gamma \approx 2$, followed by a hardening until $\approx$1200, 23 June, after this the spectrum softens to $\gamma \approx 3.5$ on 5 July some 3-days after the reverse shock of the CIR. The low-energy ions show clearly the "soft" component with $\gamma \approx 2$, followed by

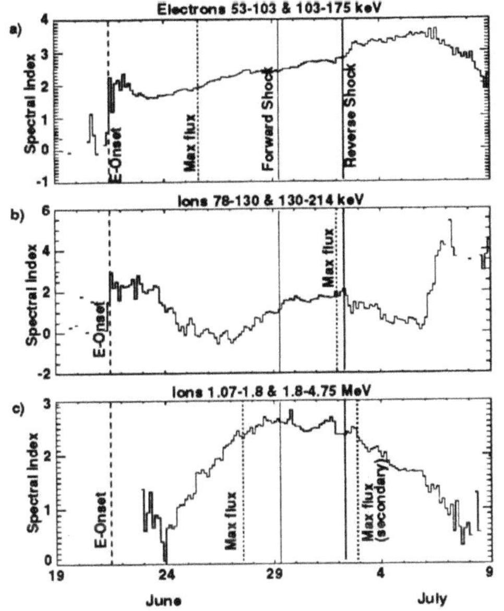

Fig 2 Spectral indices of flux channels from the LEMS30 head, (a) DE2 and 3, (b) P2 and 3 and (c) P7 and 8

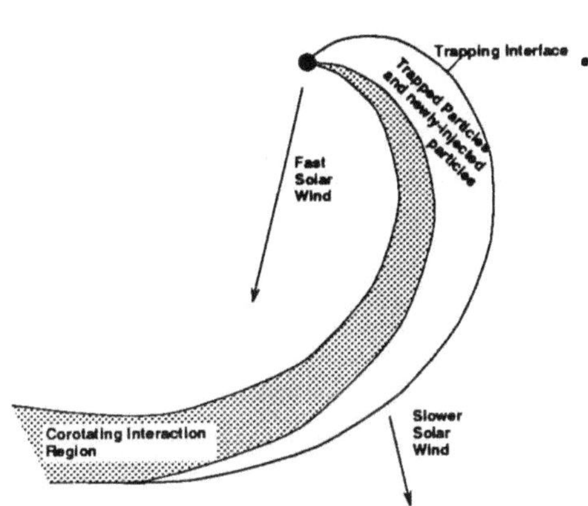

Fig 3 Conjectured solar-wind configuration giving rise to the events described

a hardening to an inverted spectrum ($\gamma \approx -0.5$) late on 26 June when the velocity-dispersed particles in this energy range begin to arrive. The spectrum then softens to $\gamma \approx 2$ at the time of maximum flux which occurs just before the reverse shock, this is followed by a hardening to around 0.5 at the time of softest electron spectra, after which the spectrum becomes very soft (but with poor statistics). For the higher energy ions where the "soft" component is absent we see an initial hard (about 0.3) spectrum on 23 June when the statistics first become usable this then softens to about 2.5 at the time of the CIR forward shock, after which there is a slow hardening until the statistics expire around 7 July. The other heads behave in a similar way.

3. Further data and Discussion

It appears that we are looking at an injection of particles which arrived at Ulysses shortly prior to the arrival of the corotating interaction region. There were no reports of flares or impulsive radio activity in *Solar Geophysical Data* (1993) in the appropriate time range.

It appears that we are seeing particles injected at the base of the solar wind without a flare or any other obvious manifestation on the solar surface. No major flares were seen for some 10 days prior to the injection (during all of which time the injection site was visible from Earth) which would seem to rule out a storage and release mechanism such as those suggested by Roelof *et al.* (1992). It is equally clear from the spectrum and intensity profiles that we are not seeing shock accelerated particles. An active region (7525) emerged on 17 June, which was consistently detected from 20 June onwards. Early on 21 June, this region was at S15-20, W20; thus it appears as a very strong candidate for the source of the particles observed at Ulysses, especially as it was the only significant AR in the Earthward half of the southern hemisphere at the time.

The ion and electron onsets are inconsistent with a single δ-function-like injection which would cause a sudden rise of flux to a maximum value at a time dependent on energy, followed by a steady decay. Therefore there must be sustained particle input at the Sun. The low-energy ions present somewhat of a problem as the obvious explanation would be that they were injected at some earlier time and then trapped in the region ahead of the forward shock of the CIR. But in the absence of a pre-existing active region it is hard to know whence they might have come.

We believe that we have identified unequivocally the propagation of solar energetic particles to 4.6 au at a latitude of 33°S, during a period otherwise dominated by shock-accelerated particles originating from a long-lived stable corotating system. The configuration shown in Fig. 3 could have given rise to these events.

Acknowledgements

We thank our HI-SCALE team colleagues for their support in this work. We also acknowledge the Ulysses shock list circulated at the meeting by A. Balogh & J. Phillips.

References

Bame, S.J., Goldstein, B.E., Gosling, J.T., Harvey, J.W., McComas, D.J., Neugebauer, M., and Phillips, J.L.: 1993, *Geophys. Res. Lett.* **20**, 2323

Lanzerotti, L.J., Gold, R.E., Anderson, K.A., Armstrong, T.P., Lin, R.P., Krimigis, S.M., Pick, M., Roelof, E.C, Sarris, E.T., Simnett, G.M. and Frain, W.E.: 1992, *Astron. Astrophys. Suppl.* **92**, 349

Roelof, E.C., Gold, R.E., Simnett, G.M., Tappin, S.J., Armstrong, T.P., Lanzerotti, L.J.: 1992, *Geophys. Res. Lett.* **19**, 1243

Scholer, M., Hovestadt, D., Kleckler, B. and Gloeckler, G.: 1979, *Astrophys. J.* **227**, 323

Simnett, G.M., Sayle, K.A., Roelof, E.C. and Tappin, S.J.: 1994, *Submitted to GRL.* ,

Tappin, S.J., Roelof, E.C. and Lanzerotti, L.J.: 1994, *Astron. Astrophys.* **Submitted**,

STUDY OF DISTRIBUTION FUNCTIONS OF INTERPLANETARY PARTICLES ACCELERATED AT CO-ROTATING INTERACTION REGION AT ~5A.U.

L. J. LANZEROTTI and C. G. MACLENNAN

AT&T Bell Laboratories, Murray Hill, NJ 07974 USA

Abstract. We report the results of a study of the 12-hour average distribution functions of high Z nuclei as measured over a 10-day interval on the Ulysses spacecraft at a helioradius of ~5.2 A.U. We use the good time and atomic mass resolution of the composition aperture of the HI-SCALE instrument to determine the form of the distribution functions for C, O, Ne, Mg, Si, and Fe over the energy range ~0.5 – 16 MeV/nucl. We find that the distribution functions of these ions can be organized by characteristic velocities that have values ranging between ~2000 – 3500 km/sec over this interval.

1. Introduction

The acceleration of interplanetary ions at co-rotating interaction regions (CIRs) has long been recognized as an important astrophysical process, one that can shed understanding on central issues in space and astrophysical plasmas (e.g., McDonald *et al.*, 1976; Barnes and Simpson, 1976; Palmer and Gosling, 1978; Pesses *et al.*, 1978; short summary in Richardson *et al.*, 1993). The acceleration at CIRs has been found to be more intense in the outer (order few A.U.) solar system than in the inner (e.g., Marshal and Stone, 1978; Mewaldt *et al.*, 1978; Christon, 1981; Zwickl and Roelof, 1981). As such, the inner solar system ion population can be augmented by these accelerated particles (e.g., Gloeckler *et al.*, 1979; Christon, 1981).

The spectra of interplanetary ions can be represented in several functional forms. Gloeckler *et al.* (1979) showed that a velocity distribution function of the form $f_0 \exp(-v/v_0)$ could organize multiple ion spectra that they measured near 1 A.U. They point out that this distribution function (particle number densities in phase space) has several important properties related to particle propagation and acceleration. Richardson *et al.* (1993) displayed spectra summed over about 46 CIR events measured near and at 1 A.U. in the declining phase of solar cycle 21. They showed that for nuclei in the energy range ~2 – 20 MeV/nucl the spectra could be fit reasonably well by power laws in energy as well as by exponentials in momentum (rigidity) and in the velocity distribution function. They did not show spectra normalized in these representations.

This paper presents the normalized velocity distribution spectra determined during an interval consisting of a solar particle event followed by a CIR and coronal mass ejection as measured at ~5 A.U. Twelve-hour average spectra are examined to determine how the distribution function might change with time during an event. Gloeckler *et al.* (1979) found an increase with time of v_0 at 1 A.U. and suggested that this may indicate an increase of the efficiency of the acceleration process with increasing helioradius. The location of Ulysses at ~5.2 A.U. allows a study of this supposition in this 10-day interval.

Space Science Reviews **72**: 335–338.
© 1995 *Kluwer Academic Publishers.*

2. Observations

The results presented herein were derived from measurements made with the Wart (W) telescope of the HI-SCALE instrument that was flown on the Ulysses spacecraft. The W aperture consists of a three-element solid state telescope of half angle 23° (Lanzerotti *et al.*, 1992). The telescope consists of a 5μ thick first detector followed by two 200μ second and third detectors. The incident particles are analyzed using two discrete energy channels and a 256-channel pulse height analyzer (PHA). A rotating four-level event selection priority scheme is used to enhance the likelihood for analyzing less-abundant species. Measurements from the PHA analysis are discussed in this paper.

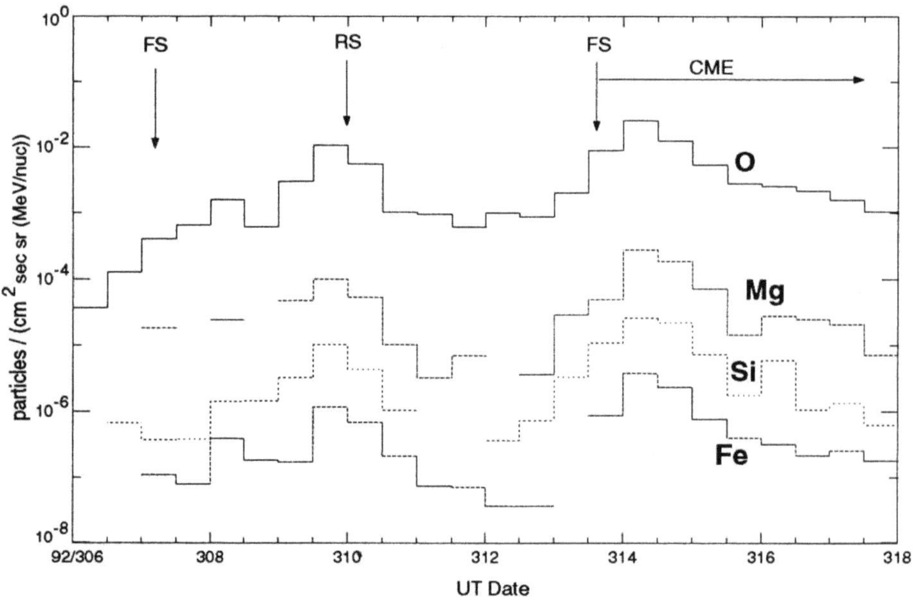

Fig. 1. 12-h averages of 1.0–8.0 MeV/nucl fluxes

Plotted in Figure 1 are 12-h averages of the spin-averaged fluxes of the 1.0 – 8.0 MeV/nucl oxygen, magnesium ($\times 10^{-1}$), silicon ($\times 10^{-2}$), and iron ($\times 10^{-3}$; top to bottom, respectively) nuclei for the interval day 306 – 318, 1992. At this time the spacecraft was at a helioradius of ~5.2 A.U. and at a southern heliographic latitude of ~20°. The times of two forward shocks (FS; days 307 and 313) and one reverse shock (RS; day 310) are indicated. Following the day 313 FS, a coronal mass ejection passed the spacecraft (information on the interplanetary events courtesy of A. Balogh and R. Forsyth, Imperial College).

The fluxes plotted in Figure 1 show the largest enhancements during the time between the FS and the RS on days 307 and 310, respectively, and just prior to and following the FS on day 313. Analysis of the intensity-time traces as a function of ion energy shows that the higher energy ions arrive first in the interval around the first FS (Brent *et al.*, 1993; Marsden *et al.*, 1993). This time dispersion indicates that this particle enhancement is a combination of solar particle event and the effects of a co-rotating interaction region (CIR).

The phase space densities (PSDs) of the measured C, O, Ne, Mg, Si and Fe ion fluxes were determined for each 12-h period during the interval shown in Figure 1. Figure 2 shows two representative PSD sets for hours 1200 – 2400 on days 309 and 314. To avoid over-crowding, only PSDs for C, O, Si, and Fe nuclei are shown. The data in the two intervals clearly indicate that each can be normalized with one characteristic velocity in the exponential function over the range from ~1 to ~$3 \cdot 10^4$ km/s.

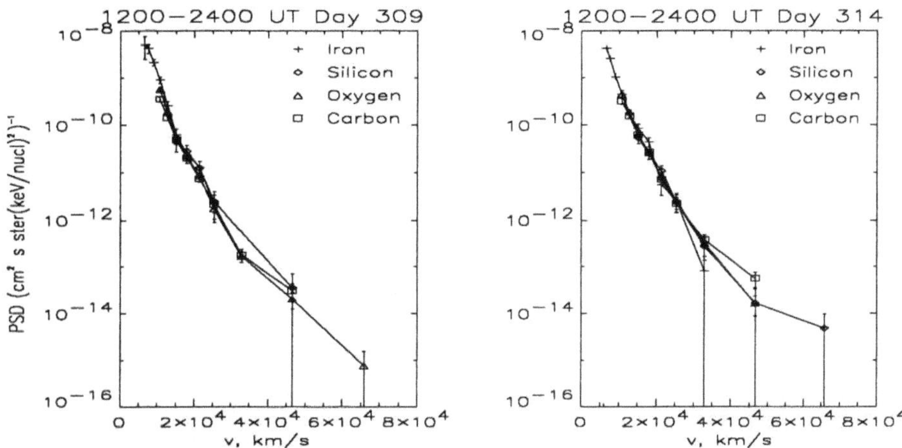

Fig. 2. 12-h average normalized distribution functions vs velocity for four species

3. Discussion

The characteristic velocities v_0 for the 12-h intervals are plotted in Figure 3. Except for the value in the first half of day 308, all of the v_0 lie within the range of ~2000 to ~3500 km/s. This range is similar to that reported by Gloeckler *et al.* (1979) at 1 A.U. The values ~4000 km/s and ~1800 km/s occur in association with the 12-h intervals following the RS on day 310 and the FS on day 313, respectively.

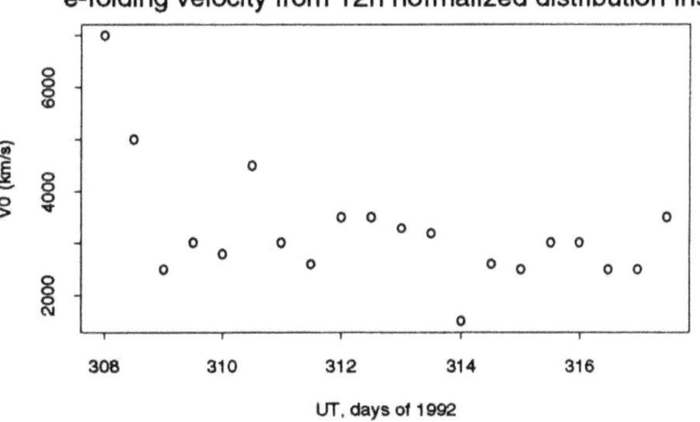

Fig. 3. Characteristic velocities from HI-SCALE energetic ions

No noticeable increase in the values of v_0 are seen during the event. Following the logic of Gloeckler *et al.* (1979), this might be ascribed to the fact that the Ulysses data are acquired in the heliosphere region where the acceleration process from the seed particle population is most effective.

Gloeckler *et al.* (1979) note that this distribution/spectral form is "compatible with a statistical mechanism as the dominant accelerating process operating in the CIR". That is, the acceleration would be a diffusion in momentum space, yielding a power law in velocity. The enhancements in the abundances occur in the intervals around the shocks on days 310 and 313, suggesting that the acceleration associated with the shocks have a statistical component to them.

One of the more significant acceleration results from Ulysses to date is the growing understanding of the importance of anomalous component species as seed populations for the low energy particles measured by the HI-SCALE detector complement (G. Gloeckler and E. C. Roelof, private communications, 1994). Additional and future work will concentrate on understanding this seed particle acceleration to the spectral forms discussed herein for a range of interplanetary abundances, not all of which may have anomalous component counterparts.

Acknowledgment

We thank our HI-SCALE colleagues for discussions and suggestions.

References

Barnes, C.W., and Simpson, J.A.: 1976, 'Evidence for Interplanetary Acceleration of Nucleons in Corotating Interacting Regions', *Ap.J.Lett.*, **210**, L91.

Brent, H.D., Lanzerotti, L.J., Maclennan, C.G., and Forsyth, R.J.: 1993, 'Interplanetary Field Control of Heavy Ion Abundances at ~5.2 A.U.', *EOS Trans. AGU,* **74**, Abs.#SH21A-2, 482.

Christon, S.P.: 1981, 'On the Origin of the MeV Energy Nucleon Flux Associated with CIRs', *J. Geophys. Res.,* **86**, 8852.

Gloeckler, G., Hovestadt, D., and Fisk, L.A.: 1979, 'Observed Distribution Functions of H, He, C, O, and Fe in Corotating Energetic Particle Streams: Implications for Interplanetary Acceleration and Propagation', *Ap.J.,* **230**, L191.

Lanzerotti, L.J., *et al.*: 1992, 'Heliosphere Instrument for Spectra, Composition and Anisotropy at Low Energies', *Astron. Astrophys.*, **92**, 349.

Marsden, R.G., Sanderson, T.R., Forsyth, R.J., and Goldstein, B.E.: 1993, 'Ulysses COSPIN Observations of a Composition Flare/CIR Event at 5.2 AU', *EOS Trans. AGU,* **74**, Abs.#SH12A-1, 480.

Marshall, F. and Stone, E.C.: 1978, 'Characteristics of Sunward Flowing Protons and Alpha Particle Fluxes of Moderate Intensity', *J. Geophys. Res.,* **83**, 3289.

McDonald, F.B., Teegarden, B.J., Trainor, J.H., von Rosenvinge, T.T., and Webber, W.R.: 1976, 'The Interplanetary Acceleration of Energetic Nucleons', *Ap.J.Lett., 203*, L149.

Mewaldt, R.A., Stone, E.C., and Vogt, R.E.: 1978, 'The Radial Diffusion Coefficient of 1.3-2.3 MeV Protons in recurrent prton Streams', *Geophys. Res. Lett.,* **5**, 965.

Palmer, I.D., and Gosling, J.T.: 1978, 'Shock-associated Energetic Proton Events at Large Heliocentric Distances', *J.Geophys.Res.,* **83**, 2037.

Pesses, M.E., Van Allen, J.A., and Goertz, C.K.: 1978, 'Energetic Protons Associated with Interplanetary Active Regions 1-5 AU from the Sun', *J.Geophys.Res.,* **83**, 553.

Richardson, I.G., Barbier, L.M., Reames, D.V., and von Rosenvinge, T.T.: 1993, 'Corotating MeV/amu Ion Enhancements at ≤ 1 AU from 1978 to 1986', *J.Geophys.Res.,* **98**, 13.

Zwickl, R.D. and Roelof, E.C.: 1981, 'Interplanetary Propagation of <1 MeV protons in Non-impulsive Energetic Particle Events', *J.Geophys.Res.,* **86**, 5449.

THE ELEMENTAL COMPOSITION IN ENERGETIC PARTICLE EVENTS AT HIGH HELIOSPHERIC LATITUDES*

M. FRÄNZ, E. KEPPLER, N. KRUPP and M.K. REUSS
Max-Planck-Institut für Aeronomie, Katlenburg-Lindau, Germany

and

J.B. BLAKE
The Aerospace Corporation, Los Angeles, USA

Abstract. The relative abundances of low energy ions (0.6-2.0 MeV/n) in solar energetic particle (SEP) and corotating interaction region (CIR) events have been measured by the EPAC experiment aboard Ulysses since launch in October 1990 until the present time. We give an overview of the abundances of heavy ions (He, C, Ne, Fe) relative to oxygen during energetic particle events lasting longer than 5 days during the in- and out-of-ecliptic phase of the mission. While the period Oct. 1990 to Aug. 1992 was dominated by high solar activity the Ulysses out of ecliptic passage at solar latitudes up to 45° went parallel to the declining phase of solar activity. Thus a very clear structure of corotating interaction regions was observed. While the in-ecliptic composition is in general agreement with measurements made near the Earth, the development of the CIR-composition shows two phases: From Aug. 1992 to May 1993 the C/O-ratio is 0.55-0.70, afterwards it increases to 0.8-0.9. This increase is correlated to the disappearance of the current sheet at 30° solar latitude reported by Smith *et al.*(1993).

Key words: Energetic particles – Ulysses – Composition.

1. Introduction

Energetic particles of solar origin in the energy range (0.6-2.0 MeV/n) observed in the heliosphere are accelerated from the high energy tail of the plasma velocity distribution in different locations and by different processes (Gloeckler *et al.*, 1984): (1) at interplanetary shocks caused by coronal mass ejections (CMEs), (2) in impulsive solar flares, (3) at the shockfronts of corotating interaction regions (CIRs) between low and high speed solar wind. Depending on the different source regions and acceleration mechanisms different chemical abundances of the respective energetic particle populations are observed (Reames *et al.*,1992,1994). During the minimum of the solar activity cycle the extension of the polar coronal holes to low heliographic latitudes causes a regular pattern of low and high speed solar wind streams. Beyond a distance of $\sim$ 1 AU from the sun the overtaking of the low speed streams by the high speed streams leads to the build up of one or more CIRs which can be stable for many solar rotations (Whang, 1991). While the in-ecliptic phase of the Ulysses mission from October 1990 until February 1992 was dominated by high solar activity accompanied by various CME

Space Science Reviews **72**: 339–342.
© 1995 *Kluwer Academic Publishers.*

events (Roelof *et al.*, 1992; Krupp *et al.*, 1992; McDonald *et al.*, 1994), from July 1992 onwards Ulysses observed stable CIRs up to a southern solar latitude of 45° in November 1993, when Ulysses finally remained in the high speed stream of the southern coronal hole (Bame *et al.*, 1993; Gosling *et al.*, 1993). The local structure of the events and the composition in the intensity minima is discussed by Keppler *et al.*(1994).

2. Instrumentation and analysis

The EPAC-instrument onboard Ulysses measures ions (electrons, protons and heavier particles up to iron) in the energy-range 0.2-6.0 MeV/n using 4 identical particle telescopes (Keppler *et al.*, 1992). Each telescope consists of 3 surface-barrier-detectors of 5, 100 and 300μm Si thickness respectively allowing dE vs. E-measurements. Depending on their nuclear charge the ions (Z $\geq$ 2) accumulate along tracks in the dE vs. E matrix. To get the relative abundance of the different elements the particle counts of the energy/nucleon-range 0.6-2.0 MeV/n are summed up parallel to the tracks by overlaying a continuous field of element numbers. The resulting curve is fitted by a multiple gaussian fit to separate the distributions of the minor species . The integral over each gaussian gives the absolute sum of counts of the respective element. The standard deviations (vertical bars in fig. 1) include the width of the distribution, the absolute count error and the quality of the fit.

3. Observations

By the method just described we have analysed the chemical composition of all energetic particle events measured by the EPAC-instrument from start in Oct. 1990 until Oct. 1993, which had a duration of more than 5 days and measurable quantities of heavier ions ($Z \geq 10$) in the energy range 0.6-2.0 MeV/n. Some events at the end of 1990 and in the second half of 1991 had to be excluded due to a different operation mode of the instrument. Fig. 1 shows the abundances of the elements He(2, scaled by a factor 10^{-2}), C(6), Ne(10) and Fe(26) relative to O(8) averaged over the in-ecliptic periods marked on the X-axis. The last four columns in fig. 1 contain reference values for the SEP-corona from CME-events (Reames *et al.*, 1992), for ecliptic CIR-events measured during the last solar minimum and for slow (v < 600 km/s) and fast (v > 600 km/s) CIR-events in the energy range 1.9-2.8 MeV/n (Richardson *et al.*, 1993). We distinguish between four periods:
I:The CME-dominated period (Day 292, 1992 - Day 190, 1991): The He/O-ratio (2) is low and fluctuating (30-70), the C/O-ratio (6) is low and constant (0.4±0.2) except for a CIR-event (day 145-150, 1991). The Fe/O (6) is high and fluctuating between 0.12-0.35, the latter value is achieved during the

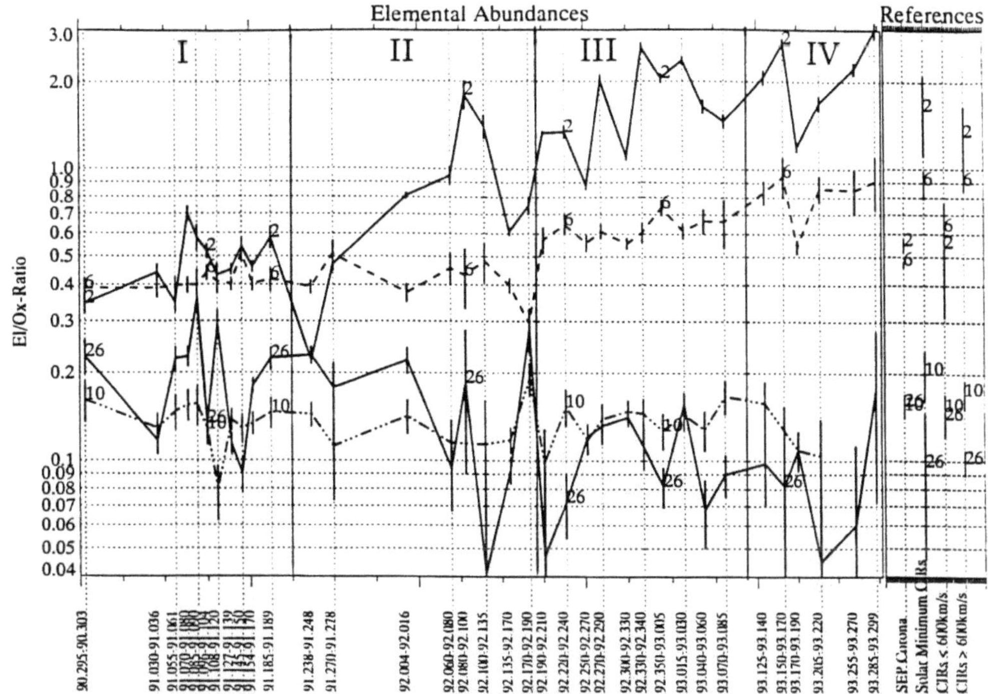

Fig 1 Chemical abundances for He(2)(scaled by 10^{-2}), C(6), Ne(10) and Fe(26) normalized to O for the energetic particle events observed by the EPAC-instrument onboard Ulysses between Oct 1990 and Oct 1993 and reference values from (Reames *et al*, 1992, Richardson *et al*, 1993)

high intensity CME-events in March 1991 (Krupp *et al.*, 1992). Ne/O (10) is rather constant (0.15±0.03) except for days 108-120, 1991.

II:The intermediate period (Day 190, 1991 - Day 190, 1992): He/O is continously rising up to 180 until day 100, 1992 and falls to 60 at the end of the period, C/O remains low with more fluctuations, Fe/O remains high (0.2) until day 16, 1992 and starts to fluctuate afterwards. Ne/O is rather constant 0.12±0.3. The last period (day 170-190, 1992) differs from all other periods observed so far by Ulysses: Fe/O $\sim$ C/O = 0.29±0.2, Ne/O=0.19±0.2. Fe/C $\sim$ 1 was only observed during the March 1991 CME-events.

III:The CIR-period below 30° (Day 190, 1992 - Day 125, 1993): He/O increases with fluctuations from 100 to 250, C/O jumps in the beginning to 0.6 and then smoothly increases to 0.7. Fe/O fluctuates between 0.04 and 0.15, Ne/O is constant 0.14±0.2. The CME-events in Nov. 1992 (day 300-315, (Reuss *et al.*, 1994)) have no significant effect on the averaged composition.

IV:The CIR-period above 30° (Day 190, 1992 - Day 125, 1993): There is one period (day 205-220, 1993) with lower He/O- and C/O-ratios. For the other events He/O fluctuates between 180 and 300 and C/O between 0.8 and 0.95. Ne/O decreases to 0.1 and Fe/O fluctuates between 0.04 and 0.18.

4. Discussion

Comparing the in- and out-of-ecliptic abundances we observe the obvious difference between the CME-dominated in-ecliptic periods and the CIR-dominated out-of-ecliptic periods showing up in the enhanced He/O and the much enhanced C/O ratios. For the second half of 1991, when Marsden *et al.*(1993) continuously observe CIR-compositions in the energy-range 4.25-6.75 MeV/n, we only see once an enhanced C/O-ratio concluding in agreement with Maclennan *et al.*(1993) that the composition at low energies is dominated by solar particle events. For the period day 170-190 (Jun 19-Jul 8), 1992 we observe a composition typical for impulsive flare events (Fe/C $\sim$1) in agreement with the values measured with the SAMPEX spacecraft at earth between Jul 17 and 20, 1992 (Mason *et al.*, 1994). At this later period we see already the CIR-composition. Whether the coincidence between the impulsive flare activity and the build-up of the stable CIR-structure is accidental must be further examined. The increase of the C/O-ratio past day 125, 1993 coincides with the disappearance of the magnetic sector structure (Smith *et al.*, 1993) and of the slow-speed solar regions between the CIRs (Bame *et al.*, 1993). On the other hand the high values agree also with the measurements made in CIRs during the previous solar minimum (Richardson *et al.*, 1993). The jump in the C/O-ratio suggests that the leaving of the region influenced by the neutral sheet is the predominant effect and not the smoothly declining solar activity.

References

Bame, S.J., et al.: 1993, *Geophys. Res. Lett.* **20**(21), pp. 2319–2326.
Gloeckler, G.: 1984, *Adv. Space Res.* **4**(2-3), pp. 127–137.
Gosling, J.T., et al.: 1993, *Geophys. Res. Lett.* **20**(24), pp. 2789–2792.
Keppler, E., et al.: 1992, *Astron. Astrophys., Suppl. Ser.* **92**, pp. 317–331.
Keppler, E., et al.: 1994, this issue.
Krupp, N., et al.: 1992, *Geophys. Res. Lett.* **19**(12), pp. 1255–1258.
Marsden, R.G., et al.: 1993, *Adv. Space Res.* **13**(6), pp. 95–98.
Mason, G.M., et al.: 1994, 'Heavy ion isotopic anomalies in ^{3}He-rich solar particle events', subm. to *Astrophys. J.*
Maclennan, C.G., et al.: 1993, *Proc. 23rd ICRC, Calgary,* **3**, pp. 330–333.
McDonald, F.B., et al.: 1994, 'The effects of the intense solar activity of March/June 1991 observed in the outer heliosphere', preprint.
Reames, D.V.: 1992, *Proc. of the First SOHO Workshop*, ESA SP-348, pp. 315–323.
Reames, D.V., et al.: 1994, *Astrophys. J., Suppl. Ser.* **90**, pp. 649–667.
Richardson, I.G., et al.: 1993, *J. Geophys. Res.* **98**(A1), pp. 13–32.
Reuss, M.K., et al.: 1994, this issue.
Roelof, E.C., et al.: 1992, *Geophys. Res. Lett.* **19**(12), pp. 1243–1246.
Smith, E.J., et al.: 1993, *Geophys. Res. Lett.* **20**(21), pp. 2327–2330.
Whang, Y.C.: 1991, *Space Sci. Rev.* **57**, pp. 339–388.

A PARTICLE EVENT AT 5 AU AND 20° SOUTHERN LATITUDE FROM MEASUREMENTS WITH THE EPAC INSTRUMENT ON ULYSSES

M. K. REUSS, E. KEPPLER, M. FRÄNZ, M. WITTE, N. KRUPP and B. WILKEN
Max-Planck-Institut für Aeronomie, Postfach 20, D-37189 Katlenburg-Lindau, Germany

A. BALOGH and R. J. FORSYTH
Blackett Lab., Imperial College, London SW7 2BZ, U.K.

and

X. MOUSSAS, J. M. POLYGIANNAKIS, A. D. KAKOURIS and A. ALEVIZOS
Astrophysics, Astronomy & Mechanics, National University of Athens, GR-15783, Greece

Abstract. In November 1992, the Ulysses spacecraft observed a multiple solar particle event and a CME event at 5.2 AU and a heliographic latitude of 20° S which were superimposed to the recurrent corotating interacting region. Distinct particle flux increases caused by these events were observed in all energy channels of the EPAC experiment. The experimental findings are discussed.

Key words: Energetic Particles – Ulysses.

1. Introduction

From mid-1992 (≈ 15° S) on until the end of 1993, Ulysses observations were dominated by regular encounters with a corotating interaction region (CIR) (cf. Bame *et al.*, 1993; Gosling *et al.*, 1993). Its periodically recurring forward and reverse shocks caused a periodic structure in energetic particle fluxes as well (cf. Keppler *et al.*, 1994) which can be seen so clearly because of the nearly vanishing background in between.

In November 1992, Ulysses being at 5.2 AU and 20° S, all channels of the charged-particle spectrometer EPAC detected enhanced particle fluxes for a full solar rotation period. These were due to a coronal mass ejection (CME) that reached Ulysses on doy 313, i. e. three days after the reverse shock (Bame *et al.*, 1993), and additionally to solar particles, as will be shown.

2. Instrumentation

The EPAC instrument consists of four telescopes which measure energetic particle fluxes by semiconductor detectors. By the E vs. dE/dx technique energies and composition can be determined. Data are sampled three dimen-

Space Science Reviews **72**: 343–346.
© 1995 *Kluwer Academic Publishers.*

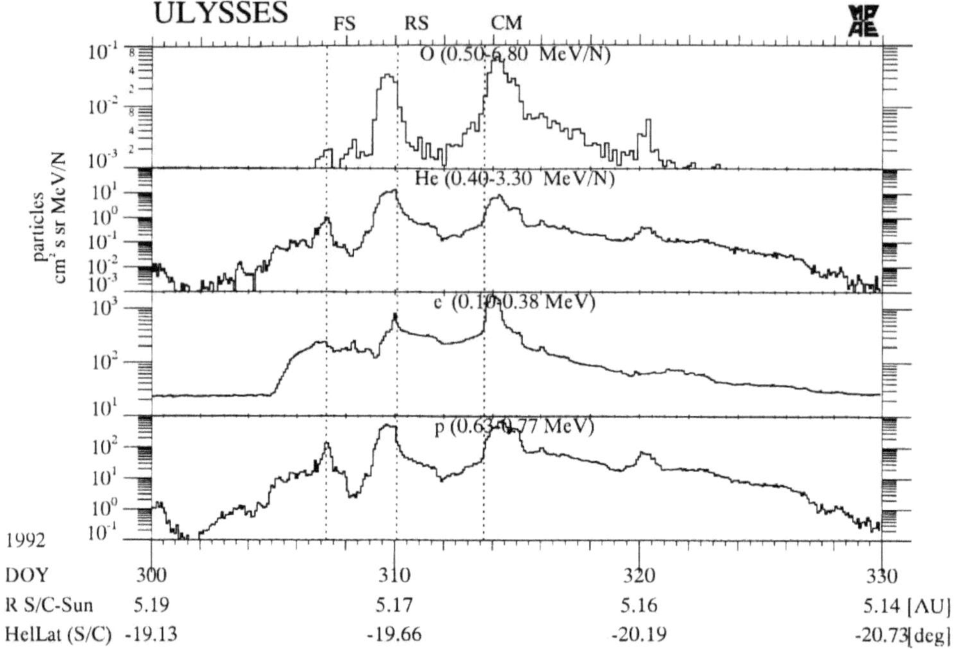

Fig. 1. Overview of particle fluxes measured by EPAC from doy 300 (October 26th) to doy 330 (November 25th). The forward (FS), reverse (RS), and CME (CM) shocks are indicated.

sionally in 32 bins on the sphere, so that anisotropies can be calculated from a spherical harmonic expansion up to the third order.

A more detailed description of the EPAC instrument is given in Keppler *et al.*, 1992; cf. also Fränz *et al.*, 1994.

3. Observations

3.1 CIR and CME particles

Fig. 1 gives an overview of the particle fluxes measured by EPAC in November 1993. Essentially, we see three peaks on doys 307, 309/310, and 313. The first two peaks are due to particles accelerated at the forward and reverse shocks of the CIR. The third peak (doy 313) is associated with the CME.

The CME probably started from an active region around 26° S heliographic latitude (NOAA no. 7321), in which several large flares occurred from October 25th to November 2nd (Solar-Geophysical Data, 1993).

From the CME shock on and during the whole post-CME decrease, composition and spectra remain nearly constant. The following table shows some element ratios at the CIR reverse shock (RS) and at the CME that have been calculated from spectral channels covering approximately the same energy/mass-range.

	Energy range (MeV/amu)	Ratio at RS	Ratio at CME
p/He	0.5 – 0.8	≤ 10	30
He/O	0.5 – 3.5	200	50 – 60
p/O	0.5 – 0.8	2000	1000 – 2000

The He/O ratio at the RS peak (doy 309/310) is consistent with the values calculated by Fränz *et al.*(1994) for the CIR-dominated out-of-ecliptic particle events, whereas it is far lower during the CME. The ratios during the CME agree with coronal elemental abundances (Reames, 1992).

The spectra of the CME particles are hard with spectral index $\gamma \approx 2$ for both protons and helium.

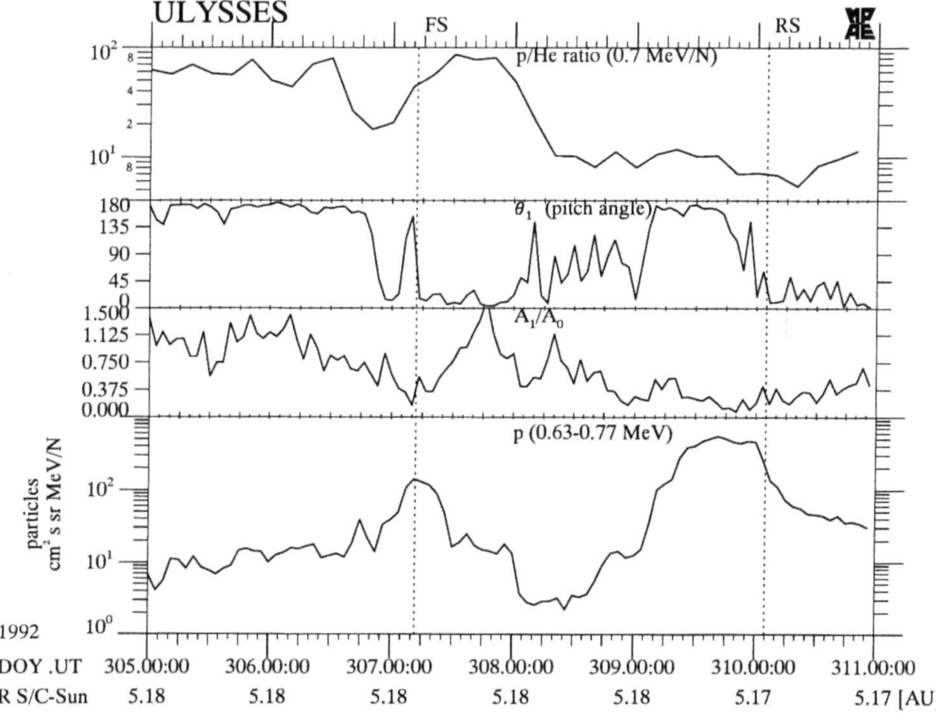

Fig. 2. The top panel shows the ratio of proton and Helium fluxes at 0.7 MeV/amu. The middle panels show the amplitude of the first order anisotropy of proton fluxes at 0.7 MeV, normalized by the omnidirectional flux, as well as the angle θ_1 between the anisotropy vector and the magnetic field. The values have been corrected for the Compton-Getting effect. The bottom panel shows the omnidirectional proton flux as a reference.

3.2 Solar particles

Solar energetic particles were also observed on Ulysses; they can be seen more clearly at higher energies, where shock accelerated particles contribute less.

If we try to trace back the Parker spiral from Ulysses to the Sun in the beginning of November, its footpoint is on the Earth-viewing side of the Sun, and it is consistent with the active region no. 7321 already mentioned. So this region is probably also the source of the solar particles. Because in the slow solar wind the angular distance between Ulysses and the footpoint is larger than 300°, Ulysses may be reached from one solar event both directly (by a CME) and via the Parker spiral (by solar particles). However, for the same reason the backmapping is very susceptible to error propagation, so that we cannot calculate the footpoint more exactly than $\pm 20°$.

In order to identify the solar component from EPAC measurements, we first look at changes in the composition (fig. 2, top panel). Before the CME shock (especially at the reverse shock) the population is α-rich ($p/\alpha \approx 10$); at the forward shock we have $p/\alpha \approx 20$, roughly corresponding to solar wind composition, whereas this ratio reaches 80 close before and after the shock. Comparing this with the fluxes (bottom panel of fig. 2), the region of low α abundance is *before* and *after* the peak of shock accelerated particles; the α particles lack the "shoulders" observed in the protons.

Furthermore, the anisotropies (fig. 2, middle panels) show that before and after the forward shock, there is a strong field aligned first order anisotropy which indicates that the observed particles come from the Sun along the Parker spiral.

These observations suggest that the particles accelerated at the forward shock are superposed to a solar population with a low α abundance.

References

Bame, S. J. *et al.*: 1993, 'Ulysses Observations of a Recurrent High Speed Solar Wind Stream and the Heliomagnetic Streamer Belt', *Geophys. Res. Lett.* **20**, 2319.

Fränz, M. *et al.*: 1994, 'The Relative Elemental Composition of Low-Energy Ions in Energetic Particle Events at high Heliospheric Latitudes', this issue.

Gosling, J. T. *et al.*: 1993, 'Latitudinal Variation of Solar Wind Corotating Stream Interaction Regions: Ulysses', *Geophys. Res. Lett.* **20**, 2789.

Keppler, E. *et al.*: 1992, 'The ULYSSES energetic particle composition experiment EPAC', *Astron. Astrophys. Suppl. Ser.* **92**, 317.

Keppler, E. *et al.*: 1994, 'Energetic Particle Observations in High Heliographic Latitudes', this issue.

Reames, D. V.: 1992, 'Energetic Particle Observations and the Abundances of Elements in the Solar Corona', in *Proc. of the First SOHO Workshop, Annapolis, Maryland, Aug. 1992*, ESA SP-348, p. 315.

Solar-Geophysical Data, 1993: 584 Part II, p. 4 – 19 and 585 Part II, p. 4 – 15, U. S. Department of Commerce, Boulder.

LATITUDE-ASSOCIATED DIFFERENCES IN THE LOW ENERGY CHARGED PARTICLE ACTIVITY AT VOYAGERS 1 AND 2 DURING 1991 TO EARLY 1994

R. B. DECKER[1], S. M. KRIMIGIS[1], R. L. MCNUTT[1],
D. C. HAMILTON[2], AND M. R. COLLIER[2]

Abstract. The Low Energy Charged Particle (LECP) instruments on Voyagers 1 (V1) and 2 (V2) measure the differential in energy fluxes and anisotropies of low energy ions ≥30 keV and electrons ≥20 keV, differential in energy ion composition ≥200 keV/nuc, and the integral rates of cosmic ray protons >70 MeV (Krimigis *et al.*, 1977). We discuss shock-accelerated ions and latitude-associated differences between V1 and V2 during 1991 to April 1994.

1. Introduction and Overview

Figure 1 is a survey of low energy ion fluxes and cosmic ray proton rates during the period of interest. A transient shock (dashed vertical line) passes V2 at ≈2300 UT on day 146 (May 26), 1991, and has no obvious counterpart at V1. A second transient

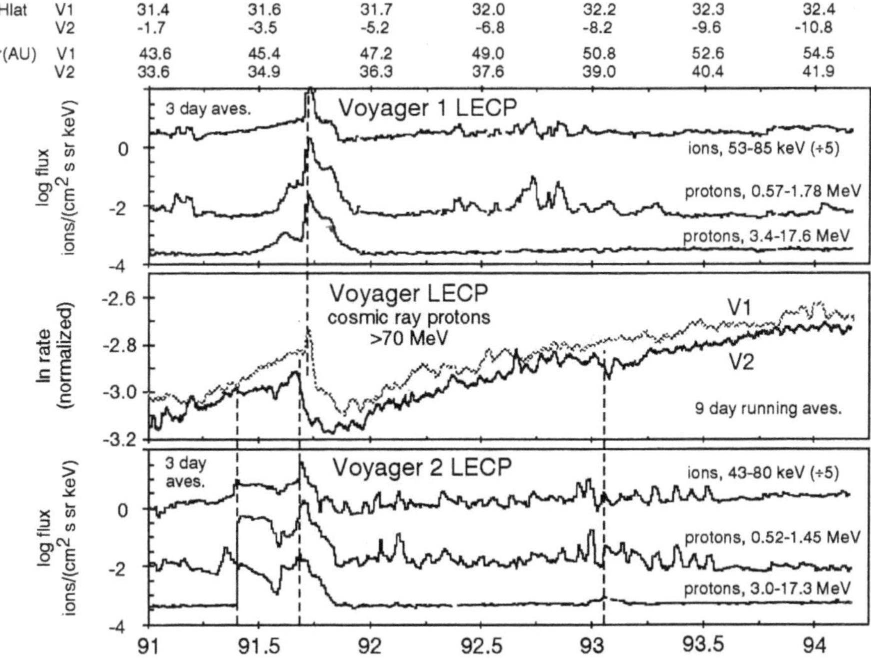

Fig. 1. V1 and V2 LECP data from 1991 to April 1994. Top and bottom panels: 3-day ave. fluxes for one ion (Z≥1) and two proton channels, with indicted energy passbands, at V1 and V2, respectively. Middle panel: 9-day running aves. of E>70 MeV cosmic ray proton rates.

Space Science Reviews **72**: 347–352.
© 1995 *Kluwer Academic Publishers.*

shock reaches V2 at ≈1200 UT on day 251 (Sept. 8) and V1 at ≈2300 UT on day 257 (Sept. 14), 1991. A third transient shock probably passed V2 in early 1993, as revealed by the local peaks in all three ion channels, particularly the 3.0-17.3 MeV protons, and the cosmic ray Forbush decrease. Figure 2 shows the same period, with the ≈1 MeV proton rates replotted in the upper panel. Daily averages of the solar wind radial velocity from the V2 Plasma Science (PLS) experiment plotted in the lower panel.

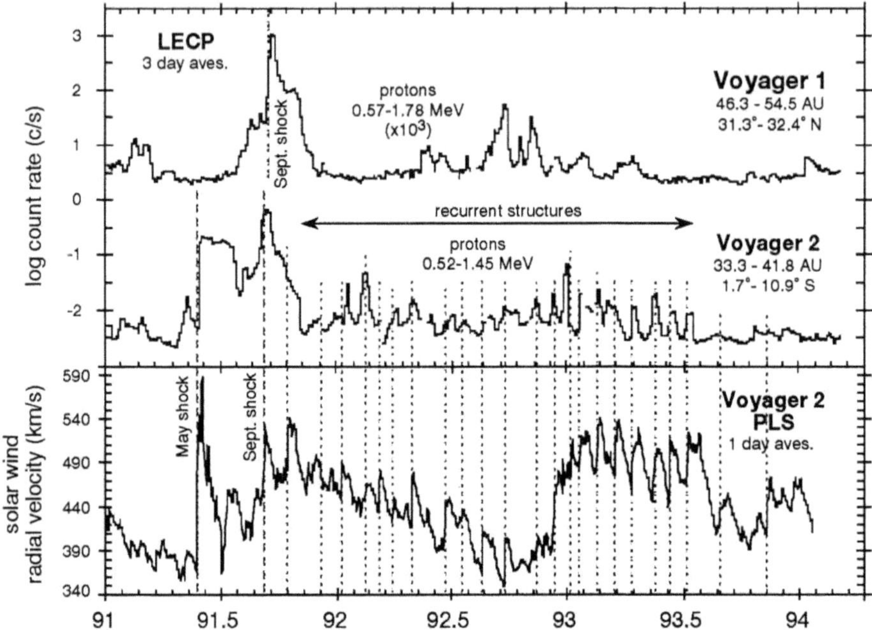

Fig. 2. Top panel: Count rates of ≈1 MeV LECP proton channels replotted from Fig. 1 (top of Fig. 1 gives V1 and V2 radii and heliographic latitudes). The ≈18 month interval marked "recurrent structures" at V2 denotes period when fluxes of ≈30 keV to ≈1 MeV ions were dominated by local acceleration at CIR shocks. Bottom panel: Daily aves. of the solar wind radial velocity from PLS instrument on V2. The 1991 transient shocks at V2 are indicated by dashed lines. Rapid increases in solar wind velocity (dotted lines) and associated recurrent ion structures in 1992 to mid-1993 are consistent with a series of CIR shocks at V2.

2. Survey of Ion Events in 1991-94

2.1. MAY 1991 SHOCK AT V2

Figure 3 shows hourly-averaged rates from eight ion channels for a 10 day period about the May 1991 shock at V2. Both before and roughly a day after the shock passage there was no response above background for ion channels with energies E>500 keV; however, early (≈0400 UT) on day 147 these channels increase rapidly over only a few hours. The ion flux from 28 to 3500 keV during day 147 is well fit by $dj(E)/dE \propto E^{-\gamma}$, with γ=1.70.

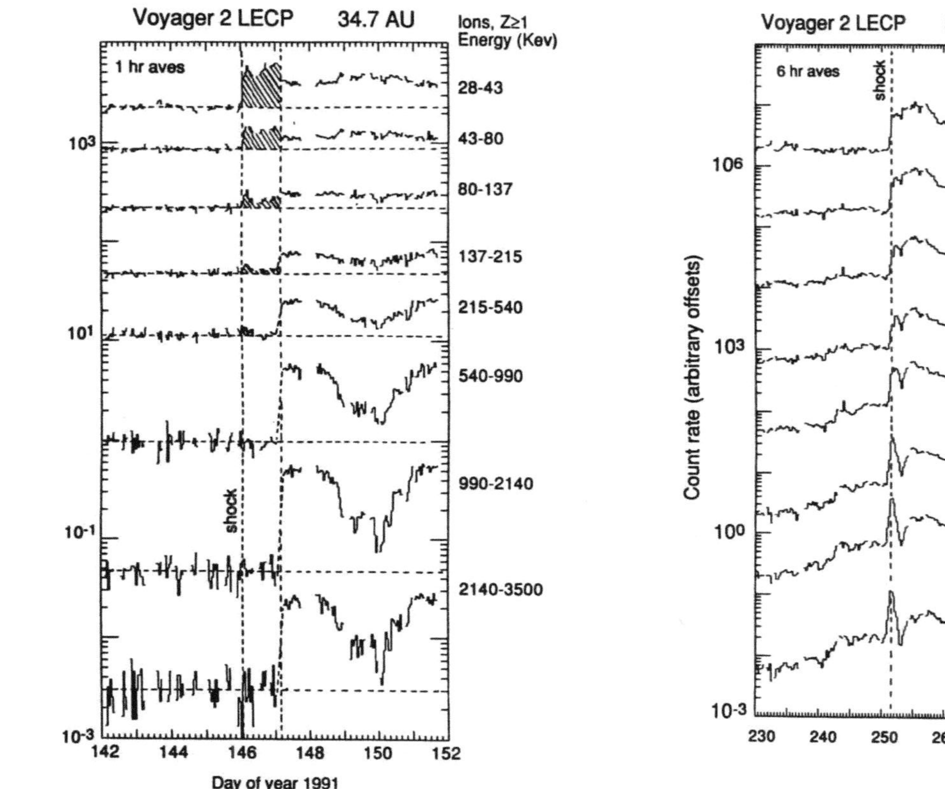

Fig. 3. 1-hour ave. count rates from 8 ion channels for a 10 day period about May 1991 shock at V2 (Decker and Krimigis, 1993a). The shock passage (first dashed line) is at ≈0230 UT on day 146. Early (≈0400 UT) on day 147 these channels increase rapidly over only a few hours when the spacecraft enters the magnetically confined driver gas.

Fig. 4. 6-hour ave. count rates of low energy ions at V2 and V1 during 50 day period about Sept. 1991 shock. Decker *et al.* (1993b) estimate average shock speed to V2≈700 km/s and to V1≈840 km/s. The ≈140 km/s decrease over 35° latitude is attributed to a disturbance that originated from flare at ≈32° N solar latitude, and drives a non-spherical shock whose speed decreases away from the central flare latitude.

This event has also been discussed by Selesnick *et al.* (1991), based on V2 data from the Cosmic Ray Subsystem (CRS). They report on the response of protons >500 keV, and associate the abrupt increase of the energetic ion flux (and plasma density) on day 147 with entry into the magnetically confined driver gas of a coronal mass ejection. Because of the similar response of the plasma and energetic ions, they suggest that the energetic ions originate not by local shock acceleration, which is clearly ineffective for ions >500 keV, but rather in the corona. The LECP data for ions >500 keV are consistent with those from the CRS, and at least partially support this hypothesis. However, it is clear from the four or five lowest energy channels (28 to ≈500 keV) that the shock does leave its signature in the shell of relatively small intensity increases confined between the shock and the plasma driver (i.e., throughout day 146). These ions also show abrupt intensity changes across the magnetic discontinuity early on day 147.

One unusual aspect of this event is that the shock is so ineffective, at least locally, in accelerating low energy ions. Upstream of the shock the level of ≈1 MeV protons was above background, providing a seed particle population available for shock acceleration (Fig. 2). Also, this event is unusual in the absence of comparable features at V1, ≈35° north of V2. This is a clear example where a transient interplanetary disturbance and associated complement of energetic particles remain confined to a limited latitudinal range well into the outer heliosphere. This disturbance is likely to have formed from the ejecta of one or more flares that occurred on the sun's southern hemisphere (≈20°-30° S) during mid to late March (e.g., Roelof *et al.*, 1992), consistent with a latitudinally-restricted disturbance being seen at V2 (3° S) but not at V1 (32° N).

2.2. SEPTEMBER 1991 SHOCKS AT V1 AND V2

Figure 4 shows a comparison of ≈30 to ≈4000 ions at V2 and V1 for a 50 day period about the Sept. 1991 shock. Despite the inferred spatial variation in shock speed and the 36 AU separating the two spacecraft, the ion profiles near the shock are remarkably similar at V1 and V2, particularly in the lower energy channels, which show step-like increases across the shock. The higher energies have localized peaks centered close to the shock passage, i.e., shock-spike events, with the intensities higher at V2. The gradual (tens of days) increases upstream of the shock at V2 is most likely comprised of ambient SEPs. At V1 and V2 the shock-accelerated proton spectrum extends to at least 30 MeV. In the intensity plateaus downstream of the shock, the fluxes are well fit by $dj(E)/dE \propto E^{-\gamma}$ with γ=1.5 for V2 (days 254-258) and 1.8 for V1 (days 262-268).

Figure 5 shows angular distributions for the ≈1 MeV protons. For a convected proton distribution isotropic in the flow frame, in the spacecraft frame the angular distribution varies as the cosine of scan angle, with anisotropy amplitude $\xi=(j_{max}-j_{min})/(j_{max}+j_{min})=2(1+\gamma)(V/v)$, to first order in V/v, where j=particle flux, V=solar wind flow speed, v=particle speed, and γ=slope of particle energy spectrum. For example, in the broad downstream intensity plateaus $\gamma\approx1.5$-2, and for V≈500-600 km/s, $\gamma\approx0.2$-0.3, so $j_{max}/j_{min}\approx1.5$-1.9, with j_{max} from the sunward direction.

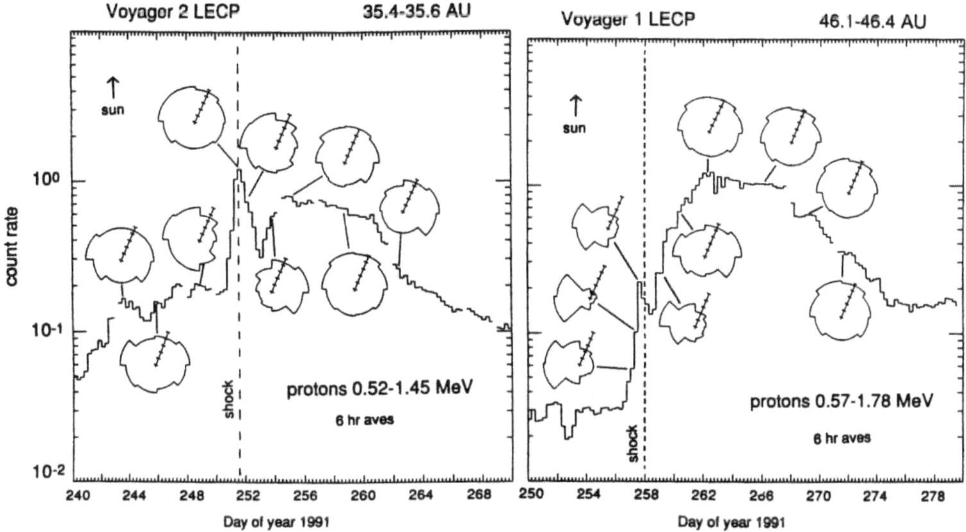

Fig. 5. Angular distributions in the spacecraft frame of ≈1 MeV protons during Sept. 1991 event (Fig. 4). The LECP instrument steps through 8 45° sectors in a scan plane tangent to a cone with vertex at the sun and opening angle relative to the solar spin axis equal to the spacecraft colatitude. During this period the LECP was scanning at the rate of 1 step/192 sec, yielding a full scan through 360° in 25.6 min. The "pie plots" in this figure represent 6-hour averages and consist of 14 full scans. In these angular plots count rates are normalized to the peak sector and are plotted on a linear scale from 0 to 1. The sun is at the top of the figure.

There are examples in Figure 5 of convected isotropic distributions consistent with the anisotropy amplitudes estimated above, particularly in the downstream plateaus. However, there are also extended regions with distinctly anisotropic features, including (i) the large, azimuthally-directed unidirectional anisotropies during the rapid pre- and post-shock rises at V1, and in the falling and rising portions of the post-shock "notch" at V2, and (ii) the bidirectional anisotropies well upstream of the shock at V2 and in the post-shock rise at V1. The angular plots show that even over radial distances ≈50 AU and latitudes ≈30° above the heliographic equator, large, non-convective anisotropies continue to be generated in low energy ion distributions, particularly at sites of local acceleration such as interplanetary shocks.

Pulse height analysis (PHA) of data during the May (doy 145-200) and Sept. (doy 200-300) 1991 events at V2 yields a ratio C/O≅0.30±0.10 for both periods. In a survey of abundance ratios of SEP nuclei from large solar flares observed at 1 AU, Mason *et al.* (1980) found that at ≈1 MeV/nuc, C/O=0.5 on average, with 2/3 of the cases falling in the range 0.40-0.59. Hamilton *et al.* reported an average C/O≅0.68 for three CIR events measured near 4 AU by the V2 LECP. Recent analyses by Lanzerotti *et al.* (1992) of SEPs 0.5-2.0 MeV/nuc observed by the HI-SCALE instrument on Ulysses (2.5 AU) during a large flare in late March 1991, most likely the major source of SEPs and solar wind disturbances later identified as the May 1991 event at V2, yielded C/O≅0.52±0.08.

Evidently, between the sun and ≈35 AU there occurred a ≈40% decrease in C/O at ≈1 MeV/nuc. One plausible interpretation is that at V2 the O is an admixture of SEPs and interstellar pick-up O ions that have been accelerated to ≈1 MeV/nuc by, e.g., transient and recurrent interplanetary shocks. Similar analysis of the V1 LECP data is under way.

2.3. IONS 43-3500 KEV AT V2 DURING CIR SHOCKS IN 1992 TO MID-93

Figures 1 and 2 show large latitudinal differences between V1 and V2 in the 1992 to mid-1993 period. There are sporadic increases from 30 keV to ≈2 MeV ions at V1. The absence of a response in 3.4-17.6 MeV protons suggests these increases arise most likely by local acceleration at CIR shocks. The differences between ion data at V1 and V2 is evidently caused by marked differences in the stream structures in the northern and southern solar hemispheres during 1992-93. It is revealing that recurrent structures such as those at V2 were observed also in low energy ion and electron data from the HI-SCALE instrument on the Ulysses spacecraft (at 4 AU and ≈50° S latitude by late 1993). A gradual intensity diminution of these events began at Ulysses by mid-1993 (Simnett et al., these proceedings). The abrupt disappearance of recurrent structures at V2 in mid-1993 reflects the response in the outer heliosphere to temporal changes in stream structures over a wide latitudinal range in the southern solar hemisphere.

Acknowledgments

This work was supported by NASA under Task I of Navy contract N00039-91-C-001. We thank the PLS team for providing V2 plasma data used in Figure 2, and L. F. Burlaga for discussions regarding the magnetic field observations and superthermal particle behavior associated with the May 1991 shock at V2.

References

Decker , R. B., and Krimigis, S. M.: 1993a, *Proc. 23rd ICRC*, **3**, 310.
Decker, R. B., Krimigis, S. M., and Venkatesan, D.: 1993b, *Proc. 23rd ICRC*, **3**, 481.
Hamilton, D. C., et al.: 1979, *Proc. 16th ICRC*, **5**, 363.
Krimigis, S. M., et. al: 1977, *Space Sci. Rev.*, **21**, 329.
Lanzerotti, L. J., et al.: 1991, *Geophys. Res. Lett.*, **19**, 1251.
Mason, G. M., et al.: 1980, *Astrophys. J.*, **239**, 170.
Roelof, E. C., et al.: 1992, *Geophys. Res. Letts.*, **19**, 1243.
Selesnick, R. S., et al.: 1991, *EOS Proc. Am. Geophys. Union*, **72**, 386.

[1]The Johns Hopkins University Applied Physics Laboratory, Laurel, Maryland, 20707-6099, USA.
[2]Department of Physics and Astronomy, University of Maryland, College Park, Maryland, 20742, USA.

LATITUDINAL AND RADIAL VARIATION OF SHOCK ASSOCIATED ≥30 KEV ION SPECTRA AND ANISOTROPIES AT VOYAGERS 1 AND 2

M. KANE, R. B. DECKER, B. H. MAUK, AND S. M. KRIMIGIS
The Johns Hopkins Univ. Applied Physics Lab., Laurel, MD 20707-6099, U.S.A.

Abstract. The spectra and anisotropies of ions ≥30 keV have been measured by the Low Energy Charged Particle experiment on Voyagers 1 and 2 in the vicinity of interplanetary shocks between radial distances of 1-55 AU and heliographic latitudes 11° S-32° N. The spectra and anisotropies associated with a recent corotating (CIR) event at low latitude observed at Voyager 2 (36.6 AU, -9°) are similar to those of another event at high latitude observed at Voyager 1 (49.8 AU, 33.5°). An earlier CIR event observed at Voyager 2 (14 AU) associated with the previous solar cycle produced spectra and anisotropies remarkably similar to the more recent events. The anisotropies are used to calculate the solar wind velocity downstream of shocks where possible using the Compton-Getting effect, allowing the determination of previously unknown velocities at the locations of Voyager 1. For the large shock event observed at Voyagers 1 (38 AU, 30°) and 2 (29 AU, 3°) in mid-1989, the postshock spectra and anisotropies are well described by convected power law distributions. The Voyager 1 and 2 postshock spectra ~4 days after the shock passage are nearly identical. The preshock anisotropies at low energy are similar, despite differences in the magnetic field orientation and the low energy spectrum. We find that the ≥30 keV ion anisotropies are generally well described by convective distributions downstream but not in the upstream region for this shock and many other shock events at Voyagers 1 and 2.

1. Introduction and Overview

The Voyager 1 and 2 spacecraft have made comprehensive measurements of the ion activity associated with numerous shock events with the Low Energy Charged Particle (LECP) detector. The LECP data contains good energy discrimination and also provides the critical angular information necessary for determination of the convection velocity and pitch angle structure (with the addition of the magnetic field data from the MAG experiment) of shock associated activity. The spectra and anisotropies of ions 30 keV-5 MeV in the vicinity of several interplanetary shock events are considered here. The Voyager 2 spacecraft has, until recently, remained at low Heliographic latitudes and is therefore used to compare similar or corresponding events detected at the higher latitude positions of Voyager 1. Since the Voyager 2 spacecraft has been in operation since 1977, it has been sampling data throughout the complete 11 year solar cycle, allowing comparison of similar events in different cycles. We focus on a large shock in mid-1989 observed at both Voyagers (Decker *et al.*, 1991; Kane *et al.*, 1993) to illustrate the latitudinal effects associated with the passage of this shock. We also consider the latitudinal and radial dependence of the spectra and anisotropies associated with corotating interaction regions (CIR's) that have been observed at both Voyager spacecraft

Space Science Reviews **72**: 353–358.
© 1995 *Kluwer Academic Publishers.*

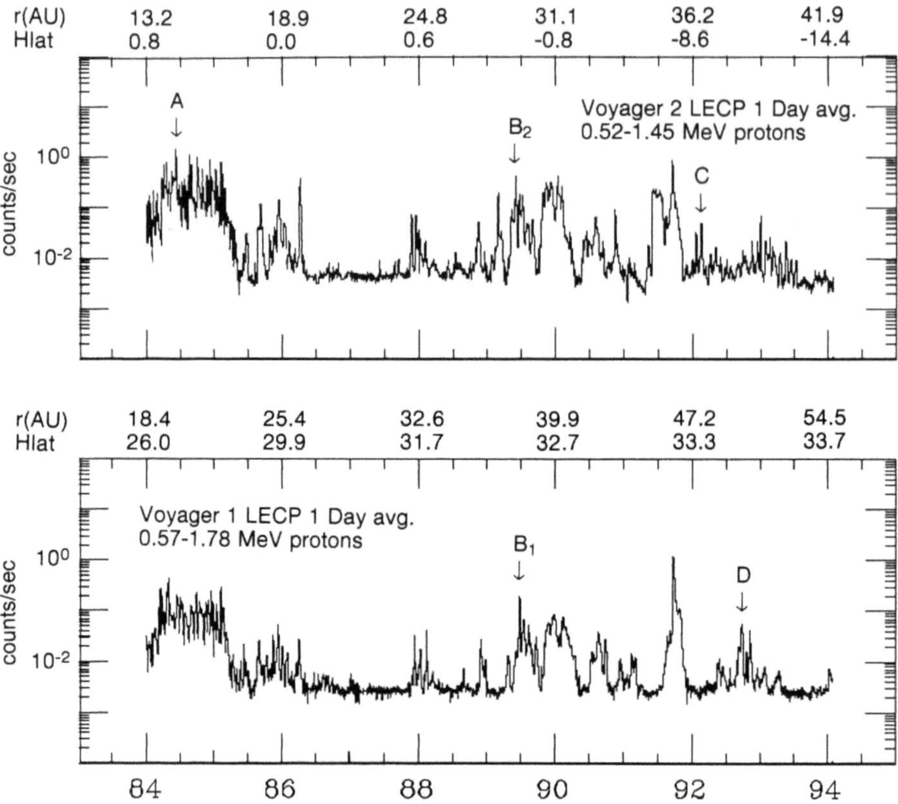

Figure 1. Overview of ~1MeV proton activity at Voyagers 1 and 2.

after 1991 (Decker *et al.*, 1993) and in the previous solar cycle prior to 1986 (Gold *et al.*, 1988).

The Voyager 1 and 2 spacecraft have detected numerous shock related enhancements of ion (primarily proton) activity since 1977. A 10 year period of such activity from 1984 to 1994 is presented in Figure 1. The activity is corrected for background, 1 day averaged, angular averaged, and is from the ~1 MeV LECP proton channels. The distance from the sun and Heliographic latitude is given above each panel for the even numbered years. Event A in the upper panel (Voyager 2) is a CIR associated with many similar adjacent events in the region (Gold *et al.*, 1988). The ion spectrum and anisotropies of event A will be compared to those of Event C, another CIR and also at low latitude, seen during early 1992 at Voyager 2. Event B_2 is the large increase associated with the mid-1989 shock at Voyager 2. This shock also has a corresponding signature at Voyager 1 (lower panel) and is labeled B_1. The preshock and postshock spectra, anisotropies, and pitch angle structure will be considered. Event D is a CIR at high latitude that will be compared with the Event C at low latitude.

The large scale pattern that emerges from the 1 MeV proton data is an early period of recurrent structures (84-86), followed by a quiet period (86-88), then a more active period as solar maximum is approached (88-92). This is then followed by a return to the recurrent structures (92-94) and perhaps a quiet period in the present (94-_). There is not necessarily a one to one correspondence between recurrent structures at low and high latitude. The events C and D were chosen because they are large enhancements. The background rates in the lowest energy ion channels of the LECP detector (<100 keV) swamp the foreground so that only the largest events could be considered at low energies.

2. CIR Spectra and Anisotropies

Event A of Figure 1 is the largest CIR in the early period and is compared to Event C in the left panel of Figure 2. Here the angular averaged flux during 1984:158 is over an order of magnitude more intense than that of Event C during 1992:47. The spectra are, however, similar despite the intensity shift. Both spectra exhibit the characteristic rolloff at high energy seen in many other CIR's. Attempts were made (Gold *et al.*, 1988) to fit the spectra to various functions but without success. It should be noted that the data cover only about an order of magnitude in velocity so that the functional dependence might not be well constrained. The two events are separated by a solar cycle but are both at low latitude. Thus, clearly the processes involved in accelerating seed particles in the CIR's to LECP energies are recurrent with the solar cycle. The spectra of Events C and D in the right panel of Figure 2 are remarkably similar, despite the large latitudinal separation. One difference that is noticable is the rolloff at low energy of the Voyager 2 spectrum relative to the Voyager 1 spectrum. The cause is not known, and was not observed in the spectra of CIR's in 1985 reported by Gold *et al.* (1988). It does appear in the nearly power law spectrum associated with the mid-1989 shock event presented in Figure 3 in both the preshock and postshock regions.

The angular anisotropies are shown above the spectra of Figure 2 for the 3 CIR's considered. The LECP detector rotates into 8 sectors, and the rates are normalized to the largest rate and proportional to the radial extent of the 45° slices of the pie. Sector 8 is at the 12 o'clock position and measures background but has been filled. All the low energy pie plots are convective, with sector 1 in the 10:30 o'clock position viewing nearly into the solar wind. The higher energy angular distribution associated with the 1984 event appears to have a symmetry rotated ccw with respect to the other distributions. Lack of magnetic field data precludes an explanation. The angular and spectral information may be used to calculate the convective flow velocity of the ions by use of the Compton-Getting effect. Analysis of the anisotropies during 1984:159 yielded a velocity of 450 km/sec. For the 1992 CIR's, the same analysis yielded 360 km/sec at Voyager 2 during 1992:48 and 490 km/sec at Voyager 1 during 1992:267. The higher velocity at the higher latitudes of Voyager 1 implies a non-spherical shock propagation. This was seen in earlier data reported by Gold *et al.* (1988) as the explanation of the shifted delay times of the onset of CIR's at the two Voyagers.

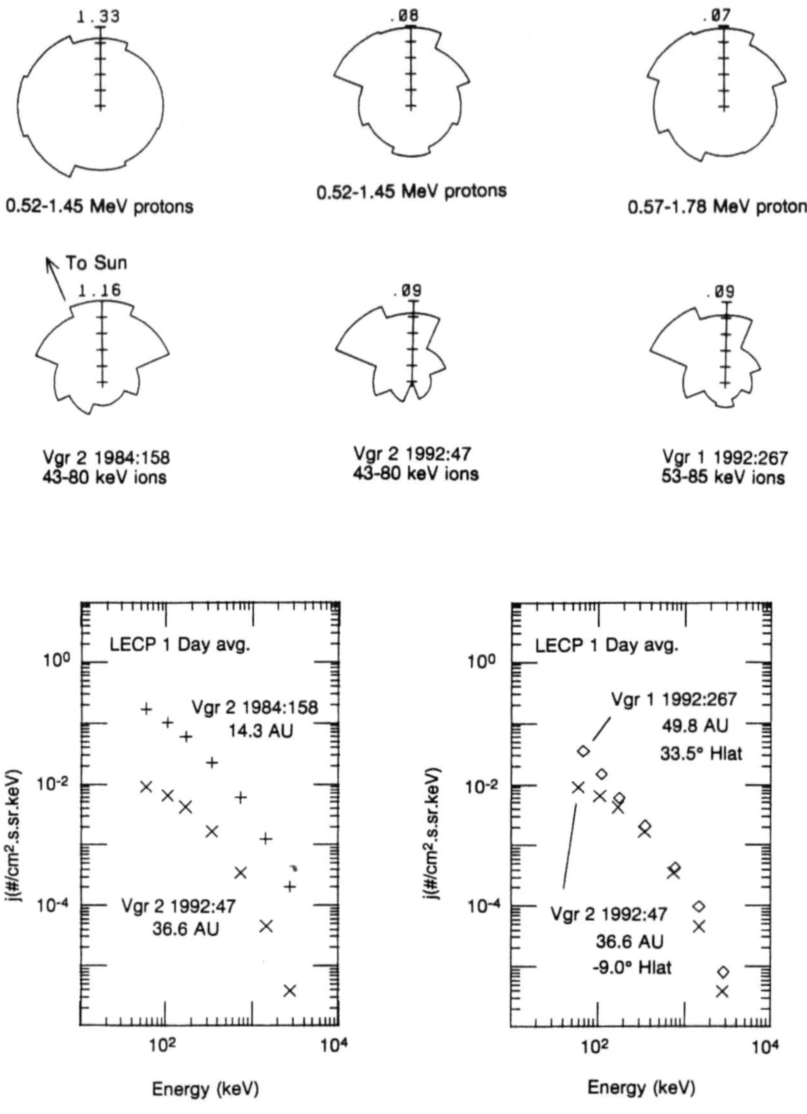

Figure 2. CIR spectra and angular anisotropies of hot ions at Voyagers 1 and 2.

3. Shock of mid-1989

We now consider the mid-1989 event seen at both Voyagers. This event was produced during a March 1989 solar flare (Venkatesan *et al.*, 1990). The shock speeds at Voyagers1 and 2 have been calculated based on IMP 8-Voyager transit times for this event and are about 600 km/sec. The shock parameters at Voyager 2 have been determined by Kane *et al.* (1993) using the PLS and MAG data and by Burlaga (1994).

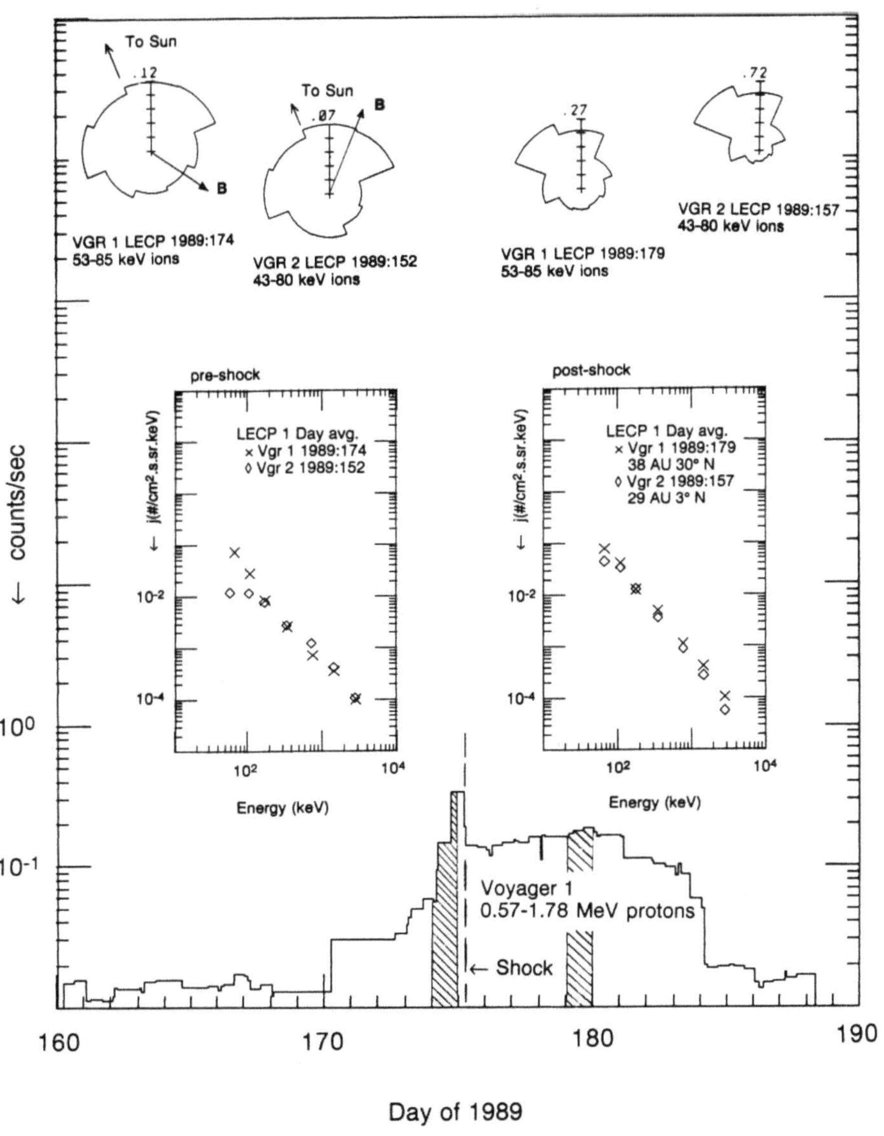

Figure 3. Spectra and anisotropies of the mid-1989 shock at Voyagers 1 and 2.

We focus here on the nature of the anisotropies, spectra, and the implied latitudinal differences between Voyager 1 and 2 for the preshock and postshock periods. The ~1 MeV proton activity is shown for Voyager 1 in the bottom of Figure 3. The dashed bars are the days where the spectra and anisotropies (above the trace) were derived. The Voyager 2 activity is not shown, but the spectra and anisotropies from Voyager 2 were obtained at the same times relative to the shock passage at Voyager 2 as those of Voyager 1. Both the preshock (left panel) and postshock spectra are nearly power law, in contrast to that of CIR's. Despite the difference of latitude of nearly 30° between the two Voyagers, the postshock spectra are nearly identical. The preshock spectra are also quite similar, with the low energy dropoff apparent in the Voyager 2 spectrum as yet unexplained. We note here that the background levels are quite high at the low energies, although the background statistics are good.

The postshock anisotropies (upper right, main panel) are convective throughout the energy range of the ions considered here. This allowed the use of a convected power law distribution to obtain the distribution functions and hence solar wind velocities of the plasma. For Voyager 2, we obtained v=564 km/sec. The addition of the magnetic field data allowed the determination of the preshock pitch angle structure (upper left of outside panel). For the low energy ions, the shape of the distribution at Voyagers 1 and 2 are remarkably similar, despite the difference in latitude and the varying orientation of the magnetic field vector. This similarity did not extend to higher energies, however.

4. Conclusions

The spectra and anisotropies of two types of interplanetary shocks seen at Voyagers 1 and 2 have been considered in the context of the radial and latitudinal differences in the locations of the observations. The CIR shocked events are evidently recurrent with solar cycle. Their spectra and anisotropies are remarkably similar both with the changing solar cycle, radial distance, and Heliographic latitude, although their occurrence frequency appears to be a function of latitude. The mid-1989 solar flare event, like many other similar events studied in the interplanetary medium at Voyagers 1 and 2, contain power law spectra and postshock convective distributions.

Acknowledgements

The authors would like to thank L. F. Burlaga for the use of the magnetic field data for the 1989 event. This work was supported by NASA under Task I of contract N00039-91-C-001 between The Johns Hopkins University and the Department of the Navy.

References

Burlaga, L. F.: 1994, J. Geophys. Res., 99, 4161.
Decker, R. B., Gold, R. E., and Krimigis, S. M.: 1991, Proc. 22nd ICRC,3, 296.
Decker, R. B., and Krimigis, S. M.: 1993, Proc. 23rd ICRC, 3, 310.
Decker, R. B., Krimigis, S. M., and Venkatesan, D.: 1993, Proc. 23rd ICRC,3, 481.
Gold, R. E., *et al.*: 1988, J. Geophys. Res., 93, 991.
Kane, M., R. B. Decker, and Krimigis, S. M.: 1993, Proc. 23rd ICRC, 3, 302.
Venkatesan, D., *et al.*, 1990, Proc. 21st ICRC, 6, 247.

ULYSSES-VOYAGER STUDY OF ENERGETIC PARTICLES ASSOCIATED WITH THE INTENSE SOLAR ACTIVITY OF MARCH/JUNE 1991

F. B. McDONALD[1], E. KEPPLER[2] and M. K. REUSS[2]

[1]*Institute for Physical Science and Technology, University of Maryland, College Park, MD USA*

[2]*Max Planck Institute für Aeronomie, Katlenburg-Lindau, Germany*

Abstract. The intense solar activity centered in March and June 1991 produced some of the largest interplanetary disturbances over the past several solar cycles. For these events the Ulysses EPAC energetic particle observations near 3 AU are compared with those of the Voyager 2 CRS experiment near 35 AU. At Voyager 2 there is a single long-lived event extending over a period of some 6 months while the Ulysses data shows the imprint of individual events as well as the formative stages of the longer lived structure. The average intensity gradient is $\sim -17\%$ AU between the 2 spacecraft. At both locations the energy spectra can be represented by an exponential in momentum. The characteristic momentum for protons, $(P_o)_H$ is on the average 4-5 times larger at 35 AU than at 3 AU and there is a significant change in the $(P_o)_{He}/(P_o)_H$ ratio. However the average H to He ratio is in the range 20-25 for both sets of measurements.

1. Introduction

Solar/interplanetary energetic particle (S/IP) events evolve with increasing heliocentric distance; inside some 5 AU they retain the characteristics of individual solar particle events; (Hamilton, 1977; Zwickl and Webber, 1977), by 10 AU the particle enhancements may extend over a solar rotation and display more pronounced shock-associated features; (Dröge et al., 1992). Beyond some 25 AU the individual increases in particle intensity observed in the inner heliosphere have lost their identity and there emerges a new type of particle event that represents the evolution and superposition of multiple energetic particle events from the inner heliosphere that have occurred over several solar rotations and the continuing acceleration of these ions by large scale interplanetary disturbances known as Global Merged Interaction Regions (GMIRs) (Van Allen and Mihalov, 1990; McDonald and Selesnick, 1991; Decker et al., 1991). The energetic particle events reflect the evolution and development of these large scale disturbances in the outer heliosphere.

Originally identified by Burlaga et al. (1984, 1985, 1993) GMIRs evolve with increasing heliocentric distance through the coalescence of interplanetary shocks produced by CMEs and high speed solar wind streams. These systems are a major element in producing the long term 11 year cosmic modulation (le Roux & Potgieter, 1993; McDonald et al., 1993; Burlaga et al. 1993) and they take on added importance with their role as the probable trigger for the strong outburst of low frequency ($\sim$2-3 K Hz) radio emission recently observed by the Voyager 1/2 Plasma Wave System (PWS) (Gurnett et al. 1993). A key question is whether the MeV ions and electrons observed in the outer heliosphere several months after an outburst of solar activity are the solar energetic particles accelerated in the inner heliosphere and partially re-energized by the shocks within the GMIR or are they more recently accelerated by these shocks from a seed population of superthermal

Space Science Reviews **72**: 359–364.
© 1995 *Kluwer Academic Publishers.*

solar wind ions.

In March 1991, some 20 months after the period of maximum solar activity there was a new increase in activity with 35 major flares (x-ray classification M-5 or higher) during the month. The great majority of these events occurred in the southern hemisphere of the Sun. This activity decreased over the next several months but in June the premier active region of cycle 22 appeared, producing in a single passage 6 X-10 (or greater) flares, all of them in the northern hemisphere. The relative temporal isolation of this intense outburst of solar activity and its marked north-south asymmetry combined with the data from Ulysses in the inner heliosphere and Voyager 2 in the outer heliosphere provide a unique opportunity to study the generation of a GMIR and the acceleration and transport of low-energy ions over a wide range of heliocentric distances.

In this paper the low energy H and He data from the Ulysses EPAC experiment near 3 AU and the Voyager 2 CRS experiment near 35 AU are used to study the evolution from many events in the inner solar system to a single event lasting almost 6 months in the outer heliosphere. Special emphasis is placed on both the spectral information and the H/He ratio measured at the 2 locations as a possible tracer of the origin of the ions observed near 3 AU.

2. Observations

The time history of 0.5-1.0 MeV protons (24 hr avg) from the Ulysses EPAC experiment for the period 1990.9 to 1991.9 are shown in Figure 1 along with the Voyager 2 data near 35 AU for 1.8-2.6 and 4.2-6.2 protons for a time period that has been displaced by 0.3 years (this would correspond to a plasma transit velocity of ~550 km/s between the 2 spacecraft). At V-2 the abrupt increase in both energy intervals on 1991.4 is associated with a strong shock as revealed in the solar wind [Belcher et al., 1993] and interplanetary magnetic field data. These MeV ions are clearly under the control

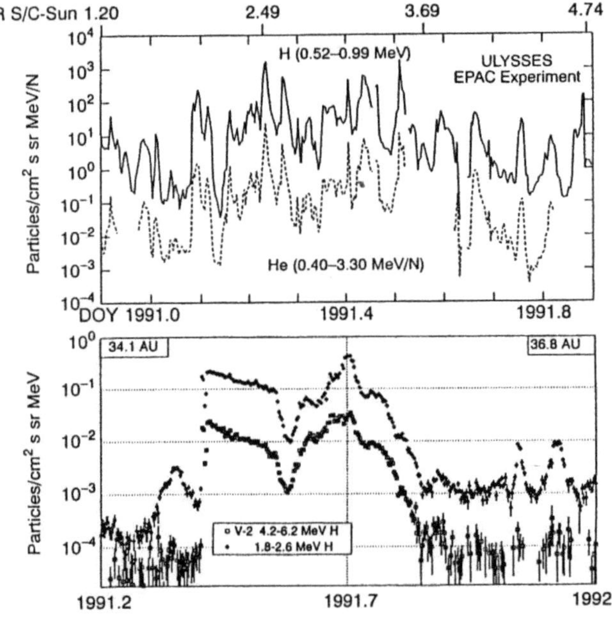

Fig. 1: **(a)** Intensity of 0.52-0.99 MeV H and 0.4-3.30 MeV/n He measured by the Ulysses EPAC experiment between 1.2 and 4.74 AU. (24 hr Averages); **(b)** Intensity of 1.8-2.6 and 4.2-6.2 MeV H measured by the Voyager 2 CRS Experiment between 34.1 and 36.8 AU

of the large-scale disturbances in the interplanetary medium. The 3 small increases in 1.8-2.6 MeV H at 1991.35, 1992.05 and 1991.12 appear to be produced by co-rotating interaction regions. In the inner heliosphere the Ulysses 0.5-1.0 MeV proton data show both the individual solar events and the nascent structure of the long-lived event observed in the outer heliosphere. The two periods in the EPAC data that extend from (a) 1991.15-1991.29 and (b) 1991.35-1991.47 reflect the intense activity centered around March and June of that year. Previously Dröge et al. (1992) had observed the intermediate state of these events at heliocentric distances out to 30 AU.

The main diagnostics available from these joint data sets for understanding the interplanetary transport and acceleration processes are the H and He energy spectra and the H/He ratio and their changes with time and with heliocentric distance. At V-2, the H and He spectra were best fit by either an exponential in momentum/nucleon (i.e., velocity at these low energies) or in kinetic energy (McDonald et al., 1994). The characteristic values for $(P_o)_H$ and $(P_o)_{He}$ (generally averaged over a 10 day period) are plotted in Figure 2 along with the H/He ratio for 1.8-2.6 MeV/n ions. The representation as a function of momentum used in this plot has the interesting feature that $(P_o)_H \approx (P_o)_{He}$ immediately at or just after the passage of 2 of the 3 largest interplanetary disturbances in this GMIR (Fig. 2). Over the rest of the event, the ratio of $(P_o)_{He}/(P_o)_H = 0.67$ which is very close to the average value of 0.75 reported by Mewaldt (1979) for a number of CIR events at 1 AU. [Mewaldt et al. expressed their ratios in terms of magnetic rigidity, R with $(P_o)_{He}/(R_o)_H \approx 1.5$ which is equivalent to

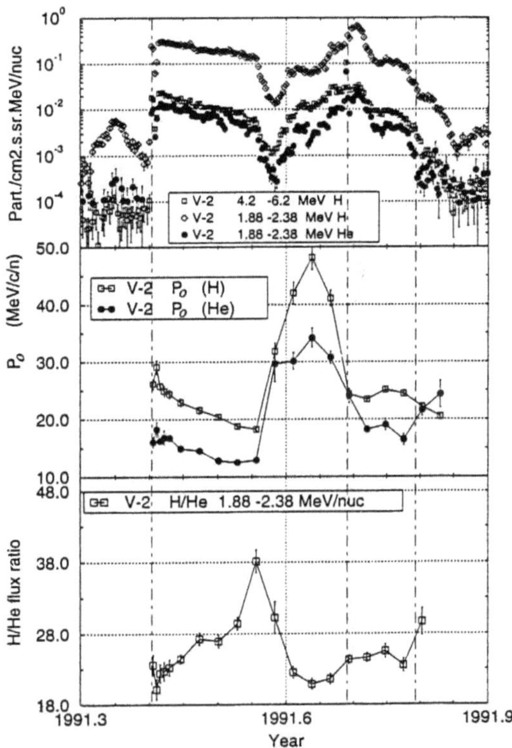

Fig. 2.: The characteristic values $(P_o)_H$ and $(P_o)_{He}$ obtained by fitting the V-2 H and He data as an exponential in momentum/nucleon (i.e., proportional to particle velocity at these energies). The averages are generally taken over a 10 day interval. Also shown for reference is the time history of 1.88-2.38 MeV/n H and He and of 4.2-6.2 MeV H and the H/He ratio for 1.9-2.4 MeV/n ions. The vertical – - – lines mark the passage of the 3 largest interplanetary shocks in the GMIR.

$(P_o)_{He}/(P_o)_H = 0.67$ since $(R_o)_{He} = 2(P_o)_{He}$]. The H/He ratio in Figure 2 is generally in the range 19-28 (except at the time of the intensity minimum in 1991.5), with an average value of 25.3.

At Ulysses the characteristic spectral values and the H/He ratios are obtained from 24 hr averages. (Figure 3). This value of $(P_o)_H$ for period a is 4.6 MV and

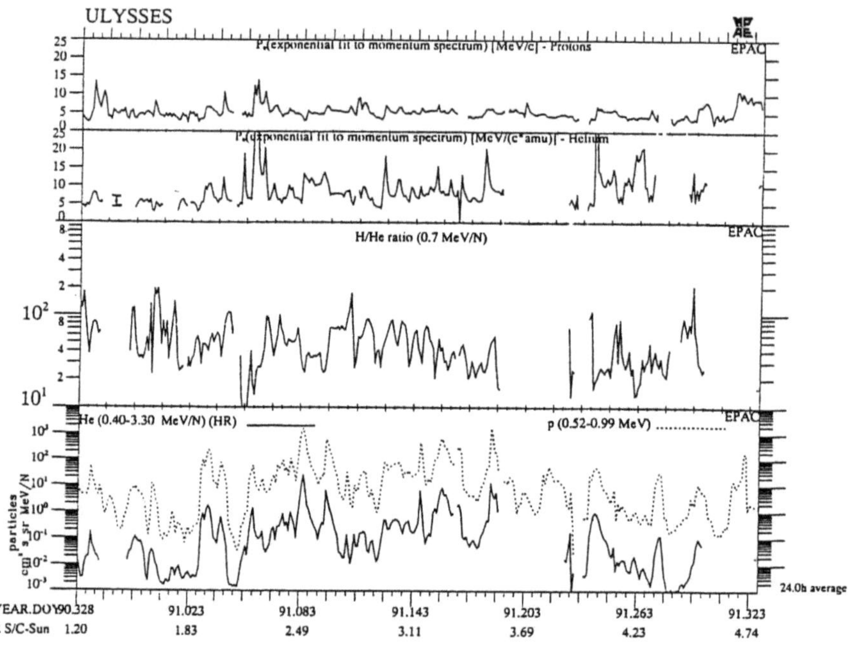

Fig. 3: The characteristic values $(P_o)_H$ and $(P_o)_{He}$ obtained by fitting the Ulysses EPAC data as an exponential in momentum/nuc (24 hr averages) also plotted is the H/He ratio at 0.7 MeV/n and the time history of 0.52-0.99 MeV H and 0.4-3.30 MeV/n He.

5.6 MV for period b which is a factor of 4-5 smaller than that measured at V-2. The average value of the ratio of $(P_o)_{He}/(P_o)_H = 2.22$ and 1.45 for the 2 periods is very different from the ratio of 0.67 observed at 35 AU. However, at Ulysses the H/He ratio over the energy range 0.5-1 MeV/n is 21.4 and 25.4 for the 2 periods which is in good agreement with the Voyager rate of 25.3.

To obtain an average intensity gradient, G_r, between the 2 spaceacraft, it is necessary to extrapolate the 1.8-2.4 MeV intensity at V-2 to 0.5-1.0 MeV assuming an exponential in momentum for the spectral form with $\bar{P}_o = 24$ MV. Averaging the V-2 data over the 2 peak regions and the Ulysses data over the a and b periods gives a value of $\bar{G}_r$ between 3 and 35 of -17%/AU. This intensity gradient is comparable to the average of -45% AU between 1 and 10 AU and -15% AU from 10-30 AU for 11-21 MeV ions measured by Dröge et al. (1992) for a series of earlier S/IP events.

3. Discussion

At 35 AU, the interplanetary disturbances associated with the March/June 1991 solar activity have formed a Global Merged Interaction Region whose properties are reflected in the long-lived energetic particle event as observed for 2.5 and 5 MeV protons (Fig. 1). In the inner heliosphere the Ulysses EPAC data show both the individual solar energetic particle events as well as the formative stages of the long-lived event. There are four principle features that stand out in comparing the properties of the energetic particles at the 2 heliocentric distances:

(i) There is a strong negative gradient of ~ 17% AU for 0.5 - 1.5 MeV H

(ii) The momentum spectra for both H and He become increasingly harder at the larger heliocentric distances with $(P_o)_H$ increasing by a factor of 4-5 between 3 and 35 AU.

(iii) Except in the spatial region near shocks $(P_o)_{He} \cong 0.67(P_o)_H$ at 35 AU. However at Ulysses $(P_o)_{He}$ ranges from 1.45 to 2.2 $(P_o)_H$.

(iv) Over most of this period the H/He ratio $\approx$ 25 at both spacecraft.

Understanding the first 2 factors will require detailed modeling of the acceleration and transport processes. The increase in $(P_o)_H$ is very striking and provides evidence that these ions are continuing to be accelerated.

Of particular importance is the observation that $(P_o)_H \approx (P_o)_{He}$ at V-2 near the time of the passage of 2 of the 3 strongest shocks in the GMIR. Such a velocity dependence is consistent with the predictions of the model of Fisk and Lee (1980) who considered shock acceleration in the forward and reverse shocks of co-rotating interaction regions. The principal loss process in this model are adiabatic energy losses in the interplanetary medium and the radial diffusion coefficient, K_r, was assumed to be of the form $K_o vr$ where K_o is a constant, v is the particle velocity and r is the heliocentric distance. This leads to spectra of the form,

$\exp - \dfrac{6K_o v\ f(\beta)}{V_{sw}}$ (where $f(\beta)$ is a function of the shock parameters), which is in agreement with the observed velocity dependence near the shock.

Previously, particle acceleration by the major interplanetary shocks passing Ulysses in the last half of March 1991 had been extensively studied using the EPAC data (Krupp et al. 1992; Naidu et al. 1992). In a detailed analysis of the March 23rd event Naidu et al. obtained a value of $(P_o)_H$ = 3.8 MV and a diffusion mean free path of 0.29 AU at 700 k ev.

There is a growing consensus that the solar energetic particles produced in large solar events are accelerated by shock waves generated by coronal mass ejections with the seed population being the ambient coronal plasma above the active region (Reames et al. 1991; Reames 1992). The Ulysses EPAC observations in March clearly establish that the CME driven shocks are still accelerating particles and at 35 AU this process continues.

It is interesting to note that the ratio of $(P_o)_{He}/(P_o)_H$ and of H\He closely resemble that of CIR events (Mewaldt et al., 1979; McGuire et al., 1978). Previous measurements of the H/He ratio for solar energetic particles at higher energies had shown a large variation from event to event. However recent studies in the 1-4

MeV/n range (Mazur 1991; Reames 1992) have found a much smaller variation and Reames finds a weighted mean average of 21.1 $\pm$ 1.6.

The most plausible explanation for the Ulysses/Voyager observation is that the 0.5-1 MeV and even lower energy ions observed near 3 AU are the seed population for the MeV ions at Voyager and this seed population is continually being replenished at least in the inner heliosphere. The observed $(P_o)_{He}/(P_o)_H$ ratio at 35 AU would appear to require that additional loss processes be included in the Fisk-Lee (1980) model and these loss processes are similar for CIRs in the inner heliosphere and GMIRs in the outer heliosphere.

References

Belcher, J. W., A. J. Lazarus, R. L. McNutt, Jr., and G. S. Gordon: 1993, **13**(6), 41.

Burlaga, L. F., F. B. McDonald, N. F. Ness, R. Schwenn, A. J. Lazarus, and F. Mariani: 1984, *J. Geophys. Res.*, 89, 6579-6587.

Burlaga, L. F., F. B. McDonald, M. L. Goldstein, and A. J. Lazarus:1985, *J. Geophys. Res.*, **90**, 12,027–12,039.

Burlaga, L. F., F. B. McDonald and N. F. Ness: 1993, *J. Geophys. Res.*, **93**, 1-11.

Decker, R. B., R. E. Gold and S. M. Krimigis: 1991, *Proc. 22nd Int'l Cosmic Ray Conf. (Dublin)*, 3, 296, 1991.

Dröge, W., R. Müller-Mellin and E. W. Cliver: 1992, *Astrophys. J.*, **387**, L97.

Fisk, L. A. and M. A. Lee: 1980, *Astrophys. J.* **237**, 620, 1980.

Gurnett, D. A., W. S. Kurth, S. C. Allendorf, and R. L. Poyntner: 1993, *Science*, *262*, 199.

Hamilton, D. C.: 1977, *J. Geophys. Res.*, **82**, 2157.

Krupp, N., E. Keppler et al.: 1992, *Geophys. Res. Letts.*, **19**, 1255.

le Roux, J. A. and M. S. Potgieter: 1993, *Advances in Space Research*, **251**, Vol. 13.

Mazur, J. E.: 1991, *Ph.D. Thesis*, Univ. of Maryland.

McDonald, F. B. and Selesnick, R. S.: 1991, *Proc. 22nd Int'l Cosmic Ray Conf. (Dublin)*, **3**, 189-192.

McDonald, F. B., N. Lal and R. E. McGuire: 1993, *J. Geophys. Res.*, **98**, 1243.

McDonald, F. B., A. Barnes, L. F. Burlaga, P. Gazis, J. Mihalov, and R. S. Selesnick: 1994, accepted for publication in *J. Geophys. Res.*

McGuire, R. E., T. T. von Rosenvinge, and F. B. McDonald: 1978, *Astrophys. J. Lett.*, **224**, L87.

Mewaldt, R. A., E. C. Stone, and R. E. Vogt: 1979, *Geophys. Res. Lett.*, **6**, 589.

Naidu, K., J. J. Quenby, et al.: 1992, *Geophys. Res. Letts.*, **19**, 1259.

Pyle, K. R., J. A. Simpson, A. Barnes and J. D. Mihalov: 1984, *Astrophys. J.*, **282**, L107.

Reames, D. V.: 1992, *Proceedings of the First Soho Workshop*, **ESA SP-348**, 315.

Reames, D. V., H. V. Cane, et al.: 1991, *Proc. 22nd Int'l Cosmic Ray Conf. (Dublin)*, **3**, 319.

Van Allen, J. A., and J. D. Mihalov: 1990, *Geophys. Res. Lett.*, **17**, 761-764.

Zwickl, R. D., and W. R. Webber: 1977, *Solar Physics.* **54**, 457.

SESSION 4

COSMIC RAYS IN THE HELIOSPHERE

COSMIC RAY MODULATION IN THE 3-D HELIOSPHERE

R. B. MCKIBBEN, J.J. CONNELL, C. LOPATE, J. A. SIMPSON, and M. ZHANG
Enrico Fermi Institute, University of Chicago, Chicago, IL, U.S.A.

Abstract: The basic physical processes that lead to the long-term modulation of cosmic rays by the solar wind have been known for many years. However our knowledge of the structure of the heliosphere, which determines which processes are most important for the modulation, and of the variation of this structure with time and solar activity level is still incomplete. Study of the modulation provides a tool for probing the scale and structure of the heliosphere. While the Pioneer and Voyager spacecraft are surveying the radial structure and extent of the heliosphere at modest heliographic latitudes, the *Ulysses* mission is the first to undertake a nearly complete scan of the latitudinal structure of the modulated cosmic ray intensity in the inner heliosphere (R<5.4 AU). *Ulysses* will reach latitudes of ~80°S in September 1994 and ~80°N in July 1995 during the approach to minimum activity in the 11 year solar cycle. We present a first report of measurements extending to latitudes of ~52°S, which show surprisingly little latitudinal effect in the modulated intensities and suggest that at this time modulation in the inner heliosphere may be much more spherically symmetric than had generally been believed based upon models and previous observations.

1. Introduction

Forbush (1954) was the first to report that the intensity of the cosmic radiation measured at Earth varies systematically in anti-correlation with the solar activity during the 11 year solar activity cycle. As shown in Figure 1, which contains the record of the smoothed sunspot number and the variation in the cosmic ray intensity as recorded by the Climax neutron monitor since 1951, we have now accumulated observations over almost four complete solar cycles. We have come to understand that the intensity variations are caused by variations in the effectiveness of the solar wind in excluding cosmic rays from the inner solar system as a result of changes in the structure of the solar wind and interplanetary magnetic field with the level of solar activity, and there is by now a basic understanding of, and some rather detailed models for, the physical processes that produce modulation of the cosmic ray intensity by the solar wind. However, as we still have only partial knowledge of the structure of the heliosphere and of its variations during the solar cycle, our picture of modulation is not yet complete.

We present in this paper a brief review of the current understanding of modulation in the heliosphere, focussing on the long-term effects governed by global heliospheric structure rather than on short-term effects that reflect primarily local conditions, and on issues particularly relevant to observations to be obtained from the *Ulysses* mission at high latitudes. We also present a first look at cosmic ray variations up to >50°S latitude during the current approach to solar minimum as measured by the COSPIN High Energy Telescope (HET) on *Ulysses*. The HET has been described by Simpson et al., (1992).

Space Science Reviews **72**: 367–378.
© 1995 *Kluwer Academic Publishers.*

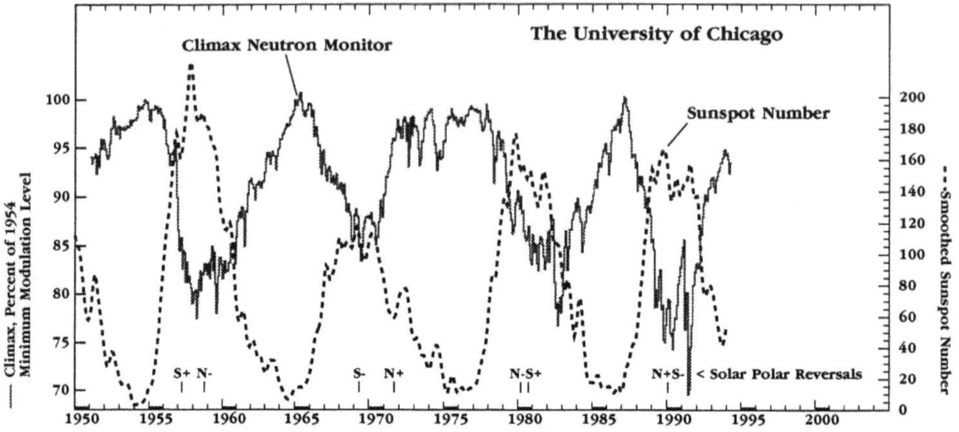

Figure 1: Smoothed sunspot number and monthly average Climax Neutron Monitor intensities.

2. Review and Discussion of Modulation Physics

The heliosphere is the region of space dominated by the solar wind. First suggested to explain ground-based observations of solar energetic particles from the February 1956 solar flare (Meyer, Parker, and Simpson, 1956) the heliosphere was originally pictured as a spherical shell of tangled magnetic fields surrounding the sun at a radius of a few AU. Since then our understanding of the heliosphere and its structure has developed greatly as spacecraft have pushed our observational knowledge to larger radii (currently to ~60 AU with Pioneer 10) and higher latitudes (currently ~60° S with *Ulysses*) and as observations have accumulated over four solar cycles with increasingly sophisticated instrumentation (see review by Simpson, 1989). While results from Pioneer 10 have shown that the solar wind flow and the solar modulation of cosmic rays persist at least to 56 AU from the sun (Lopate and Simpson, 1993), at some large radius from the sun, estimated by pressure balance with the interstellar medium to be of order ~100 AU (*eg.* Axford 1990), it is believed that the solar wind must undergo a shock transition to subsonic flow. After passing through a transition region where it may be further decelerated by external pressures, the shocked, subsonic solar wind must eventually be entrained with the interstellar medium which flows past the heliosphere at a velocity of ~25 km/s (Bertin *et al.*, 1993; Witte et al., 1993). While at its average velocity of ~400 km/s near the ecliptic the solar wind propagates from the sun to 100 AU in ~1.2 years, at ~25 km/s it takes ~19 years for the sun to progress 100 AU through the interstellar medium. Thus the transition region most likely contains at all times solar wind emitted over more than a solar cycle. As a result, while there may be some depression of the cosmic ray intensity below interstellar levels in the transition region, the time scale for significant change in the accumulated solar wind plasma beyond the termination shock is so

TABLE 1

LONG-TERM MODULATION PROCESSES

	Term	Process	Spatial Variation
$\frac{\partial U}{\partial t} =$	$\nabla \cdot [\kappa \cdot \nabla U]$	Diffusion	$\kappa_{\perp}$ dominant near ecliptic beyond ~5 AU $\kappa_{\parallel}$ dominant near poles and inside ~5 AU Varies with B and in regions of enhanced solar wind turbulence
	$-V_{sw} \cdot \nabla U$	Convection	V_{sw} higher at high latitudes
	$-V_d \cdot \nabla U$	Drift	$V_d = V_d(r,\theta)$ throughout the heliosphere
	$+\frac{1}{3}\nabla \cdot V_{sw}(\alpha TU)$	Adiabatic Deceleration	$\nabla . V_{sw}$ larger in high velocity solar wind
	$+\Phi(r,t)$	Source	Uniform on boundary for GCR Unknown for Anomalous CR

long that the 11 year variations in modulation must be produced almost entirely in the supersonic region of the solar wind inside the termination shock.

The basic physics of solar modulation is well known and has been the subject of several recent reviews (*eg.* Kota, 1989; McKibben, 1988, 1990; Potgieter, 1994, Venkate-

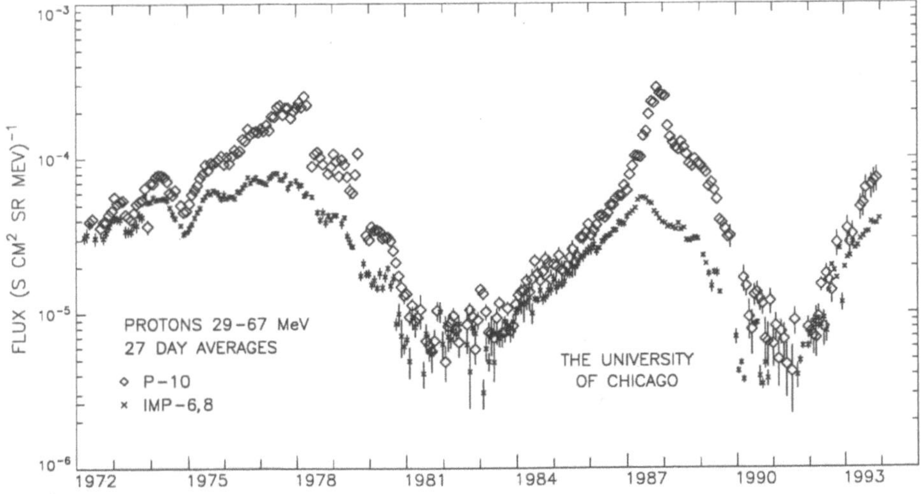

Figure 2: Modulated intensities of 29-67 MeV galactic cosmic ray protons from University of Chicago experiments on Pioneer 10 and the IMP 6 and 8 spacecraft at 1 AU.

san and Badruddin, 1990) which provide a discussion far more detailed than is possible here because of space limitations. The solar modulation results from the interaction of the cosmic rays with the interplanetary magnetic field, which has its source in the corona and is frozen into and drawn out by the outflowing solar wind plasma. The principal physical processes that describe this interaction, first identified by Parker (1965), are listed in Table 1. They consist of outward convection by the solar wind, adiabatic deceleration from the divergence of the spherically expanding solar wind, inward diffusion through the irregular interplanetary field in response to the gradient set up by the convection and deceleration, and gradient and curvature drifts as a result of the global gradients in magnetic field strength and the curvature resulting from the overall spiral pattern imposed on the magnetic field by the rotation of the sun. The modulated cosmic ray intensity at any point results from the integrated effect of all of these processes on cosmic ray nuclei and electrons as they travel inward to the point of observation from the modulation boundary of the heliosphere, which, as argued above, is most likely nearly coincident with the solar wind termination shock.

Early models of the modulation considered only diffusion, convection, and adiabatic deceleration, assumed quasi-steady conditions in a spherically symmetric heliosphere, and postulated only simple variations of the parameters (*eg.* diffusion coefficients) within the heliosphere. Since the intensity at any one point depends only on the integral of the modulation effects along the path of the particles, these models gave useful first-order results for understanding the modulation as observed at Earth. As spacecraft began to travel ever further from Earth in the heliosphere, however, such models became inadequate to describe the increasingly detailed observations of the temporal and spatial dependence of the cosmic ray intensity. Among the first discoveries when spacecraft left the vicinity of Earth were that the modulation region in the heliosphere is very big, and that changes in modulation propagate outwards at the solar wind velocity. This is illustrated in Figure 2, which shows the record of the modulated intensity of ~30-70 MeV protons as observed at Earth by the IMP spacecraft and simultaneously by Pioneer 10 as it travelled outwards through the heliosphere to its present radius of ~60 AU. The outward propagation is particularly clear for the maximum intensities in the solar minimum of 1987, which occurred about 5 months later at Pioneer 10 than at IMP, corresponding to propagation of the solar wind from 1 AU to Pioneer 10 at 42 AU at an average velocity of ~450 km/s. As a result of this propagation of modulation, the phase of the solar cycle modulation may be very different in different regions of the heliosphere. For example, in 1987 the increased modulation from the new solar cycle had already begun at earth while the strength of modulation was still declining at Pioneer 10. Therefore modulation must be treated as a dynamic, intrinsically time and space dependent process, and spherically symmetric quasi-steady models are completely inadequate to provide real understanding. From Pioneer 10 data the scale of the heliosphere is clearly much larger than ~60 AU in 1993. Recent estimates based on radial gradients of the anomalous components suggest that the boundary location lies in the range ~60 to ~90 AU and that its location may vary during the solar cycle (*eg.* Cummings *et al.*, 1993; Lopate and

Simpson, 1993). Furthermore, since the solar activity level changes significantly on time scales of order 1 year, comparable to the solar wind propagation time to the boundary, the modulation process is never in equilibrium.

While drifts had been neglected in early models, recognition of the existence of the large scale magnetic structure of the heliosphere, consisting of hemispheres of opposite magnetic polarity separated by a wavy current sheet that is near-equatorial at solar minimum, but that is deeply convoluted and extends to high latitudes at solar maximum, led to the suggestion that the gradient and curvature drifts in the interplanetary magnetic field could play an important role in guiding the propagation of cosmic rays through the heliosphere (Jokipii, Levy, and Hubbard, 1977), and thus in modulation. Since they depend upon the 3-dimensional magnetic structure of the heliosphere, inclusion of drifts requires abandonment of spherically symmetric models of modulation. In the drift mod-

TABLE 2

SELECTED EXPERIMENTAL EVIDENCE RELATED TO ROLE OF GRADIENT AND CURVATURE DRIFTS IN SOLAR MODULATION OF GALACTIC COSMIC RAYS

Measured Quantity	Prediction/Observation	Reference
Latitude Gradients for nucleons	Opposite sign in successive solar minima	
	Low rigidities	
	qA>0 ('74-'75) Positive	McKibben (1989)
	qA<0 ('86-'88) Negative	McDonald *et al.* (1992)
	High rigidities (1936-1988)	Bieber & Chen (1991)
Radial Gradients for nucleons	Smaller for qA>0 than for qA<0	
	NOT CONFIRMED	Chen & Bieber (1993)
Correlation with Current Sheet Tilt	Nucleons: Stronger for qA<0 than for qA>0	Smith (1990)
Shapes of Solar Minima	Nucleons: Broad for qA>0, Narrow for qA<0	Garcia-Munoz *et al.* (1991)
	Electrons: Opposite behavior	
	NOT CONFIRMED	Garcia-Munoz *et al.* (1991)
Nucleon/Electron Ratio	22-year cyclic variation	Garcia-Munoz *et al.* (1991)
Diurnal Anisotropies	Phase more radial for qA>0	Chen & Bieber (1993)

els, special importance is assigned to the high latitude regions where the spiral interplanetary field lines are nearly radial and to the current sheet that separates the opposing magnetic polarities. Outside of the polar and equatorial regions transport via drifts is primarily in the latitudinal direction, so that observations from *Ulysses* can be especially important for assessing the role of drifts in modulation. While their inclusion in modulation models was controversial at first, drift models make specific predictions concerning the behavior of cosmic ray nuclei and electrons during the 22 year solar magnetic cycle. In Table 2 we list a number of these specific predictions with selected references to reported experimental tests that, with only the two exceptions marked, have confirmed the predictions. Thus it is generally conceded today that realistic models of modulation must include the effects of drifts.

Most recently the importance for the solar cycle modulation of large discrete propagating disturbances in the solar wind has been recognized (*eg.* Webber and Lockwood, 1993, McDonald *et al.*, 1993, and included refs.). Discovered in the outer heliosphere by the Voyager and Pioneer spacecraft, these disturbances - termed Merged Interaction Regions (MIRs), and, when of global scale, Global MIRs (GMIRs) - result from the dynamical merging at large radii of interaction regions formed between solar wind streams of differing velocities, by solar-flare-induced shock waves, and by coronal mass ejections. Simple models of the response of the cosmic rays to decreased diffusion coefficients in the such structures have had remarkable success in reproducing the observed time-intensity behavior of the cosmic rays over the solar cycle by using the observed interplanetary magnetic field strength as input to the models (*eg.* Perko, 1993; Burlaga, McDonald, and Ness, 1993).

The current view of modulation accepted by most investigators (McDonald, Lal, and McGuire, 1993; Potgieter, 1994) is that all of the above processes are important, but that their relative importance varies throughout the solar cycle. In the period near solar minimum, when the magnetic structure of the heliosphere is particularly simple and approximates the ideal of two hemispheres of opposite magnetic polarity separated by an equatorial current sheet, drifts may play an important role in the transport of cosmic rays through the heliosphere. Even here, though, as pointed out by Jokipii and Kota (1989), irregularities in the polar fields may become important enough at large radii to significantly diminish the effectiveness of drifts. In the complex magnetic structure characteristic of solar maximum, it is likely that drifts play only a small role and that modulation is dominated by large scale disturbances in the solar wind.

During the present approach to solar minimum, observations with *Ulysses* should reflect a heliosphere where the importance of drifts for the modulation of cosmic rays is steadily increasing. For the current sign of the solar magnetic dipole (qA>0) drift models predict an increase in the intensity of positively charged particles towards high latitude (*cf.* Table 2), and such an increase was observed up to a latitude of 17° in 1974-75 (McKibben, 1989). Furthermore, even in the absence of drifts, the increase in solar wind velocity at high latitudes (*eg.* Schwenn, 1994; McComas *et al.*, 1994) and the looser winding of the spiral magnetic field lines (Forsyth, 1994) might be expected to

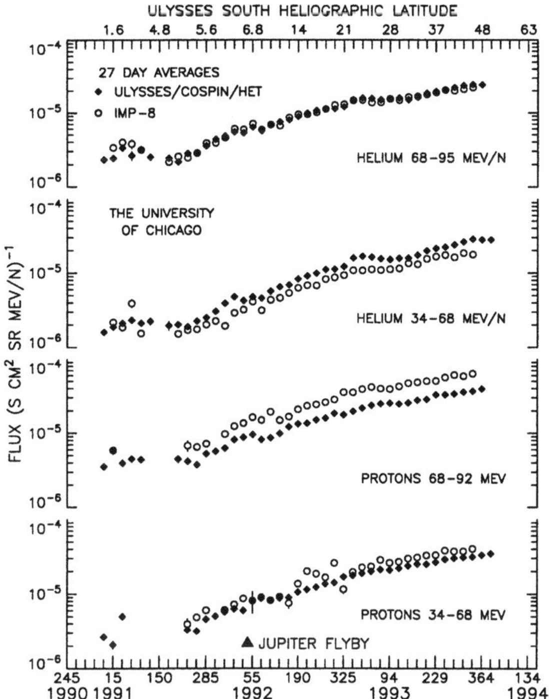

Figure 3: 27 day average fluxes of protons and helium nuclei from the *Ulysses*/COSPIN/HET and from IMP 8 in Earth orbit. Periods with significant contributions from solar energetic particles are not included.

produce significant differences in modulation at high latitudes. Jokipii and Kota (1994) and Potgieter and Haasbroek (1994) discuss model predictions in relation to *Ulysses* observations elsewhere in these proceedings.

3. The *Ulysses* Observations of Modulation

After its launch in October of 1990, *Ulysses* flew out to Jupiter in a period of high solar activity during which the modulated intensity of cosmic rays was the lowest ever recorded by the Climax neutron monitor (*cf.* Fig. 1). Since its flyby of Jupiter in February 1992, it has been climbing to steadily higher southern heliographic latitudes while slowly returning inwards towards the sun. At the same time solar activity has been decreasing rapidly to the low levels characteristic of solar minimum, and the cosmic ray intensity has been increasing correspondingly. As of the middle of April 1994, *Ulysses* was at a radius of ~3.3 AU at a latitude of ~60°S. Of the charged particle instruments on *Ulysses*, the best suited for studies of the solar modulation of the cosmic rays are the High Energy Telescope (HET) and the Kiel Electron Telescope (KET), which are part of the COSPIN Collaboration experiment on the spacecraft. These instruments have been described by Simpson *et al.* (1992). Results from the KET are reported separately in these proceedings (Heber *et al.*, 1994; Kunow *et al.*, 1994) and are entirely consistent with the results from the HET described below.

Since modulation has been changing in time at the same time that *Ulysses* has been rising in latitude, to separate temporal from spatial variations requires a reference detector that remains at a constant location in the heliosphere and that measures cosmic rays in energy ranges comparable to those available from *Ulysses*. For this study we have used the University of Chicago telescope on IMP 8 (Garcia-Munoz, Mason, and Simpson, 1975) in Earth orbit, which provides measurements very well matched to those available from the *Ulysses* HET.

An overview of the modulated intensity measured for two energy ranges each of protons and helium nuclei by IMP 8 and *Ulysses* since the launch of *Ulysses* is shown in Figure 3. 27-day averages have been used to remove to first order the effects of corotating features in the modulation such as those discussed elsewhere in this volume by Kunow *et al.* (1994) and McKibben *et al.* (1994), and thus to concentrate on the long-term modulation that is not dependent on strictly local conditions. These data have also been selected to remove contamination by solar energetic particles. The requirement that contamination be negligible at both IMP and Ulysses, coupled with the high level of solar activity during the early months of the Ulysses mission accounts for the very sparse measurements obtained from 1 to ~4 AU while Ulysses was outbound to Jupiter. Since data for each particle species and energy range were selected individually, the set of available measurements is different for each energy range and species. The lower energy helium interval (34-68 MeV/n) may contain some contribution from the anomalous helium component (*cf.* Klecker, 1994), but the higher energy helium and protons in both energy intervals consist only of galactic cosmic ray nuclei. The approximately constant offsets between fluxes measured by IMP and *Ulysses* in several of the energy intervals result from systematic errors of so-far-unknown origin in calculating fluxes from the two spacecraft. Efforts are continuing to identify and remove these errors. Until they are understood the results presented below should be considered preliminary since they

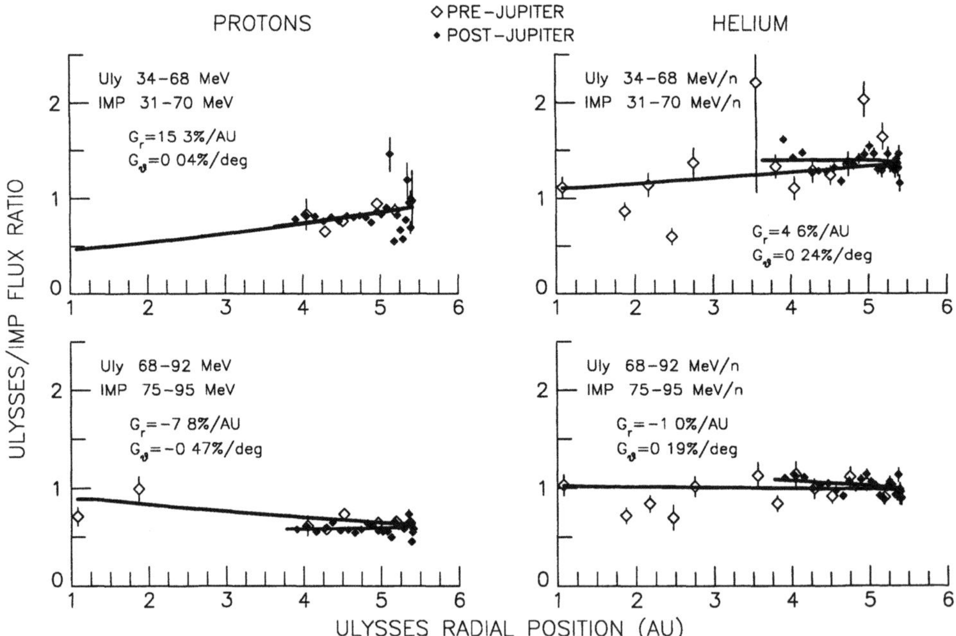

Figure 4: *Ulysses*/IMP flux ratios as a function of *Ulysses* radial position. Open symbols represent measurements near the ecliptic before Jupiter flyby. Solid symbols are used for measurements after Jupiter flyby when *Ulysses* was leaving the ecliptic plane. Solid lines are best fits to functions of the form Ratio = $A*\exp(G_r*R+G_\theta*|\theta|)$ where r and θ are the radius and heliographic latitude of *Ulysses*

rely on the assumption that the systematic errors are independent of time and intensity level.

The clear impression from Figure 3 is that the modulation at *Ulysses* has been almost identical to that observed at IMP 8 throughout the run of observations presented, which extend to latitudes for *Ulysses* of ~52° S. A more direct way of visualizing the dependence of the ratios of the fluxes at *Ulysses* and IMP on the position of *Ulysses* is shown in Figure 4, where the ratios of the fluxes presented in Figure 3 are plotted as a function of *Ulysses* radial position. During its flight from Earth to Jupiter, *Ulysses* did not depart significantly from the ecliptic plane, and thus any variation of the ratio with position is a measure of the radial gradient. After its Jupiter flyby, *Ulysses* left the ecliptic plane while coming back inward towards the sun. Thus the signature of a latitude gradient, assuming a time-independent radial gradient, would be a departure of the post-Jupiter ratio vs. radius relation from that measured prior to Jupiter flyby. In the figure, solid lines represent the unweighted best fits to functions of the form

$$\text{Ratio} = A^*\exp(G_r{}^*R + G_\theta{}^* |\theta|) \quad (1)$$

where R and θ represent the heliocentric radius and heliographic latitude of *Ulysses*, G_r and G_θ represent the radial and latitudinal gradients, and A is a relative normalization determined by the fitting procedure. Clearly there is as yet no compelling evidence for significant latitudinal gradients for protons and helium in these energy ranges. Even radial gradients are uncertain given the high level of solar activity and consequent scarcity of measurements during *Ulysses* outbound trajectory to Jupiter.

Less affected by solar activity are heavy ion intensities that are measured by both *Ulysses* and IMP 8, although in less well-matched energy intervals. Figure 5 shows daily averages of heavy ion ($Z \geq 6$) counting rate ratios for a differential energy window centered around 100 MeV/n (for Carbon) and for the integral intensity corresponding to

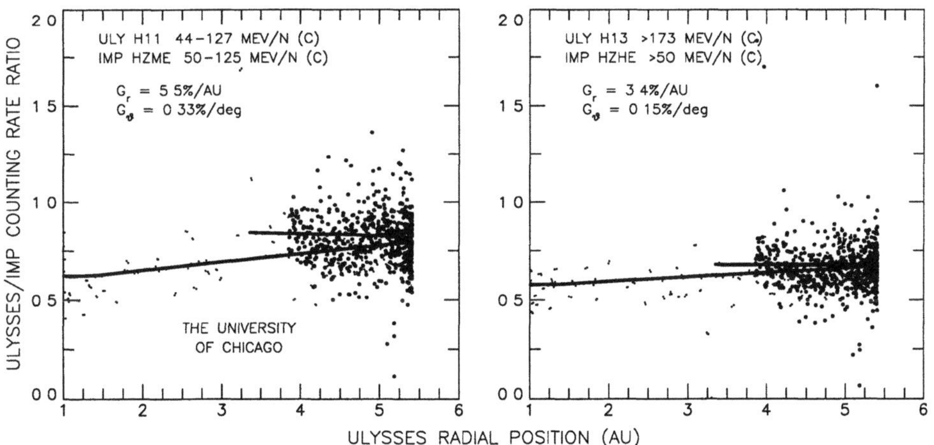

Figure 5: Daily average ratios of heavy ion counting rates from IMP 8 (HZME and HZHE) and from the *Ulysses*/COSPIN/HET (H11 and H13) through the end of 1993. No quiet selection has been performed, but best fit gradients have been derived using an iterative procedure to delete outlying points from the fit. Heavy dots are used for post-Jupiter flyby observations.

TABLE 3

SUMMARY OF LATITUDE GRADIENT MEASUREMENTS

	1975-76 (qA>0) *McKibben, 1989*		1985-86 (qA<0) *Cummings et al., 1987*		1992-94 (qA>0) *This work*	
SPECIES	ENERGY (MeV/n)	G_θ (%/degree)	ENERGY (MeV/n)	G_θ (%/degree)	ENERGY (MeV/n)	G_θ (%/degree)
Protons	29-67	0.2±0.1			34-68	-0.04±0.4
					68-92	-0.47±0.16
			130-210	-0.9±0.1		
	>70	0.01±0.04	>70	-0.34±0.08		
Helium	11-20	1.5±0.2	10-22	-2.2±0.7		
	29-67	1.2±0.1	30-56	-1.6±0.3	34-68	0.24±0.12
					68-92	0.19±0.10
Hi-Z					44-127 (C)	0.33±0.04
(Z>5)					>127 (C)	0.15±0.03
S/C	P-11, IMP 8		P-10, Vgr-1/2		*Ulysses*, IMP 8	
Latitude	$-7 \le \theta \le 16°$		$0° \le \theta \le 27°$		$-58° \le \theta \le 6°$	
Radius	R≅4 AU		R≅25 AU		R≅4-5 AU	

all particles above a certain threshold energy (50 MeV/n for Carbon at IMP, 173 MeV/n at *Ulysses*). Also shown are the best fits to a relation of the form (1). No quiet time selection has been done for these data, but the fits were performed in an iterative manner to successively discard outlying points until a stable fit was achieved. Observations in this figure extend through the end of 1993 when *Ulysses* was at 48°S latitude. The low energy (~100 MeV/n) heavy ions in this figure provide the strongest evidence we have so far from *Ulysses* data for the existence of a latitude effect on the galactic cosmic rays.

A summary of the results of fits to the *Ulysses* data for all particle species measured (energies correspond to the Ulysses energy ranges) is given in Table 3, together with measurements reported over smaller latitudinal ranges near previous solar minima in 1974-75 and 1985-86. The current results for latitudinal gradients are much smaller than previous measurements. Based on the best information available last year, Potgieter and Haasbroek (1993) made predictions for the variations of spatial origin that would be observed during the *Ulysses* mission for protons of 220 MeV and 1.2 GeV. 220 MeV protons have a rigidity of 679 MV, which is comparable to that of the ~100 MeV/n C (889 MV) included in the lower energy range in Figure 5.

Figure 6 shows the *Ulysses* observations compared to the prediction for 220 MeV protons, where the apparent "non-local radial gradient", G_r, is computed as

$$G_r = [\ln(\textit{Ulysses}/\text{IMP}) - \ln(A)]/\Delta r$$

where the relative normalization, A, was determined from the first 30 days of observations available from *Ulysses* when the spacecraft was very near 1 AU, and Δr is the radial interval from IMP to *Ulysses*. While there is a large scatter in the data, by the end of 1993 it appears that there was a clear discrepancy between the observations and the prediction. It is also clear, however, that the major effect is expected in 1994. Close examination shows a slight upturn in the apparent radial gradient at the end of 1993, perhaps corresponding to the beginning, somewhat late, of the expected effect.

4. Summary

As *Ulysses* began its climb to high solar latitudes in 1992 both previous experience in earlier solar minima and the current ideas concerning modulation would have predicted with some confidence that latitudinal effects in the modulation would produce intensity variations of order 1%/degree at low energies (<~100 MeV/n). This prediction has not been borne out by *Ulysses* observations up to beyond 50°S latitude. Why this should be so is not clear. The most clearly measured latitude effects in the past were those measured by Voyager in the outer heliosphere (R ≅ 25 AU) in 1985-86. However Pioneer 11 also appeared to observe a clear effect in 1974-75 at R ≅ 4 AU, in the same radial range as Ulysses. A more significant difference may be the phase of the solar cycle, since Voyager measurements were made in a period spanning solar minimum and Pioneer measurements were made in a period of extended near-minimum modulation, , while *Ulysses* is making measurements in a period of rapidly decreasing modulation.

Ulysses will reach its maximum latitude of ~80°S in September of 1994, and will execute another high latitude pass in the northern hemisphere in 1995. Current modulation models predict large effects at the highest latitudes, even though we will not yet have reached the minimum level of solar activity in this cycle. Observations during the next year will provide a strong test of our understanding of the

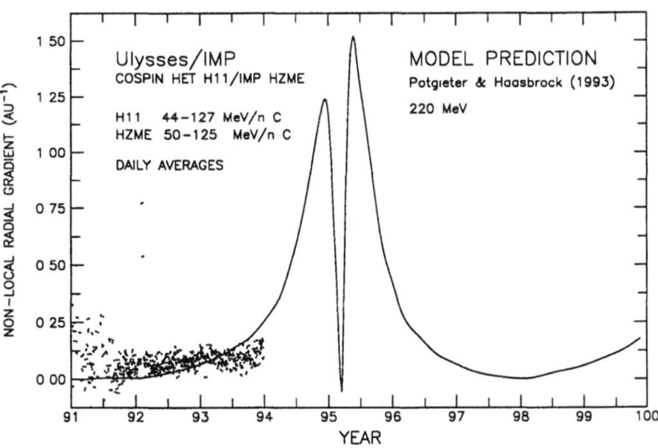

Figure 6: Comparison of apparent radial gradients for heavy nuclei (~900 MV rigidity) with model predictions for the *Ulysses* orbit for 220 MeV protons (~700 MV).

three-dimensional aspects of modulation and of the propagation of charged particles through the heliosphere. Evidence so far suggests that modulation in the inner heliosphere is far more spherically symmetric than expected from current modulation models.

Acknowledgments: This work was supported in part by NASA/JPL Contract JPL-955432, by NSF Grant ATM92-1522, and by NASA Grants NAG5-706 and NAG2-380

References

Axford, W. I.: 1990, in *Physics of the Outer Heliosphere, (Proceedings of the 1st COSPAR Colloquium)*, S. Grzedzielski and D. E. Page (eds), Pergamon Press, p. 7.
Bertin, P., *et al.*: 1993, *J. Geophys. Res.*, **98**, 15193.
Bieber, J. W., and Chen, J.: 1991, *Astrophys. J.*, **372**, 301.
Burlaga, L. F., McDonald, F. B., and Ness, N. F., 1993:, *J. Geophys. Res.*, **98**, 1.
Chen, J. and Bieber, J. W.: 1993, *Astrophys. J.*, **405**, 375.
Cummings, A. C., Stone, E. C., and Webber, W. R.: 1987, *Geophys. Res. Lett.*, **14**, 174.
Cummings, A. C., Stone, E. C., and Webber, W. R.: 1993, *J. Geophys. Res.*, **98**,15165.
Forbush, S. E.: 1954,*J. Geophys. Res.*, **59**, 525.
Forsyth, R. J.: 1994, these proceedings.
Garcia-Munoz, M., Mason, G. M., and Simpson, J. A.: 1975, *Ap. J. (Letters)*, **201**,L145.
Garcia-Munoz, M., *et al.*: 1991, *Proc. 22nd Int'l Cosmic Ray Conf. (Dublin)*, **3**, 497.
Heber, B., *et al.*: 1994, these proceedings.
Jokipii, J. R., and Kota, J.: 1989, *Geophys. Res. Letters*, **16**, 1.
Jokipii, J. R., and Kota, J.: 1994, these proceedings.
Jokipii, J. R., Levy, E. H., and Hubbard, W. B.: 1977, *Astrophys. J.*, **213**, 861.
Klecker, B.: 1994, these proceedings.
Kota, J: 1989, in *Physics of the Outer Heliosphere, (Proceedings of the 1st COSPAR Colloquium)*, S. Grzedzielski and D. E. Page (eds), Pergamon Press, p. 119.
Kunow, H., *et al.*: 1994, these proceedings.
Lopate, C., and Simpson, J. A.:, 1993, *Proc. 23rd Int'l Cosmic Ray Conf. (Calgary)*, **3**, 415.
McComas, D. J., *et al.*: 1994, these proceedings.
McDonald, F. B., Moraal, H., Reinecke, J. P. L., Lal, N., and McGuire, R. E.: 1992, *J. Geophys. Res.*, **97**, 1557.
McDonald, F. B., Lal, N., and McGuire, R.E.: 1993, *J. Geophys. Res.*, **98**, 1243.
McKibben, R. B.: 1988, in *Solar Wind 6*, NCAR Tech. Note-306+Proc., **2**, 615.
McKibben, R. B.: 1989, *J. Geophys. Res.*, **94**, 17021.
McKibben, R. B.: 1990, in *Physics of the Outer Heliosphere, (Proceedings of the 1st COSPAR Colloquium)*, S. Grzedzielski and D. E. Page (eds), Pergamon Press, p. 107.
McKibben, R. B., *et al.*: 1994, these proceedings.
Meyer, P., Parker, E. N., and Simpson, J. A.: 1956,*Phys. Rev.*, **104**, 768.
Parker, E. N.: 1965, *Planet. Sp. Sci.*, **13**, 9.
Perko, J. S.: 1993, *J. Geophys. Res.*, **98**,19027.
Potgieter, M. S.: 1994, *Proc. 23rd Int'l Cosmic Ray Conf. (Calgary)*, to be published.
Potgieter, M. S., and Haasbroek, L. J.: 1993, *Proc. 23rd Int'l Cosmic Ray Conf. (Calgary)*, **3**, 457.
Potgieter, M. S., and Haasbroek, L. J.: 1994, these proceedings.
Schwenn, R: 1994, these proceedings.
Simpson, J. A.: 1989, *Adv. in Sp. Res.*, **9** (#4), 5.
Simpson, J. A., *et al.*: 1992, *Astron. Astrophys.*, **92**,365.
Smith, E. J.,: 1990, *J. Geophys. Res.*, **95**, 18731.
Venkatesan, D. and Badruddin: 1990, *Sp. Sci. Rev.*, **52**, 121.
Webber,. W. R. and Lockwood, J. A.: 1993, *J. Geophys. Res.*, **98**, 21095.
Witte, M. et al.: 1993, *Adv. in Sp. Res.* **13** (No. 6), 121.

THREE-DIMENSIONAL COSMIC-RAY SIMULATIONS: HELIOGRAPHIC LATITUDE AND CURRENT-SHEET TILT

J.R. JOKIPII and J. KÓTA
The University of Arizona, Tucson, AZ 85721, U.S.A.

Abstract. We present the results from a study of the variations of the cosmic-ray intensity with time, heliographic latitude, and longitude, and for varying interplanetary conditions, using our three-dimensional, time-dependent computer code for cosmic-ray transport in the heliosphere. Our code also produces a solar-wind and interplanetary magnetic field (IMF) configuration which is compared with observations. Because of the fully three-dimensional nature of the model calculations, we are able to model time variations which would be expected to be observed along Ulysses's trajectory as it moves to high latitudes. In particular we can model the approximately 13- and 26-day solar-rotation induced variations in cosmic rays, solar wind and IMF, as a function of increasing heliographic latitude, as one moves poleward of the interplanetary current sheet. Our preliminary model results seem to be in general form quite similar to published data, but depend on the physical parameters used such as cosmic-ray diffusion coefficients, boundary conditions, and the nature of the solar wind and IMF and current sheet.

Key words: Cosmic-Rays – Heliosphere – Modulation

1. Introduction.

The modulation of cosmic rays by the solar wind involves a number of physical effects - diffusion, convection, guiding-center drifts and energy change. Global modelling suggests that the drift terms play the most important role in the large-scale, eleven-year variation (see, e.g., Kóta and Jokipii, 1983, Potgieter, et al, 1987, and Jokipii 1985, 1988, Kóta, 1990, Jokipii and Kóta, 1991). Notable successes have been the prediction of various observed 22-year solar magnetic cycle effects and the correlation between the magnetic field and rate of change of the cosmic-ray intensity associated with co-rotating interaction regions.

In the present paper we examine the latitudinal and longitudinal behavior of a global, three-dimensional, time-dependent model which contains a realistic, co-rotating solar-wind configuration. Emphasis is in the 13-day and 26-day variations and how they vary with latitude and current-sheet tilt.

2. The Simulation Model

The model is the three-dimensional, time-dependent model used by Kóta and Jokipii (1991, see also Kóta and Jokipii, 1983 and Jokipii and Kóta, 1989) We use Parker's (1965) general equation which includes convection, anisotropic diffusion, drifts, and both adiabatic cooling and shock acceler-

Space Science Reviews **72**: 379–384.
© 1995 *Kluwer Academic Publishers.*

ation. An external boundary condition is imposed, which is taken to be a power law in total energy with an exponent -2.65.

The large-scale magnetic field (e.g., Smith, et al, 1985) consists of a Parker Archimedean spiral, with the field directed in one direction in the northern heliospheric hemisphere and in the opposite direction in the southern. The fields are separated by a thin current sheet. At sunspot minimum, the current sheet is nearly equatorial. The structure near sunspot maximum is not as simple.

Observations (e.g., Newkirk and Fisk, 1985; Gazis et al, 1988, Bame, et al, 1993) show that the solar-wind velocity increases with heliomagnetic latitude λ_{mag}, from 350-400 km/sec near the current sheet, to some 800 km/sec for $\lambda_{mag} \gtrsim 20^o$. Solar rotation results in regions of fast solar wind overtaking previously-emitted slower solar wind, which then form CIR's, or shocks.

Here we assume that the current sheet at the sun is a plane inclined at an angle α to the equatorial plane. The wind is radial near the sun and varies as $V_w = V_{pole} - (V_{pole} - V_{eq})/[1 + (sin(\lambda_{mag})/sin(\lambda_{1/2}))^2]$, where the V_{pole}, V_{eq} and $\lambda_{1/2}$ are as in the previous paragraph. Since transverse stresses are small, we neglect the transverse velocities and solve the radial momentum equation with a frozen-in magnetic field, including gas pressure and magnetic forces. Cosmic rays are treated as test-particles and the calculation is followed until it settles into a corotating pattern.

The structure beyond the solar-wind termination shock is uncertain, and we simply assume that there is a shock transition at some specified radius R_{sh}. Beyond this the radial velocity drops by 1/4 and falls off as $1/r^2$, out to an outer, spherical boundary where the cosmic-ray distribution function takes on the specified "interstellar" value.

We use the same form for the diffusion tensor as that in our previous modelling efforts, which is $\kappa_\parallel = \kappa_0 p^{1/2} \beta B_E/B$ and $\kappa_\perp = \eta\kappa_\parallel$, where B is the local magnetic field and B_E is the magnetic field at the Earth. If the momentum p is expressed in GV/c then κ_0 is typically 10^{22} to yield $\kappa_\parallel$ in cm^2/sec. The normalization of $\kappa_\parallel$ by the magnetic field is to scale the parallel mean free path with the local particle gyro-radius.

3. Discussion of Results

First we calculate the solar wind and frozen in magnetic field from a rotating tilted dipole as described in Kóta (1992). Figure 1 shows snapshots illustrating the typical evolution of the solar wind speed, plasma density, and temperature for a tilt angle of 30^0 along two different radial lines. At 10^0 latitude, close to the heliospheric equatorial plane (left panel) there are two velocity minima in each solar rotation which results in the formation of two CIRs. At higher latitudes (30^0, right panel), on the other hand, there is

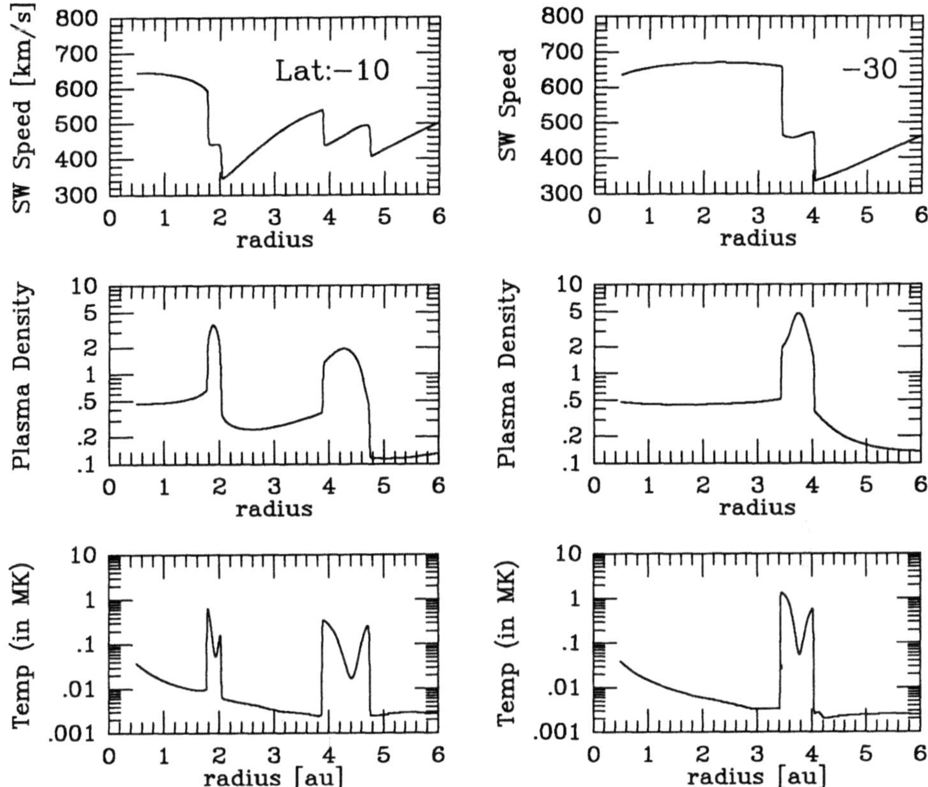

Fig. 1. A typical snapshot of solar wind speed, normalized plasma density (nr^2, in arbitrary units), and temperature as obtained for a tilt angle of 30^0 along two radial lines, at latitude 10^0 (left) and 30^0 (right).

only one single CIR leading to a 26-day wave (instead of the 13-day variation close to the equator). Shocks form typically at 2-3 a.u. from the sun, this distance is expected to increase as one moves toward higher latitudes. The forward and reverse shocks can clearly be identified.

A slowly moving observer, like Ulysses, will see quasi-periodic 26-day variations due to the rotation of the sun. During the period from 1992 to 1994 Ulysses moves poleward while the tilt angle of the current sheet decreases in the declining phase of solar activity. Figure 2. (left panel) shows a numerical simulation for the solar wind speed and magnetic field along the trajectory of the spacecraft as obtained from our tilted dipole model with a tilt angle slowly changing from 45^0 in 1992 to 20^0 in 1994 (Hoeksema, private communication). Since the change of tilt is slow, our steady-state solutions should apply as reasonable approximations. These results indicate that one may expect large-amplitude well-defined 26-day waves in 1993 when the latitude of Ulysses is comparable to the maximum excursion of the current sheet.

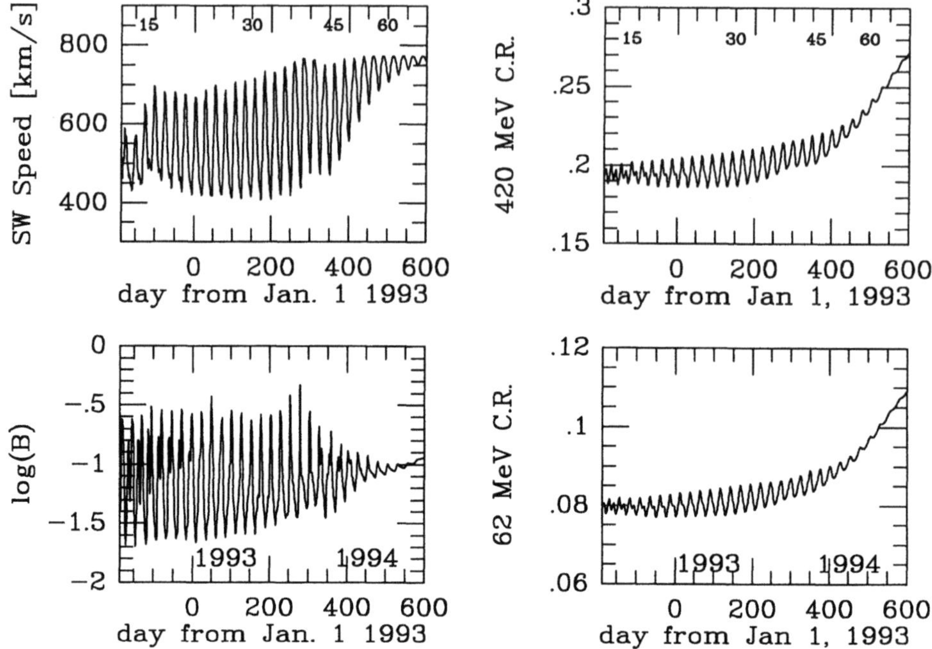

Fig. 2. Synthesized solar wind speed and magnetic field strength (left) along the trajectory of Ulysses obtained for the period of mid-1992 to mid-1994. The tilt angle is slowly changed from 45^0 in 1992 to 20^0 in 1994. The right panels show the predicted variation of 62 MeV and 420 MeV cosmic-ray proton fluxes. Small numbers on the top indicate the latitude of Ulysses.

In 1994, when Ulysses is already constantly below the sheet, and moves poleward, the wind speed should level out at a constant high value. Because the variation in the wind speed becomes smaller, CIR-s will form farther and farther out, thus steady recurrent shocks are not expected to be seen.

For mid-1992 and before, when the tilt angle was large and Ulysses was still close to the heliospheric equator one may expect 13-day waves to appear. In reality, of course, this clean 13-day pattern will seriously be modified by asymmetries and transients at the sun. The two peaks of a 26-day rotation become asymmetric as the spacecraft moves away from the equatorial plane and gradually turn into a single peak per rotation structure.

The simple tilted-dipole model with a plane current sheet is overly simplified and would predict high average speeds at the equator for the period of high solar activity when the tilt angle is large. To account for the warps and complexity of the current sheet during solar maximum, we included multiple current sheets for the cases of large ($> 40^0$) tilt angles. As expected, and seen in Figure 2, this modification yields lower average speeds and smaller 26-day and 13-day waves for the period of mid 1992 and earlier.

The 3-D structure of solar wind and magnetic fields provide the background for the transport of cosmic rays. In the present solar cycle the IMF is directed outward above the current sheet and inward below it ($A > 0$). Galactic cosmic-ray ions drift into the heliosphere primarily from the polar regions thus they are relatively insensitive to the structure of the current sheet (Kóta and Jokipii, 1983). The cosmic-ray flux should increase toward the poles, the rate of this increase should depend on the effectiveness of drifts and cross-field diffusion; smaller perpendicular diffusion results in larger latitudinal gradients. Furthermore cosmic-ray intensities at the poles may also be greatly influenced by the structure of the polar magnetic field. If the field is weak and radial at large heliospheric distances as predicted by the regular Parker-spiral field then cosmic rays may have an easy access to the inner heliosphere, resulting in little modulation at the poles. A modification of the polar field by transverse components as proposed by Jokipii and Kóta (1989) may, on the other hand, impede cosmic-ray transport in the polar regions. To account for these transverse fields, we reduced the diffusion coefficients at the poles. As a consequence, drifts, instead of being concentrated at the poles, will be more evenly distributed over a larger area.

Figure 2. (right panel) shows the simulated variations of 62 MeV and 420 MeV proton fluxes predicted (for $\eta = 0.05$) along the trajectory of Ulysses assuming slowly decreasing tilt angle from 1992 to 1994. The average intensity tends to increase as the spacecraft moves poleward. Enhanced 26-day variations are predicted for the year of 1993 when Ulysses is close to the maximum excursion of the current sheet. The short-term variations of cosmic-ray flux is primarily due to the increased magnetic field and smaller diffusion in the CIR-s (Kóta and Jokipii, 1991). The variation of magnetic field strength and cosmic-ray flux are not in phase, the strength of the magnetic field correlates with the rate of decrease in the cosmic-ray flux. The rate of the poleward increase of cosmic-ray flux, as well as the magnitude of the predicted 26-day and 13-day variations, will strongly depend on the perpendicular diffusion. Reducing the value of $\kappa_{\perp}/\kappa_{\|}$ from 0.05 (Figure 2.) to 0.02 would result in about a twofold increase in both the global latitudinal gradient and the amplitude of the 26-day waves.

4. Summary and Conclusions

We have presented a 3-D numerical simulation for the solar wind, IMF, and cosmic-ray fluxes along the trajectory of Ulysses from 1992 to 1994. The simple rotating tilted-dipole model has been somewhat modified to account for the warps and complexity of the current sheet during high solar activity. The simulated results reproduce a number of the features observed (Smith et al, 1993). In addition to the global changes due to the poleward trajectory of the spacecraft, quasi-periodic 13-day and 26-day variations are predicted.

The 13-day (2 peaks per rotation) variation, which is characteristic for the equatorial region, turns gradually into a 26-day pattern as the spacecraft moves away from the equator. The 26-day waves are most pronounced in the period when the spacecraft is close to the maximum excursion of the current sheet and are expected to diminish afterwards.

A recurrent 26-day wave activity is predicted in the cosmic-ray fluxes. These waves are associated with CIR-s in the solar wind, and their amplitudes should depend on the cross-field diffusion. This may offer a diagnostic tool for estimating the perpendicular diffusion. More numerical modelling is required before a quantitative conclusion can be drawn.

Acknowledgements

The authors are grateful to Dr. T. Hoeksema for providing data on the tilt angles. This work was supported by NASA/JPL as part of in interdisciplinary investigation on the Ulysses mission, JPL contract 959376. In addition, this work was also supported by the NSF under Grant ATM-9307046 and by NASA under Grants NAGW-2549 (JK) and NAGW 1931 (JRJ).

References

Bame, S.J., V.E. Goldstein, J.T. Gosling, J.W. Harvey, D.J. McComas, M. Neugebauer and J.L. Phillips: 1993, *Geophys. Res. Let* **20**, 2323

Gazis P.R., A. Barnes and A.J. Lazarus: 1988, 'Intercomparison of Voyager and Pioneer Plasma Observations' in Proceedings of the Sixth International Solar Wind Conference, ed(s)., *V. J. Pizzo, T. E. Holzer, and D. E. Sime*, NCAR:Boulder, II-563

Jokipii, J.R.: 1985, 'Effects of Three-Dimensional Heliospheric Structures on Cosmic-Ray Modulation' in The Sun and Heliosphere in Three Dimensions, ed(s)., *R. Marsden*, Reidel:Dordrecht, 375

Jokipii, J.R.: 1988, *Adv. Space Res.9* **12**, 105

Jokipii, J.R. and J. Kóta: 1989, *Geophys. Res. Lett.* **16**, 1

Kóta, J.: 1992, 'A Numerical Model of the Large-Scale Solar Wind in the Outer Heliosphere' in E. Marsch and R. Schwenn, ed(s)., *Solar Wind Seven*, ,, Pergamon:Oxford 205

Kóta, J.: 1990, 'Diffusion, Drifts, and Modulation of Galactic Cosmic Rays in the Heliosphere' in S. Grzedzielski and D.E. Page, ed(s)., *Physics of the Outer Heliosphere*, Pergamon:Oxford, 119

Jokipii, J.R., and J. Kóta: 1991, *Geophys. Res. Lett.* **18**, 1797

Kóta, J. and J.R. Jokipii: 1983, *Astrophys. J.* **265**, 573

Kóta, J. and J.R. Jokipii: 1991, *Geophys. Res. Lett* **18**, 1797

Newkirk, G. and L.A. Fisk: 1985, *J. Geophys. Res.* **90**, 3391

Parker, E.N.: 1965, *Planet. Space Sci.* **13**, 9

Potgieter, M.S., J.A. LeRoux and R.A. Burger: 1987, *Proc. 20th Intl Cosmic Ray Conf.* **3**, 287

Smith, E.J.,J.A. Slavin, and B.T. Thomas: 1985, 'The heliospheric current sheet: 3-dimensional structure and solar-cycle changes' in The Sun and Heliosphere in Three Dimensions, ed(s)., *R. Marsden*, Reidel:Dordrecht, 375

Smith, E.J. M.Neugebauer, A. Balogh, S.J. Bame, G. Erdős, R. J. Forsyth, B.E. Goldstein, J.L. Phillips , and B. Tsurutani: 1993, *Geophys. Res. Lett.* **20**, 2327

THE MODULATION OF COSMIC RAYS IN THE HIGH LATITUDE HELIOSPHERE: A COMPUTER SIMULATION

L. J. HAASBROEK and M. S. POTGIETER
Space Research Unit, Department of Physics, Potchefstroom University for CHE, 2520 Potchefstroom, South Africa

Abstract. Modulation models based on the numerical solution of Parker's transport equation for galactic cosmic rays in the heliosphere make clear predictions about modulation in the high latitude heliosphere. However, for these predictions certain assumptions have to be made, for example, what the heliospheric magnetic field (HMF) looks like above the solar poles and what the spatial dependence of the diffusion coefficients are. For this presentation the general predictions of a standard drift model for the modulation of cosmic rays in the high latitude heliosphere, in particular predictions for the Ulysses trajectory, are discussed and critically reviewed. Preliminary results from Ulysses show a significant increase in the solar wind speed towards higher latitudes. The effects of this strong latitudinal dependence together with different modifications of the HMF at these high latitudes on the apparently too large diffusion and drifts predicted by current models are also shown.

1. Introduction

The network of interplanetary spacecraft has contributed tremendously towards a better understanding of modulation mechanisms and modulation in general during the past decades. Voyager 1, currently at ~25 AU above the ecliptic, made significant contributions towards our understanding of mid-heliolatitude modulation phenomena. Other spacecraft with mainly equatorially bounded trajectories gave precious insight concerning this region of the heliosphere. To advance our understanding of the high latitude heliosphere, the Ulysses spacecraft started its journey in October 1990 which took it in a gravity assist manoeuvre around Jupiter towards higher latitudes where it will traverse the southern high latitude regions of the heliosphere at ~2.2 AU during mid 1994 and the northern regions at nearly the same radial distance a year later. Ulysses will spend a total of ~8 months at latitudes above 70° and promises with the aid of a variety of cosmic-ray related instruments to put these regions into a new perspective. The purpose of this paper is to show *what current modulation models predict for the high latitude heliosphere and what the effects of more realistic representations of the HMF and solar wind might have in these regions* – first, for modulation in general, and second, along the Ulysses trajectory.

2. The Modulation Model

A two-dimensional, steady-state drift model based on the numerical solution of the Parker transport equation (TPE) and incorporating a simulated wavy

Space Science Reviews **72**: 385–390.
© 1995 *Kluwer Academic Publishers.*

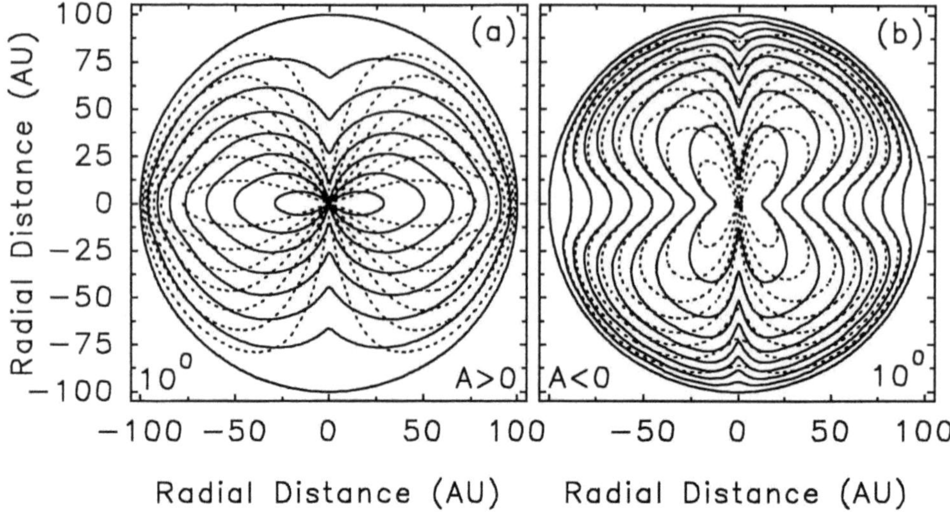

Fig. 1. The intensity contours for 220 MeV protons during (a) A > 0 and (b) A < 0 polarity epochs with a 10° tilt angle. The dashed lines represent the reference solution with a standard Parker HMF and the solid lines a JK-modified HMF (see text). The intensity decreases from a 100% in steps of 10% as an observer moves from the outer boundary at 100 AU towards the Sun.

neutral sheet (HNS) was used (Potgieter and Moraal, 1985). The diffusion coefficients parallel and perpendicular to the background HMF, and the "drift coefficient" were chosen as

$$K_{\|} = K_0\,\beta(K_P)\left(\frac{B_0}{B}\right), \qquad K_{\perp} = 0.01\,K_{\|}, \qquad K_T = \frac{\beta P}{3B} \tag{1}$$

with β the ratio of particle speed to that of light, P the particle rigidity with $K_P = (P/P_0)$ when $P \geq 0.4$ GV and $K_P = (0.4\ \mathrm{GV}/P_0)$ when $P < 0.4$ GV; $P_0 = 1$ GV; $B_0 = 6$ nT which is the HMF magnitude at Earth. Here $K_0 = 2.2 \times 10^{22}\ \mathrm{cm^2.s^{-1}}$. The modulation boundary was assumed as 100 AU. For the reference solutions of the TPE shown in Fig. 1 and 2, B represents the HMF magnitude as given by the *standard* Parker HMF but with a latitude dependent solar wind speed; in Fig. 3, $V = 450\ \mathrm{km.s^{-1}}$. The dashed lines in Fig. 1 show the intensity contours for 220 MeV protons and represent what may be considered a standard drift solution. Note (i) the extremely effective diffusion and drifts occurring inwards through the polar regions during the A > 0 cycle and (ii) the effects of the wavy HNS in the equatorial regions during the A < 0 cycle.

3. Model Modifications

To obtain a more "realistic" representation of two major modulation agents, i.e. the HMF and solar wind, we used three different modifications to the

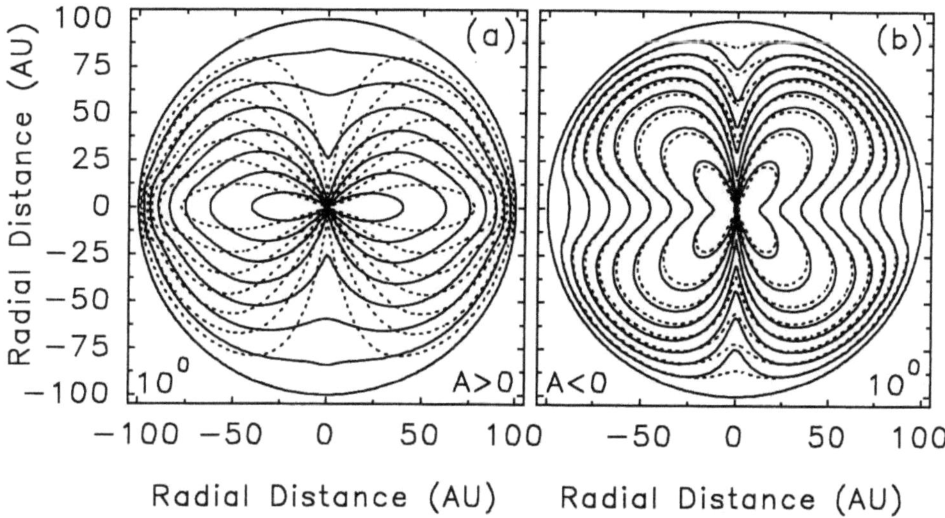

Fig. 2. The intensity contours for 220 MeV protons resulting from the SBM modification (solid lines) compared with the standard solution (dashed lines). The allocation of contours are the same as before.

standard modulation model. The first and second being modified representations of the HMF and the third a latitudinal dependent solar wind.

Jokipii and Kóta modification (JKM). Jokipii and Kóta (1989) suggested that the magnetic field at high latitudes might be significantly larger than the Parker HMF due to randomly orientated transverse magnetic field components present near the surface of the Sun. This component causes the HMF magnitude to become proportional to $1/r$ whereas the Parker HMF magnitude is proportional to $1/r^2$, implying a larger HMF magnitude at high latitudes, with $r \gg 1$ AU, but with the average direction the same as for the Parker HMF.

Smith and Bieber modification (SBM). Smith and Bieber (1991) examined measurements of the HMF made by various interplanetary spacecraft and found that the spiral structure of the field at Earth is nearly always more tightly wound that what the Parker theory predicts. To address this, they suggested that differential solar rotation might justify a modified expression for the standard HMF spiral angle ψ. Since $\tan\psi \propto \sin\theta$, the Parker theory predicts a radially dominated HMF at high latitudes (small θ) with an ever decreasing azimuthal component that is zero at the polar axis. The SBM added a term to the ψ-dependence of the HMF that is proportional to r and implies a non-zero azimuthal component around the polar axis. This should have a major effect on the inward transport of cosmic rays compared to the zero azimuthal component and the consequent easy inward transport predicted by the standard model.

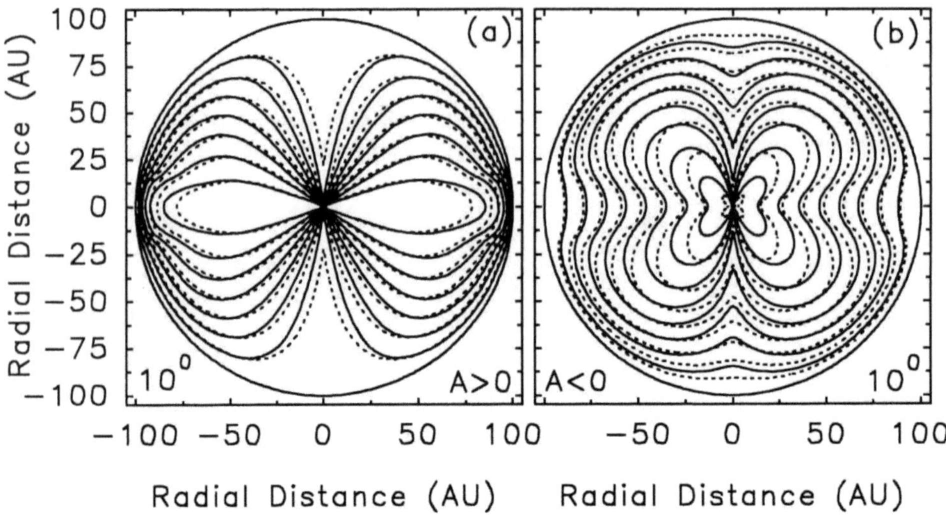

Fig. 3. The intensity contours for 220 MeV protons resulting from a solution with a latitudinal dependent solar wind profile (solid lines) compared with the standard solution (dashed lines) where $V = 450$ km.s^{-1}. The allocation of contours are the same as before.

Latitudinal dependent solar wind. It is known from observations made by existing spacecraft that the solar wind speed increases with latitude (Barnes, 1989). Recent data from Ulysses also showed a large increase in the solar wind speed at higher latitudes (Bame *et al.*, 1993). To determine the effects of such a profile on modulation, the following modified solar wind profile was used: (i) It expands radially at 450 km.s^{-1} in the equatorial plane. (ii) It increases to a maximum of 900 km.s^{-1} at the poles. (iii) This maximum speed is reached at a latitude of 30^0 North and South. (iv) It increases exponentially from zero at the inner boundary to the respective speeds over ~0.3 AU. This profile can be seen as an extreme case and enables one to study the effects of such a modification.

4. Results and Discussion

The influence of the JKM on modulation contours is portrayed by the solid lines in Fig. 1. The major differences between the JKM and the reference solution can be summarized as: (i) A significant reduction of inward diffusion and drifts in the polar regions during $A > 0$ cycles [Fig. 1(a)] due to the $1/B$ dependence of the diffusion coefficients in (1). (ii) The radial gradients in the inner heliospheric regions increased near the equatorial plane but decreased over the poles. The general radial dependence in the equatorial plane – an increase towards the outer boundary – remained the same. (iii) The JKM has no effect close to the equatorial regions during $A < 0$ cycles but do have an effect in the polar regions.

The effect of the SBM on modulation can be seen in Fig. 2. Compared with Fig. 1, the major differences are: (i) For the respective magnitudes of the modifications used, the SBM decreased the inward propagation of particles in the polar regions during A $>$ 0 cycles considerably more than the JKM. This greater effect can be contributed to the fact that the SBM modifies the HMF direction and implicitly also the HMF magnitude whereas the JKM only modifies the latter. (ii) The radial dependence of the radial gradients are qualitatively the same as for the JKM. (iii) The SBM has a relative small effect on modulation during A $<$ 0 cycles. Taken together these features make the SBM the more suitable modification.

The results of the latitudinal dependent solar wind profile (solid lines) compared with a standard solution with $V = 450$ km.s^{-1}, are shown in Fig. 3. The major differences are: (i) The modified solar wind *increases* inward drift and diffusion during A $>$ 0 cycles. It increases outward convection in the high latitude regions, but the HMF spiral is simultaneously drawn open so that inward diffusion also increases. Furthermore, since the radial component of the particle drift velocity, v_r, $\propto \partial V/\partial\theta$ and $v_\theta \propto V$, the latitudinal gradient established in the solar wind causes increased drift effects (see also Reinecke and Moraal, 1993). (ii) Effects during A $<$ 0 cycles are similar. Compared to the effects of the HMF modifications, the latitudinal dependent solar wind *enhances* both diffusion and drifts in the polar regions and therefore reduces modulation. This effect is undesirable from a modelling point of view, but has to be incorporated because it is more realistic than what was used before. However, combined with a SBM for the HMF, it has a minor effect on modulation in general as was shown in Fig. 2.

5. Implication for Ulysses

The observations awaited from Ulysses may be the first to show to what extent the modifications are justified. If the results in Fig. 1, 2 and 3 are shown along the trajectory of Ulysses for the A $>$ 0 cycle, which is currently the case, the results in Fig. 4 follow (see also Potgieter and Haasbroek, 1993). The major implication of the HMF modifications is a smaller difference (by a factor of $\sim$2) between modulation in the polar and equatorial regions since the JKM and SBM tend to lessen the differences in the level of modulation between these two regions. The maximum intensity should however still be reached during the two polar passes in 1994 and 1995, unless the inner heliosphere is much more spherical symmetric than we anticipated.

6. Conclusions

Drift models based on the diffusion coefficients in (1) predict considerable less modulation in the polar regions of the heliosphere compared with the

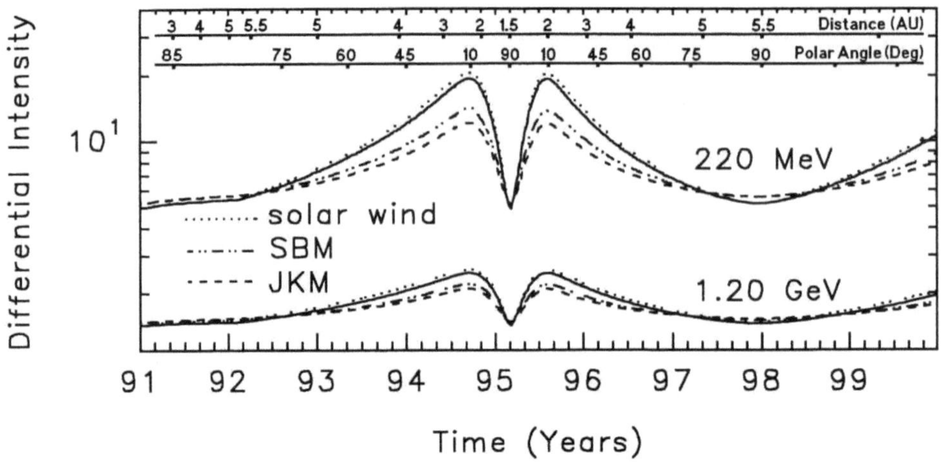

Fig. 4. The simulated intensity in units of particles.m^{-2}.sr^{-1}.s^{-1}.MeV^{-1} for 220 MeV protons along the Ulysses trajectory from 1991 to 1999 as given by the standard solution with $V = 450$ km.s^{-1} (solid line), the JKM, the SBM and the modified solar wind profile.

equatorial plane (factor of ~4 for 220 MeV protons) during A > 0 cycles (e.g. 1970–80; 1991–?). This causes e.g. very small radial gradients in the equatorial plane during these cycles (Potgieter *et al.*, 1989). In this paper, modifications to the standard drift model, with a Parkerian HMF and diffusion coefficients proportional to $1/B$, have been considered: (i) Increasing the latitude dependence of the solar wind speed made the situation worse. (ii) Modifying the direction and the magnitude of the HMF, especially over the poles, according to Smith and Bieber (1991) improved the situation significantly despite of (i) above. This modification increased the modulation in the polar regions and subsequently produced what we consider more realistic radial and latitudinal gradients. The observations from the Ulysses mission may be the first to show to what extent these modifications to the standard Parker model are justified. A second phase of this study will be devoted to models where $K_{||}$, for instance, is not proportional to $1/B$.

References

Bame, S.J., Goldstein, B.E., Gosling, J.T., Harvey, J.W., McComas, D.J., Neugebauer, M. and Phillips, J.L.: 1993, *Geophys. Res. Let.* **20**, 2323.

Barnes, A.: 1989, in S. Grzedzielski and D.E. Page (eds.), 'Physics of the outer heliosphere', Pergamon Press, Oxford, pp. 235–240.

Jokipii, J.R. and Kóta, J.: 1989, *Geophys. Res. Let.* **16**, 1.

Potgieter, M.S. and Haasbroek, L.J.: 1993, *Proc. 23rd Int. Cosmic Ray Conf.* **3**, 457.

Potgieter, M.S. and Moraal, H.: 1985, *Astrophys. J.* **294**, 425.

Potgieter, M.S., Le Roux, J.A. and Burger, R.A.: 1989, *J. Geophys. Res.* **94**, 2323.

Reinecke, J.P.L. and Moraal, H.: 1993, *Proc. 23rd Int. Cosmic Ray Conf.* **3**, 473.

Smith, C.W. and Bieber, J.W.: 1991, *Astrophys. J.* **370**, 435.

HIGH ENERGY COSMIC RAY NUCLEI RESULTS ON ULYSSES: 1. MISSION OVERVIEW

B. HEBER, A. RAVIART[2], C. PAIZIS[3], W. DRÖGE[1], R. DUCROS[2], P. FERRANDO[2], H. KUNOW[1], R. MÜLLER-MELLIN[1], C. RASTOIN[2], K. RÖHRS[1], H. SIERKS[1] and G. WIBBERENZ[1]

[1] *Institut für Kernphysik, Universität Kiel, 24118 Kiel, Germany*
[2] *CEA, Service d'Astrophysique, C.E.Saclay, 91191 Gif-sur-Yvette, France*
[3] *Instituto Fisica Cosmica CNR, Universitá di Milano, 20133 Milano*

Abstract. The cosmic ray flux observed with the Kiel Electron Telescope on board the ULYSSES spaceprobe varies with solar activity as well as with heliospheric position. Determination of the latitudinal gradients requires a careful analysis of the influences of the current sheet tilt angle, the number of major solar flares, interplanetary shocks and interaction regions evolving in the expanding solar wind. In this paper we concentrate on nuclei with rigidity above 1 GV. We discuss the effects of the variable solar activity in the declining phase of the present solar cycle and the variation with radial distance as a basis for separating latitudinal effects. We show that during this phase of the solar cycle modulation of GV nuclei is ordered by temporal evolution, radial distance and negligible latitudinal effects even at latitudes between 30° and 50° South.

Key words: heliosphere – solar modulation – Ulysses

1. Introduction

The ULYSSES spacecraft was launched on October 6, 1990 in the declining phase of solar cycle 22. In this paper we report observations taken from launch until March 1994, when the maximum latitudinal excursion was 55° South. The Kiel Electron Telescope (KET) is part of the COsmic and Solar Particle INvestigation [1]. In order to separate spatial from temporal variations in the galactic cosmic ray flux, we use measurements from the Chicago particle instrument on board IMP-8 (protons with energies > 106 MeV, mean rigidity 2 GV, [2]), from the Kiel Neutron Monitor KNM and Hermanus Neutron Monitor HNM (mean rigidities 17.1 and 19.4 GV).

2. Observations

Fig. 1 shows daily averaged counting rates registered by KET in two selected nuclei channels, electrons will be discussed in detail in [3]. The proton channel 34-125 MeV (400 MV) in the upper panel is a good indicator for solar activity showing increases due to individual energetic particle events. This channel has been used to select quiet periods when no solar particles contaminate the galactic flux. The next panel shows counting rates for protons of 250 to 2200 MeV with a mean rigidity of 1.7 GV. Plotted are only

Space Science Reviews **72**: 391–396.
© 1995 *Kluwer Academic Publishers.*

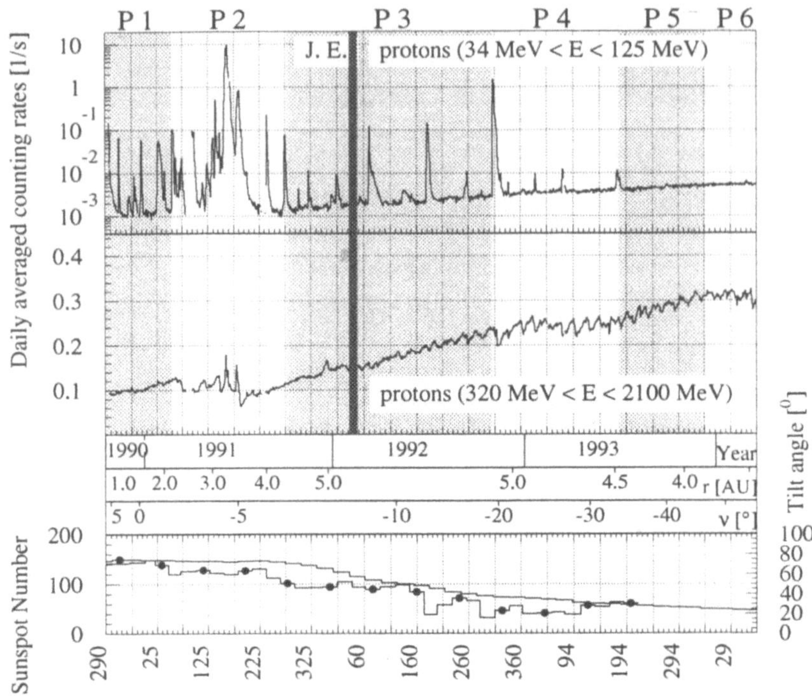

Fig. 1. Daily averaged countıng rates for KET (cf. Table 1), and 27 day averaged Sunspot number and Tilt angle (dots) J E marks the Jovian Encounter ın Feb 1992

time periods when KET is working nominally. So for example the periods of the March event 1991 and the Jovian Encounter (J.E.) have been omitted, because KET was partly saturated. The lowest panel shows the 27 day averaged measured and predicted sunspot numbers published by NOAA and the tilt angle indicating the evolution of solar activity. From Fig. 1 we can distinguish three different categories in the counting rates of GV nuclei:
(A) Monotonic increases in the count rates of high energy nuclei are characteristic for Period 1 (P1) from launch to Feb. 1991, P3 from Oct. 1991 to Oct. 1992 and P5 from Jul. 1993 to Dec. 1993. Solar activity decreases from a moderate to a low level. During P1 and P3 ULYSSES is moving close to the ecliptic plane within 16° heliographic latitude, while in P5 ULYSSES is climbing up from 30° to 50° South. During P1 ULYSSES is moving outward from 1 AU to 2 AU radial distance, whereas during P3 the radial distance is nearly constant between 5 and 5.4 AU. In P5 ULYSSES is moving inward from 4.5 to below 4 AU.
(B) For the 1.7 GV protons the continuous increases from P1, P3 and P5 respectively were inhibited during P2 (Feb. 1991 to Oct. 1991), P4 (Oct.

1992 to Jul. 1993) and P6 (Dec. 1993 to Mar. 1994). The suspension in P2 is due to the high solar activity [4] (see also the increases in the lower energy proton channel (Fig. 1)); this is unlike for P4 and P6.
(C) The counting rates of the high energy nuclei show a 26 day modulation from Aug. 1992 until Mar. 1994 (P3-P6). This kind of modulation is discussed in detail in [5]. For our further analysis in this paper we use only the time periods P1, P3 and P5, when the counting rates of the high energy nuclei are continuously increasing.

3. Data Analysis

To distinguish between temporal and spatial influences in two dimensions we have to compare the fluxes measured with identical instruments on three spacecraft separated in radial distance *and* heliographic latitude. Analysing data from two spacecraft separated in radial distance and heliographic latitude we calculate the temporal and spatial variations under the assumptions that the radial and latitudinal gradients are not varying significantly in time and space. As a 1 AU reference we use the daily averaged KNM and IMP 8 counting rates. From the selected time periods P1 and P3 and the special trajectory of ULYSSES we are able to separate temporal and radial variations. Extending this method to P5 gives an estimate for latitudinal effects.

3.1 Temporal Variation and Radial Gradients

Until the end of 93 the fluxes of galactic cosmic rays are increasing with time. To decide whether the observed increases are due to temporal or spatial variation, we analysed the daily averaged counting rates $R_N(t)$ measured by the KNM and HNM in P1 and P3 (cf. Fig. 2). The counting rates of the KNM and the HNM show qualitatively the same behaviour as the high energy nuclei measured by the KET: there are continuous increases in P1 and P3 and a drastic decrease in P2. After the disturbed period, the intensities start to increase again with essentially the same rate as before, but on a reduced overall level. We approximated $R_\mathrm{N}(t)$ in the marked two periods by an exponential law in time with different offsets in P1 and P3:

$$R_\mathrm{N}(t) = R_{\mathrm{N,0,i}} \cdot \mathrm{e}^{\gamma(P)\cdot t} \tag{1}$$

with $\gamma = 1/R_\mathrm{N} \cdot \partial R_\mathrm{N}/\partial t$ (%/year). The result of the fit gives the superposed line in Fig. 2. We conclude tentatively (i) that for fixed radial distance r (ULYSSES) the increase γ is nearly the same in both periods of the analysis and the time dependence is given by eq. 1, and (ii) that the radial gradient $G_r(P) = 1/R \cdot \partial R/\partial r$ and the temporal rate of increase γ do not vary significantly with time t or radial distance r, but depend only on rigidity. With these assumptions we can use the following approximation for the

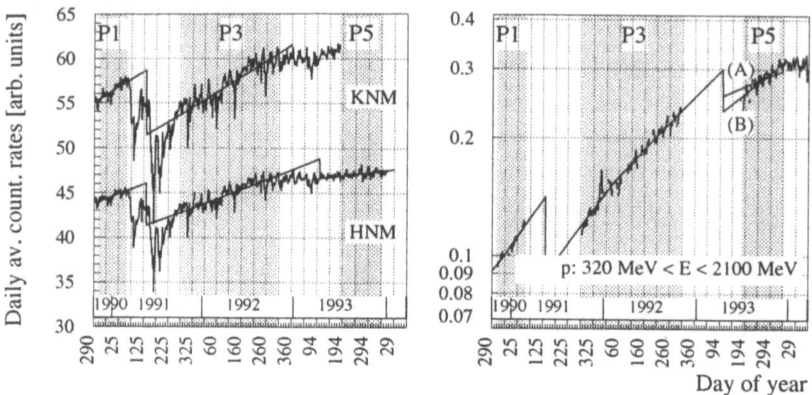

Fig 2 Quiet time counting rates at 1 AU (left) and at ULYSSES with approximation by eq 1, 2 and 3

counting rate $R_i(P,t,r)$ of the KET nuclei channel (p, α-particles) in Pi:

$$R_i(P,r,t) = R_{i0} \cdot e^{G_r(P)\,r} e^{\gamma(P)\,t}. \tag{2}$$

Since the radial distance r is not linearly related to time t (see Fig. 1), we can fit the data according to eq. 2 which yields unambiguously values of $\gamma(P)$ and $G_r(P)$. Here γ is determined for the period from November 1990 to November 1992. Fig. 2 (right) shows as an example the behaviour of the 1.7 GV quiet time proton fluxes together with a fit according to eq. 2. Fig. 3 (left) shows the temporal rate of increase $\gamma = 1/R \cdot \partial R/\partial t$ of the daily averaged counting rates as a function of rigidity. For nuclei > 1 GV rigidity the temporal variation decreases with increasing rigidity. The calculated radial gradients G_R are shown in Fig. 3 (right). The radial gradient can also be calculated directly using the counting rate ratio of IMP 8 (E>106 MeV) to KET (E>125 MeV). This ratio, determined until September 1992, is fitted well by a single value of $3.2 \pm 0.8\%/\mathrm{AU}$ for the integral gradient. Within uncertainties this value is in good agreement with our determination by usage of eq. 2.

3.2 Latitudinal Effects

Simply extending our method to P5 is not possible because the HNM data (see Fig. 2) show a smaller rate of increase γ than in the two periods P1 and P3; this can also be estimated from the IMP 8 data. Nevertheless, the HNM data show that we can approximate the counting rates by an exponential law in time with a smaller rate of increase during P5. Because we do not know $\gamma(P)$ for every rigidity P and the heliographic latitude is continuously increasing during P5 we cannot fit R by an exponential law

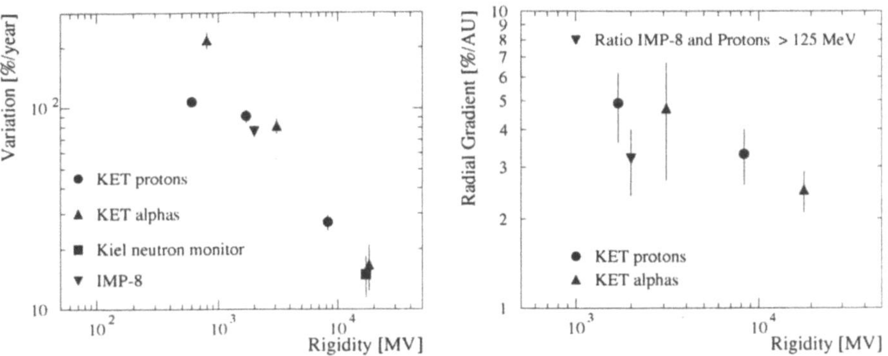

Fig 3 Left Variation γ in %/year of daily averaged counting rates and right Radial gradient G_R in %/AU

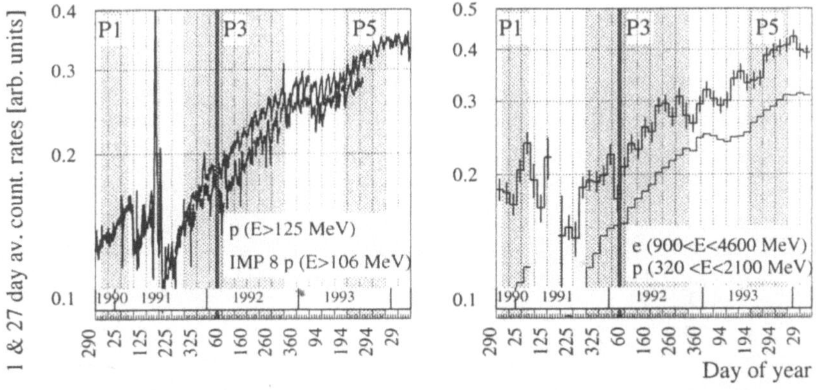

Fig 4 Left Daily averaged counting rates measured by KET and IMP 8 Right 27 day averaged KET 1 7 GV-proton and electron quiet times counting rates

in time, radial distance and heliographic latitude with unambiguous values for the temporal and spatial variations. To give an estimate for latitudinal effects, we assume: (i) the radial gradient to be the same as in P1 and P3 and (ii) that the temporal rate of increase is scaled by a factor f_i as estimated from the HNM and the IMP 8 data. With these assumptions we use the following approximation for the counting rate $R_i(P,t,r,\theta)$ of the KET nuclei channel in Pi:

$$R_i(P,r,t) = R_{i0} \cdot e^{G_r(P)\, r} \; e^{G_\theta\, \theta} \cdot e^{\gamma(P)\, t\, f_i} \quad . \tag{3}$$

Herein G_θ is the latitudinal gradient and $f_i = 1$ for P1 and P3 and varying in P5 between 50% (HNM) and 70% (IMP 8). Fig. 2 (right) shows the counting rates of 1.7 GV protons in the three selected time periods and the approximation by eq. 3 for (A) fixed $f_i = 0.5$ and assuming no latitudinal gradient and (B) with an appropriate G_θ of 0.6%/°. Fitting the whole data set with eq. 3 leads to latitudinal gradients G_θ between 0.1%/° ($f_i = 0.7$) and 0.6%/° ($f_i = 0.5$). These values are comparable to values found by McKibben [6] with Pioneer 10 and 11 and IMP 8 from 1975 to 1976. Our result using only KET nuclei data and the assumptions listed above is confirmed by comparing KET protons above 125 MeV and IMP 8 protons above 106 MeV. In Fig. 4 we show the two corresponding proton channels (left). We also compare in Fig. 4 (right) the 27 day averaged counting rates of protons and electrons with rigidity of 1.7 GV. We find that - in contrast to predictions from drift models - the result for the two particle types are surprisingly similar (see also [3] for discussion).

4. Conclusions

During two selected time periods from 1990 to 1992 the flux variation of nuclei > 1 GV rigidity can be fitted with a model assuming exponential variations with time and radial distance. The annual variation for nuclei decreases with increasing rigidity. The radial gradient for nuclei in the GV range from the data analysed so far shows a slight decrease with rigidity. The average value of about 3.3%/AU determined directly for relativistic particles is in good agreement with integral gradients from previous missions (for discussion see [7]). For a selected time period in 1993 when ULYSSES is moving from 30° to 50° South we are able to estimate a latitudinal gradient of 0.1%/° to 0.6%/°. So we find that the modulation processes during the declining phase of the solar cycle are characterized by large temporal and small radial variations with negligible latitudinal effects.

Acknowledgements

We are grateful to B. McKibben, Prof. P.H. Stoker and T. Hoeksema for making available IMP-8 particle data, Hermanus data and magnetic tilt angles. This work was partly supported by Deutsche Agentur für Raumfahrtangelegenheiten (DARA) under grant No. 50 ON 9106.

References

[1] Simpson J.A. et al.: 1992 *Astron. Astrophys. Suppl.* **92**, 365
[2] Garcia-Munoz M., Mason G.M. and Simpson J.A.: 1975, *Ap. J.* **201**, L145
[3] Raviart A. et al.: 1994 this volume
[4] Müller-Mellin R. et al.: 1993 *Adv. Space Res.***13** 86
[5] Kunow A. et al.: 1994 this volume
[6] McKibben R.B.: 1989, *JGR* **A12**, 17021-17033
[7] Flückiger E.O.: 1991, *Proc. 22nd Int'l. Cosmic Ray Conf. Dublin* 5, 273

HIGH ENERGY COSMIC-RAY NUCLEI RESULTS ON ULYSSES: 2. EFFECTS OF A RECURRENT HIGH-SPEED STREAM FROM THE SOUTHERN POLAR CORONAL HOLE

H. KUNOW, W. DRÖGE, B. HEBER, R. MÜLLER-MELLIN, K. RÖHRS, H. SIERKS and G. WIBBERENZ
Institut für Kernphysik, Universität Kiel, 24118 Kiel, Germany

R. DUCROS, P. FERRANDO, C. RASTOIN and A. RAVIART
CEA, Service d'Astrophysique, C.E. Saclay, 91191 Gif-sur-Yvette, France

and

C. PAIZIS
Instituto Fisica Cosmica del CNR, Universitá di Milano, 20133 Milano, Italy

Abstract. Since June 1992 the Kiel Electron Telescope on board ULYSSES measures 26-day variations of the order of 6% in the fluxes of high energy H and He. In May 1993 ULYSSES entered into the unipolar region of the southern polar coronal hole, but continued to observe similar effects: increases in the MeV proton channels due to acceleration near the shocks of the corotating interaction region and decreases in the intensity of galactic nuclei associated with the same region. Amplitude variations are presented for different magnetic rigidities and the effects are discussed in view of corotating shock development in a 3-dimensional heliospheric structure.

Key words: Heliosphere – solar modulation – Ulysses – Corotating Interaction Region

1. Introduction

On its way to the southern solar polar region during the declining phase of solar cycle 22 the ULYSSES spacecraft encountered a stable long-lasting corotating high-speed solar wind stream from June 1992 to March 1994. The underlying heliospheric current sheet is described by Hoeksema (1994), the development of the interplanetary structure by Balogh et al. (1993), Smith et al. (1993), Bame et al. (1993), and Gosling et al. (1993a). Heber et al. (1994) present an overview of energetic particle measurements observed by the Kiel Electron Telescope on board ULYSSES, and Sanderson et al. (1994) describe the corotating events in the MeV range.

Effects of recurrent high-speed solar wind streams on energetic and cosmic ray particles as observed in the plane of the ecliptic since several decades (e.g. Venkatesan and Badruddin (1990) and Kunow et al. (1991)) are:

1. Acceleration of suprathermal particles to energies up to several tens of MeV: Corotating Interaction Region events (CIR events), first described by Barnes and Simpson (1976), and
2. Temporary decreases of galactic cosmic rays which can be described as Recurrent Forbush Decreases (RFD).

Space Science Reviews **72**: 397–402.
© 1995 *Kluwer Academic Publishers.*

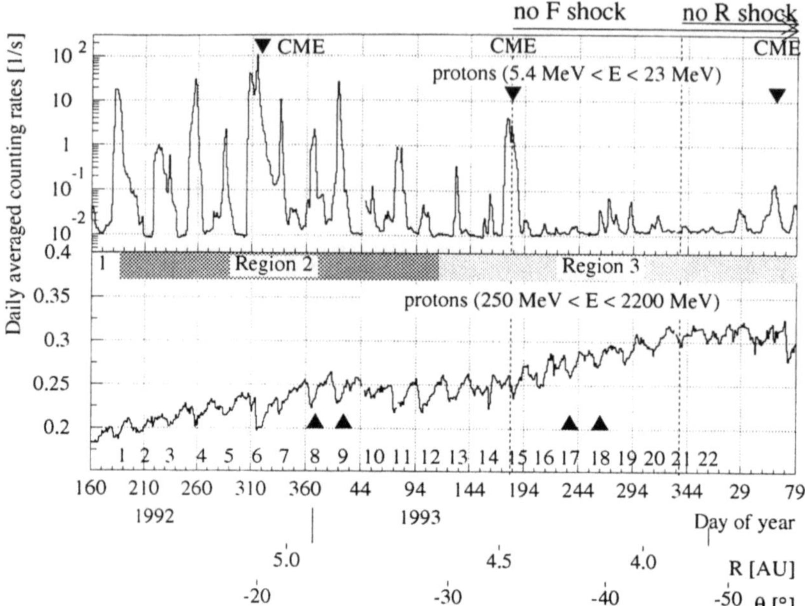

Fig. 1. Daily averaged counting rates for COSPIN Kiel Electron Telescope. Downward pointing triangles: CME, upward pointing triangles: Events shown in Fig. 2 in detail.

We describe the effects of this recurrent high-speed stream on energetic and cosmic ray particles with increasing heliographic latitude.

2. Observations

Observations are based upon the ULYSSES COSPIN Kiel Electron Telescope (Simpson et al. (1992), Kunow et al. (1991a)). Fig. 1 provides an overview of the ascending phase of ULYSSES from June 8, 1992, through March 20, 1994, equivalent to heliographic latitudes from $-13°$ to $-55°$. The bottom panel displays daily averaged counting rates of protons (250–2200 MeV) showing a general increase due to the long-term recovery in the declining phase of solar cycle 22, discussed in Heber et al. (1994). The rates of this channel are modulated by fairly regular decreases with a recurrence time scale of about 26 days, equivalent to the sidereal rotation of the sun.

The 5.4–23 MeV proton response (top panel Fig. 1) shows increases of up to 4 orders of magnitude in coincidence with the modulation depressions. Amplitudes are highest within region 2, although some are associated with CMEs as marked in Fig. 1. These corotating events seem to disappear slowly within region 3, although a recurrence tendency is still observed.

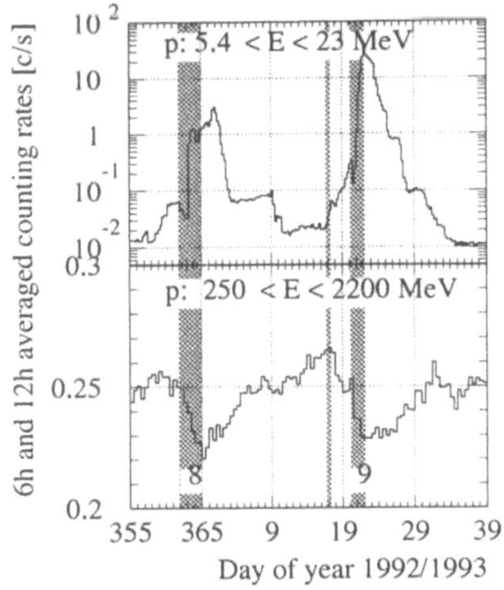

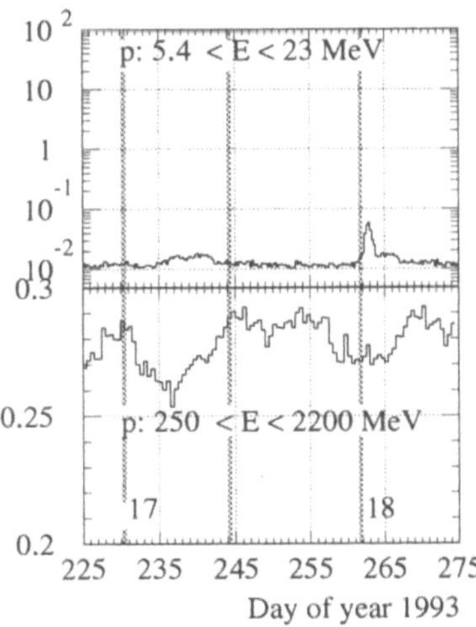

Fig. 2. Selected CIR events on expanded scale. Left: Events No. 8 and 9 of region 2. The shaded areas mark the CIR bounded by forward and reverse shock passage. Right: Events No. 17 and 18. Vertical lines mark identified shocks. Event numbers as defined in Fig. 1.

3. Analysis

The onset of the recurrent high-speed stream occurs at a solar latitude of $-13°$ and marks the exit from the streamer belt dominated region (Bame et al. 1993). Region 2 is determined by the heliospheric current sheet and covers the latitude range from $-13°$ through $-29°$. Up to $-33.6°$ ULYSSES observes a well developed CIR bounded by a forward and a reverse shock (Phillips et al. (1994)). Region 3 beyond $-29°$ is dominated by a high-speed stream from the southern polar coronal hole. No more current sheet crossings are observed (Smith et al. (1992)). Beyond $-33.6°$ the forward shock and beyond $-45.5°$ the reverse shock disappears as well suggesting that the CIR is locally no longer present. We discuss in the following section the influence of the CIR in terms of the heliospheric latitude structure.

3.1 CIR Effects within Latitude Region 2

Fig. 2 shows the effects of the CIR on energetic and cosmic ray particles on an expanded scale for two typical events in region 2 (left) and region 3 (right). For most CIR events in region 2 we found an acceleration of particles up to some tens of MeV while protons above some hundreds of MeV show RFD's. For accelerated protons the times of maximum are correlated with the reverse shock passage (Fig. 2, left) whereas the forward shock has a much smaller effect. The increase starts already three days prior to the forward shock passage, and the decay upstream the reverse shock lasts about 3–

4 days. The RFD's show a different time dependence: The decay starts at the passage of the forward shock, reaches the maximum of the depression with the reverse shock passage, and has a recovery phase of approximately one week.

3.2 Effects of the CIR beyond −29°

Polewards of −33.6° the forward shock of the CIR has disappeared while a backward shock persists up to −45.5°. Beyond −45.5° also the reverse shock disappears. With the disappearance of the forward shock we observe a dramatic decrease in the accelerated particle amplitude. However, the RFD's still persist with about the same amplitude. The disappearance of the reverse shock causes the amplitude of the accelerated particles to decrease even further. The modulation of the high energy particles, however, persists undiminished. The rigid time relation between maxima of accelerated particles and minima of modulated particles is no longer obvious (Fig. 2, right). The increase is delayed by several days with respect to the minimum of the modulation minima.

4. Discussion

The observations described above can be extended to other particle species and energy intervals with similar acceleration increases (e.g. Sanderson et al. (1994)) and modulation decreases. The rigidity dependence of the recurrent modulation (Fig. 3, left) shows a general decrease of the modulation effect with increasing rigidity above 1 GV although the dependence is not as convincing as for the long-term modulation described by Heber et al. (1994).

The time (or latitude) development seems to provide more insight. The accelerated proton peaks (Fig. 3, right), defined by the enhancement above the local background rates, are largest for events No. 4–8 during times when a fully developed CIR was observed bounded by strong forward and reverse shocks (for definition of event numbers see Fig. 1). The amplitude decreases by two orders of magnitude with the disappearance of the forward shock (event No. 14 in Fig. 3) and continues to decrease even beyond the disappearance of the reverse shock. The magnitude of the modulation depression in the 250–2200 MeV proton channel rises until event No. 11 and falls afterwards when the reverse shock disappears. Compared to that the variation of the >2.2 GeV protons remains about constant over the whole time period with a magnitude of about 6%.

Three questions remain to be answered after analyzing the data:

1. How can acceleration occur in absence of both the reverse and the forward shock?
2. How can the modulation persist even beyond the latitude where the CIR affects the large scale structure?

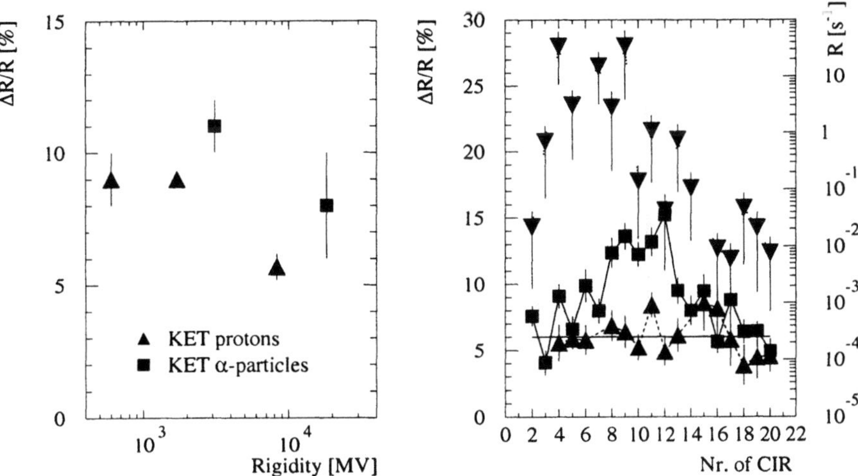

Fig. 3. Left: Rigidity dependence of CIR induced intensity variations Δ R/R. Right: Time dependence (event numbers as defined in Fig. 1) of acceleration increases respectively modulation decreases for different proton energies.(Downward pointing triangles: 5 – 23 MeV, right scale; squares: 250 – 2200 MeV, right scale; upward pointing triangles: >2.2 GeV

3. How can the delay of the accelerated particle maxima with respect to the modulated particle minima after the disappearance of the reverse shock be explained?

In the following we will suggest answers to these questions.
1. Inside the orbit of earth HELIOS observed accelerated particles from CIR's even though neither the forward nor the reverse shock had developed locally (Kunow et al. (1991)). The maximum intensity of the corotating event was observed on a field line which crosses the reverse shock at a distance about 3–5 AU away from the observer. The particles were accelerated at the reverse shock and travelled about 5 AU along the interplanetary fieldline to the observer inside 1 AU. Similar conditions are observed with ULYSSES: Gosling et al. (1993a) observed upstream suprathermal electrons both from the forward and from the reverse shock. They described a poleward motion of the reverse shock with increasing solar distance. One consequence of this behaviour is the detection of the reverse shock of the CIR up to −45.5° although the neutral sheet disappeared already at −29° and the forward shock at −33.6°. The possible connection to the reverse shock even if locally absent is explained in Fig. 4 where interplanetary fieldlines allow propagation of particles from the poleward progressing reverse shock at larger distance from the sun to ULYSSES where the reverse shock is no longer observable. In analogy to the observations in the inner solar system

ULYSSES observes particles accelerated at the corotating reverse shock on fieldlines which encounter the shock at a distance of several AU.

2. The reason for the persistence of the modulation is increased scattering between interstellar space and ULYSSES happening predominantly inside the CIR. As this is an integral effect occasional or final disappearance of the reverse shock has not changed the conditions markedly resulting in an unchanged time dependence.

3. The delay of the accelerated particle maximum with respect to the modulation minima can also be explained from Fig. 4: no reverse shock capable to accelerate particles is observed at the latitude and solar distance of ULYSSES. With increasing solar distance the reverse shock, however, has moved eastward and poleward (Gosling et al. (1993)). Only fieldlines east of the extrapolated shock passage fieldline can therefore encounter the reverse shock at larger distances of several AU These fieldlines are swept later over ULYSSES causing a delay of the maximum of accelerated corotating particles relative to the RFD.

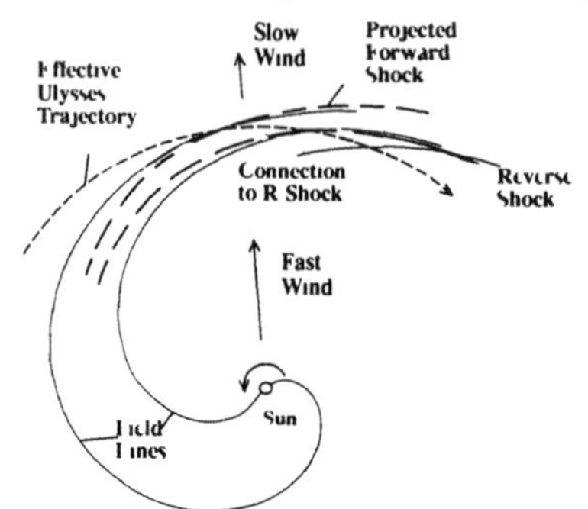

Fig 4 Interplanetary field configuration, extrapolated shock locations and ULYSSES trajectory to explain reverse CIR shock accelerated particle observations after disappearance of shock (after Gosling et al (1993))

Acknowledgements

This work was partly supported by Deutsche Agentur fur Raumfahrtangelegenheiten (DARA) GMBH under grant No. 50 ON 9106.

References

[1] Balogh A et al 1993, *Geophys Res Lett*, 20(21) 2331
[2] Bame S J et al 1993, *Geophys Res Lett*, 20(21) 2323
[3] Barnes C W and J A Simpson 1976, *Astrophys J*, 207 977 1976
[4] Gosling J T and B E Goldstein 1993, *Geophys Res Lett*, 20(21) 2335
[5] Gosling, J T et al 1993a, *Geophys Res Lett*, 20 2789
[6] Heber B et al 1994, *this volume*
[7] Hoeksema, J T 1994, *submitted to Space Science Rev*
[8] Kunow H et al 1991, in *Physics of the Inner Heliosphere (R Schwenn and E Marsch Edts)* Springer-Verlag, Berlin Vol 2, p 243 342
[9] Kunow H et al 1991a, *Proc 22 Int Cosmic Ray Conf Dublin* Vol 2 503 506
[10] Phillips et al 1994, *Geophys Res Lett, in press*
[11] Sanderson T et al 1994, *Geophys Res Lett, in press*
[12] Simpson J A et al 1992, *Astron and Astrophys Suppl* 92 365
[13] Smith E J et al 1993, *Geophys Res Lett*, 20(21) 2327
[14] Venkatesan D and Badruddin 1990, *Space Science Rev* 52 121 194

May 26, 1994

ULYSSES OUT-OF-ECLIPTIC OBSERVATIONS OF "27-DAY" VARIATIONS IN HIGH ENERGY COSMIC RAY INTENSITY

R.B. MCKIBBEN, J.A. SIMPSON AND M. ZHANG
Enrico Fermi Institute, The University of Chicago, Chicago, IL 60637, U.S.A.

S. BAME
Los Alamos National Laboratory, Los Alamos, N.M., U.S.A.

A. BALOGH
Blackett Laboratory, Imperial College of Science, Technology and Medicine, UK

Abstract. We report the discovery that for latitudes above ~ 40°S, the observed recurring modulation of cosmic rays and anomalous nuclei occurs without the detection by *Ulysses* of the solar wind velocity and magnetic field recurring enhancements that have, heretofore at lower latitudes, defined corotating interaction regions — i.e., the mechanism producing the recurring intensity variations > 40°S appears to be located beyond the radial range of *Ulysses*

1. Introduction

Following its encounter with Jupiter in February 1992, where it was deflected from the ecliptic towards the heliographic south pole, the *Ulysses* instruments observed almost continuously a recurrent variation of cosmic ray intensity with approximately the period of solar rotation (the "27-day variation") up to the heliographic latitude of 56°S (as of April 1994), confirming that corotating interaction regions extend beyond the maximum tilt angle of the heliographic current sheet, as reported for *Pioneer-11* at > 16°N in the period near the solar minimum of 1976-77 (Bastian et al., 1979; Simpson, et al., 1986). In this paper we also show for latitudes $\gtrsim$ 33°S that, although the 27-day modulation of galactic cosmic ray and anomalous component intensities persist to the highest latitudes reached so far by *Ulysses*, the observations of solar wind velocity and magnetic field intensity variations which are characteristic of CIRs gradually disappear above ~ 35-40°S. Furthermore, we found that this cosmic ray recurrent modulation is observed with the same phase at Earth extending in magnetic rigidity to > 13 GV. These new observations have important implications for the origin of the recurrent cosmic ray modulation.

The charged particle measurements were obtained from the High Energy Telescope (HET) channel H9 and the recent report of the Low Energy Telescope (LET Sanderson et al., 1994). The solar wind and magnetic field data are from the plasma (SWOOPS) and vector helium magnetometer instruments, respectively. For details of the instruments and data analyses, see Astronomy and Astrophysics Supplement Series special issue (Vol. 92, No. 2, January 1992). Simultaneous measurements in the ecliptic at 1 A.U. were obtained by the Climax and Maui neutron monitors which observed the recurrent

Space Science Reviews **72**: 403–408.
© 1995 *Kluwer Academic Publishers.*

modulation of cosmic ray nucleons above the magnetic rigidities of ~ 3 GV and ~ 20 GV, respectively.

2. Observations

As pointed out by Sanderson et al. (1994), and as evident from the solar wind velocity (V_{sw}) panel in FIGURE 1, *Ulysses* so far has encountered three different regions of solar wind structures. In the following we divide our analyses into these three regions:

1. In the latitude region less than ~ 13°S (before July 1992), the spacecraft was immersed in the solar wind streamer belt where the velocity fluctuations are small and irregular but where the magnetic field displayed recurrent passages of compressed magnetic field. Since the cosmic radiation average intensity was increasing over the years 1992 to the present due to relaxation of the 11 year solar cycle modulation as shown in the third panel of FIGURE 1, we detrended this long term intensity change in panels four and five for analysis of the shorter term recurring intensity variations. From the HET channel H9 we obtain the integral intensity of protons with energies > 92 MeV which is equivalent to an average proton energy of ~ 1 GeV. The ΔH9 panel shows the detrended recurring intensity variations expressed as percent variations from the "27-day" averages. The maximum amplitude is approximately 4 percent.

In the same period, neutron monitors at Earth observed the "27-day" variations extending beyond ~ 3 GV (Climax) and ~ 13 GV (Maui) magnetic rigidity. Taking into account the changing longitude difference between the *Ulysses* and Earth and the time for solar wind structure to propagate at ~ 450 km/s from the orbit of Earth to *Ulysses*, we found that the cosmic ray intensity variations - both at Earth and the spacecraft were produced by the *same* stream from the Sun. In order to compare measurements at Earth with *Ulysses* we indicated with "+" signs the times when the spacecraft or Earth crossed a constant heliographic longitude. For example, in the period 92.30-92.35 the Climax intensity peaks led the ΔH9 intensity peaks by 23 days. This was consistent with ~ 100° (7 days) longitude lag of the spacecraft relative to Earth, plus the propagation lag from 1 A.U. to 5 A.U. (~ 17 days). The same kind of analysis was carried out for CIR's beyond 16°N in 1976-1977 with *Pioneer-11* (Bastian et al., 1979).

2. In the region of middle latitudes between ~ 13°S and ~ 33°-35°S (July 1992 - July 1993) the velocity of the solar wind passing the spacecraft exhibited a large periodic variation from ~ 400 km/s to >750 km/s. In this region, as shown by the example in FIGURE 2A, at the leading edges of the fast streams and strongly compressed magnetic fields the high energy cosmic ray intensity (H9) decreased rapidly reaching intensity minima as the compressed magnetic field passed the spacecraft. Concurrently, intensity increases of LET protons with energies 0.9-1.2 MeV (cf. Figure 2A) were shown by Sanderson et. al. (1994) to be proton acceleration associated with forward and reverse corotating shocks — observations consistent with corotating shocks produced by Corotating Interaction Regions (Barnes and Simpson, 1976).

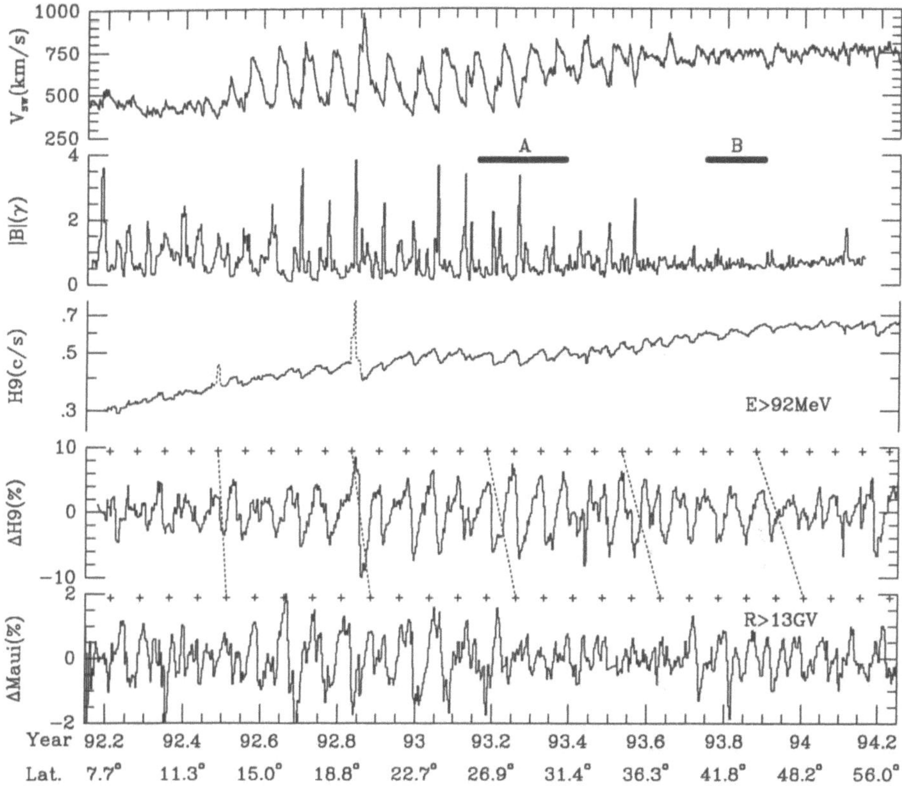

Fig.1. The bars A and B correspond to Figure 2.

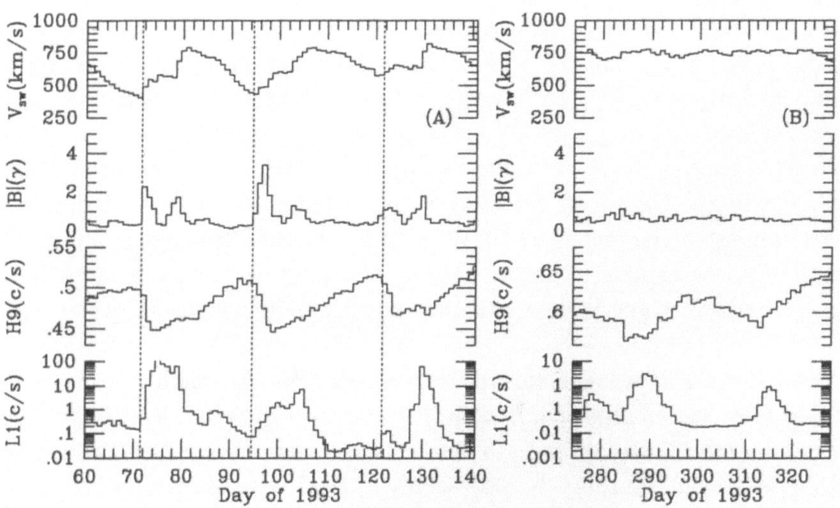

Fig. 2. Time variations at middle latitudes (A) and two rotations at high latitudes (B).

Examination of the synoptic charts of solar Fe XIV emission (*Solar-Geophysical Data*, Vols. 577–590, 1992-1993) revealed that the longitude position of the southern coronal hole at the latitude equivalent to that of the spacecraft did not vary much during July 1992 to July 1993. This is probably the reason that the variations in the solar wind velocity and in the intensity of cosmic rays were very regular as indicated by the good agreement between the "+" signs and the peaks of ΔH9 in Figure 1.

The amplitude of high-energy cosmic ray intensity variations was the largest in middle-latitude regions. The maximum amplitude of ΔH9 was about 8%. The appearance of the maximum variation at middle latitudes was a latitude effect and not the result of the time variation of solar activity, since the Maui neutron monitor did not observe the strongest variations in the same period (see Figure 1). This suggests that the strongest recurrent modulation region lies in the middle latitudes.

By using the same method as for region 1, we found that the "27-day" cosmic ray modulations both at the spacecraft and at Earth were produced by the same solar wind structure from the solar corona which was in phase with the low latitude recurring variation. The mechanism producing this type of coincidence is not clear. Future studies involving solar wind measurements in the ecliptic may help to resolve this question.

3. We were surprised to find that, although all evidence for CIR's had ceased to exist at all latitudes $\gtrsim$ 33°-35°S, the recurrent modulation of the galactic cosmic rays continued to the highest southern latitudes so far observable (~ April 1994). In the range of heliographic latitudes greater than ~ 33°S the spacecraft was continuously immersed in a high speed stream at ~ 750 km/s from the southern coronal hole. Although there were small magnetic field variations, as shown in FIGURES 1 and 2B, there were *no* identifiable compressed field regions in the passing solar wind. Furthermore, there was essentially no phase shift of recurrent modulation with respect to the lower latitudes. The amplitude of the recurring intensity modulation decreased with increasing latitude throughout this period by a factor ~ 2 by 1994.

The "27-day" variations appeared not only in high-energy cosmic ray protons (>92MeV, H9; >3GV, Climax; >15GV, Maui), but also in other cosmic ray elements and the anomalous helium component at lower energies. Anomalous (i.e., singly charged) helium recurring intensity variations are larger than for protons in the 38-50 MeV/nucleon energy range since the rigidity of anomalous He is higher than that of 600MeV protons. The strong variation in the helium is consistent with the variation in the H9 protons of energies > 92 MeV/nucleon. These observations suggest that the strength of the variations is not a function of energy per nucleon but a function of magnetic rigidity, which indicates that modulation is by particle diffusion and particle drift.

FIGURE 3 shows a magnetic rigidity spectrum for the amplitude of the "27-day" variation. Here the amplitude was defined to be the difference of averaged percentage variation from the 27-day averages of fluxes for the time when ΔH9>0 and for the time when ΔH9<0. Data from July 1, 1992 to July 1, 1993 were used in the calculation. We assumed that the helium (points with the symbol "b") was composed entirely of anomalous helium. The data points indicate the lower limits for the amplitude of the "27-day" variation at those rigidities. The amplitudes of variation for heavy cosmic rays nuclei (Z>3, mostly C and O) are plotted with the symbol "c". Two proton flux channels are indicated by the symbol "a". Two data points (with "d") are from the neutron monitor observations at Earth. The magnetic rigidity dependence for the "27-day" variations from Climax (~3 GV) to Huancayo (~20 GV) was first reported by Fonger (1953) and Simpson (1954) and is in approximate agreement with these new results. It appears that the maximum amplitude of the "27-day" modulation is somewhere near 600-1000 MV. This must be the result of the balance between the particle diffusion scale inside the CIRs and recovery time scale after the passage of the CIRs. For very-high-rigidity particles, the CIR modulation decreases with increasing rigidity because of their large gyroradius. On the other hand, for low-rigidity particles, their transport is slow and they cannot recover within 27 days; their intensity stays constantly at low levels.

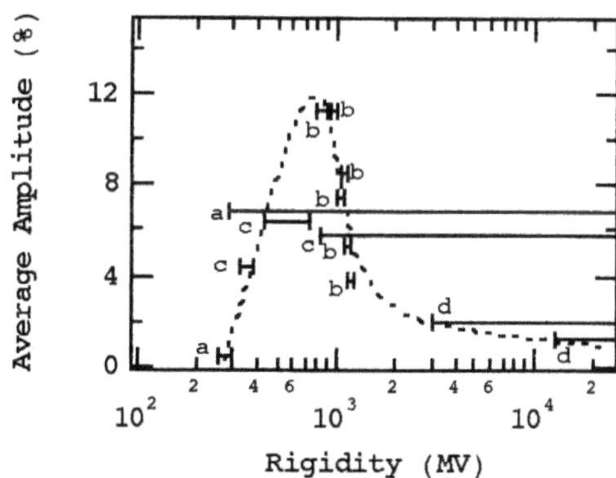

Fig. 3 Rigidity spectrum of amplitude of the recurrent modulation of galactic cosmic rays.

3. Summary and Conclusions

The principal new findings derived from our *Ulysses* investigations of corotating interaction regions and the recurring modulation of galactic cosmic rays and anomalous ^{4}He nuclear components are:

First, the discovery that, for latitudes in excess of ~ 40°S, the local recurring modulation of charged particles occurs without the detection by *Ulysses* of the solar wind velocity and magnetic field enhancements that have, heretofore, defined CIRs such as observed at lower latitudes - i.e., the observed particle recurring intensity variations are not a local modulation phenomenon.

Second, the phases of the recurring modulation at *Ulysses* and at Earth, as measured by neutron monitors, are the same and are consistent with having a common solar source for the solar wind stream and magnetic field which generates the CIR.

Third, by comparing the amplitudes of recurring modulation of the galactic cosmic rays and anomalous He nuclei, we find that the modulation mechanism and particle propagation probably is a function of particle magnetic rigidity, rather than energy per nucleon.

Fourth, since the in-phase modulation extends in magnetic rigidity from ~ 1 GV (*Ulysses*) to beyond 13 GV (Maui neutron monitor) the recurring modulation mechanism is a large-scale, three-D phenomenon in the heliosphere.

Previous models for the "27-day" recurring intensity variations relied upon the observed coexistence of the particle intensity variations and solar streams or, later, CIRs over low and intermediate latitudes. However, the new results we report extend recurrent modulation in three dimensions place new requirements on models which could explain their origin.

Recently, a three-dimensional magnetohydrodynamic numerical mode developed by Pizzo (1994) to compute the properties of "tilted dipole" CIRs in the distant solar wind can be extended to provide some insight into the origin of the discoveries we have reported in this paper. For example, Pizzo's modeling shows that a forward front of a CIR should propagate preferentially towards and across the equator, whereas reverse fronts should travel towards the poles (in the inertial or *Ulysses* frame) This model explains the persistence of reverse shocks to high latitudes observed by Sanderson et al (1994). This model should permit the formation of a modulating CIR beyond the radial range of *Ulysses* at high latitudes. The modulated particles could then propagate inward by fast diffusion and gradient drift (Jokipii and Kopriva, 1979) to the location of *Ulysses* beyond 40°S where the CIR had not yet formed In the months following this report we expect to more clearly interpret these new observations as *Ulysses* approaches the heliographic pole in September 1994.

4. Acknowledgments

We thank K.R. Pyle for providing the Climax and Maui neutron monitor data This work was supported in part at the University of Chicago by NASA/JPL Contract 955432 and NSF Grant 9215122.

5. References

Bastian, T S , McKibben, R B , Simpson, J A 1979, 'Recurrent modulation of galactic cosmic rays and the anomalous helium', *Proc. 16th Int'l Cosmic Ray Conf.*, **12**, 324

Barnes, C W , Simpson, J A 1976 'Evidence for interplanetary acceleration of nucleons in corotating interaction regions', *Astrophys. J. (Letters)*, **210**, L91

Fonger, W H 1953, 'Cosmic ray intensity-time variations and their origin, II energy dependence of 27-Day variations', *Phys. Rev.* **91**, 351

Jokipii, J R , Kopriva, D A 1979, 'Effects of particle drift on the transport of cosmic rays', *Astrophys. J.*, **234**, 384

Pizzo, V 1994, 'Global quasi-steady dynamics of distant solar wind 2, Deformation of the heliospheric current sheet', *J. Geophys. Res.*, **99**, 4185

Ibid, 'Origin of north-south flows in the outer heliosphere', 4173

Sanderson, T R , et al 1994, 'Ulysses high-latitude observations of ions accelerated by corotating interaction regions', *J. Geophys. Res*, in press

Simpson, J A 1954, 'Cosmic radiation intensity-time variation and their origin III the origin of 27-day variations', *Phys. Rev.*, **94**, 425

Simpson, J A , Smith, E J , Tsurutani, B T 1986, 'Shock acceleration of nucleons at >16° solar latitude associated with interplanetary corotating interaction regions', in R Marsden (ed), The Sun and the Heliosphere in Three Dimensions , Reidell, p 319

VARIATIONS OF THE HIGH ENERGY ELECTRON FLUX ALONG THE ULYSSES TRAJECTORY

RAVIART A.[1], DUCROS R.[1], FERRANDO P.[1], RASTOIN C.[1], HEBER B.[2], KUNOW H.[2], MÜLLER-MELLIN R.[2], SIERKS H.[2], WIBBERENZ G.[2], and PAIZIS[3] C.

[1]*Service d'Astrophysique, C.E. Saclay, 91191 Gif-sur-Yvette, France*
[2]*Institut für Kernphysik, University of Kiel, 2300 Kiel, Germany*
[3]*Instituto Fisica Cosmica del CNR, Universita di Milano, Italy*

Abstract. We describe for the first time the analysis of high energy electrons (above 240 MeV) from the COSPIN/KET experiment onboard Ulysses. The electron time profiles in four energy windows are presented from Oct. 90 to the end of March 94, up to a maximum heliographic latitude of 57 °S. The recovery rates we derived for the electrons are compared to the recovery rates of positively charged particles with the same rigidity.

1. Introduction

The understanding of the solar modulation process is one of the major aims of the Ulysses mission. The current solar modulation models, which include large scale drift effects in the heliosphere, predict major differences between electrons and protons because of their opposite charges. At the time of the symposium, Ulysses will be over 60° S of heliographic latitude, being in a good position for checking and constraining the modulation models.

We present for the first time the detailed data analysis of the highest electron energy channel of the KET experiment onboard Ulysses. Four energy windows, with average rigidities of about 1.1, 1.8, 2.3 and 5.0 GV were defined. In order to look for charge-dependent effects on the recovery of solar modulation, we compare solar rotation averaged time profiles of these four electron energy windows to those of positively charged particles with similar rigidities.

2. KET Data Analysis

A detailed description of the KET (Kiel Electron Telescope), including the definition of the coincidence channels, has been given in Kunow *et al.* (1991) and Simpson *et al.* (1992). The energy ranges for protons and alpha particles given in the above publications are valid for the spare model, which was initially selected for flight. Shortly before launch, the spare and flight models were exchanged. We have now completely analysed our calibration data for the KET model onboard Ulysses, and found differences in the energy ranges between flight and spare models, as detailed in Sect. 2.1. Sect 2.2 addresses the high energy electron analysis.

Space Science Reviews **72**: 409–414.
© 1995 *Kluwer Academic Publishers.*

2.1. PROTON AND ALPHA ENERGY WINDOWS

We are interested here in the energy proton and alpha coincidence channels listed in the table below. The corresponding energy windows are determined by the thresholds of the Cerenkov detector C1, the calorimeter C2, and the penetration scintillator S2 (see Simpson *et al.*, 1992). These thresholds have been measured with KET in the flight configuration (detectors and electronics) during calibration runs of the instrument at CERN, SATURNE, and ALS accelerators. We give below the energy ranges for these channels, as derived from the calibration analysis of the flight model.

KET channel	particle	Emin MeV/n	Emax MeV/n	mean Energy MeV/n	mean Rigidity MV
P116	proton	125	250	190	630
P190	proton	250	2200	1100	1750
P4000	proton	> 2200		7400	8300
A190	alpha	250	2200	900	3100
A4000	alpha	> 2200		8200	18200

We have also indicated in this table the average energy and the average rigidity for each channel. They were numerically calculated assuming typical shapes of the interplanetary proton and alpha spectra (Webber and Golden, 1987; Potgieter and Burger, 1990; Seo et al., 1991; Webber, 1993). The exact interstellar shape and modulation level have a very minor effect on the P116 average energy and rigidity. On the contrary, they result in a non negligible uncertainty on these values for the other channels, estimated to be around ± 10 %.

2.2. THE ELECTRONIC COMPONENT

We present here the first analysis of the E300 channel, corresponding to electrons passing through all detectors of the KET instrument. The bidimensional analysis of the signal amplitudes in the calorimeter C2 and the penetration detector S2 allows : i) a rejection of protons triggering an E300 coincidence down to a negligible level compared to the electron counts, and ii) an energy evaluation.

Figure 1 shows the bidimensional C2-S2 distributions obtained during accelerator calibrations with

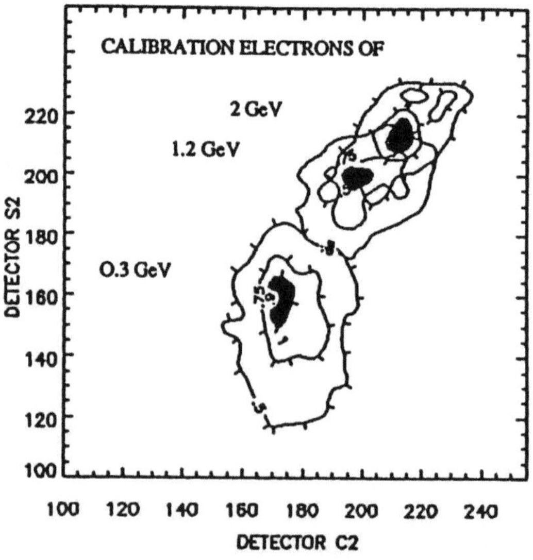

Fig 1: C2-S2 channel-number bidimensional distributions. Contour levels are plotted at 0.9, 0.75 and 0.5 of the maximum values.

300, 1200, and 2000 MeV electron beams. The poor C2 resolution and the S2 inhomogeneity prevent a direct measurement of the electron energy. However, Fig.1 indicates that it is possible to define reasonably well separated energy windows in the E300 channel.

We defined in C2-S2 four cuts (or "boxes") which were extensively characterized using calibration runs at ORME (7.5 to 150 MeV), BONN (50 to 1200 MeV) and DESY (200 to 5000 MeV) accelerators. The boxes are defined in such a way that they have reasonably large efficiencies with limited energy overlap. The relative efficiency of each box is derived from the accelerator runs, and combined to the Monte-Carlo (GEANT) calculated geometric factor of the E300 channel. This results in the energy-dependent geometric factors for each box, shown in Fig.2. The curves above 5 GeV are simple extrapolations of the calibration data. The FWHM limits of these curves are given in the following table.

KET E300-box	Emin MeV	Emax MeV	Rigidity MV
E300-1	240	1550	1100
E300-2	470	2960	1800
E300-3	880	4650	2300
E300-4	2420	52500	5000

As for protons, we derived the average rigidity for each box. The interplanetary electron spectrum we used gives a good fit to the box counting rates and is in full agreement with published electron spectra at high energy. The uncertainty on the average rigidities is estimated to be around ± 10 %.

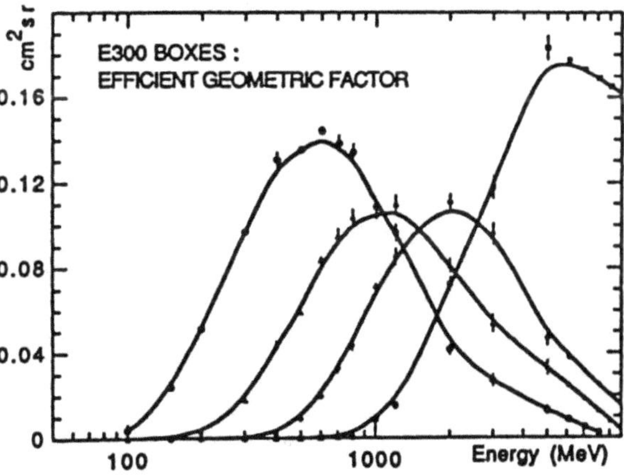

Fig 2: Energy-dependent geometric factors for the four electron boxes.

3. Observations

Figure 3 shows the time profiles of the 4 electron boxes counting rates, and of the P4000 channel, averaged over 104 days to minimise the CIR related effects observed at Ulysses, even at very high latitude (Kunow *et al.*, this conf.). Electrons, protons and alphas have similar time profiles. The counting rates increase from the beginning until the very large Forbush of mid 1991. This is followed by a regular increase up to the very beginning of 93, when hadron fluxes stay at a plateau value, or even decrease, during about 5 solar rotations (Heber *et al.*, this conf.). Although the statistics are poorer for electrons, this plateau is also visible in the electron channels. Then all particle fluxes increase again, with the appearance of a second plateau starting at the end of 93.

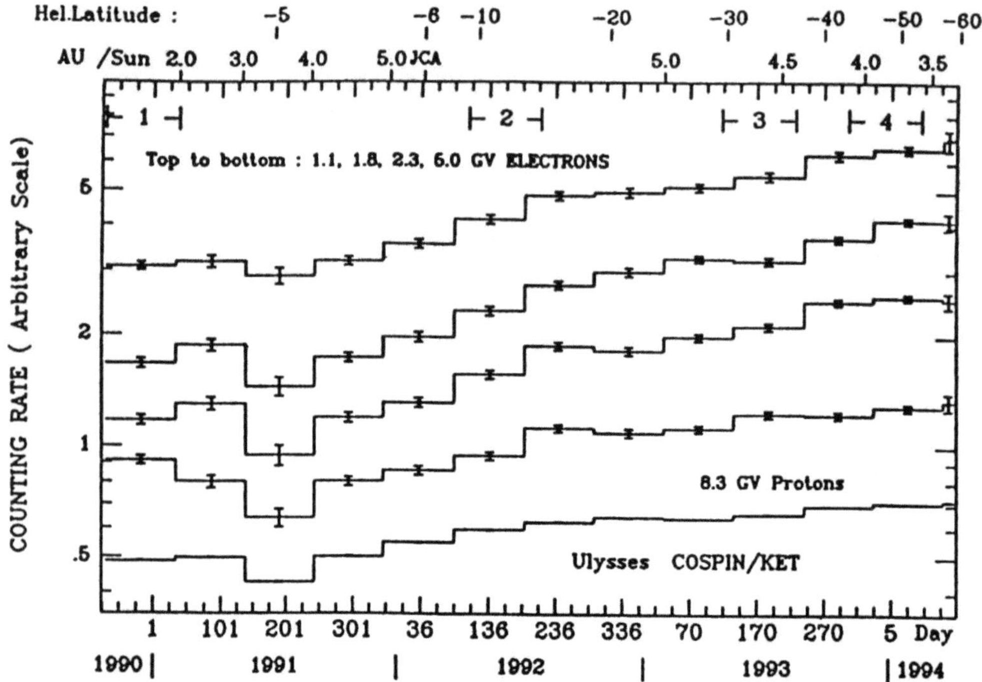

Fig 3: Time profiles of the four electron boxes and of the P4000 channel with 104-day averages. Periods of solar activity have been excluded.

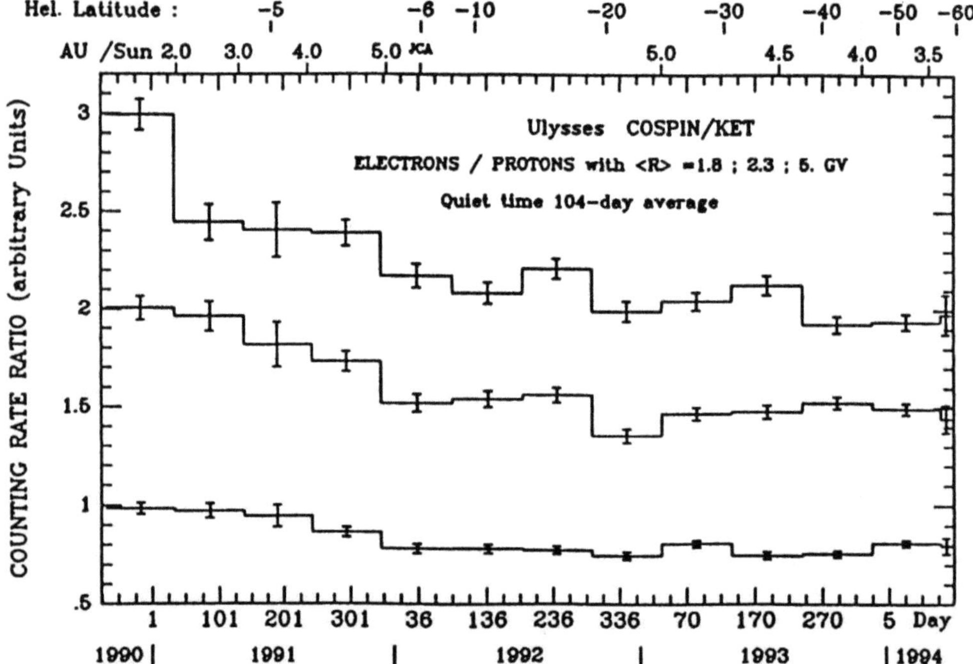

Fig 4: Time profiles of flux ratios of electrons with rigidities of 1.8, 2.3 and 5. GV (from top to bottom) to hadrons of the same rigidity. All channels are 104 day averaged. Ulysses distance from the Sun (in AU) and heliographic latitude are indicated above.

Figure 4 shows the counting rate ratios of electrons of 1.8, 2.3, and 5.0 GV to protons and alphas of corresponding rigidities. Hadron fluxes at 2.3 and 5.0 GV are computed by mixing channels P190, A190, and P4000. Fig. 4 gives evidence that the increase rate from launch to middle of 92 is much larger for protons and alphas than for electrons. Later, when Ulysses was above 20° of latitude, the protons, alphas and electrons at these rigidities have clearly the same rate of increase. There is thus no obvious effect of a charge dependence in this phase of the solar cycle. In particular, according to modulation models including drifts (Kota and Jokipii, 1983; Potgieter and Moraal, 1985), we should expect latitudinal gradients with opposite signs for electrons and positively charged particles. For this given cycle, and in an equilibrium case, this would result in electron/proton (alpha) ratios decreasing with increasing latitude. This is not observed. However, the heliosphere is not yet in equilibrium, the fluxes of cosmic rays are still rising, so that our observations do not invalidate the above predictions of the drift models. We note also that there is still no latitudinal gradient detected for protons at Ulysses (Heber *et al.*, this conf.).

Moreover, we defined four quiet time periods of 104 days each, as shown at the top of Fig.3, in order to investigate the rigidity dependence of the recovery rates.

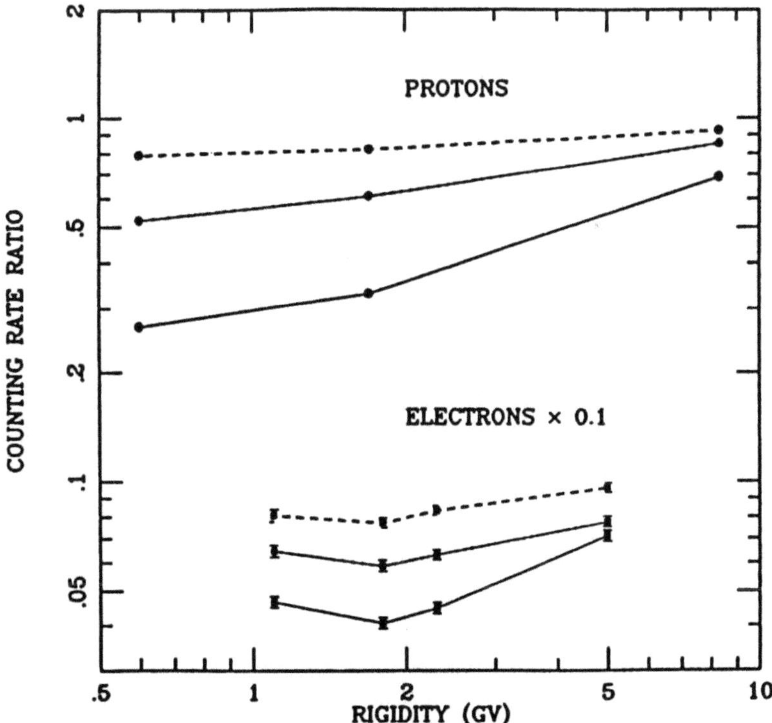

Fig 5: Proton and electron fluxes of time periods 1,2 and 3 relative to the flux of period 4 (see Fig. 3) versus the channel mean rigidity.

The fluxes in periods 1, 2 and 3 are compared to the temporary maximum flux at Ulysses, defined to be period 4. Figure 5 represents the ratios of fluxes of the periods 1, 2, 3, to that of period 4 for all the channels considered. Protons show a rather smooth rigidity dependence for the modulation recovery. The three higher energy electron channels also follow a similar trend. However, the lower energy electron channel does show a much smaller rate of increase than the rate extrapolated from the higher rigidity points. Indeed, rate ratios at 1.1 GV are even larger that those found for the three periods at 1.8 GV. We have currently no explanation for this effect.

References

Heber B. *et al.* : 1994, 'High-Energy Cosmic-Ray Nuclei Results on Ulysses: 1. Mission overview', this conf.

Kota J. and J.R. Jokipii : 1983, 'Effects of Drift on the Transport of Cosmic-Rays. VI. A three-dimensional Model Including Diffusion', Ap. J. **265**, 573.

Kunow H. *et al.* : 1991, 'The Kiel Electron Telescope onboard Ulysses', Proceedings of the 22nd Int. Cosmic Ray Conf., Dublin, **1**, 372.

Kunow H. *et al.* : 1994, 'High-Energy Cosmic-Ray Nuclei Results on Ulysses: 2. Effects of a Recurrent High-Speed Stream from the Southern Coronal Hole', this conf.

Potgieter M.S. and H. Moraal : 1985, 'A Drift Model for the Modulation of Galactic Cosmic-Rays', Ap. J. **294**, 425.

Potgieter M.S. and R.A. Burger : 1990, 'The Modulationic-Ray Electrons, Positrons and Helium Nuclei as Predicted by a Drift Model with a Simulated Wawy Neutral Sheet', Astron. Astroph. **233**, 598.

Seo E.S., *et al.* : 1991, 'Measurements of Cosmic-Ray Proton and Helium Spectra during the 1987 Solar Minimum', Ap.J. **378**, 763.

Simpson J.A., *et al.* : 1992, 'The Ulysses Cosmic-Ray and Solar Particle Investigations', A&A Suppl. **92**, 365.

Webber W.R, *et al.* : 1987, 'Cosmic-Ray Proton and Helium Spectra from 5-200 GV measured with a Magnetic Spectrometer', Proc. 20th Int. Cosmic Ray Conf., Moscow, **1**, 325.

Webber W.R. : 1993, private communication, interstellar spectrum resulting from a standard galactic propagation calculation.

POLAR RAIN ENTRY OF GALACTIC ELECTRONS INTO THE INNER HELIOSPHERE?

E. R. CHRISTIAN
USRA, Laboratory for High Energy Astrophysics, Code 661
NASA Goddard Space Flight Center, Greenbelt, Maryland 20771, U.S.A.

J. F. COOPER
Hughes STX Corporation, National Space Science Data Center
NASA Goddard Space Flight Center, Greenbelt, Maryland 20771, U.S.A.

Abstract. The understanding of the relative intensity variations in cosmic ray ions and electrons with respect to solar modulation is a grand challenge for cosmic ray modulation theory. Although effects of the heliospheric neutral sheet, gradient-curvature drifts, and merged interaction regions provide qualitative explanations for observed solar cycle variations of high energy protons and ions, these effects do not account for the anomalously high intensities of high energy galactic electrons at 22-year intervals of the solar magnetic solar cycle. From the similar modulation responses of protons and heavy ions it does not appear that cosmic ray pressure effects, dominated by protons, can account for the charge-sign asymmetry of cosmic ray modulation. External factors including modulation in the heliosheath and polar linkage to the interstellar magnetic field are examined as potential causes of symmetry breaking for electron modulation with respect to the solar magnetic polarity at solar minimum.

1. Introduction

Near solar minimum, the large-scale order of the heliospheric magnetic field influences the access of charged particles to the inner heliosphere via gradient-curvature drifts of ions (electrons) down from the polar heliosphere for outward (inward) magnetic fields at the northern solar pole and from the outer heliosphere at low latitude for the opposite solar polarity (Jokipii *et al.* 1977). McDonald *et al.* (1993) have reported that only for the so called A<0 solar minima, when the northern magnetic field is inward toward the sun (e.g., 1965, 1987) are the cosmic ray nucleon intensities significantly affected by drifts; more important at all other times are collective global effects of merged interaction regions from fast solar wind streams overtaking slower streams and from interplanetary shocks often associated with coronal mass ejections.

A significant limitation of using cosmic ray particle observations to study heliospheric dynamics is that the interstellar cosmic ray pressure, mostly from protons, may significantly contribute to the hot plasma and magnetic field configuration of the outer heliosphere where most of the observed modulation

Space Science Reviews **72**: 415–418.
© 1995 *Kluwer Academic Publishers.*

presumably occurs (e.g., Lee and Axford, 1988). This problem can only be avoided by using observations for cosmic ray particles acting as test particles with intensities (i.e., partial pressures) too low to significantly affect the heliospheric boundary configuration. Cooper and Baker (1993) have noted that a analogy to polar rain of superthermal solar wind electrons in the Earth's magnetosphere may be useful in placing constraints on the direction of the magnetic field in the Very Local Interstellar Medium (VLISM) within 10^3 A.U. of the Sun and outside the heliosphere. Galactic electron intensities at high energies (i.e., GeV) are an extremely small fraction (10^{-3}) of the galactic ion intensity.

2. Observations

Tuska *et al.* (1991) have compiled high energy ion and electron observations over more than two solar cycles and found that GeV electrons were relatively enhanced in intensity during the inward polarity solar minima of 1965 and 1987 when drifts would have provided access to the inner heliosphere from high latitudes. Although conventional modulation theory was adequate to describe modulation of protons and helium for data in the two solar cycles spanning the years 1965-1984, the onset of relatively higher electron intensities after solar polarity reversals in 1980 was difficult to understand in view of the absence of a corresponding increase in the ions during the opposite polarity and extended solar minimum in the mid-1970's. The final conclusion of Tuska *et al.* was that charge asymmetry of modulation arose from saturation of the proton component during the interval of preferred proton access from the polar heliosphere after the solar polarity reversals of 1969 (south pole) and 1971 (north pole).

If the conclusion of Tuska *et al.* (1991) was correct, we would then expect that ions heavier than protons and helium would show large increases relative to protons and electrons during the 1970's. The data from IMP-8, Pioneer 10, ISEE 3/ICE, and the Climax neutron monitor in Figure 1 show that this was not the case. All ion profiles for electrons, protons, and heavier ions through the CNO group were similar before 1981, and only the electron profile diverged thereafter. Since the CNO abundance in cosmic rays is two orders of magnitude below that of protons, it is unlikely that either component was near saturation at any time, contrary to Tuska *et al.*'s theory.

The observed trend of the charge-sign asymmetry effect continues into the 1990's, when Ulysses electron and ion data from Heber *et al.* (1993) show similar time profiles in the recovery from maximum solar activity in 1989-1991. After the next solar polarity reversals at year 2001-2002 the electron intensities would then show a relative increase. Such an increase might be observed near the end of a proposed extended mission for Ulysses as well as by continuing earth-based measurements with balloon and satellite experiments. A magnetic spectrometer measurement would be desirable to separate the electron and positron components.

3. Models

In this paper we propose that the galactic magnetic field in the immediate vicinity of the heliosphere may be the cause of the observed differences in electron and ion modulation. The direction of the very local galactic magnetic field is not known, but the large scale field has been determined. The local (< 3 kpc) magnetic field, as measured with pulsar rotation measures (Rand and Kulkarni 1989) points roughly in the direction of galactic rotation (~96° galactic longitude). In inertial heliographic coordinates this direction is northward 67° from the solar equator and 108° in east longitude from the direction of the nose of the heliosphere. Due to the local fluctuations about the measured mean direction of the galactic field, however, the direction of the local field just outside the heliosphere may be very different, especially given our location within a few parsecs of a small interstellar cloud (Bertin *et al.* 1993).

Solar magnetic field and solar wind flow out to the solar wind termination shock is qualitatively understood in the context of a Parker spiral. The region beyond the termination shock is not as well understood since there are no direct observations of this region. The level of cosmic ray modulation caused by this region is also not well modeled, although Quenby *et al.* (1990), and others have shown that modulation beyond the shock can be substantial. It is in this heliosheath that any connection between the solar and galactic magnetic fields would take place. In earth's magnetosphere, only one polarity of the solar wind field induces a "polar rain" of electrons; in the heliosphere the A<0 cycle of the sun may "short circuit" modulation of cosmic ray electrons in the heliosheath and bring higher fluxes into the solar poles.

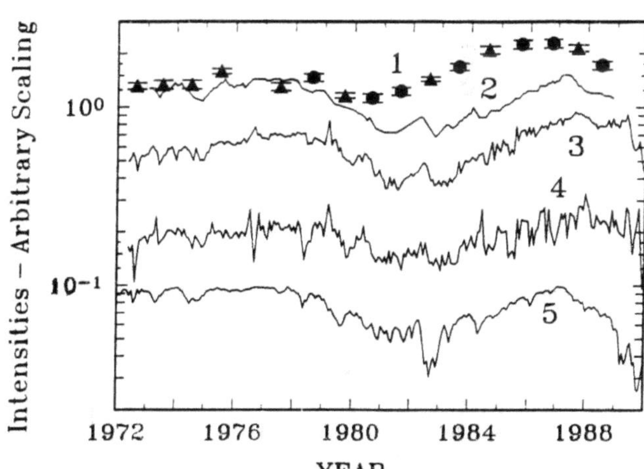

Fig. 1. Electron fluxes (1) at 5-6 GeV were reported by Tuska et al. (1991). Curves (2)-(5) show 27-day averages (flares removed) for >60 MeV protons (IMP-8 CPME experiment), >70-100 MeV/nucleon Z≥1 and Z≥5 nuclei (Pioneer 10 Charged Particle Instrument), and the Climax neutron monitor, respectively. IMP-8 and Pioneer 10 data are from the OMNI and COHO databases at NSSDC, respectively. Climax data are from the Univ. of Chicago.

Nerney *et al.* (1993) have modeled the heliosheath region, and have concluded that solar-VLISM field connection would occur only near the nose of the heliosphere where the heliopause breaks up into strips of alternating magnetic polarity due to the oscillation of the neutral current sheet at heliographic latitudes above and below the stagnation point of the inflowing VLISM gas. In this case there would be no difference in heliosheath modulation for solar cycles of opposite magnetic polarity. However, during solar minimum the tilt of the current sheet is of the same order as the offset of the stagnation point from the heliographic equator. This can result in large unipolar strips which map to the downstream magnetopause and heliosheath near each solar minimum.

Possibly, the heliosheath and VLISM fields are best connected during the A<0 phase. The heliosheath region may be too turbulent to maintain the smooth structures modeled by Nerney et al., so the external field might leak through the heliosheath and connect to the internal field at a rate common for planetary magnetospheres (e.g., ten percent). Since electron propagation along unipolar field lines can be very fast as compared to transport in fields of mixed polarity, even a small connection rate could significantly affect total modulation of electrons. Even if few field lines connect directly, the large scale geometry of fields penetrating into the polar heliosheath may enhance electron access via drifts for A<0 solar cycles.

4. Conclusion

The solar magnetic cycle of galactic electron modulation, as inferred from electron increases in 1965 and 1987, may have provided a critical new clue to the understanding of solar modulation and a sensitive probe for remote sensing of the heliospheric configuration beyond the termination shock. Theories of Quenby *et al.* (1990), and others, that a significant fraction of cosmic ray modulation probably takes place beyond the termination shock, need modification to allow for magnetic field configurations, including solar-VLISM field connection and drifts, which might account for the observed magnetic cycle effect on VLISM electron modulation.

References

Bertin, P., *et al.*: 1993, *J. Geophys. Res.* **98**, 15193.

Cooper, J. F., and Baker, D. N.: 1993, *Adv. Sp. Res.* **13(6)**, 103.

Heber, B., *et al.*: 1993, *Proc. 23rd Intern. Cosmic Ray Conf., Calgary* **3**, 461.

Jokipii, J. R., *et al.*: 1977, *Astrophys. J.* **213**, 861.

Lee, M. A., and Axford, W. I.: 1988, *Astron. Astrophys.* **194**, 297.

McDonald, F. B., *et al.*: 1993, *J. Geophys. Res.* **98**, 1243.

Nerney, S., Suess, S.T., and Schmahl, E.J.: 1993, *J. Geophys. Res.* **98**, 15169.

Quenby, J. J., Lockwood, J. A., and Webber, W. R.: 1990, *Astrophys. J.* **365**, 365.

Rand, R.J., and Kulkarni, S.R.: 1989, *Astrophys. J.* **343**, 760.

Tuska, E., Evenson, P., and Meyer, P.: 1991, *Astrophys. J.* **373**, L27.

THE ANOMALOUS COMPONENT OF COSMIC RAYS IN THE 3-D HELIOSPHERE

B. KLECKER
Max-Planck-Institut für extraterrestrische Physik, D 85740 Garching, Germany

Abstract More than 20 years ago, in 1972, anomalous flux increases of helium and heavy ions were discovered during solar quiet times. These flux increases in the energy range < 50 MeV/nucleon showed peculiar elemental abundances and energy spectra, e.g. a C/O ratio ≤ 0.1 around 10 MeV/nucleon, different from the abundances of solar energetic particles and galactic cosmic rays. Since then, this "anomalous" cosmic ray component (ACR) has been studied extensively and at least six elements have been found (He, N, O, Ne, Ar, C) whose energy spectra show anomalous increases above the quiet time solar and galactic energetic particle spectrum. There have been a number of models proposed to explain the ACR component. The presently most plausible theory for the origin of ACR ions identifies neutral interstellar gas as the source material. After penetration into the inner heliosphere, the neutral particles are ionized by solar UV radiation and by charge exchange reactions with the solar wind protons. After ionization, the now singly charged ions are picked up by the interplanetary magnetic field and are then convected with the solar wind to the outer solar system. There, the ions are accelerated to high energies, possibly at the solar wind termination shock, and then propagate back into the inner heliosphere. A unique prediction of this model is that ACR ions should be singly ionized. Meanwhile, several predictions of this model have been verified, e.g. low energy pick-up ions have been detected and the single charge of ACR ions in the energy range at ≈10 MeV/nucleon has been observed. However, some important aspects such as, for example, the importance of drift effects for the acceleration and propagation process and the location of the acceleration site are still under debate. In this paper the present status of experimental and theoretical results on the ACR component are reviewed and constraints on the acceleration process derived from the newly available ACR ionic charge measurements will be presented. Possible new constraints provided by correlative measurements at high and low latitudes during the upcoming solar pole passes of the ULYSSES spacecraft in 1994 and 1995 will be discussed.

1. Introduction

About 20 years ago anomalous flux increases in the low energy spectra of helium (< 50 MeV/nuc), oxygen, and nitrogen (< 20 MeV/nuc) were observed in near-Earth interplanetary space by several groups (Garzia-Munoz, Mason, and Simpson, 1973, Hovestadt et al, 1973, McDonald et al, 1974). The composition and energy spectra showed for instance a helium / proton ratio > 1 at 30 MeV/nucleon and carbon / oxygen ratio < 0.1 at 10 MeV/nuc, not compatible with solar and galactic cosmic ray abundances. Subsequently the elements neon (von Rosenvinge and McDonald, 1975, Klecker et al, 1977), Argon, and small amounts of carbon (C/O ≈1%, Cummings and Stone, 1987) have been detected in this "anomalous" cosmic ray (ACR) component. The apparent overabundance of elements with large first ionisation potential led to the

Space Science Reviews **72**: 419–430.
© 1995 *Kluwer Academic Publishers.*

hypothesis that the ACR component originates from interstellar neutral particles, penetrating into the inner heliosphere, being ionized by solar UV and by charge exchange with solar wind ions, convected into the outer heliosphere, and accelerated there to energies of 10 to ≈ 100 MeV/nuc. Being predominantly singly ionized, these ions would be much less modulated than galactic or solar cosmic rays of the same velocity and thus could be observed in the inner heliosphere (Fisk, Kozlovsky, and Ramaty, 1974). This model provided a qualitative explanation of the ACR component and much effort in the last 20 years was devoted to the quantitative modelling of the physical processes involved: (1) the penetration of interstellar neutrals into the heliosphere, (2) the ionization (pickup) of neutral particles in the solar wind, (3) losses due to charge exchange and adiabatic deceleration, (4) acceleration of pick-up ions from ≈ 4 keV/nuc to > 10 MeV/nuc, and (5) the modulation and propagation into the inner heliosphere. With the hypothesis of interstellar neutral particles being the source, the model also provided a crucial check, because ACR ions should be singly ionised.

2. Key Observations and their Implications

2.1 THE DETERMINATION OF THE IONIC CHARGE OF ACR IONS

The direct determination of the ionic charge in the energy range of tens of MeV/nucleon is still beyond the possibilities of current spacecraft experiments. However, a number of methods have been used over the last 20 years to infer the ionic charge of ACR ions. One method relies on arguments based on the rigidity dependence of the modulation in the heliosphere and uses rigidity dependent hystheresis effects or the shape of ACR spectra to infer the ACR ionic charge. It has been shown, for instance that observed phase lag effects were compatible with ACR helium being singly ionized (McKibben, 1977) and ACR oxygen having low charge states < 3 (Klecker et al., 1980). From the effect of modulation on the ACR energy spectra Cummings et al. (1984) inferred that ACR He, N, O, and Ne ions are singly ionized. Their finding was based on the assumption that the modulation depends on particle velocity and rigidity and that spectral features such as e.g. the flux maxima of different species show up at energies where the value of the diffusion coefficient is the same.

A different method with the advantage of not being dependent on propagation models in the heliosphere, has become available with ACR measurements onboard spacecraft in low altitude, high latitude orbits. This method utilizes the Earth's magnetic field as a magnetic spectrometer. The ionic charge can then be inferred from a comparison of the interplanetary ACR flux with the integrated low altitude flux (integral measurement) or by computing the trajectories of individual ions in the Earth's magnetic field (trajectory tracing method). From integral measurements of ACR oxygen onboard Skylab 3 Biswas et al. (1977) inferred that ACR oxygen is only partially ionized and recently Adams et al. (1991) derived a mean ionic charge $\langle Q \rangle = 0.9\ (+0.3/-0.2)$. For a recent review see e.g. Biswas et al. (1993).

The trajectory tracing method was first applied to ACR measurements obtained with experiments onboard Spacelab 1 and Spacelab 3. Because of limited counting statistics the ionic charge could be derived for a few particles only, however, singly ionized oxygen (Oschlies et. al., 1989) and singly ionized O, N, and Ne (Singh et al., 1991, Dutta et al., 1993) were observed. Since the successful launch of the Solar, Anomalous, and Magnetospheric Particle Explorer (SAMPEX) in July 1992 (Baker et al., 1993) with its advanced instrumentation optimized for the observation of ACR ions (Klecker et al., 1993) the trajectory tracing method can now be applied for a large number of ACR ions. With the much improved accuracy in the determination of observation time, position, and direction of particles, SAMPEX also provides both much improved counting statistics and better accuracy for the ionic charge determination. Using SAMPEX data obtained in December 1992, it was shown by Klecker et al., (1993) that all oxygen, nitrogen, and neon ions in the energy range 8 - 28 MeV/nuc observed at $L < 3$ (≈ 100 particles) were singly ionized. In summary, there is good evidence now that the ACR component is singly charged. However, a small contribution of higher charge states cannot be excluded yet and needs further investigation.

2.2 INTERSTELLAR NEUTRALS AND PICKUP IONS

According to the model of Fisk, Kozlovsky, and Ramaty (1974), interstellar pickup ions are the seed particles for anomalous cosmic rays. Thus, the knowledge of the distribution of neutral particles and pickup ions in the heliosphere is highly important for the understanding of ACR ions. The major supply of neutral particles in the heliosphere is the local interstellar medium (LISM) which enters the heliosphere at a relative speed of about 25 km/s. Evidence of interstellar hydrogen and helium in the heliosphere was provided by measurements of resonantly scattered solar UV already before the discovery of the ACR component (Bertaux and Blamont, 1971, Thomas and Krassa, 1971) and the existence of pickup ions had been inferred from these measurements (e.g. Holzer, 1977, and references therein). More recently, interstellar neutral helium has been measured directly with advanced instrumentation on the Ulysses spacecraft (Witte et al., 1993). Pickup He^+ ions have been detected with the AMPTE / IRM spacecraft at 1 AU in 1985 (Möbius et al., 1985). Other pickup ions, namely H^+, N^+, O^+, and Ne^+ were only recently discovered on the Ulysses deep space probe (Gloeckler et al., 1993, Geiss et al., 1994). Another important question is whether the composition of pickup ions in the heliosphere is similar to the composition of the LISM. As pointed out by e.g. Fahr (1990, 1991), charge exchange reactions in the heliospheric interface plasma could significantly modify the composition of the neutral LISM species H, O, N, whereas neutral He and Ne would be much less affected. Recent results of atomic abundances derived from the measured pickup ion abundances by Geiss et al. (1994) indicate, however, that these changes are small and very likely less than a factor of two as can be seen from He/O abundance in Fig 1. Another important recent finding from Ulysses is the observed acceleration of pickup H^+ and He^+, as well as of solar wind particles at the passage of a corotating interaction region (Gloeckler et al., 1994). These measurements are interesting in two respects: (1) they provide new insight into the injection process and (2) they show that a pre-accelerated component with energies of ≈ 100 keV/nuc may be available in the outer heliosphere for the acceleration to ACR energies.

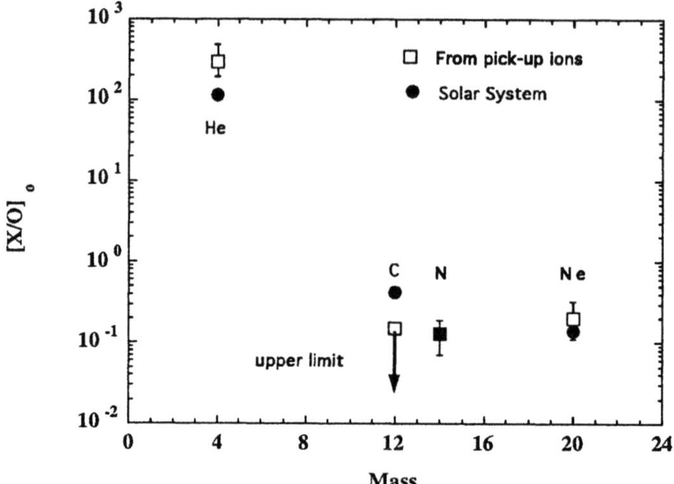

Fig. 1 Atomic abundances relative to oxygen derived from pickup ion observations (from Geiss et al., 1994), and solar system abundances from Anders and Grevesse, 1988.

2.3 LONG-TERM MODULATION

Measurements over the last two solar cycles showed that the anomalous component is especially sensitive to solar modulation. The flux of ACR oxygen at ≈ 10 MeV/nuc in the inner heliosphere varied for instance by a factor ≥ 100 and was observable at 1 AU only for solar minimum conditions. The maximum intensity shows a pronounced asymmetry in consecutive solar cycles, similar to the time profiles of galactic cosmic rays, indicative of the importance of the direction of the solar magnetic field, i.e. the importance of drift effects for the propagation in the heliosphere (Mewaldt et al., 1993).

2.4 COMPOSITION AND ENERGY SPECTRA

In addition to ACR He, O, N, and Ne discovered in the inner heliosphere, measurements in the outer heliosphere showed that also Ar and, to a much lesser extent, C (≈ 1%) exhibit anomalous flux increases above the galactic cosmic ray spectrum (Cummings and Stone, 1987). Possibly H has also been detected (Christian, Cummings and Stone, 1988). However, the decomposition of the observed H spectrum into an ACR and GCR component depends highly on the assumed GCR spectrum, and, as pointed out by McDonald et al. (1992), reduced adiabatic energy losses in the outer heliosphere might also account for the observations.

The energy spectra of ACR ions, after correction for solar and galactic cosmic ray contributions, exhibit a characteristic maximum, scaling with the mass of the ion (Cummings and Stone, 1987). The energy of the flux maximum depends on the acceleration and modulation processes and thus provides a sensitive boundary condition for the models. Significant (factor ≈ 2) differences of the energy of the flux maximum were reported for the ACR oxygen spectrum in 1985/1986 when compared with ACR fluxes during the previous solar minimum (Cummings and Stone, 1986, see also discussion in Fisk, 1986) However, subsequent measurements reported by Cummings and Stone (1990) show that these differences of the ACR oxygen spectra are much smaller for the actual solar minimum conditions reached in 1987. Only a very small shift of ≈ 30% of the energy of the flux maxima was also observed in 1987 for ACR helium (McDonald et al., 1992).

2.5 RADIAL AND LATITUDINAL GRADIENTS

The radial and latitudinal gradients also place important constraints on the acceleration and propagation models of ACR ions in the heliosphere. A parameter particularly sensitive to the model calculations is the latitude gradient. If drift effects are important then the latitude gradient is predicted to change sign in two consecutive solar cycles, i.e. with the reversal of the polarity of the solar magnetic field. This sign reversal was observed, indeed, in the last two solar cycles as shown in Fig. 2 (McKibben, 1989, Cummings et al., 1987). Radial and latitudinal gradients and new results obtained with experiments

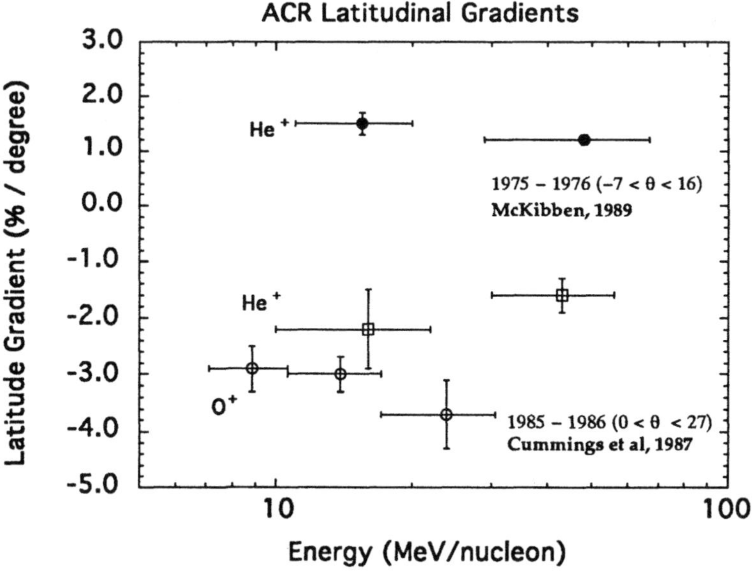

Fig 2 Latitudinal gradients in two consecutive solar cycles with different magnetic field polarity.

onboard the Ulysses spacecraft at latitudes < 60° are discussed in detail in another paper of this volume (McKibben, 1994). These results indicate that the latitudinal gradient in the present solar cycle so far is very small (≈ 1% per degree) and positive, in qualitative agreement with the model predictions. It will be very interesting to see the latitudinal gradient in the upcoming solar pole passes of the Ulysses spacraft in 1994 and 1995.

3. Acceleration and Modulation of ACR Ions

With the detection of pickup He, O, N, and Ne between 1 and 5 AU the seed particle population of ACR ions has been identified. The measured pickup energy distribution exhibits a typical cut-off at twice the solar wind velocity (Möbius et al., 1985, Gloeckler et al., 1993), thus the pickup ions need to be accelerated from ≈ 4 kev/nuc by a factor > 1000 to account for the observed energy of ACR ions. Whereas the origin of the ACR component now appears to be reasonably well established, the location of the acceleration site and the acceleration mechanism are still under debate. The observational constraints, i.e. the positive radial gradient observed out to ≈ 40 AU and the correlation of the modulation with the modulation of GCR, place the acceleration region into the outer heliosphere. So far essentially three theories have been developed for the acceleration of the anomalous component.

3.1 TRANSIT-TIME DAMPING

Acceleration by large scale magnitude fluctuations of the magnetic field in the outer heliosphere, generated by e.g. stream-stream interaction regions in the azimuthal field configuration of the low latitude solar wind has been proposed by Fisk (1976a). Using the magnitude and spectrum of these fluctuations as free parameters Klecker (1977) was able to reproduce the intensity and spectra of ACR oxygen, nitrogen, and neon reasonably well in a spherically symmetric, steady state model for the acceleration and modulation. However, whether the interaction of ACR ions with these fluctuations can be approximated by a linear theory is not completely clear (see also discussion in Fisk, 1986). With the deep space probes coming closer to the termination shock an experimental test may become available, because the energy spectra expected from transit-time damping (see e.g. Fisk, 1976b, Klecker, 1977) would not show the power law energy dependence at low energies as predicted by diffusive shock acceleration discussed below.

3.2 ACCELERATION AT THE TERMINATION SHOCK OVER THE SOLAR POLES

A qualitative model for the acceleration of ACR ions at the quasi-parallel termination shock over the solar poles, including also the effects of gradient and curvature drifts, was proposed by Pesses, Jokipii, and Eichler (1981). Particles can be accelerated at this shock by being scattered back and forth across the shock front, and gain energy by the converging solar wind flow (diffusive shock acceleration, also called the first-order Fermi mechanism), much as it occurs at interplanetary shocks or at the Earth's bow shock. A quantitative model of this acceleration process, including the effects of diffusion, convection, adiabatic deceleration, and drift and assuming particle injection at small polar angles was presented by Jokipii (1986). An important result of the drift models is the prediction of a 22 year cycle which apparently is observed for the modulation and latitudinal gradient as discussed above. The spectral differences in two consecutive solar cycles as initially predicted by the model can qualitatively be understood from the drift path of positive ions: for time periods with the solar magnetic field pointing outward in the northern hemisphere ($qA>0$, 1972 - 1975 solar minimum) positive ions injected near the poles spend considerably less time near the termination shock than during time periods with $qA<0$ (1987 solar minimum), when these particles drift to larger values of the polar angle when they gain energy. However, as the results for solar minimum conditions in 1987 show, the measurements now indicate that the spectral differences are small. This could be due to drift effects being less important or due to the assumption on the injection region. However, it could also be due to the polar magnetic field configuration: if the polar magnetic field is significantly larger than the nominal Archimedian spiral field as proposed by Jokipii and Kota (1989), drift effects would be reduced considerably and the spectral differences would be reduced by about an order of magnitude (Jokipii, 1990).

3.2 ACCELERATION AT THE TERMINATION SHOCK IN THE ECLIPTIC

Acceleration at the quasi-perpendicular termination shock near the equatorial plane has been suggested by Fisk (1986). In this nearly azimuthal field configuration there is a potential problem with the injection as also pointed out by Pesses et al. (1981): for the diffusive acceleration process to work, the injected particles have to travel back upstream. However, large directional fluctuations of the on average nearly azimuthal field have been observed with the Pioneer 11 spacecraft at distances of $\approx$ 32 AU (Smith, 1993). Using the width of the measured distribution ($\Phi_{FWHM} \sim 35°$) and assuming that the distribution will be similar near the termination shock, the percentage of time when the angle between the magnetic field direction and the shock normal (Θ_{BN}) is e.g. less than 70°, is $\sim$ 20%, i.e. injection at the equatorial plane seems not to be a problem. In fact, the injection may take place at all polar angles.

In the next paragraph time scales for different acceleration and loss processes will be investigated in order to obtain constraints for the acceleration mechanism and acceleration site.

4. Time Scales for Acceleration and Loss Processes

As pointed out by Jokipii (1992), the time scales of loss processes acting on ACR ions in the outer heliosphere impose severe observational constraints on possible acceleration mechanisms. Relevant loss processes are (1) losses due to charge exchange reactions and (2) energy loss due to adiabatic deceleration. The loss time scales τ_{loss} can be compared with acceleration time scales τ_{acc} and only mechanisms with $\tau_{acc} < \tau_{loss}$ will provide efficient acceleration. The present estimates for the distance of the termination shock range from $\approx$ 60 to $\approx$ 90 AU (Potgieter, 1993), therefore for the following estimate of acceleration and loss time scales a termination shock distance of 80 AU will be assumed.

4.1 LOSS TIME SCALES

Losses by charge exchange reactions of O^+ ($O^+ \rightarrow O^{n+}$) in the local interstellar medium have been investigated by Adams and Leising (1991). Using a local interstellar neutral hydrogen density of $n_H = 0.1 / cm^3$ and an upper limit of $\langle Q \rangle < 1.6$ for the mean charge state of oxygen as reported by Adams et al. (1990), they inferred a maximum propagation distance d_{max} of 0.2 pc for O^+ ions at 10 MeV/nuc and concluded from this result that the ACR source must be local. Applying this maximum propagation distance to the outer heliosphere, it can be converted to a loss time scale of $\tau_{loss} = d_{max} / v = 4.6$ years.

At low energies (< 0.1 MeV/nuc), the dominant loss process is charge pick-up. With typical values of $\lambda_{ec} \approx 10^{-16}$ cm^2 (e.g. Spjeldvik, 1979) for the electron capture cross section of O^+ at 0.1 MeV/nuc and $n_H = 0.1 / cm^3$ we obtain $\tau_{loss} = 1 / (v \lambda_{ec} n_H) = 0.73$ years.

With particle momentum p, heliospheric radial distance r and solar wind velocity V_{SW}, the time scale for adiabatic deceleration is $\tau_{ad} = (1/p \; dp/dt)^{-1} = 3r / 2V_{SW}$. Using 350 km/s as an average solar wind velocity, this results in $\tau_{ad} = 1.63$ y at 80 AU. Note, that near the termination shock the energy loss rate may be reduced because particles are alternating between the upstream and downstream region, as pointed out by Jokipii (1992).

4.2 ACCELERATION TIME SCALES

In the following paragraph we will compare the loss time scales derived above with acceleration time scales for the first- and second order Fermi process and transit-time damping.

The second order Fermi acceleration is generally slow , it scales as $\tau_{F2} \approx 2 \kappa / V_A^2$ (e.g. Ip and Axford, 1986), where κ is the diffusion coefficient and V_A the Alfvén velocity. A lower limit for the acceleration time can be obtained by using the Bohm limit (κ_B) for the diffusion coefficient, $\kappa_B = v \, r_g / 3$), with particle velocity, v, and particle gyroradius r_g. For the following discussion a nominal Archimedian spiral magnetic field configuration is assumed with typical solar wind and magnetic field parameters at 1 AU (v_A = 50 km/s, B = 5 nT, V_{SW} = 350 km/s). The resulting acceleration times for 10 MeV/nuc O^+ at 80 AU and 0° and > 80° latitudes are 60 and > 300 years, respectively. It is evident from this result that the acceleration time scale for second order Fermi acceleration is much too long to account for the observed energies.

The acceleration rate for transit-time damping (τ_{ttd}) depends on the correlation lengths scales, on the magnitude fluctuations of the magnetic field ($\eta = <(\delta B)^2>/ B_o^2$) and on the power spectrum index , q, of the fluctuations, and scales approximately as $\tau_{ttd} \sim \eta \; v \; r_g^{q-2}$, provided the gyroradia are small compared to the correlation lengths of the fluctuations. (Fisk, 1976a, Klecker, 1977). Typical values obtained for the acceleration rate of He^+ at 10 MeV/nuc and 50 AU using η=0.1, and typical solar wind parameters are $2*10^{-8}$ sec^{-1} (Fisk, 1976a, Klecker, 1977). Using the simple scaling law given above, this value corresponds for q=1 to an acceleration rate of $5*10^{-9}$ sec^{-1} for O^+ at 10 MeV/nuc, i.e. to an acceleration time scale $\tau_{ttd} \sim 3$ years . This is somewhat larger than the time scale for adiabatic energy losses and comparable with charge exchange losses (see also Fig 3). However, considering that this statistical acceleration also results in considerable energy spreading, transit-time damping cannot be excluded on the basis of the constraints posed by the loss processes discussed.

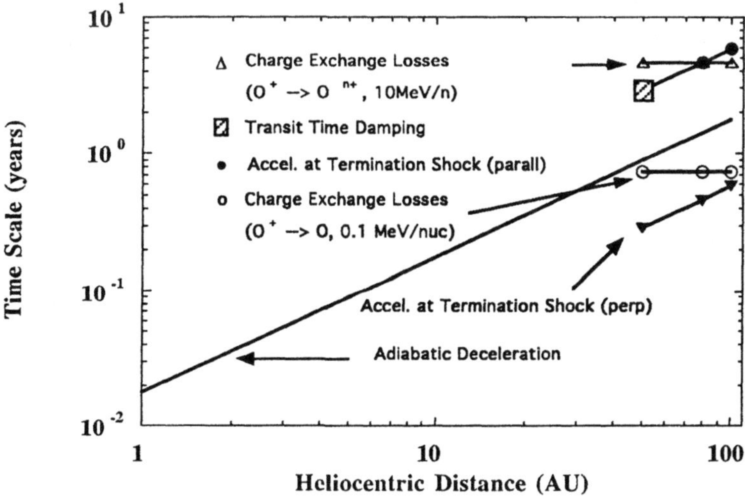

Fig. 3 Acceleration and loss time scales at 0° latitude in the outer heliosphere, for details see paragraph 4.

The acceleration time scales for first order Fermi acceleration (τ_{F1}) at parallel and perpendicular shocks have been reviewed by Jokipii (1987) and were subsequently used by Jokipii (1992) for an estimate of the acceleration time scales of ACR oxygen. Using again the Bohm limit for the diffusion coefficient, a lower limit of the acceleration time scale at a quasi-parallel shock is given by : $\tau_{F1||} = 2\, r / (r-1)\; v\; r_g / V_{sh}^2$ (Jokipii, 1992), where r is the shock compression ratio and V_{sh} the shock velocity. This time scale is plotted in Fig. 3 for 10 MeV/nuc O^+ and 0° latitude as a function of heliocentric distance and in Fig 4 as a function of latitude at 80 AU. In order to approximate the influence of the high solar wind velocity in the polar coronal holes a linear increase of the solar wind velocity from 350 km/s at 45° to 700 km/s at 90° latitude has been assumed in Fig. 4. Considering that these values of $\tau_{F1\parallel}$ are a lower limit and that at low latitudes quasi-parallel conditions will only exist for a small fraction of the time, diffusive shock acceleration seems to be too slow to compete with the losses at 10 MeV/nuc. However, with $\tau_{F1\parallel} \sim v\, r_g$ the acceleration time scale for 0.1 MeV/nuc O^+ is a factor of $\sim$ 100 smaller than the values shown in Fig. 3 and 4, i.e. the charge exchange loss time scale of 0.73 years at 0.1 MeV/nuc does not seem to pose a serious problem.

Much smaller time scales can be obtained for acceleration by a quasi-perpendicular shock, because here κ_{perp} is more relevant than $\kappa_{\parallel}$ and κ_{perp} is not limited by the gyroradius. The acceleration time scales $\tau_{F1,perp}$ obtained with typical values of κ as used in modulation model calculations ($k_{\parallel} = 5\ 10^{21}\ R^{0.5}\ \beta\ B_E/B$ and $kperp = 0.03^* k_{\parallel}$, Jokipii, 1992) is also shown in Fig. 3 and Fig. 4. These time scales are considerably smaller than the loss time scales , i.e. diffusive shock acceleration at a perpendicular shock would satisfy the observational constraints for a large range of latitudes.

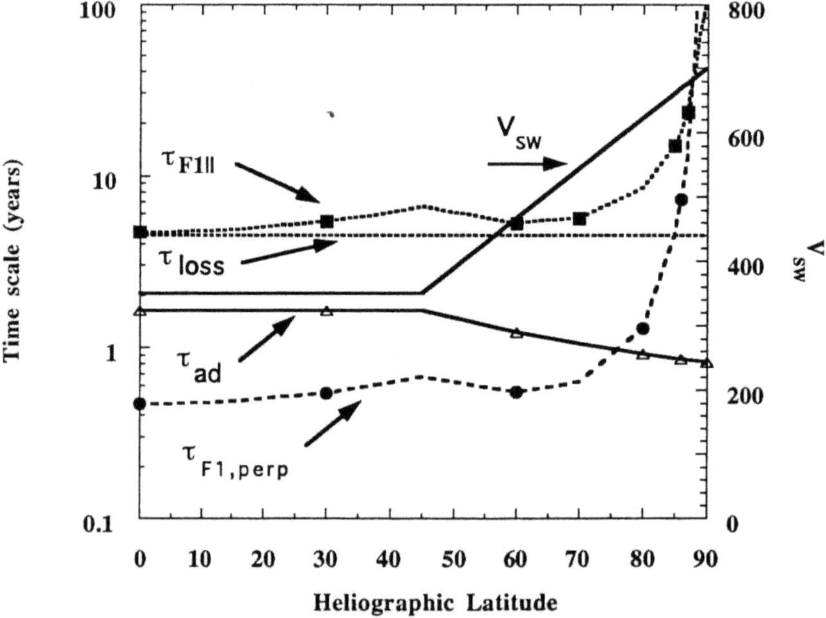

Fig. 4 Time scales for acceleration and loss processes at 80 AU, for details see paragraph 4.

5. Summary

There is now compelling evidence that the source of the anomalous component of cosmic rays is indeed interstellar neutrals, ionized and accelerated in the heliosphere as already proposed 20 years ago. In particular, the ionic charge of ACR ions at $\approx$ 10 to 30 MeV/nuc has been determined and singly charged O, N, Ne have been unambiguously identified. In addition to interstellar pickup He^+, the pickup ions H^+, O^+, N^+, and Ne^+ have been detected with the novel plasma composition experiments onboard the Ulysses spacecraft. This enables us now to measure the seed population of ACR ions directly.

Much less clear is the question of the acceleration mechanism and the acceleration site. Comparing acceleration and loss time scales it is evident that a second order Fermi process is too inefficient to compete with charge exchange losses and energy loss by adiabatic deceleration. Transit-time damping at low latitudes cannot be ruled out, however, it remains still to be shown that the magnitude fluctuations of the magnetic field do have the required fluctuation levels and scale sizes. Acceleration at a quasi-parallel shock at high latitudes as originally proposed does not have the required small acceleration time scale if a nominal Archimedian spiral field is assumed. However, an increased polar field as also discussed in connection with propagation models could change this result significantly. The shortest acceleration times are obtained with diffusive shock acceleration at a quasi-perpendicular shock. In fact, this acceleration time scale satisfies the constraints imposed by the loss processes over a large range of latitudes.

So far Ulysses provided already important new results on the seed particle population of ACR ions. The comparison of ACR fluxes and energy spectra obtained during the polar passes of Ulysses in 1994 and 1995 with the data from near-Earth and interplanetary spacecraft can be expected to provide a new and unique opportunity for the test of the acceleration site, the acceleration mechanism and the propagation of the anomalous component in the heliosphere.

References

Adams, J.H., et al.: 1991, *Astrophys. J.* **375**, L45.

Adams, J. H., and Leising, M.D.: 1991, *Proc 22nd Int. Cosmic Ray Conf.*, **3**, 304.

Anders, E., and Grevesse, N.: 1988, *Geochim. Cosmochim. Acta* **53**, 197.

Baker, D.N., Mason, G.M., Figueroa, O., Colon, G., Watzin, J.G., and Aleman, R.M.: 1993, *IEEE Trans. Geosci. Remote Sensing* 31, 531.

Bertaux, J.L., and Blamont, J.E.: 1971, *Astron. Astrophys.* **11**, 200.

Biswas, S., Durgaprasad, N., Nevatia, J., Sarkar, S., and Vekatavaradan, V.S.: 1977, *Proc 15th I nt. Cosmic Ray Conf.*. **2**, 327.

Biswas, S., Durgaprasad, N., Mitra, B., and Dutta, A.: 1993, *Space Science Rev.* **62**, 3.

Christian, E.R., Cummings, A.C. and Stone, E.C.: 1988, *Astrophys. J.* **334**, L77.

Cummings, A.C, and Stone, E.C: 1984, *Astrophys. J.*, **287**, L99.

Cummings, A.C, and Stone, E.C.: 1986, *J. Geophys. Res.* **91**, 2896.

Cummings, A.C, and Stone, E.C.: 1987, *Proc 20th Int. Cosmic Ray Conf.*. **3**, 413.

Cummings, A.C., Stone, E.C., McDonald, F.B., Webber, W.R., and Lal, N: 1987, *Proc 20th Int. Cosmic Ray Conf.*. **3**, 417.

Cummings, A.C., and Stone,E.C.: 1990, *Proc 21st Int. Cosmic Ray Conf.*. **6**, 202.

Dutta, A., Goswami, J.N., Biswas, S., Durgaprasad, N., Mitra, B., and Singh, R.K.: 1993, *Astrophys. J.* **411**, 418.

Fahr, H.J.: 1990, in S. Grzedzielski and D.A. Page (eds), *Physics of the Outer Heliosphere*, Pergamon, New York, pp 327-343.

Fahr, H.J.: 1991, *Astron. Astrophys.* **241**, 251.

Fisk, L.A., Kozlovsky, B., and Ramaty, R.: 1974, *Astrophys. J.* **190**, L35.

Fisk, L.A.: 1976a, *J. Geophys. Res.* **81**, 4633.

Fisk, L.A.: 1976b, *J. Geophys. Res.* **81**, 4641.

Fisk, L.A.: 1986, in R.G. Marsden (ed), *The Sun and the Helisphere in Three Dimensions*, Reidel Publishing Company,, Dordrecht, pp 401-412.

Garcia-Munoz, M., Mason, G.M., and Simpson, J.A.: 1973, *Astrophys. J.* **182**, L81.

Geiss, J., Gloeckler, G., Mall, U., von Steiger, R., Galvin, A.B., and Ogilvie, K.W.: 1994, *Astron. Astrophys.* **282**, 924.

Gloeckler, G., et al.: 1993, *Science* **261**, 70.

Gloeckler, G., et. al.:1994, *J. Geophys. Res.* (accepted for publication).

Holzer, T.E.: 1977, *Rev. Geophys. Space Phys.* **15**, 467.

Hovestadt, D., Vollmer, O., Gloeckler, G., and Fan, C.Y.: 1973, *Phys. Rev. Letters* **31**, 650.

Ip, W.H., and Axford, W.I.: 1986, *Planet. Space Sci.* **34**, 1061.

Jokipii, J.R.:1986, *J. Geophys. Res.* **91**, 2929.

Jokipii, J.R.:1987, *Astrophys. J.* **313**, 842.

Jokipii, J.R., and Kota, J.: 1989, *Geophys. Res. Lett.* **16**, 1.

Jokipi, J.R.: 1990, in S. Grzedzielski and D.A. Page (eds), *Physics of the Outer Heliosphere*, Pergamon, New York.

Jokipii, J.R.:1992, *Astrophys. J.* **393**, L41.

Klecker, B.: 1977, *J. Geophys. Res.* **82**, 5287.

Klecker, B., Hovestadt, D., Gloeckler, G., and Fan, C.Y.: 1980, *Geophys. Res. Letters* **7**, 1033.

Klecker, B., et al.: 1993, *IEEE Trans. Geosci. Remote Sensing* **31**, 542.

Klecker, B., Hovestadt, D., Scholer, M., Blake, J.B., Mabry, D., and Mason, G.M.: 1993, *Proc 23. Int. Cosmic Ray Conf.*. **3**, 427.

McDonald, F.B., Teegarden, B.J., Trainor, J.H., and Webber, W.R.: 1974, *Astrophys. J.* **187**, L105.

McDonald, F.B, Moraal, H, Reinecke, J.P.L., Lal, N., and McGuire, R.E.: 1992, *J. Geophys. Res.* **97**, 1557.

McKibben, R.B.: 1977, *Astrophys. J.* **217**, L113.

McKibben, R.B.:1989, *J. Geophys. Res.* **94**, 17021.

McKibben, R.B.:1994, this volume.

Mewaldt, R.A., et al.: 1993, *Geophys. Res. Lett.* **20**, 2263

Möbius, E. , Hovestadt, D. Klecker, B., Scholer, M., Gloeckler, G., and Ipavich, F.M.: 1985, *Nature* **318**, 426.

Oschlies, K., Beaujean, R, and EngeW.: 1989, *Astrophys J.* **345**, 776.

Pesses, M.E., Jokipii, J.R., and Eichler, D.: 1981, *Astrophys. J.* **246**, L85.

Potgieter, M.S.: 1993, in D.A. Leahy, R.B. Hicks, and D. Venkatesan (eds), *Invited, Rapporteur & Highlight Papers of the23. Int. Cosmic Ray Conf.*, World Scientific, Singapore, p 213-242.

Singh, R. K., et al.: 1991, *Astrophys. J.* **374**, 753.

Smith, E. J.: 1993, *Adv. Space Res.* **6**, 5.

Spjeldvik, W.N.: 1979, *Space Sci. Rev.* **23**, 499.

Thomas, G.E., and Krassa, R.F.: 1971, *Astron. Astrophys.* **11**, 218.

von Rosenvinge, T.T., and McDonald, F.B.: 1975, *Proc 14th Int. Cosmic Ray Conf.*. **2**, 792.

Witte, M., Rosenbauer, H., Banaszkiewicz, M., and Fahr, H.: 1993, *Adv. Space Res.* **13**, 121.

THE EVOLUTION OF THE ANOMALOUS COMPONENT OF THE COSMIC RAYS DURING THE ULYSSES MISSION

K. J. TRATTNER, R. G. MARSDEN, T. R. SANDERSON and K.-P. WENZEL
Space Science Dept. of ESA. ESTEC, Noordwijk, The Netherlands

June 29, 1994

Abstract.
We have studied the evolution of the anomalous cosmic ray component at ULYSSES using observations of the quiet time helium fluxes obtained with the COSPIN Low Energy Telescope during the years from launch in October 1990 to the latest data in 1994. Shown are preliminary results of measurements in the energy range from 4.0 to 19 MeV/n. Following the drop in solar activity in late 1991, the low energy cosmic ray helium flux started to increase and continue to rise until the end of 1993, signifying the appearance of the anomalous component at ULYSSES.

1. Introduction

Interplanetary space is rich in particle phenomena up to several GeV. Since the beginning of the space age it has been known that there are at least two main sources of energetic particles: from outside the solar system the so-called Galactic Cosmic Rays (GCR), and solar energetic particles (SEP) associated with transient events on the sun. However, about twenty years ago a third component of energetic particles, the anomalous component of cosmic rays (ACR) was discovered by instruments onboard Pioneer-10, IMP-5 and IMP-7. This third component shows an anomalous increase, compared to the other two components, of species having high first ionization potentials in the energy range below 50 MeV/n (e.g., Hovestadt *et al.*, 1973; McDonald *et al.*, 1974). While carbon and oxygen in SEP and GCR are almost equally abundant, oxygen exceeds carbon in abundance by 30 to 1 in the ACR. There are also more alpha particles than protons in the ACR, while in GCR and SEP protons are typically $\geq$10 times more abundant than alpha particles.

It is generally believed that the source of the ACR is the neutral interstellar gas. These particles are unaffected by the interplanetary magnetic field and drift into the inner heliosphere where a fraction becomes singly ionized by solar UV or by charge exchange with the solar wind. The ions are picked-up by the solar wind and convected into the outer heliosphere where they gain energy of 10 to 100 MeV/n due to shock acceleration at the solar wind termination shock (e.g., Fisk *et al.*, 1974; Pesses *et al.*, 1981). This model explains also the unusual composition of ACR which is caused by the different ionization potentials of the atoms in the interstellar medium.

Space Science Reviews **72**: 431–434.
© 1995 *Kluwer Academic Publishers.*

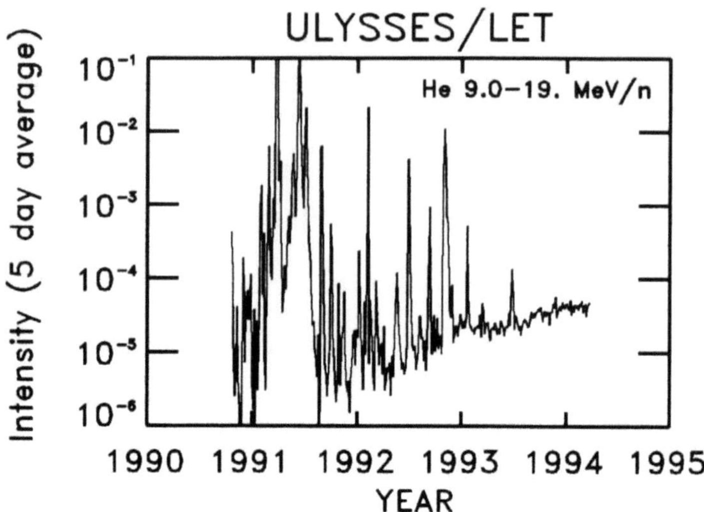

Fig. 1. The alpha particle intensity in the energy range from 9 MeV/n to 19 MeV/n during the complete ULYSSES mission. The data are averaged over five days.

Atoms with higher ionization potentials like He, N, O and Ne are predominantly neutral in the interstellar medium and are able to enter the inner heliosphere more easily than atoms with lower ionization potentials (e.g., C, Mg, Si and Fe).

The intensity of the ACR is strongly modulated by the solar activity. Similar to the intensity variations of the GCR (about a factor of 10), the ACR intensity is inversely correlated to the 11-year solar activity cycle and shows variations between solar maximum and minimum of about 100. Due to these strong variations the ACR at 1 AU is only observable at solar minimum (see e.g., Mewaldt *et al.*, 1993).

The trajectories of energetic particles in the interplanetary medium depend on diffusion, convection with the solar wind and drift processes (curvature and gradient drift) in the magnetic field. The direction of the drift of energetic particles in the interplanetary medium (towards the pole or from the polar region) depends on the magnetic topology which also reverses in a 11 year cycle as the solar magnetic field reverses (e.g., Jokipii and Davila, 1981). We expect for the current decrease in solar activity a similar behaviour to 22 years ago with an increase of the ACR intensity towards the solar pole.

2. Observations

The following analysis is based on data from the Low Energy Telescope (LET) of the COSPIN (Cosmic Ray and Solar Particle Instrumentation)

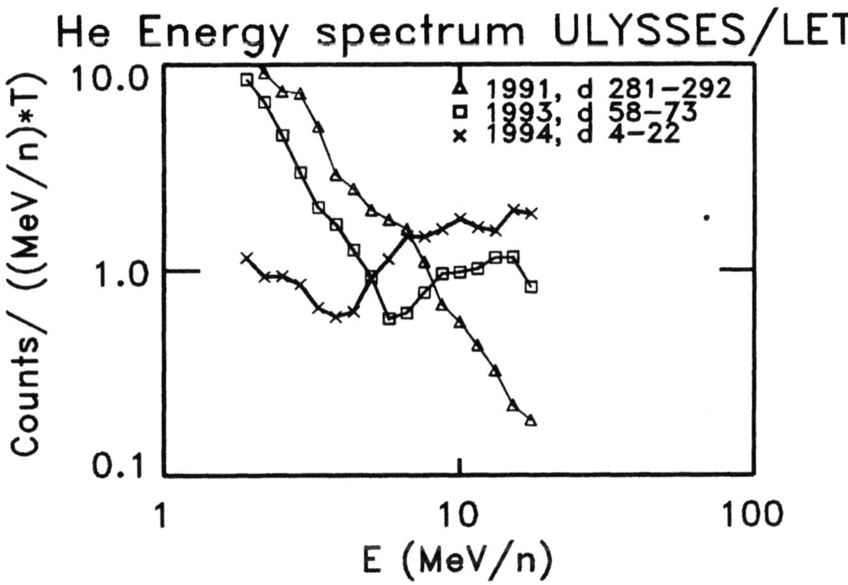

Fig. 2. Alpha particle energy spectra compiled from ULYSSES LET data for three quiet time periods in 1991 (days 281–292), 1993 (days 58–73) and 1994 (days 4–22).

instrument on board the ULYSSES spacecraft. This instrument is able to measure protons, alphas and heavy ions over a range of energies from ~1 to 50 MeV/n. For this preliminary investigation we use alpha particle data in the energy range from 4 to 19 MeV/n. A complete description of the COSPIN instrument can be found in Simpson *et al.* (1992).

Figure 1 shows five day average values of the alpha particle intensity in the energy range from 9 to 19 MeV/n where ACR should be detectable. Plotted are data from the complete ULYSSES mission up to March 24th, 1994, covering a latitude range of 0° – 50° S. As mentioned above, the intensity of the ACR depends strongly on solar activity and was masked during solar maximum in 1991 by the high intensity of solar energetic particles accelerated by flare events (see Figure 1). Solar activity started to decrease in late 1991 but even in this declining phase there are still spikes visible in the data profile mentioning bursts of energetic particles accelerated by interplanetary traveling shocks and/or corotating interaction regions. However, these active periods become progressively weaker and rarer while the intensity in quiet periods, dominated by the ACR, shows an increase till at least the end of the available data in 1994.

Figure 2 shows alpha particle energy spectra computed from ULYSSES LET data for three quiet time periods in 1991 (days 281–292), 1993 (days 58–73) and 1994 (days 4–22). Plotted is the number of counts per energy bin per day versus energy per nucleon. The first spectrum was derived in 1991 during a time where the general solar activity was still very high. This spectrum shows the typical slope known for solar energetic particles in this energy range. The other two spectra were evaluated during the declining phase of the solar activity. Therefore the number of counts below about 10 MeV/n decreases, compared to the 1991 spectrum, and the decreasing modulation of ACR can be seen as a strong increase of the flux above 10 MeV/n.

3. Summary

By using data from the COSPIN LET instrument on board ULYSSES we have shown a significant increase of the intensity of the ACR during the declining phase of solar activity. In summary we find the following:

1. The drop in solar activity in late 1991 and thereafter has allowed the detection of the anomalous component of cosmic rays in the inner heliosphere by the LET instrument. We report here preliminary observations of the anomalous helium component at radial distances $\sim 3 - 5$ AU from the sun and over a heliographic latitude range $0^o - 50^o$ S.

2. The energetic alpha particle intensity observed at ULYSSES started to rise in early 1992 and has continued to rise till the end of the available data in March, 1994.

There are two possible reasons for the increase of the anomalous component: the decrease in solar activity and a positive latitudinal gradient. Data at higher latitudes will hopefully allow us to distinguish between these.

References

Fisk, L.A., Kozlovsky, B., and Ramaty, R.: 1974, An interpretation of the observed oxygen and nitrogen enhancements in low-energy cosmic rays, *Astrophys. J.*, **190**, L35.

Hovestadt, D.O., Vollmer, O., Gloeckler, G., and Fan, C.Y.: 1973, Differential energy spectra of low energy (<8.5 Mev per nucleon) heavy cosmic rays during solar quiet times, *Phys. Rev. Lett.*, **31**, 650.

Jokipii, J.R., and Davila, J.M.: 1981, Effects of drift on the transport of cosmic rays IV, More realistic diffusion coefficients, *Astrophys. J.*, **248**, 1156.

McDonald, F.B., Teegarden, J.B., Trainor, J.H., and Weber, W.R.: 1974, The anomalous abundance of cosmic ray nitrogen and oxygen nuclei at low energies, *Astrophys. J.*, **185**, L105.

Mewaldt, R.A., Cummings, A.C., Cummings, J.R., Stone, E.C., Klecker, B., Hovestadt, D., Scholer, M., Mason, G.M., Mazur, J.E., Hamilton, D.C., von Rosenvinge, T.T., and Blake, J.B.: 1993, The return of the anomalous cosmic rays in 1 AU in 1992, *Geophys. Res. Lett.*, **20**, 2263.

Pesses, M.E., Jokipii, J.R., and Eichler, D.: 1981, Cosmic ray drift, shock wave acceleration, and the anomalous component of cosmic rays, *Astrophys. J.*, **246**, L85.

Simpson, J.A.: 1992, The Ulysses cosmic ray and solar particle investigation, *Astron. and Astrophys.*, **92**, 365.

ON THE OFF-ECLIPTIC DISTRIBUTIONS OF ANOMALOUS COSMIC RAYS AND THEIR RELATION TO THE LARGE-SCALE GEOMETRY OF THE HELIOSPHERIC SHOCK

HORST FICHTNER
Max-Planck Institut für Radioastronomie, Auf dem Hügel 69, D-53121 Bonn, Germany

HANS-JÖRG FAHR
Institut für Astrophysik and Extraterrestrische Forschung der Universität Bonn, Auf dem Hügel 71, D-53121 Bonn, Germany

and

S. RANGA SREENIVASAN
Department of Physics and Astronomy, University of Calgary, 2500 University Drive N.W., Calgary, Alberta T2N 1N4, Canada

Abstract. The scenario explaining the origin of the anomalous component of cosmic rays (ACR) implies a close relation between these high energy particles and the solar wind termination shock representing their main acceleration region. Consequently, one should expect the ACR distributions in the heliosphere to reflect some information about the structure as well as the large-scale geometry of the shock. We study the influence of a non-spherically symmetric heliospheric shock on the off-ecliptic - i.e. high latitude - ACR distributions using a two-dimensional model including their anisotropic diffusion and drift in the heliospheric magnetic field as well as a solar wind flow dependent on the heliographic latitude. The model calculations are used to investigate the probability of a possible polar elongation of the heliospheric shock from observations of the distributions of the ACR at high latitudes during solar minimum conditions.

Key words: Anomalous Cosmic Rays, Heliospheric Shock

1. Introduction

The exploration of the high-latitude heliosphere with the ULYSSES satellite will complement our knowledge about the three-dimensional structure of the heliosphere (reviewed in e.g. Fahr and Fichtner, 1991) with most valuable in-situ measurements which can be expected to have considerable impact on the future modeling of numerous subjects of heliospheric physics, e.g. the latitudinal variation of the solar wind (e.g. Witt et al., 1979), the latitudinal variation of the heliospheric magnetic field (HMF) and, in particular, its structure close to the symmetry axis (Jokipii and Kota, 1989), as well as the modulation of spectra of galactic (GCR) and anomalous cosmic rays (ACR) (e.g. Potgieter and Moraal, 1988; Jokipii, 1990; McDonald et al., 1992).
The ACR turned out to be of great value for the investigation of the heliospheric structure since their distribution in the heliosphere is determined

Space Science Reviews **72**: 435–440.
© 1995 *Kluwer Academic Publishers.*

by their diffusion as well as non-negligible drifts in the HMF, and their convection by the solar wind. Moreover, infering from the single ionization of the ACR a local production (Adams et al., 1991; Adams and Leising, 1991) which most probably takes place at the solar wind termination shock (Jokipii, 1992), these particles represent, in contrast to GCR, an intrinsic heliospheric phenomenon. The connection between the ACR and the heliospheric shock (HS) gives rise to the suspicion that their distributions might reflect useful information about the structure and geometry of the HS.
Although a non-spherical geometry of the HS was expected earlier by Dessler (1967) and this expectation is frequently mentioned in the literature (e.g. Holzer, 1989; McKibben 1990; Suess, 1993), quantitative investigations of its consequences for the heliospheric structure were performed only very recently (Chalov, 1993; Fahr et al., 1993; Fichtner et al., 1993).
In contrast to our earlier studies of upwind-downwind signatures in the ACR distributions (Fahr et al., 1992; Fichtner et al., 1993; 1994a,b), we concentrate here on the consequences of a polar extension of the HS for the high-latitude distributions of ACR. Combining an observation of the increase of the solar wind velocity towards higher latitudes (Kakinuma, 1977) with the derived mass fluxes from polar coronal holes (Lallement and Bertaux, 1985) during periods of minimum solar activity, we might conclude that there is evidence that the momentum flux of the wind above the poles is two times that in the ecliptic implying that the heliocentric distance of the HS above the solar poles might be as large as $\sqrt{2}$ times its distance in the ecliptic (Suess, 1993).
If this is true, the assumption of a spherically symmetric HS significantly underestimates the heliocentric distance of the ACR source at polar regions presently assumed to be nearly identical with the HS and, consequently, leads to an overestimation of the ACR pressure values at high latitudes. Therefore, we will incorporate explicitly such a shock asymmetry into a model describing the diffusion, drift and convection of ACR.
The study might serve as a useful tool to understand and interpret the forthcoming in-situ observations of the high-latitude ACR with ULYSSES.

2. The model

2.1 The cosmic ray transport equation

Excluding the acceleration process and the details of the energy spectra of ACR, we base our study on the time-independent and momentum-averaged fundamental equation for the transport of cosmic rays (Axford et al., 1977; Drury and Völk, 1981):

$$\nabla\left[\frac{\gamma_{acr}}{\gamma_{acr}-1}\left(\mathbf{v}_{sw}+<\mathbf{v}_{dr}>\right)p_{acr}-\overleftrightarrow{\kappa}\,\nabla\left(\frac{p_{acr}}{\gamma_{acr}-1}\right)\right]=\mathbf{v}_{sw}\cdot\nabla p_{acr},\quad(1)$$

i.e. a second order partial differential equation for the ACR pressure $p_{acr}(\mathbf{r}) = \frac{4\pi}{3}\int_0^\infty f(\mathbf{r},\mathbf{p})wp^3dp$ defined as the second order moment of the ACR kinetic distribution function $f(\mathbf{r},\mathbf{p})$. It represents an integral of the differential intensities $p^2 f(\mathbf{r},\mathbf{p})$ and is motivated by the pressure definition in statistical mechanics (see Drury and Völk, 1981). $\mathbf{p}$ without a subscript denotes the particle momentum, w its velocity. Methods to obtain p_{acr} from ACR data are discussed in Fichtner et al. (1994b). $4/3 < \gamma_{acr} < 5/3$ is a polytropic index, $\mathbf{v}_{sw} = 400(1+\epsilon\ cos^2\theta)km/sec\ \mathbf{e}_r$ is the latitude-dependent solar wind velocity with $\epsilon \in [0,1]$ and θ being the co-latitude, $< \mathbf{v}_{dr} >$ denotes a momentum averaged value of the ACR drift velocity $\mathbf{v}_{dr} = \frac{wpc}{3q}\nabla \times \left(\frac{\mathbf{B}}{B^2}\right)$ with c and q being the speed of light and the particle charge, respectively. Since the drift can be treated as off-diagonal elements of the diffusion tensor $\overleftrightarrow{\kappa}$ (e.g. Potgieter and Moraal, 1988), its momentum average can be obtained via that of the latter: $\overleftrightarrow{\kappa} = \kappa_\perp(1-\hat{b}\hat{b}) + \kappa_{||}\hat{b}\hat{b}$ with $\hat{b} = \mathbf{B}/B$. $\kappa_{||} = 10^{20}|B_0|/B\ cm^2/sec$ and $\kappa_\perp = 0.03\kappa_{||}$ (Kota and Jokipii, 1993) are the diffusion coefficients parallel and perpendicular to the HMF $\mathbf{B} = AR\left(\frac{r_0}{r}\right)^2\left(\mathbf{e}_r - \frac{\Omega(r-r_\odot)}{v_{sw}}sin\theta\mathbf{e}_\phi\right)$ representing with $A = \pm 10\mu G$ and $\Omega = 2.9\cdot 10^{-6}s^{-1}$ Parker's spiral field modified with a factor $R(r,\theta)$ describing a non-radial field on and close to the symmetry axis (for details see Jokipii and Kota, 1989). The index '0' refers to values taken at $r = 1AU$. The momentum-averaging of $\overleftrightarrow{\kappa}$ is discussed in detail in Fichtner et al. (1994b).

2.2 The geometry of the heliospheric shock

Since the acceleration of the ACR is excluded from the explicit description, we have to consider the HS as a source surface for ACR (as done similarly by Reinecke et al., 1993). As explained above, the HS is expected to have a polar elongation during solar minimum periods. Neglecting any additional deviation from spherical symmetry - in particular any upwind-downwind asymmetry - a plausible representation of the HS is an ellipsoid axisymmetric with respect to the sun's rotation axis. Assuming for all configurations a shock distance of $r = 100AU$ in the ecliptic, we may distinguish three cases: a spherically symmetric shock S1, and two polewardly elongated ellipsoidal shocks S2, S3 with polar shock distances of 115 and 140 AU, respectively.

2.3 The source distributions

To obtain boundary conditions for the elliptical partial differential equation (1) we prescribe the ACR pressure at the shock surface. To this end, based on the results given by Steenkamp and Moraal (1993) for anomalous oxygen (O^+), we employ a latitudinally varying normalized source distribution

$$p_{acr} = \begin{cases} 1 - 0.2\cdot sin(\pi/2(1-\theta/15^o)); & 0^o \leq \theta \leq 15^o \\ 1 - 0.5\cdot sin(\pi/2(1-(\theta-15^o))/75^o); & 15^o < \theta \leq 90^o \end{cases} \tag{2}$$

The symmetry with respect to the ecliptic plane yields the source for $90^o < \theta \leq 180^o$. A location of the source maximum at $\theta \approx 15^o$ was also found earlier by Jokipii (1986) and is valid for $A > 0$ conditions. Thus, here we restrict the analysis to O^+ during the $\sim$, 1998 minimum period relevant for the ULYSSES observations. This boundary condition is supplemented by the requirement that p_{acr} vanish very close to the sun ($p_{acr}(\mathbf{0}) = 0$).

3. Results

We have solved equation (1) for O^+ with assumptions described in the previous section using an integration algorithm developed by Houstis et al. (1985). A more detailed description with a test of the model including an investigation of the radial as well as the latitudinal gradients and an application to the case $A < 0$ will be presented in a forthcoming publication. Here we concentrate solely on a comparison of the distributions resulting from the spherical shock S1 and the ellipsoidal shocks S2, S3 for solar minima characterized by $A > 0$ and the influence of drift motions.
Fig.1 shows the resulting (normalized) O^+ distribution for (a) S1 and (b) the extreme case S3. These distributions derived within the framework of a hydrodynamic theory are comparable to those found by Jokipii (1990) and Potgieter (1994) who used kinetic models. We consider this as an *a posteriori* justification of the use of equation (1). The effect of a shock-asymmetry is obvious: as a consequence of a more distant source above the solar poles the O^+ pressure at a given heliocentric distance is lower for S3 compared to that for S1. This is illustrated in Fig.1(c) giving the ratio $p_{acr}(S3)/p_{acr}(S1)$ in the region enclosed by the spherical shock S1 and, revealing that the pressure for the case S3 is up to 17% lower than that for S1. Fig.1(d) shows the same ratio when S2 instead of S3 is considered. The effect is still evident although reduced approximately by a factor of two, i.e. if the shock distance above the poles is 15 AU farther than that in the ecliptic plane the O^+ density is $\sim$ 8% lower than the one obtained with a spherical shock.
Notice that the use of the source function (2) for non-spherical shocks represents an upper limit since the production efficiency of ACR tends to decrease with increasing heliocentric distance as a consequence of lower conversion rates of pick-up ions (see Chalov and Fahr, this conference) regarded as the seed population of the ACR. Due to this upper limit the derived numbers can be seen as lower limits for pressure reduction, since a decreased source strength would yield even lower values for the cases S2 and S3.
Finally, we studied the importance of drifts by computing a non-drift solution for S1 and S2. The comparison shows that the absence of drifts leads to an even stronger pressure decrease. This can easily be understood by considering the orientation of the drift motion during a period characterized by $A > 0$. During such periods the ACR particles drift from the HS

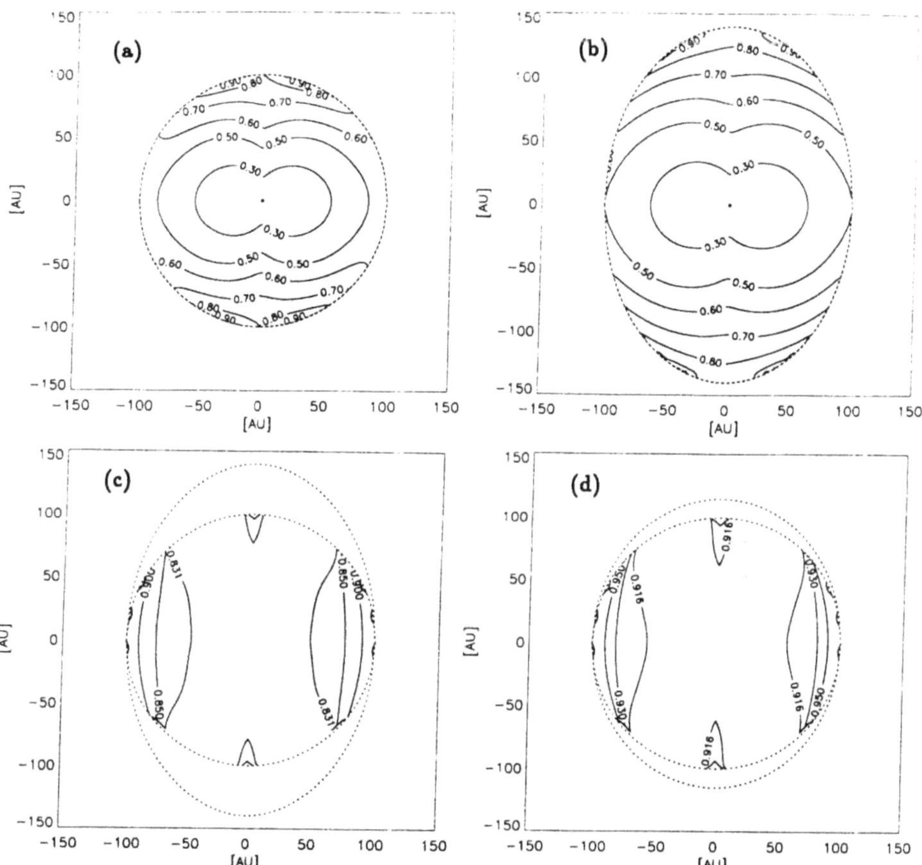

Fig. 1. (a) The pressure distribution of O^+ for a spherically symmetric heliospheric shock S1 (dashed line) for $A > 0$ conditions. (b) Same as (a) for an elongated shock S3 (dashed line) with a distance of $140AU$ above the solar poles. (c) The ratio $p_{acr}(S3)/p_{acr}(S1)$ of the pressure fields shown in (a). The dashed lines indicate the two shocks. (d) The ratio $p_{acr}(S2)/p_{acr}(S1)$ of the pressure fields for the cases S1 and S2. The shock S2 has a polar distance of $115AU$. The dashed lines indicate the two shocks.

poles inward, along the ecliptical current sheet towards the HS and along the shock surface poleward. Thus, compared to the non-drift case, the ACR are transported more efficiently inward resulting in higher pressures.

In the outer heliosphere, for distances $r \geq 50AU$, the pressure reduction resulting from the shock asymmetry is an off-ecliptic effect for all cases discussed. However, for $r < 50AU$ the ACR pressures are reduced also in the ecliptic plane. This effect should be included in model calculations aiming at a determination of the ecliptic distance of the HS from the measured radial variation of the ACR pressures (as reported e.g. by Cummings and Stone, 1987; McDonald et al., 1992; Reinecke et al., 1993) in order to avoid overestimation of the shock distance in the ecliptic plane.

4. Conclusions

We have shown that a polar extension of the heliospheric shock does influence the distributions of ACR. The ACR pressures to be measured by ULYSSES during its polar passages might be significantly lower than those predicted with the assumption of a spherically symmetric shock. Considering, as an example, anomalous oxygen we have found depending on the polar extension of the shock and on the importance of drift motions, $8 - 17\%$ lower pressures where the smaller values correspond to less pronounced shock asymmetry and significant drifts. The detection of such 'reduced' ACR pressures would represent the first clear indication of a non-spherically symmetric heliospheric shock.

References

Adams, J.H. and Leising, M.D.: 1991, Proc. 22. Int. Cosm. Ray Conf., Dublin, SH 4.2.8
Adams, J.H. et al.: 1991, *Astrophys. J.* **375**, L45
Axford, W.I., Leer, E. and Skadron, G.: 1977, Proc. 15. Int. Cosm. Ray Conf., Plovdiv, 11, 132
Chalov, S.V. 1993: *Planet. Space Sci.* **41**, 133
Cummings, A.C. and Stone, E.C.: 1987, Proc. 20. Int. Cosm. Ray. Conf., Moscow, 3, 421
Dessler, A.J.: 1967, *Rev. Geophys.* **5**, 1
Drury, L.O'C. and Völk, H.: 1981, *Astrophys. J.* **248**, 344
Fahr, H.J. and Fichtner, H.: 1991, *Space Sci. Rev.* **58**, 193
Fahr, H.J., Fichtner, H. and Grzedzielski, S.: 1992, *Solar Phys.* **137**, 355
Fahr, H.J., Fichtner, H. and Scherer, K.: 1993, *Astron. Astrophys.* **277**, 249
Fichtner, H., Fahr, H.J. and Sreenivasan, S.R.: 1993, Proc. 23rd Int. Cosm. Ray Conf., Calgary, Vol. **3**, 423
Fichtner, H. et al.: 1994a, *Astrophys. J. Suppl.* **90**, 595
Fichtner, H. et al.: 1994b, *Astron. Astrophys.* **284**, 599
Holzer, T.E.: 1989, *Ann. Rev. Astron. Astrophys.* **27**, 199
Houstis, E.N., Mitchell, E.F. and Rice, J.R.: 1985, *ACM Trans. Math. Soft.* **11**, 413
Jokipii, J.R.: 1986, *J. Geophys. Res.* **91**, 2929
Jokipii, J.R.: 1990, '' in S. Grzedzielski and D.E. Page, ed(s)., *Physics of the outer heliosphere*, Pergamon Press: Oxford, 169
Jokipii, J.R.: 1992, *Astrophys. J.* **393**, L41
Jokipii, J.R. and Kota, J.: 1989, *Geophys. Res. Lett.* **16**, 1
Kakinuma, T.: 1977, '' in Shea, M.A., Smart, D.F. and Wu, S.T., ed(s)., *Study of Travelling Interplanetary Phenomena*, Reidel: Dordrecht,
Kota, J. and Jokipii, J.R.: 1993, *Adv. Space Res.* **13**, 257
Lallement, R. and Bertaux, J.L.: 1985, *J. Geophys. Res.* **90**, 1413
McDonald, F.B. et al.: 1992, *J. Geophys. Res.* **97**, 1557
McKibben, R.B.: 1990, '' in S. Grzedzielski and D.E. Page, ed(s)., *Physics of the outer heliosphere*, Pergamon Press: Oxford, 107
Potgieter, M.S. and Moraal, H.: 1988, *Astrophys. J.* **330**, 445
Potgieter, M.S.: 1994, Proc. 23. Int. Cosm. Ray Conf., Calgary; Invited, Rapporteur and Highlight Papers, 213
Reinecke, J.P.L., Moraal, H. and McDonald, F.B.: 1993, *J. Geophys. Res.* **98**, 9417
Steenkamp, R. and Moraal, H.: 1993, Proc. 23rd Int. Cosm. Ray Conf., Calgary, Vol. **3**, 419
Suess, S.T.: 1993, *J. Geophys. Res.* **98**, 15147
Witt, N., Ajello, J.M. and Blum, P.W.: 1979, *Astron. Astrophys.* **73**, 272

SIMULATIONS OF PICKUP-ION ACCELERATION AT QUASI-PERPENDICULAR SHOCKS

J. GIACALONE and J. R. JOKIPII
Department of Planetary Sciences, University of Arizona, Tucson, AZ 85721

Abstract. We present results from hybrid simulations (kinetic ion/fluid electron) of the interaction of interstellar pickup ions with collisionless shocks. Since cross-field transport is unphysically suppressed in the one-dimensional geometry used here, an ad hoc scattering algorithm is used to model this effect. This is a necessary step to accelerate the pickup ions from their initial low energies at quasi-perpendicular shocks to the high energies which are often observed associated with traveling interplanetary shocks by Ulysses.

Key words: shock acceleration – interstellar pickup ions – interplanetary shocks

1. Background

Collisionless shocks are known to be efficient accelerators of charged particles. It is generally believed that the orientation of the average upstream magnetic field with respect to the shock-normal direction is of paramount importance. For instance, quasi-parallel shocks, which are those in which the average upstream magnetic field and shock normal direction are nearly aligned, are regarded as efficient accelerators of low-energy (near the thermal energy of the plasma) particles; but that high-energy (cosmic ray) particles are not as effectively accelerated as their quasi–perpendicular counterparts. The reason for this is that the dominant acceleration mechanism in quasi-parallel shocks is Fermi acceleration, whereas, at perpendicular shocks it is drift acceleration which occurs on a much shorter time scale. A key issue in shock acceleration physics is whether quasi-perpendicular shocks are also efficient at accelerating low-energy particles.

The SWICS instrument on board the Ulysses satellite is capable of making detailed measurements of the interstellar pickup ion distribution function (Gloeckler *et al.*, 1993; Geiss *et al.*, 1994). These were originally neutral particles that were ionized on approach to the sun and subsequently swept up by the solar wind. Their energy shortly after ionization is the solar wind ramming energy, making them exceptional candidates for the study of the injection problem at quasi-perpendicular shocks.

Here we present numerical simulation results of the interaction of pickup H^+ with quasi-perpendicular shocks. The well known hybrid computational method is used in which the ions are treated kinetically and electrons as a charge-neutralizing fluid. For computational reasons, the model assumes that the plasma density, currents and fields depend on only one spatial variable. Hence, the motion of charged particles across the magnetic field

Space Science Reviews **72**: 441–446.
© 1995 *Kluwer Academic Publishers.*

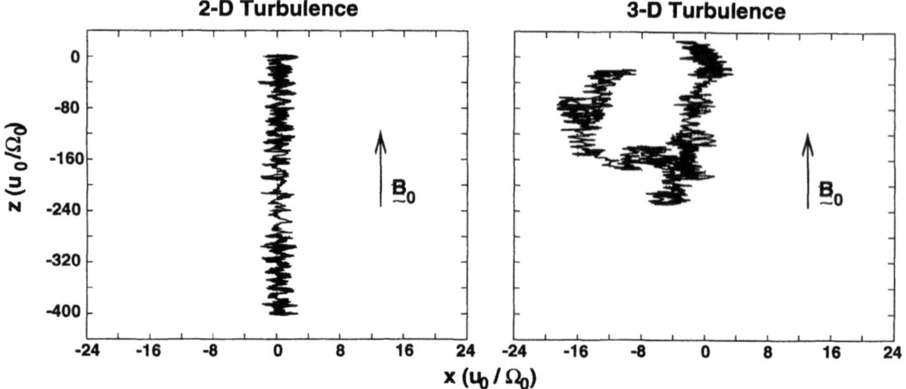

Fig. 1. Sample particle trajectories of a charged particle moving in an irregular magnetic field projected onto the x-z plane. The initial speed of the particle (the same in each case) is u_0 and has a gyroradius u_0/Ω_0 (one unit on this scale).

is unphysically suppressed (Jokipii *et al.*, 1993) and an ad hoc scattering mechanism must be employed.

2. Cross-Field Diffusion

Jokipii *et al.* (1993) demonstrated that when the electric- and magnetic-field vectors are independent of at least one spatial variable, charged particles are unphysically forced to remain within a Larmor radius of the field line on which it started. On the other hand, if the field is fully three-dimensional, the particles readily diffuse across the field. This is illustrated in Figure 1, which is taken from Giacalone and Jokipii (1994). Plotted here is the trajectory of a charged particle in an irregular magnetic field which is two-dimensional (upper panel) and three-dimensional (lower panel).

Giacalone and Jokipii (1994) calculated a cross-field diffusion coefficient by directly computing the Fokker-Planck coefficients based on the motions of an ensemble of test particles. The ratio of the diffusion coefficients is in reasonable agreement with the hard-sphere scattering result

$$\kappa_\perp = \frac{1}{3}\frac{\lambda_\parallel/r_g}{1+(\lambda_\parallel/r_g)^2}\frac{u_0^2}{\Omega_0} \qquad [1]$$

where $\kappa_\perp$ and $\kappa_\parallel$ are the diffusion coefficients across and along the average magnetic field, respectively, $\lambda_\parallel$ is the parallel mean free path, r_g is the Larmor radius, and Ω_0 is the gyrofrequency..

In our model, we have chosen an ad hoc scattering mechanism, in which after a time τ_i, a particle is scattered by re-randomizing the particle's gyrophase, while conserving it's pitch angle and energy in a frame co-moving with the plasma. τ_i is chosen from an exponential distribution with a mean scattering time τ. This mechanism gives rise to a cross-field diffusion coefficient given by (Giacalone *et al.*, 1994)

$$\kappa_\perp = \frac{1}{4}\frac{u_0^2}{\Omega_0^2 \tau} \qquad [2]$$

Recent self-consistent hybrid simulations without ad hoc scattering (Giacalone *et al.*, 1993) suggests that the parallel mean free path of an ion moving at the plasma velocity in a frame moving with the plasma (i.e. similar to a pickup ion) is about $18r_g$. Inserting this into Equation 1 gives $\kappa_\perp \approx \frac{1}{54}\frac{u_0^2}{\Omega_0}$. We see by comparison with Equation 2 that our choice of scattering mechanism is reasonable provided $\Omega_0\tau > 10$. We note that, although our approach is somewhat crude, it is clearly better than having none at all. Short of prohibitively costly three-dimensional simulations, this is the best method we have found to evaluate acceleration in this regime.

3. Shock Model

The shock is generated by having the convecting plasma collide with a rigid, perfectly conducting boundary. The subsequent perturbation forms into a shock which moves away from the boundary. The fields, density and currents are functions of x, which is the direction in which the shock propagates. In the simulations discussed here, the flow speed in the wall frame is $5v_A$, where v_A is the Alfvén speed. This gives rise to a shock frame Alfvén Mach number of 7.2. The upstream ion and electron plasma beta's are 0.5. The total simulation domain is large enough so that the shock can propagate well into the upstream medium and accelerate particles for several scattering times (the total simulation time is $120\Omega_0^{-1}$). The mean scattering time is $\tau = 20\Omega_0^{-1}$ (when it is incorporated). To model the pickup ions, the trajectories of an additional population, whose plasma frame speed is $7.2v_A$, are also integrated and the density and currents due to this population are coupled with the thermal plasma to be used in solving for the fields. The ratio of the unshocked pickup-ion density to the unshocked solar-wind density is 0.01.

We report on three simulations. In the first two (one with ad hoc scattering, one without), the upstream magnetic field is perpendicular to the shock's propagation direction. For the third, the shock normal angle is 75°.

4. Results and Discussion

Plotted in Figure 2 are the distribution functions of pickup ions both ahead and behind the shock front at $t = 120\Omega_0^{-1}$. The solid circles are for the case of no ad hoc scattering, while the open circles are for $\Omega_0\tau = 20$. The distributions are spatial averages over the intervals $-150v_A/\Omega_0 < x - x_{shock} < -50v_A/\Omega_0$ and $50v_A/\Omega_0 < x - x_{shock} < 150v_A/\Omega_0$, respectively.

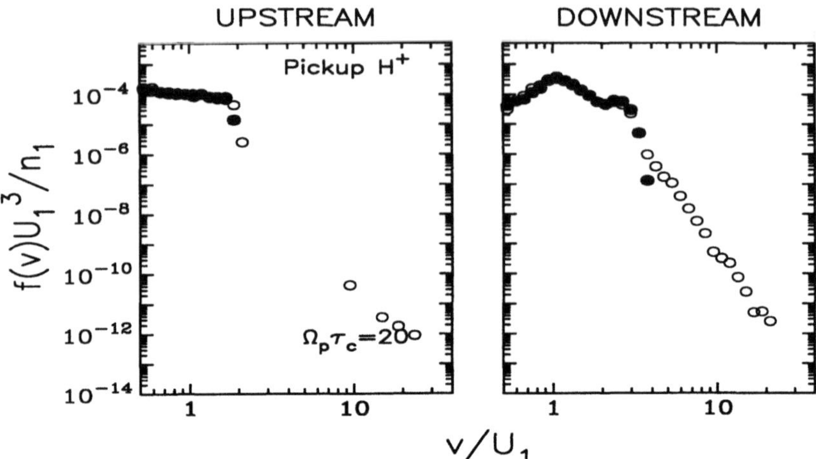

Fig. 2. Upstream and downstream velocity distribution functions of pickup protons for a case with no scattering (solid circles) and a case with scattering using $\Omega_0\tau = 20$.

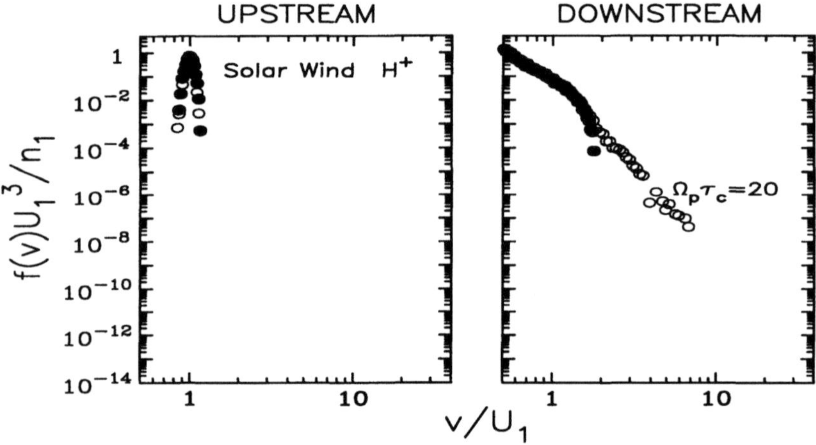

Fig. 3. Upstream and downstream velocity distribution functions of solar protons for a case with no scattering (solid circles) and a case with scattering using $\Omega_0\tau = 20$.

It is clear that the ad hoc scattering has lead to a high energy tail in the distributions shown in Figure 2. This is consistent with Ulysses observations of pickup ions (Gloeckler *et al.*, 1994) and agrees with Ogilvie *et al.* (1993). Note that when no scattering is used, there is no high energy tail.

The fact that the pickup ions have been accelerated so readily is somewhat surprising since the process is not a diffusive one. This is easily seen by defining a quantity Υ, which is the ratio of the flow velocity to the maximum

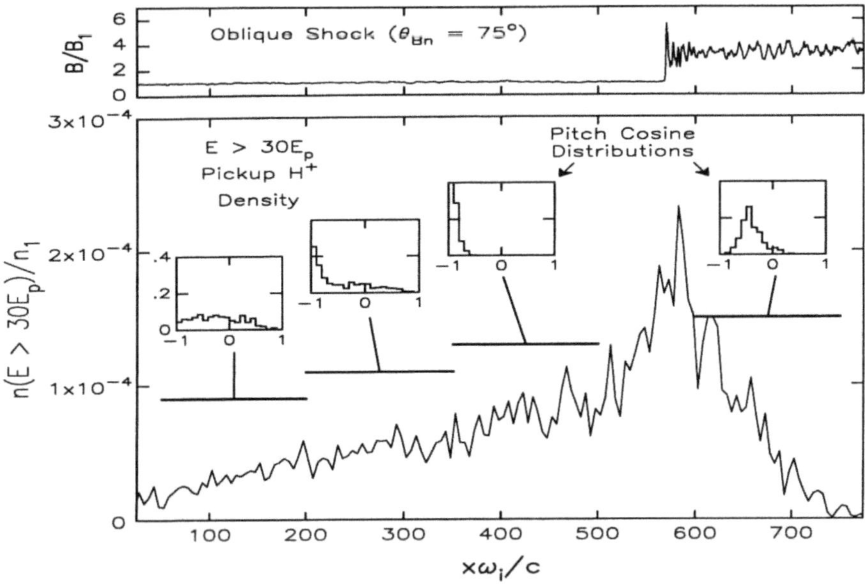

Fig. 4. Total magnetic field (upper panels) and density of mildly energetic ions (lower panels) at $t = 120\Omega_0^{-1}$ as a function of position for an oblique shock ($\theta_{Bn} = 75°$) using a scattering time $\Omega_0\tau = 10$. Displayed in the sub–panels are pitch angle cosine distributions.

"diffusive" velocity (Jokipii, 1991). From Equation 2, this is

$$\Upsilon = \frac{U_1 r_g}{\kappa_\perp} = 4\frac{U_1}{u_0}\Omega_0\tau \qquad [3]$$

When the pickup ions first encounter the shock $u_0 = U_1$, where U_1 is the shock frame plasma velocity, and for $\Omega_0\tau = 20$, this gives $\Upsilon = 80$ which is well outside the diffusive regime ($\Upsilon \gg 1$ implies that the particles are convecting with the flow more rapidly than they can diffuse against it).

Figure 3 is the same as Figure 2 except that the distributions are of solar wind ions. Contrary to the pickup ions, the solar wind ions are not efficiently accelerated to high energies. We have argued that this is also due to the one-dimensionality of the model (Giacalone *et al.*, 1994). Although, it is reasonable to expect that the thermal solar wind ions will not be as efficiently accelerated as the pickup ions.

The lower panel of Figure 4 displays the density of mildly energetic ($E > 15mU_1^2$) pickup ions at $t = 120\Omega_0^{-1}$. The inserts display the plasma frame pitch angle cosine distributions indicating that the ions are highly field-aligned immediately ahead of the shock which is in agreement with observations upstream of interplanetary shocks (e.g. Armstrong *et al.*, 1970). The upper panel displays the total magnetic field strength indicating the location of the shock. The explanation for the existence of these field aligned ions is that they have been drift accelerated, and reflected upstream of the shock on a time scale which is shorter than the mean scattering time. Note that far upstream, the distributions are much more isotropic.

5. Summary

We have reported on results from a new model for ion injection and acceleration at quasi–perpendicular shocks. Our model, although somewhat crude, contains many of the necessary physical effects to effectively describe this process. We have used the one–dimensional hybrid simulation (kinetic ions/ fluid electrons) and have placed an additional assumption on the model which allows for cross-field scattering of the ions which is otherwise suppressed in typical one- and two-dimensional simulations.

We conclude that provided some scattering exists, pickup ions can be injected and accelerated to large energies, even at purely perpendicular shocks. The acceleration is very rapid. For conditions typical of the solar wind at 5 AU, we have found that pickup ions can be accelerated from their initial energy of $\sim$ 1 keV to many tens to a few hundred keV in 10–20 minutes. Moreover, our system size is small ($\sim 10^5$km $\sim 10^{-3}$AU) and it is probable that interplanetary shocks are more efficient, even though they are considerably weaker than the shocks we have simulated.

6. Acknowledgments

This work was supported by NASA/JPL as part of an interdisciplinary Investigation on the Ulysses mission, JPL contract number 959376. The research was also supported in part by the NSF under Grant ATM 9307046 and by NASA under Grants NAGW-1931 and NAG52251 (JRJ).

References

Armstrong, T. P., S. M. Krimigis, and K. W. Behannon: 1970, *J. Geophys. Res.* **75**, 5980

Geiss, J., G. Gloeckler, U. Mall, R.von Steiger, A. B. Galvin, K. W. Ogilvie: 1994, *Astron. Astrophys.* **282**, 924

Giacalone, J., D. Burgess, S. J.Schwartz, and D. C. Ellison: 1993, *Astrophys. J.* **402**, 550

Giacalone, J., J. R. Jokipii, and J. Kóta: 1994, *J. Geophy. Res.* , in press

Giacalone, J., and J. R. Jokipii: 1994, *Astrophys. J. Lett.* , in press

Gloeckler, G., J. Geiss, H. Balsiger, L. A. Fisk, A. B. Galvin, F. M. Ipavich, K. W. Ogilvie, R.von Steiger, B. Wilken: 1993, *Science* **261**, 70

Gloeckler, G., J. Geiss, E.C. Roelof, L. A. Fisk, F. M. Ipavich, K. W. Ogilvie, L.J. Lanzerotti, R.von Steiger, B. Wilken: 1994, *Geophys. Res. Lett.* , submitted

Jokipii, J. R., J., Kóta, and J. Giacalone: 1993, *Geophys. Res. Lett.* **20**, 1759

Jokipii, J. R.: 1971, 'Diffusive shock acceleration: acceleration rate, magnetic-field direction and the diffusive limit' in G. P. Zank and T. K. Gaisser, ed(s)., *Particle Acceleration in Cosmic Plasmas*, AIP: New York, 137

Ogilvie, K. W., M. Baring, R. Forsyth, and G. Gloeckler: 1994, *EOS Trans* **74**, 480

FURTHER STUDIES OF WAVES ACCOMPANYING THE SOLAR WIND PICK-UP OF INTERSTELLAR HYDROGEN

N. MURPHY, E. J. SMITH, and B. T. TSURUTANI

Jet Propulsion Laboratory, California Institute of Technology, Pasadena, CA 91109, USA.

A. BALOGH and D. J. SOUTHWOOD

Imperial College of Science, Technology and Medicine, Blackett Laboratory, London SW7 2BZ, UK.

Abstract. Data collected by the magnetometer onboard the Ulysses spacecraft are surveyed for the occurrence of waves generated during the pick-up of interstellar hydrogen. Thirty one wave events were found during a 640 day study period, between March 21, 1992 and December 20, 1993 (after the Ulysses encounter with Jupiter). It is found that observation of the waves does not depend on the magnitude of the background magnetic field, but is a strong function of the angle between the magnetic field and the solar wind flow direction, with small angles being favored.

1. Introduction

While observations of waves associated with ion pickup are well reported (e.g. at comets), hydromagnetic waves expected to accompany the pickup of interstellar hydrogen have been elusive. Observations of interstellar hydrogen ions by the Ulysses SWICS experiment (Gloeckler et al., 1993) have led to a renewed search for these waves in the Ulysses Magnetometer data. Theoretical studies of the expected wave signature of interstellar ion pick up (e.g. Lee and Ip, 1987; Williams and Zank, 1994; Wu and Davidson, 1972) suggest that the waves should be left hand polarized in the spacecraft frame of reference and exhibit a cutoff in the wave spectra at the hydrogen gyrofrequency. The first observations of such waves were reported by Smith et al., 1993, and the present work follows up this discovery with a statistical study of the occurrence of waves generated by interstellar hydrogen pick-up.

2. Data analysis

The wave events reported in Smith et al., 1993 had a duration of the order of one to several hours, and so our data analysis centred on searching for events of this duration in the large volume of Ulysses magnetometer data. Three analysis tools were used to survey the data. Initially a 'wave occurrence index', $\Delta f_{g\pm}$, was developed. The index $\Delta f_{g\pm}$ is a measure of the difference between the average slope in the spectrum of the transverse wave power and the spectral slope local to the hydrogen gyrofrequency, f_g. The index is calculated by making a least squares straight line fit to the transverse power

Space Science Reviews **72**: 447–452.
© 1995 *Kluwer Academic Publishers.*

spectrum to estimate its slope, then calculating the slope around f_g using the mean power in a small range of frequencies above and below f_g. The difference in these two slopes is then $\Delta_{fg\pm}$. A positive value of $\Delta_{fg\pm}$ then implies that there is an enhancement in wave power at frequencies above f_g, consistent with the expected signature of waves produced by interstellar hydrogen pick up. Periods during which $\Delta_{fg\pm}$ was greater than some fiducial value were then analysed using dynamic spectra (a value of 0.6 was used initially). It should be noted that the main value of using $\Delta_{fg\pm}$ is to exclude periods in which there are no waves present in the data and so reduce the volume of data requiring more intensive analysis. Figure 1 shows an example of a 20 day plot, for the period Jan. 1 to Jan. 21 1993, showing $\Delta_{fg\pm}$ compared to the magnetic field magnitude and the angle between the magnetic field vector and the radial to the sun (RTN x direction). Several peaks can be seen in $\Delta_{fg\pm}$. The sharp peaks seen on doy 11 and doy 16 1993 appear to be associated with shocks. The two peaks on doy 6 1993 (marked with a vertical arrow) are thought to be due to waves produced by interstellar hydrogen pick-up.

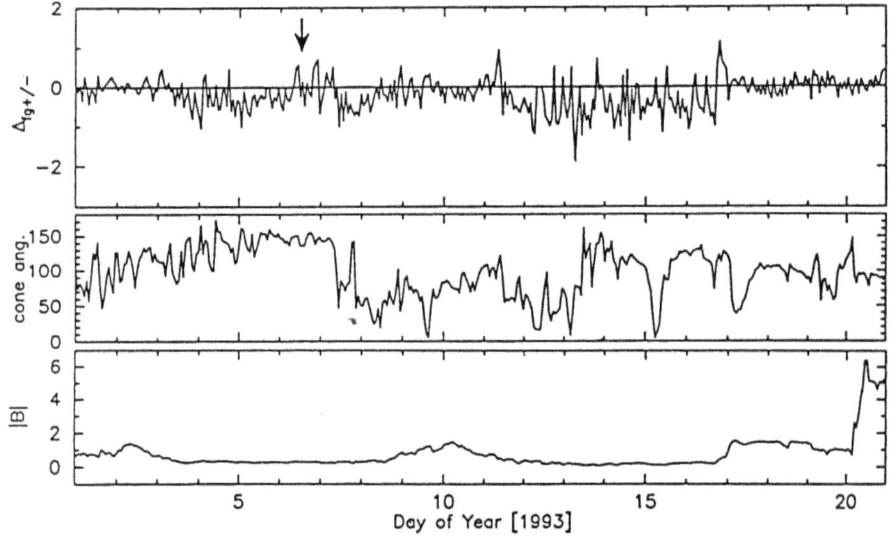

Figure 1. Plot of $\Delta_{fg\pm}$, the cone angle between the magnetic field and the radial direction to the sun (α in the text) and the magnetic field magnitude $|B|$, for the period Jan. 1 to Jan. 21 1993. The period of pick-up wave activity on doy 6, 1993 is marked by a vertical arrow.

Following the application of the above technique, dynamic spectra and dynamic polarization plots of periods containing potential wave events were produced. Events showing an enhancement in power above the hydrogen gyrofrequency and circular, or near circular polarization were then selected for further analysis. The application of this technique showed that the waves seen on doy 6 1993 were left-hand circularly polarized in the spacecraft frame of reference.

After events were identified using the above two methods, auto and cross spectra were computed over one or two hour blocks, and time series plots were examined, to characterize events more fully. It was at this stage that events were eliminated if they didn't meet the expected theoretical criteria, or they were associated with shocks in the solar wind.

Data used in this study were collected over 640 days, following Ulysses' encounter with Jupiter, and sufficiently far from Jupiter so as to rule it out as a source of waves (pick-up wave events have been found in data taken before the encounter, but they are not used in this study). During this period the spacecraft moved from 5.4 to 4.6 AU from the sun, and from -1° to -40° heliographic latitude.

3. Event example

Figure 2 shows autospectra of two hours of magnetic field data taken between 09:00 and 11:00 UT on Jan. 6 1993. Prior to the calculation of the autospectra the data are transformed into the BH coordinate system , in which the X-axis points along the mean magnetic field direction and the Y-axis is parallel to the solar equator. From left to right the three panels show the power parallel to the magnetic field and in the two directions perpendicular to the magnetic field.

From the plot it is obvious that there is considerably more power in the transverse magnetic field components, and there is a significant enhancement in power just above the hydrogen gyrofrequency. This period is taken during the wave enhancements identified in Figure 1. The enhancement in power is consistent with the expected signature of pick-up ion produced waves, however to be definitive more information is required. Examination of the cross spectrum of the magnetic field components perpendicular to the mean field (not shown) confirms that the waves seen during this event are indeed left-hand polarized in the spacecraft frame.

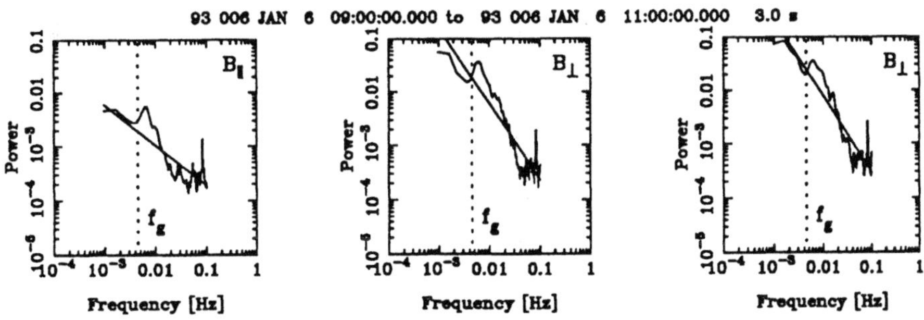

Figure 2. Autospectra for a two hour period between 09:00 and 11:00 UT on Jan. 6 1993. The vertical dotted line in each spectral plot denotes the hydrogen gyrofrequency. Prior to the calculation of the spectral estimates the data are transformed into the BH coordinate system, which has it's x-axis aligned with the mean field direction.

4. Statistical results

During the period between March 28 1992 and June 29 1993, 31 events were found which satisfy the theoretical expectation for waves produced by the pick-up of interstellar hydrogen by the solar wind. In previous studies we have found a number of examples of right hand polarized waves, with similar autospectral signatures, however in this studies no RH polarized waves were found which were not associated with shocks.

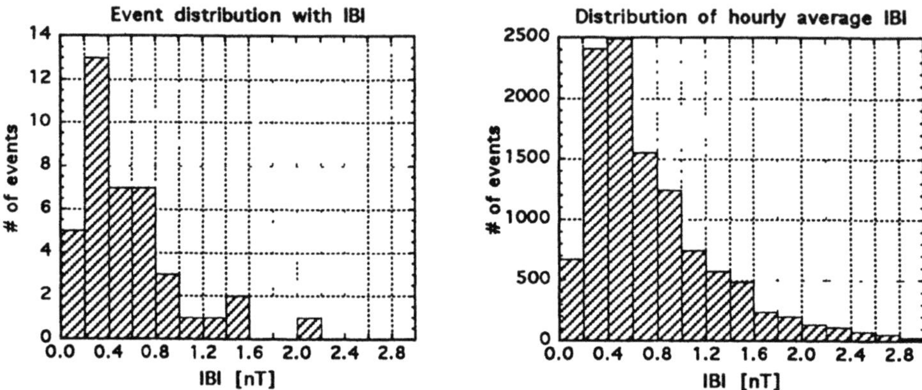

Figure 3. Event distribution as a function of magnetic field magnitude |B|, compared with the distribution of 1 hour averages of |B| for the whole study period. (March 1992 until December 1993). Bins are 0.2 nT wide.

It is interesting to note that no wave events have been found after June 29, when Ulysses passed -25° heliographic latitude. At present it is unclear whether this absence is due to the higher level of fluctuations seen in the magnetic field masking the presence of the waves, or the waves not being generated at higher heliographic latitudes.

Figure 3 shows a comparison between the distribution of events as a function of magnetic field magnitude and the distribution of one hour averages of magnetic field magnitude taken over the entire study (most of the events seen last one to several hours, and so it is appropriate to use 1 hour averages in comparisons with event distributions). The distribution on the left shows a tendency for events to be seen when the magnetic field magnitude is low, however the distribution's shape is very similar to the distribution of one hour averages on the right. This does not suggest that the occurrence of pick-up wave events is dependent of the background magnetic field strength.

In figure 4 we compare the distribution of events as a function of the angle α between the background magnetic field direction and the sun-spacecraft line (i.e. the nominal solar wind flow direction) with the distribution of one hour averages of α. Here we can see that the events are clumped around a value of 40° for α, with no events seen with α greater than 60° (α equals 90° would have the magnetic field transverse to the solar wind flow, and produce the largest VxB electric field). A comparison between thc top two panels in this figure shows that the event distribution is markedly different than the background distribution of α, which not surprisingly is peaked close to 90°. This suggests a strong tendency for events to be observed when the magnetic field is close to radial. Support for this conclusion also comes from the lower panel in figure 4, which shows a scatter plot of peak wave power seen in each event (normalized to $|B|^2$). Although there is considerable scatter in the data, there is a tendency for events with higher peak power to be seen at smaller values of α.

5. Discussion

We have surveyed 640 days of continuous magnetic field data collected by the Ulysses spacecraft, following the encounter with Jupiter. Care was taken to select a study period far enough away from Jupiter so as to rule it out as a source of the waves, and indeed, there is no evidence of increased wave activity with closeness to Jupiter in this data set.

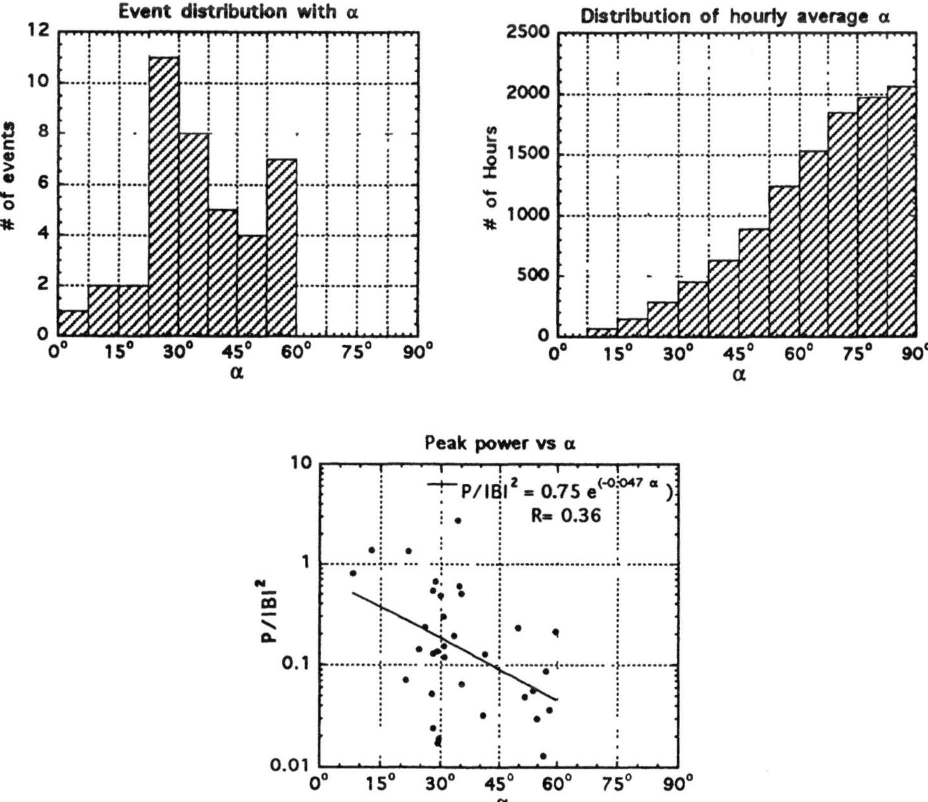

Figure 4. The upper two panels show the distributions of events as a function of the acute angle, α, between the background magnetic field direction and the sun-spacecraft line compared with the distribution of 1 hour averages of a over the whole study period (March 1992 until July 1993). The lower panel shows a scatter plot of normalized peak wave power in an event versus a. A linear fit is shown.

Previously we have compared the occurrence of wave events with energetic electron data from the COSPIN instrument on Ulysses, and found no correlation between the two. Our study shows that enhancements in wave power consistent with that expected from the pick-up of interstellar hydrogen are seen frequently in the solar wind (one event roughly every 20 days). We have found that there is no correlation between the observation of pick-up waves and the background magnetic field magnitude, as had been suggested by our preliminary studies. We do, however, find that the angle between the magnetic field and the solar wind flow direction, α, exerts a strong influence on the likelihood of observation of pick-up waves, and further that the peak power seen in the pick-up waves increases as a approaches 0°. This result suggests that the free energy needed to generate waves during the interstellar hydrogen pick-up process does not come from the VxB electric field.

Finally, although our analysis techniques were not specifically designed to detect them, several examples of wave power enhancements with a low frequency cutoff at the singly charged helium gyrofrequency were uncovered in the survey. Theory suggests that the wave power produced by helium pick-up should be significantly less that produced by the pick-up of hydrogen (Lee and Ip, 1987.). Future comparisons between hydrogen and helium pick-up generated waves should provide a useful tool to test wave generation theory.

Acknowledgments

This research was supported by the Jet Propulsion Laboratory, California Institute of Technology, under contract with the National Aeronautics and Space Administration.

References

Lee, M.A. and Ip W.-H.: 1987, 'Hydromagnetic wave excitation by ionized interstellar hydrogen and helium in the solar wind', *J. Geophys. Res.*, **92**, 11041.

Gloeckler, G. et al.: 1993, 'Detection of interstellar pick-up hydrogen in the solar system', *Science*, **261**, 70-73.

Smith, E.J. et al.:1994, 'Observation of waves generated by the solar wind pickup of interstellar hydrogen ions', submitted to *J. Geophys. Res.*

Williams, L.L. and Zank, G.P.:1994, 'An upper limit to the wave excitation of the interstellar pickup process', submitted to *J. Geophys. Res.*

Wu, C.S. and Davidson, R.C.: 1972, 'Electromagnetic instabilities produced by neutral particle ionization in interplanetary space', *J. Geophys. Res.*, **77**, 5399.

SESSION 5

INTERSTELLAR GAS AND COSMIC DUST

INTERSTELLAR GAS IN THE HELIOSPHERE AND THE SOLAR WIND ANISOTROPIES

R. LALLEMENT
Service d'Aeronomie, Paris, France

E. KYRÖLÄ AND T. SUMMANEN
Finnish Meteorological Institute, Helsinki, Finland

Abstract. A critical review of the interstellar hydrogen in the heliosphere will be presented. Recent Sun-interstellar matter interaction model improvements, a non-stationary flow and a flexible latitude dependence, will be discussed. We also consider the influence of heliospheric interface on neutral flow and the remaining refinements, which could help to better interpret the results of the SWAN experiment on board SOHO.

1. Introduction

One of the important unanswered questions in solar physics (and one of the main objectives of the SOHO mission) is the fast solar wind acceleration process. What type of energy (thermal or non-thermal) is transported from the convection zone to above the critical point (Holzer and Leer; 1980, Leer et al., 1982) and how? The classical Parker thermal model succeeds in explaining terminal velocities of up to about 250 km/s but it fails to reproduce high-speed streams. The coronal base thermal pressure alone is unable to lift up streams carrying up to 4 to 10 times more energy (velocity increase by a factor of 2 to 3.5) than the slow wind. It is well established now that fast solar wind (>400 km/s) flows from coronal holes and slow wind from elsewhere on the Sun. Polar coronal holes, sources of the high speed wind around the minimum of activity, are extended and stable for months and certainly easier to study as compared with equatorial coronal holes.

Models devoted to the study of the solar wind acceleration use the average ecliptic mass flux as a boundary condition for the expansion. However, if the high speed wind mass fluxes were significantly smaller, the dynamics in the acceleration region and the way different types of acceleration models account for it would certainly be changed. In particular, the observational evidence from white-light coronagraphy for the existence of a non-thermal energy flux emanating from the coronal base (Munro and Jackson, 1977) is strongly dependent on the polar mass flux amplitude (Lallement et al., 1986). From the observational point of view, the question of the amplitude of the mass flux from open-field regions and polar coronal holes, is still a matter of debate. The Ulysses results will provide first direct measurements of the high latitude mass flux!

Space Science Reviews **72**: 455–466.
© 1995 *Kluwer Academic Publishers.*

Here we are concerned with an indirect method which provides large scale solar wind latitudinal variations from ecliptic measurements, i.e. the use of the shape and temporal evolution of the interstellar neutral H cavity around the Sun. The Ulysses set of solar wind measurements at high latitudes is an unique opportunity to validate this method which will be used during the SOHO mission.

2. Description of the Method

The principle of the "interstellar neutral H" method is the following: the solar wind (and in minor proportions the solar EUV flux) is carving a cavity in the interstellar hydrogen flow distribution. The main solar wind effect is charge-exchange between the solar wind protons and the interstellar neutral H. Electrons play a negligible role. The shape of this cavity, deduced from the Lyman α resonance glow, allows to infer the solar wind latitudinal dependence. If the solar wind were isotropic, the H distribution would be axisymmetric around the upwind-downwind axis. The departures from axisymmetry reflect the properties of the solar wind encountered by the neutral H traveling at different latitudes. The solar rotation averages the longitudinal variations, only the latitudinal variations of the solar wind can be inferred. The accurate determination of the solar asymmetries requires a mapping of the sky at Lyman α. Schematically, the near-ecliptic upwind/downwind variations are first fitted by a model, then departures from axisymmetry are fitted by introducing a heliographic dependence of the ionization.

The characteristic scales of the phenomenon are the following: The size of the region contributing to most of the emission is about 5 AU on the upwind side, 20 AU on the downwind side, and about 10 AU in the region where the departures from the isotropic model are at maximum. The latter direction is at 45° from the downwind axis and in a plane containing the pole and the wind axis. The interstellar flow is moving at about 4 AU per year. The fact that the downwind region has been influenced by about 3-4 years, twice more than the upwind side, shows that time-dependent models are necessary, if one wants to infer solar wind time-variations with a temporal resolution finer than 2-3 years.

A last important point is the difference between the influence of the solar wind and the solar photon flux. The former is a directional effect, i.e. H atoms at a certain latitude are ionized by solar wind emitted from the same latitude. On the other hand, the same atoms are photoionized and illuminated by photons emitted from the whole solar disk. This makes photoionization much less latitude dependent. Around the solar minimum, photoionization and illumination are almost independent of latitude. At higher activity levels this is no longer true and one has to take into account the latitude dependence of photoionization and the illuminating flux (Cook and Meier, 1981; Pryor et al., 1992; Ajello et al., 1994).

The initial product of the H flow modeling and comparison with the data is the latitudinal variation of the global ionization rate of H. After subtracting ionization by solar photons we are left with the latitudinal variation of the solar wind ionizing power. A final step is then to derive the solar wind mass flux latitudinal variation: the use of a model describing the latitude dependence of the wind velocity in order to estimate the charge-exchange cross section variation with latitude. Around solar minimum the velocity field alone influences significantly the H distribution. With average velocities increasing from 350 up to 750 km/s the charge-exchange cross section decreases by 32 %. Only the remaining part of the observed ionization rate decrease is due to the mass flux variation.

3. Previous Results

Results obtained in the past from experiments on board Mariner 10, Prognoz 5-6, Voyager 1-2, and Pioneer-Venus have all shown a significant decrease of the total ionization of the neutral H from the ecliptic towards the solar poles, ranging from 20 to 40 %. After subtracting photoionization this gives 25-50% variation for the solar wind ionization. All these experiments have been analysed by using a stationary model: the derived anisotropies refer to 1-3 years period preceding the observations. An attempt to derive a solar-cycle evolution of the global ionization anisotropies has been done by Lallement (1987). The data suggest an increase of this anisotropy around minima of activity (A further analysis of 1986 Pioneer-Venus observations has finally given a 30 % contrast instead of the 25% preliminarily announced).

Now, these results have been more controversial in terms of implied solar wind latitudinal mass flux variations. It is clear that the accuracy of the mass flux determination is strongly dependent on a good knowledge of the velocity field. Only the Pioneer-Venus data have been analysed in the light of the IPS maps of Kojima and Kakinuma (1987) observed during the whole year preceding the observations. These Pioneer-Venus 85-86 observations suggest the velocity pattern as the main factor for the ionizing power decrease and they do not require any significant mass flux decrease with latitude (or less than 10%, see Lallement and Stewart, 1990). Now, when applying the same velocity field, which is in agreement with the 1975 IPS results of Coles and Rickett (1976) below 60 deg latitude, to the Prognoz 76-77 observations made just after the previous solar minimum of 1975, the data on the contrary imply a significant mass flux decrease of about 25-30% (Lallement et al., 1985). These results are somewhat contradictory, since they apply to about the same phase of the solar cycle (within one year). The inclination of the magnetic axis may explain this discrepancy or there may be odd/even cycle effects.

4. Influence of the Heliospheric Interface Neutral Perturbations

As it has been understood during the past few years, there may be substantial modifications of the neutral hydrogen flow at the heliospheric interface crossing. The reason for these changes is the coupling with the protons via charge-exchange. Before being deviated and excluded from the heliosphere, the interstellar H^+ are slowed down, heated, and compressed. After charge-exchange with them, the neutrals properties are modified toward some resemblance to them. Since the first discussion of this effect (Wallis, 1975), modeling has been carried out in several ways (Baranov and Malama 1993; Osterbart and Fahr, 1992). All models predict substantial effects, provided the heliosphere is confined within a volume of less than about 200 AU by the interstellar plasma and magnetic pressure, which indeed is the actual case, if one refers to recent interpretations of the Voyager 2-3 KHz detected radiations (Gurnett et al., 1993) and of the Anomalous Cosmic Rays behaviour (Cummings et al., 1993).

Some anomalies of the neutral H behaviour, which are possibly related to perturbations at the interface, have been reported recently. At large distances from the Sun, there appears to be some dissymmetry between the upwind and downwind side (Ganghopadhyay and Judge, 1993; Fahr and Osterbart, 1993). While this comparison suffers to some extent from the use of two different spacecrafts, strong support has been obtained by comparison between full radiative transfer models at Lyman α and the Voyager all-sky measurements (Hall et al., 1993; Lallement 1993; Quemerais 1993). They show some H density increase in the upwind direction with increasing distance from the Sun. An additional sign has come from the comparison between the interstellar H properties outside the heliosphere, as derived from both neutral helium which is unaffected by the plasma (Witte et al., 1993), and Local Interstellar Cloud measurements (Lallement and Bertin, 1992; Bertin et al., 1993), and inside the heiosphere as observed by UV space experiments and telescopes. A deceleration of 5-6 km/s at the heliospheric boundary is inferred, compatible with a small heliosphere of about 100 AU.

The question is now, how these perturbations will affect our solar wind measurements. First of all, recall that the size of the "useful" part of the neutral H ionization cavity is rather small (< 30 AU) in comparison with the size of the whole heliosphere. It is more a question of having information on what are the H flow characteristics close to the Sun. The two main questions are: What are the bulk velocity and the temperature? What are the departures from an homogeneous flow?

In principle, the "modified" bulk velocity and temperature have already been measured, in particular by Hydrogen Cells on board the Prognoz satellites (20 km/s, 8000 K). But they should be measured more accurately with the help of the SWAN Hydrogen Cell, due to a much greater Doppler shifts range for the Lyman α line scanning (see Bertaux et al, 1988). A special attention will be brought to these scans, since the found "reduced" velocity and the simultaneous apparent absence of a large heating of H are puzzling, and do not really fit the interface models predictions. But, the possible corrections to be expected should not have a significant impact on the anisotropy results. The overall ionization rate may be modified but the relative departures from axisymmetry of the distribution will not be much influenced and the derived pole-equator ratio should not change significantly. Now, once this question of a possible readjustment of V and T with better H cell observations is solved, it remains that all models used to calculate the interaction with a non-isotropic Sun start with a simple Maxwell distribution in the rest frame of the gas. The problem is that the possible effect of the plasma coupling is the introduction of some deviations from the Maxwell-distribution (see e.g. Lallement et al., 1992). It is a difficult task to estimate in which extent these deviations would affect the determination of latitudinal variations. Our feeling is that, because they are second-order effects they should be negligible, but but certainly further work is needed to answer quantitatively.

5. Comparison With Other Data

The conclusion that mass fluxes associated with fast wind would be smaller as quiet wind mass fluxes, was also inferred from the Helios data. This spacecraft has brought interesting informations on the relative mass fluxes of different types of wind around the ecliptic. Statistically the fast wind mass flux (> 600 km/s) is about 45 % less than the slow wind flux (<400 km/s) (Schwenn, 1983), suggesting a similarity between the equatorial and the polar open-field regions. Recently new results were published from an entirely new method (Woo and Goldstein, 1994). These authors infer the latitudinal variation of the solar wind from both spectral broadening and coherence bandwith of radio signals (13 cm). Used together with white-light coronograph results on the electron density latitudinal variation (decrease by a factor of 5 of the density between equator and pole), these results imply a decrease by a factor of 2.3 of the mass flux (57% reduction) from equator to pole in the 3 to 8 R_S region. While these last results are essentially dependent on both electron density and electron density fluctuations measurements, it is interesting to see that they also show a significant mass flux decrease. Now, certainly the most reliable and direct way to confirm or not the significant polar mass flux decrease is to go over the pole to measure it directly. As we emphasized in the introduction, the Ulysses results will bring important constraints.

6. Latitudinal Anisotropy of the Solar Wind

For the modeling of the latitudinal anisotropy of the solar wind we assume that the anisotropy can be considered constant. During the solar maximum also anisotropic solar Lyman α flux has to be included (Pryor et al., 1992). This is because during the solar maximum active regions cause an excess in the Lyman α intensity and the active regions are concentrated around the equator. During the solar minimum active regions are less prevalent. Therefore, the solar Lyman α flux is assumed isotropic and latitudinal effects are due to only the solar wind.

The latitudinal dependence of the ionization rate has usually been modeled with a harmonic function

$$\beta(\lambda,r)=\beta_e(1-A\sin^2\lambda)\frac{r_e^2}{r^2} \tag{1}$$

where β_e is the ionization rate at 1 AU, λ is the solar latitude, A is the anisotropy factor, r_e is 1 AU and r is the distance to the Sun. According to different spacecraft measurements the anisotropy factor is 0.2 - 0.4 (± 0.1).

The harmonic model gives a rough indication of how the solar wind proton flux changes from the equator toward the poles but it cannot describe the detailed latitudinal structure. We have used a new model (Summanen et al., 1993) where instead of the harmonic ionization rate (1) the rate is given by adjustable values at 19 different latitudes. The best ionization model is found by seeking the combination of the adjustable rates which gives the best agreement with the measured Lyman α intensity values. We have used the same Prognoz 6 measurements as in the earlier study by Lallement (1985) where the harmonic model was used. Two constraints were imposed. First, the ionization rate was taken symmetric around the equator, i.e. only 10 parameters were adjusted, and second, the rate was assumed to increase toward the equator. The fitting procedure was made to four Prognoz 6 measuring sessions.

We found out two features common to all data sets. First, the best fit was achieved using enhanced ionization near the solar equator and second, large plateaus around the latitudes 30°–70° appeared. The degree of anisotropy remained about the same as in earlier analysis based on Eq. (1) (Lallement, 1985). The average ionization rate using four different measuring sessions is shown in Figure 1.

The solar proton mass flux can be calculated from the total ionization rate. First we assume that the total ionization rate is a sum of photoionization and charge-exchange ionization. Photoionization is estimated to be 5.5 $10^{-8}s^{-1}$ over all latitudes (Ajello, 1987). We use the velocity model by Zhao and Hundhausen (1983) (model 4 of Kojima and Kakinuma (1990)) to calculate the charge exchange cross section. The solar proton mass flux at 1 AU is then

$$n(\lambda,r_e)v(\lambda)=(\beta(\lambda,r_e)-\beta_{ph}(\lambda,r_e))\,/\,\sigma(v(\lambda)) \tag{2}$$

where $n(\lambda,r)$ is the proton number density, $\beta_{ph}(\lambda,r)$ the photoionization rate, σ the charge exchange cross section, and v(λ) the solar wind proton velocity. The solar wind proton flux calculated by Eq. (2) is shown in Figure 2. The mean level is probably too small due to the lack of corrections for downwind intensity enhancements through multiple scattering resulting in underestimating the overall ionization (Quemerais, 1993), but large plateaus and a sharp change around 30° are there. These results can be compared with new Ulysses results between 0-56 heliographic latitude (Phillips et al, 1994; Goldstein et al., this issue; McComas et al., this issue). It is interesting to see that the broad plateaus at high latitudes and the narrow equatorial band of high fluxes are also present in the Ulysses data.

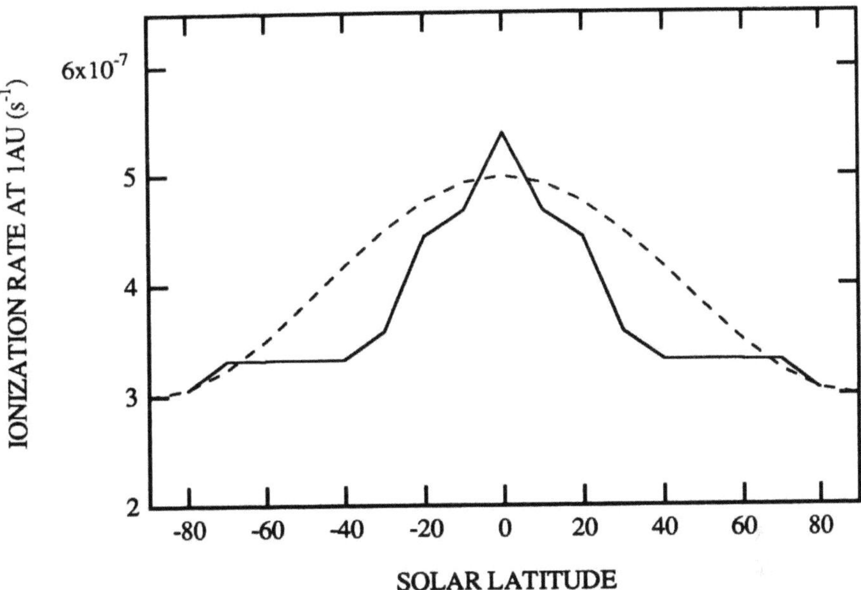

Fig. 1. The average ionization rate using four different measuring sessions of the Prognoz 6 satellite (solid line) The harmonic ionization rate (dashed line) used as the initial value in the search for a best fit is also shown.

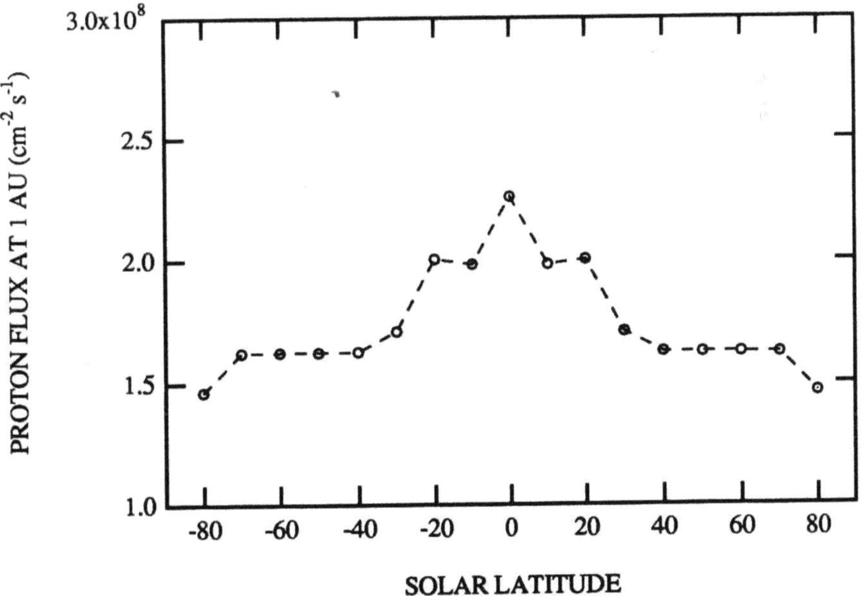

Fig. 2. The solar wind proton mass flux calculated from the average ionization rate shown in Fig. 1 using the velocity model by Zhao and Hundhausen (1983).

We want to point out that there are still many uncertainties in the method used. The final results can depend on the starting level of the ionization rate as well as on the method to find out the best ionization rate. In fact, we found out that the level of the best ionization rate changed in few cases when parameters of the model were changed. However, the two main features, broad plateaus and enhanced ionization near the equator, remained.

These phenomenological features can be verified near the solar wind minimum by in situ measurements of Ulysses and also by SWAN, which measures Lyman α all over the sky. Then, the results of the model of Summanen et al. (1993) can be used as a possible alternative when modeling hydrogen density distribution.

When the stationary model is used, it is assumed that by the use of different parameters for different periods of the solar cycle, the neutral flow can be modeled reliably. To test this assumption a time-dependent model is needed.

7. Solar Cycle Modulation

The state of the neutral hydrogen gas outside the heliosphere, i.e. in the local interstellar matter is poorly known. For modeling the interaction of the interstellar gas with the Sun we need to know how homogeneous the nearby interstellar gas flow is. The distances to the nearest stars, which could induce inhomogeneities, are order of parsecs (1 pc= 206162 AU). As the collision distance in the neutral hydrogen gas (n=0.1 cm^{-3}) is of the order of 10000 AU an adequate thermalization can be assumed. There is also a possibility that the solar system is moving across an inhomogeneous interface region of interstellar clouds or the cloud is in a turbulent state but there is no evidence to assume this to be the case. Therefore, a reasonable starting point is to assume that the interstellar gas is in a state of a thermal equilibrium and flowing in a stationary way.

The passage of interstellar matter through the heliospheric interface region is an active research area but a definite consensus does not yet exist concerning the modification of the neutral gas distribution function by the interface. If the distribution function deviates from the thermal one it must be taken into account in the modeling but it does not induce any time dependent effects. The crossing of the interface can induce time dependent variations if the structure of the interface region itself depends on time. The location of the solar wind termination shock is an obvious candidate for this kind of influence. As the location is fixed by the pressure balance between the interstellar medium (including the magnetic field and cosmic rays) and the solar wind it is susceptible to the large variations known to exist in the solar wind. According to recent studies by Barnes (1993) and Suess (1993) the variations of the location can be order of a few AU in a month. As neutrals proceed at a pace of 0.3 AU in a month they will experience the shock many times in their passage toward the Sun. As this kind of studies are still premature they have not been taken into account in modeling.

The flow of neutral gas from the heliopause toward the Sun is perturbed by interactions with the solar radiative and corpuscular emissions. Both interactions are subject to solar cycle modulation but also to modulation at shorter time scales. The solar rotation period is a typical time scale here. The variations in the radiative field are felt almost immediately in the heliosphere (100 AU in 13 hours). The solar wind propagates with speeds varying from 400 km/s to 800 km/s, i.e. 80-160 AU/year. This means that variations reach the whole heliosphere in a year.

The fast variations are not considered important when we are looking for the general behaviour of neutral gas in the heliosphere. The important temporal variations are due to the solar cycle. While the total radiative output of the Sun has been estimated to stay nearly constant in the time scales discussed here the UV radiation flux does have a strong solar cycle variation. The solar Lyman α can have 60% modulation during the solar cycle. The ionizing part of the radiation field is estimated to have up to a factor of three variations. The solar wind also, the main ionizing agent, varies according to solar cycle, resulting in up to a factor of two variations of the ionization rate in the ecliptic plane.

Let us now look how the solar cycle affects the dynamics of neutral atoms. Hydrogen atom trajectories are determined by a central force which is a sum of the gravitational force and the radiation pressure force. For hydrogen these two opposite forces are near balance but during the solar cycle the net force does show both attractive and repulsive phases. The force can be described by a formula

$$\bar{F}(r,t) = -\frac{GM}{r^2}(1 - \mu_0 - \Delta\mu \ \sin(\frac{2\pi t}{T_c}))\bar{r}_0 \tag{3}$$

As a consequence hydrogen atom trajectories tend generally to be a mixture of repulsive and attractive trajectories (see Kyrölä et al., 1994). Not only individual trajectories but the whole ensuing density distribution will be modified by the solar cycle. In the stationary case an analytical solution for the Boltzmann equation can be found (see Danby and Camm, 1957). For the time dependent case a numerical simulation must be used. We have approached the problem by using Monte Carlo technique. Particles are sent from a distant plane and they are followed through the heliosphere. The velocity and density distributions are recorded in the locations near the Sun.

The solar cycle modifications in the density distribution are most clearly seen when the temperature of the flowing gas is low. It has been shown (Blum et al., 1993; Kyrölä et al., 1993; Rucinski and Bzowski, 1994) that a time dependent central force gives rise to a constant succession of depletion areas and focusing peaks downwind from the Sun. These structures travel at the bulk velocity of the incoming gas.

When we are looking for the solar cycle effects on the interplanetary gas around the Sun we must include variations also in the ionization rate. Ionization is modeled by

$$\beta_e = \beta_{sw}(1 \pm 0.2\sin(\frac{2\pi t}{T_c})) + \beta_{ph}(1 + 0.33\sin(\frac{2\pi t}{T_c})) \tag{4}$$

where $\beta_{sw} = 5.1 \cdot 10^{-7}$ s^{-1} and $\beta_{ph} = 7.5 \cdot 10^{-8}$ s^{-1}. The two signs in the solar wind ionization correspond to the possibilities that the solar wind mass flux may be in the opposite phase compared to the radiation cycle of the Sun.

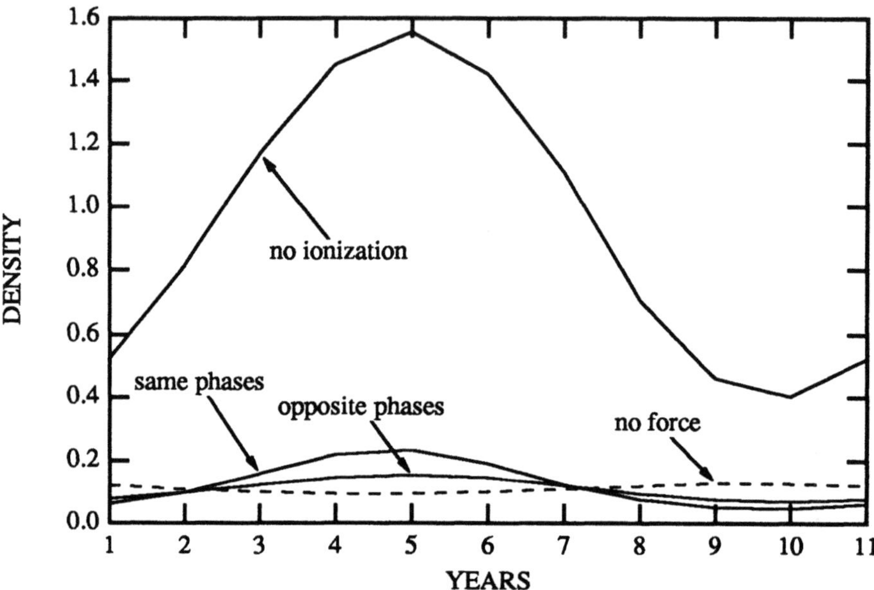

Fig. 3. Density variations at a downwind location r=10 AU when T=7500 K, bulk velocity=25 km/s, $\mu_0 = 1$ and $\Delta\mu = -0.3$. The phases refer to the two signs in Eq. (4).

An example of the results is shown in Fgure 3. The figure shows the neutral H density variations at a downwind location. At the upwind location the variations caused by the solar cycle in the force are negligible and the variation is due to the changes in ionization. The variation in the density is about 5% around the mean. In the downwind the deviations are clearly due to the force and the magnitude is about 60% around the mean when ionization is turned off. Ionization alone gives only 15% variation. Including both the variation is 80% when the force and the solar wind ionization are in the same phase and about 40 % when they are in the opposite phases.

The calculations just described have shown that solar cycle modifications do occur in the neutral interplanetary gas. How much do they affect on the measured quantities, i.e. Lyman α is still under investigation (see Rucinski and Bzowski, 1994). Models which also include time dependent latitudinal changes and the temporal modifications of the terminal shock position are also being developed.

8. Conclusion

Observing the interstellar hydrogen ionization cavity is one of the ways to infer large scale latitudinal variations of the solar wind and of its mass flux. The SOHO mission will provide ideal conditions for improving the precision of the mass flux measurements by means of: - greater precision of the data - full sky mapping every two days- better time coverage - simultaneous measurements of solar wind velocities at all latitudes through IPS -simultaneous measurements of ecliptic solar Lyman α fluxes and mapping of the full-disk active regions allowing a precise latitudinal dependence of the Lyman α fluxes illuminating the interstellar H. The question of the simultaneous variation of the velocity of SW from the open-field polar regions and the associated mass fluxes is to be addressed in the future ambitious models of the interstellar flow ionization. These models incorporate recent improvements such as time dependence (described in section 7), discretized latitudinal dependence (already proven successful, see section 6), and modeling of the illuminating Lyman α. However, at present a crucial issue in this field is the Ulysses high-latitude series of direct measurements. These new findings should allow to check and maybe better interpret the results from the neutral H indirect method. Also, assuming SOHO is launched on time, the first SOHO-SWAN data will be available for correlation with the Ulysses results obtained in the northern hemisphere in 1995-6, providing a valuable source for inter-comparison.

References

Ajello, J. M., Stewart, A. I., Thomas, G. E., and Graps, A.: 1987, *Astrophys. J.* **317**, 964.

Ajello, J. M., Pryor, W. R., Barth , C. A., Hord C. W., Stewart A. I. F., Simmons K. E., and Hall D. T.: 1994, *Astron. Astrophys.*, in press.

Baranov, V. B. and Malama, Yu. G.: 1993, *J. Geophys. Res.* **98**, A9, 15157.

Barnes, A.: 1993, *J. Geophys. Res.* **98**, 15137.

Bertaux, J. L. et al.: 1988, ESA SP-1104.

Bertin, P., Lallement R., Ferlet R., and Vidal-Madjar , A.: 1993, *J. Geophys. Res.* **98**, A9, 15193.

Blum, P., Gangopadhyay P., Ogawa H. S., and Judge D. L.: 1993, *Astron. Astrophys.* **272**, 554.

Cummings, A. C., Stone E. C., and Webber , W. R.: 1993, *J. Geophys. Res.* **98**, A9, 15165.

Cook, J. W., Meier , R. R., Bruekner , G. D., and Van Hoosier , M. E.: 1981, *Astron. Astrophys.* **97**, 394.

Danby, J. M. A., and Camm, G. L.: 1957, *Monthly Notices Roy. Astron. Soc.* **117**, 50.

Fahr, H. J.: 1971, *Astron. Astrophys.* **14**, 263.

Fahr , H. J. and Osterbart, R.: 1993, *Adv. Space Res.* **13**, 6, 159.

Ganghopadhyay, P. and Judge, D. L.: 1993, *Adv. Space Res.* **13**, 6, 173.

Goldstein et al.: 1994, *this issue.*

Gurnett, D. A., Kurth W. S., Allendorf S. C., and Poynter , R. L.: 1993, *Science* **262**, 199.

Hall, D. T., Shemansky , D. E., Judge, D. L., Gangopadhyay , P., and Gruntman , M. A.: 1993, *J. Geophys. Res.* **98**, A9, 15185.
Holzer, T. E. and Leer , E.: 1980, *J. Geophys. Res.* **85**, 4665.
Kojima, M. and Kakinuma, T.: 1987, *J. Geophys. Res.* **92**, 7269.
Kyrölä, E., Summanen, T., and Råback, P.: 1993, submitted to *Astron. and Astrophys.*
Lallement, R., Bertaux, J. L., and Kurt, V. G.: 1985, *J. Geophys. Res.* **90**, A2, 1413.
Lallement, R., Holzer, T. E., and Munro, R. H.: 1986, *J. Geophys. Res.* **91**, A6, 6751.
Lallement, R.: 1987, *Solar Wind 6*, NCAR/TN-306, 651.
Lallement, R. and Stewart, I. A.: 1990, *Astron. Astrophys.* **227**, 608.
Lallement, R., Malama, Yu. G., Quemerais, E., Bertaux, J. L., and Zaitzev, N. A.: 1992, *Astrophys.J.*, **396**, 696.
Lallement, R. and Bertin, P.: 1993, *Astron. Astophys.* **266**, 479.
Lallement, R.: 1993, *Adv. Space Res.* **13**, 6,113.
Leer, E., Holzer, T. E., and Fla, T.:, 1982, *Space Sci. Rev.* **33**, 162.
McComas et al.: 1994, *this issue.*
Munro, R. H. and Jackson, B. V.: 1977, *Astrophys. J.*, **213**, 874.
Osterbart, R. and Fahr, H.J.: 1992, *Astron. Astrophys.*,**264**, 260.
Phillips et al.: 1994, submitted to *Geophys. Res. Lett.*
Pryor, W. R., Ajello, J. M., Barth , C. A., et al.: 1992, *Astrophys. J.* **394**, 363.
Rucinski, D. and Bzowski, M: 1994, submitted to *Astron .. and Astrophysi.*
Quemerais, E.,: 1993, Paris VI University Thesis.
Schwenn, R.: 1983, *Solar Wind 5*, NASA CP-2280, 1983.
Schwenn, R. and Marsch, E.: 1990, *Physics of the Inner Heliosphere I, Large -Scale Phenomena*, p. 161.
Suess, S. T.: 1993, *J. Geophys. Res.* **98**, 15147.
Summanen, T., Lallement, R., Bertaux, J. L., and Kyrölä, E.: 1993, *J. Geophys. Res.* **98**, 13215.
Wallis, M. K.: 1975, *Nature* **254**, 202.
Witt, N., Ajello, J. M., and Blum, P. W.: 1979, *Astron. Astrophys.* **73**, 272.
Witte, M., Rosenbauer, H., Banaszkiewicz, M., and Fahr, H. J.: 1993, *Adv. Space Res.* **13**, 6, 121.
Woo, R. and Goldstein, R. M.: 1994, *Geophys. Res. Lett.* **21**, 2, 85.
Zhao, X-P. and Hundhausen, A. J.: 1983, *J. Geophys. Res.* **88**, 451.

SOLAR CYCLE MODULATION OF THE INTERSTELLAR HYDROGEN DENSITY DISTRIBUTION IN THE HELIOSPHERE

M.BZOWSKI AND D.RUCINSKI

Space Research Centre of the Polish Academy of Sciences, Bartycka 18A, 00-716 Warsaw, Poland

Abstract: A new approach taking explicitly into account time dependent variations of the solar radiation pressure and ionisation rate is developed for more realistic description of the distribution of the hot (~8000 K) interstellar hydrogen gas within the heliosphere. It is shown that the solar cycle-related variabilities cause temporal modulation of the hydrogen density patterns, manifested by propagation of density waves smearing out only beyond about 20 AU and 60 AU in the upwind and downwind directions, respectively. Density variations induce modulation of the size of hydrogen cavity by ~15% around the mean value, affect (mainly in the antiapex direction) the location of the maximum emissivity region (MER) contributing to the Lyman-α backscattered glow, and lead to a significant temporal modulation of the upwind-to-downwind Lyman-α intensity ratio as seen from the inner Solar System.

1. Introduction

It is well known that the combined effect of the solar radiation pressure and ionisation losses modifies significantly the patterns of the local hydrogen density distribution in an extended area around the Sun in comparison with its unperturbed homogeneous distribution in the Very Local InterStellar Medium (VLISM). A typical travel time of an H atom through the disturbed region is comparable with the time scale of the solar cycle. It becomes therefore apparent that the classical static "hot" model (Danby and Camm, 1957; Fahr, 1971), assuming purely stationary solar gravitational and ionisation influence on the penetrating gas, gives only a simplified description of the physical reality.

In our approach we develop a more realistic model of the hot hydrogen gas distribution taking into account solar cycle variability of the radiation pressure and ionisation rate. The adopted amplitudes of variations of the radiation pressure and total ionisation rate are presented in Fig.1 and reflect reasonably well the typical solar cycle of the canonical period of 11 years (see e.g. Ajello et al., 1987; Lean and Skumanich, 1983; Schwenn, 1983; Bruno et al., 1994). The interstellar H gas parameters v_B = 20 km/s, T = 8000 K are taken after Bertaux et al. (1985).

2. Method of calculations

An analytic link between the unperturbed distribution function of the gas in VLISM (assumed to be the "shifted" Maxwellian corresponding to the temperature T) and the distribution function at a local point $\boldsymbol{R}$ in the interplanetary space exists only for a stationary model (Fahr, 1971; Wu and Judge, 1979). Establishing such a link in the

Space Science Reviews **72**: 467–470.
© 1995 *Kluwer Academic Publishers.*

nonstationary scenario adopted in this paper requires numerical tracking of individual atoms along their trajectories. Selecting an atom arriving at a time t at a point $\boldsymbol{R}=(r,\theta)$ with a velocity $V=(v_r, v_\rho)$ and tracing it back in time sufficiently far away to be able to neglect the solar dynamic and ionisation influence, one can calculate its survival probability w_i and the probability w_{Maxw} of launching such an atom from the shifted Maxwellian population of VLISM. The two values multiplied return the actual local distribution function of the gas. The numerical tracking of the atomic trajectories and survival probability is based on the following set of equations.

$$\dot{x}=v_x;\quad \dot{y}=v_y;\quad \dot{v}_x=-\frac{GM(1-\mu(t))}{r^3}x;\quad \dot{v}_y=-\frac{GM(1-\mu(t))}{r^3}y;\quad \frac{\dot{w}_i}{w_i}=-\beta(t)\left(\frac{r_E}{r}\right)^2;\quad r=\sqrt{x^2+y^2}$$

Integration of the local distribution function computed in this way over the local velocity space returns the required density of hydrogen atoms. Iterating this scheme for a sufficiently large number of space points $\boldsymbol{R}$ and moments of time t, one is able to monitor the global evolution of hydrogen density distribution during solar cycle.

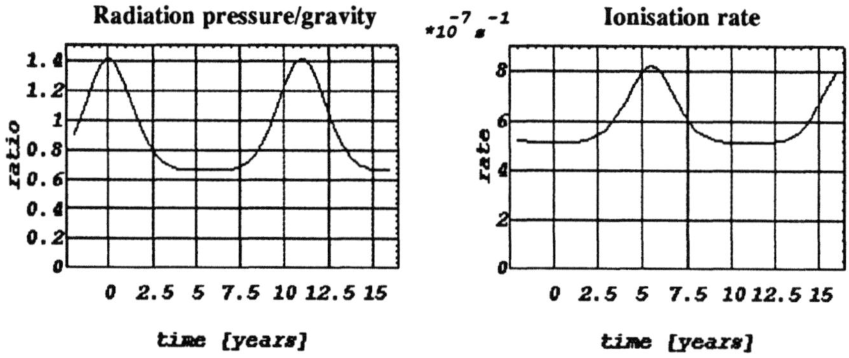

Fig.1. The courses of the ratio of radiation pressure force to solar gravity (left panel) and of the total ionisation rate (in the units of $10^{-7}s^{-1}$) at r_E = 1 AU (right panel) over the solar cycle adopted in the paper. Note that the two functions anticorrelate.

3. Discussion of the results

The results of the modeling are based on extensive calculations ($\sim 4\cdot 10^7$ trajectories tracked), providing us a very good and reliable temporal and spatial coverage of the hydrogen density distribution, guaranteeing a high accuracy and internal consistency (better than 1%) of the results. As a reference model we adopted the stationary model with μ and β values equal to the mean values (averaged over the solar cycle period) of the relevant functions presented in Fig.1. The main conclusions from the analysis of the modulation of hydrogen distribution during the solar cycle are following:

(i) the smaller heliocentric distance, the higher temporal density variations. They are especially pronounced in the downwind region. Along the upwind axis, the typical range of these variations around the mean value of the expected density is of the factor of 0.6 to 1.6 at 1 AU, and decrease to about 0.95 to 1.05 at 10 AU. In the downwind direction at 1 AU, the variability may cause in extreme cases either an enhancement by factor of 8 above the mean value or a drop to practically zero (see Fig.2). The corresponding density modulation does not exceed about 35 percent around the mean value at 10 AU;

(ii) the above-mentioned density waves propagate in a huge circumsolar region. They smear out beyond the distances of about 20 AU upwind and about 60 AU downwind, where the temporal fluctuations of the hydrogen density is well below 5% (see Fig.2);

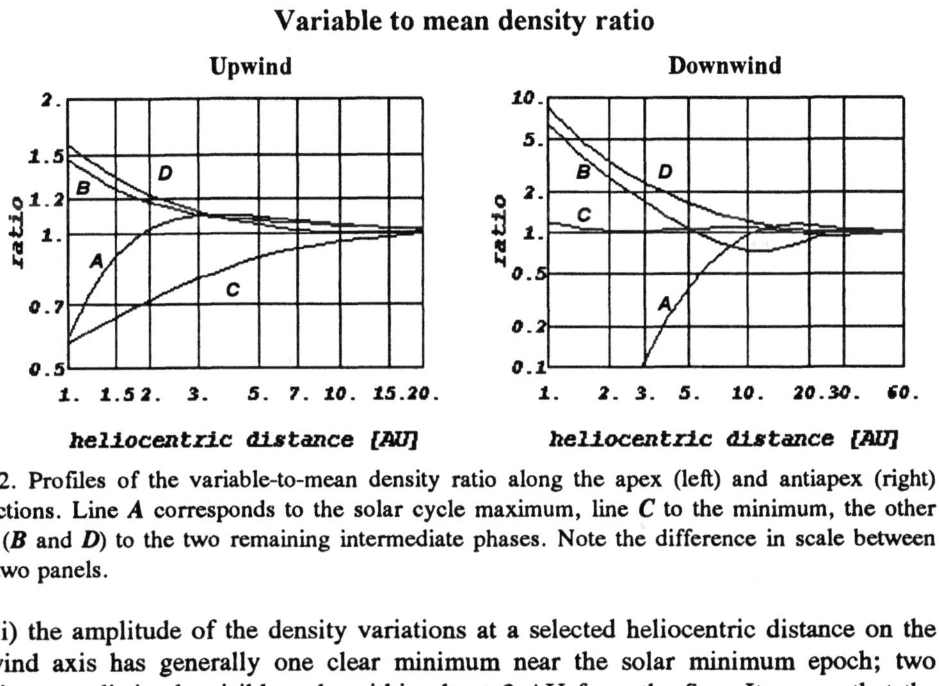

Fig.2. Profiles of the variable-to-mean density ratio along the apex (left) and antiapex (right) directions. Line ***A*** corresponds to the solar cycle maximum, line ***C*** to the minimum, the other two (***B*** and ***D***) to the two remaining intermediate phases. Note the difference in scale between the two panels.

(iii) the amplitude of the density variations at a selected heliocentric distance on the upwind axis has generally one clear minimum near the solar minimum epoch; two minima are distinctly visible only within about 2 AU from the Sun. It means that the effects of ionisation are the main factor shaping the density pattern in the upwind hemisphere. The corresponding downwind variations in the inner heliosphere show two minima separated in time by about one half of the solar cycle. They are associated with the epochs of high ionisation occurring at the minimum of the solar cycle and high radiation pressure at the solar maximum, respectively;

(iv) the hydrogen cavity region (defined here as an area where the hydrogen density is below 1/e of the unperturbed density in VLISM) also changes its size within the solar cycle. It extends up to 3.5 — 5 AU upwind and to 20 — 26 AU downwind. As a consequence of this pulsation, the total size of the cavity along the apex – antiapex direction at 1 AU varies from about 23.5 to 31 AU, the maximum coinciding with the solar minimum (i.e. maximum of ionisation) epoch;

(v) the density modulations affect the spatial location of MER contributing to the Lyman-α backscattered radiation. The position of the peak of emissivity corresponding to the maximum of $n(r,t)/r^2$ ranges from about 1.3 to 2 AU upwind and from about 3.5 to 15 AU downwind. The absolute value of the emissivity is also significantly affected.

(vi) the variations of emissivity patterns influence the upwind-to-downwind Lyman-α intensity ratio. It is shown (see Fig.3) that this ratio at 1 AU may vary from the factor of about 7 (slightly above one year after the solar maximum) to ~4.5 at the solar minimum. This noticeable temporal variations are pronounced for the observations from the inner Solar System and smear out for larger distances. Typically, at 10 AU the upwind intensity (in the units of actual solar Lyman-α flux) is within 5% from the mean stationary case, whereas downwind variations do not exceed 15%, leading to fluctuation of the above-mentioned ratio by only 10% around its mean value.

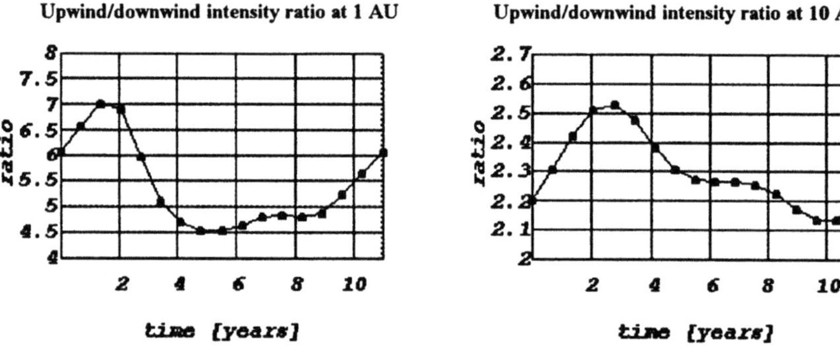

Fig.3. Solar cycle evolution of the upwind-to-downwind backscatter radiation intensity ratio at 1 AU (left) and 10 AU (right) measured in the antisolar direction.

Acknowledgements: We are indebted to Prof. G.Sitarski for drawing attention to the Recurrent Power Series method of orbits tracking and helpful discussions on that subject. This work was supported by the State Committee for Scientific Research under contract No. 2 P 304 018 04.

References

Ajello J.M., Stewart A.I., Thomas G.E., and Graps. A.: 1987, *Astrophys.J.* **317**, 964

Bertaux, J.L., Lallement, R., Kurt, V.G., and Mironova, E.N.: 1985, *Astron.Astrophys.* **150**, 1

Bruno, R., Villante, U., and Stecca, A.: 1994, *Ann.Geophys.* **12**, 105

Danby J.M.A. and Camm G.L.: 1957, *M.N.R.A.S.* **117**, 50

Fahr H.J.: 1971, *Astron.Astrophys.* **14**, 263

Fahr H.J.: 1979, *Astron.Astrophys.* **77**, 101

Lean, J.L. and Skumanich, A.: 1983, *J.Geophys.Res.* **88**, 5751

Schwenn, R.: 1983, in *Solar Wind Five*, NASA CP-2280, p 489

Wu, F.M. and Judge D.L.: 1979, *Astrophys.J.* **231**, 594

THE FLUX OF INTERSTELLAR DUST OBSERVED BY ULYSSES AND GALILEO

M. BAGUHL, E. GRÜN, D.P. HAMILTON, G. LINKERT,
R. RIEMANN and P. STAUBACH
Max-Planck-Institut für Kernphysik, P.O.Box 103980, 69029 Heidelberg, Germany

and

H.A. ZOOK
Nasa, Johnson Space Center, Houston, U.S.A.

June 6, 1994

Abstract. Interstellar dust detected by the dust sensor onboard Ulysses was first identified after the Jupiter flyby when the spacecraft's trajectory changed dramatically (Grün et al., 1994). Here we report on two years of Ulysses post-Jupiter data covering the range of ecliptic latitudes from 0° to -54° and distances from 5.4 to 3.2 AU. We find that, over this time period, the flux of interstellar dust particles with a mean mass of $3 \cdot 10^{-13}$ g stays nearly constant at about $1 \cdot 10^{-4}$ m^{-2} s^{-1} $(\pi\ sr)^{-1}$ with both ecliptic latitude and heliocentric distance.

Also presented are 20 months of measurements from the identical dust sensor onboard the Galileo spacecraft which moved along an in-ecliptic orbit from 1.0 to 4.2 AU. From the impact direction and speeds of the measured dust particles we conclude that Galileo almost certainly sensed interstellar dust outside 2.8 AU; interstellar particles may also account for part of the flux seen between 1 and 2.8 AU.

Key words: Interstellar dust – Ulysses dust measurements – Galileo dust measurements

1. Introduction

There is clear evidence that the distribution of dust in the outer solar system is quite different from the zodiacal population which dominates within a few AU of the Sun. The first dust measurements in the outer solar system were performed by the Pioneer 10 and 11 spacecraft (Humes, 1980). At a distance of 3 AU, these probes found impact rates about a factor of three lower than those observed at 1 AU. Curiously, between 3 AU and 18 AU, the impact rates stayed roughly constant despite the fact that the spacecraft reached ecliptic latitudes of 15°. Humes (1980) concluded that the observations are inconsistent with the low-eccentricity low-inclination orbits typically assumed for zodiacal dust particles. On the contrary, the observations seemed to imply a population of dust particles on randomly-inclined highly-eccentric orbits.

Grün et al. (1992a) argued that beyond about 3 AU the Ulysses dust detector also measured a non-zodiacal dust population. From the flux ratio before and after Jupiter flyby, Grün et al. (1992b) concluded that the back-

Space Science Reviews **72**: 471–476.
© 1995 *Kluwer Academic Publishers.*

ground dust flux was on retrograde trajectories. Based on the analysis of a year of Ulysses data covering the ecliptic latitude range from 0° to -16°, Grün et al. (1993 and 1994) concluded that the trajectories and impact speeds could best be explained by dust particles on hyperbolic orbits penetrating the solar system from the upstream direction of the local interstellar wind (Witte et al., 1993).

Divine (1993) developed an empirical model of the interplanetary dust cloud using impact rates observed by Pioneers 10 and 11 as well as early data from Galileo and Ulysses. Although Divine's model lacks an interstellar population, it does contain a component with randomized inclinations that, to some extent, mimics the behavior of interstellar particles. In this paper we compare data from the later phases of the Galileo and Ulysses missions with predictions from both Divine's interplanetary model and a simple model of interstellar dust.

2. Ulysses measurements from 0° to -54°

In Fig. 1 we show the flux of dust particles observed by Ulysses during the first two years after Jupiter flyby. Except for three of the latitude intervals, the flux shows no strong trend, remaining roughly constant over a large range of latitudes. Assuming that the flux is constant, we obtain a mean dust flux of $1 \cdot 10^{-4}$ m^{-2} s^{-1} $(\pi \text{ sr})^{-1}$ with a mean mass of $3 \cdot 10^{-13}$ g. Divine's model, by contrast, shows a gradual rise in the flux due to increases in both the spatial density of dust particles and in typical impact velocities as Ulysses moves inward. The data is marginally compatible with Divine's model, but also consistent with a flux of interstellar grains.

In particular, the interstellar flux is expected to be nearly constant over the timespan shown in Fig. 1. Deflection and focusing by solar gravity should cause the flux to increase by only about 10% over the Ulysses orbit from 5.4 AU to 3.2 AU. Unfortunately, these variations are too small to be easily detected in the data. In addition, the importance of sublimation of interstellar dust, which leads to a diminishing interstellar flux with decreasing radial distance, is difficult to assess. This effect depends on particle sizes and on material properties.

A better test of the predicted flux rates will occur as Ulysses continues to descend beneath the ecliptic. At extreme southern latitudes, the spatial density of Divine's interplanetary population is primarily latitude dependent and eventually drops to levels five times lower than at -45°. In the absence of sublimation, the interstellar dust flux should continue to increase to a value about 20% larger at maximum southern latitude than at -45° .

The predictions of the two models can be tested more stringently by combining the information contained in measured impact directions and speeds. This has been performed by Grün et al. (1994) for impact data until

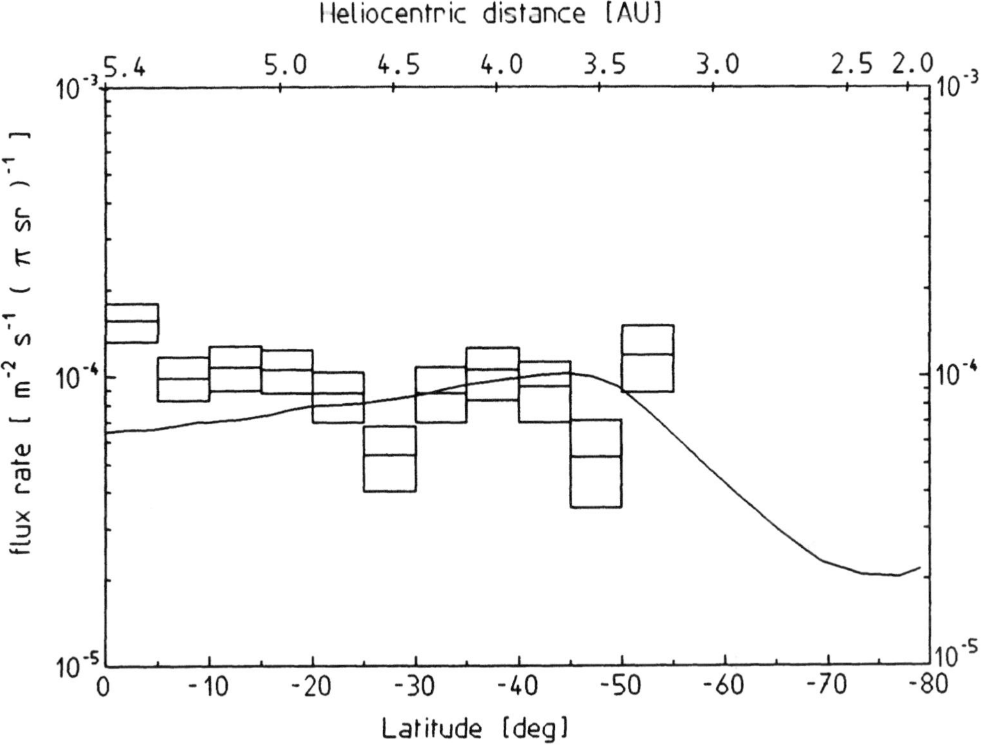

Fig. 1. Ulysses flux rate as a function of ecliptic latitude and heliocentric distance. The boxes represent the flux of particles with a mean mass of $3 \cdot 10^{-13}$ g measured after Jupiter flyby in 5° intervals of latitude and their heights correspond to statistical errors. Jovian stream particles, identified by their collimation in impact time and direction, were excluded from the displayed impact rate. The solid line is the flux predicted by Divine's model over the same time period.

the end of 1992 (ecliptic latitude from 0° to -16°). In Fig. 2 we show the impact directions of dust particles measured between 1 March 1992 and 24 April 1994 versus ecliptic latitude.

If bound orbits dominate the flux, then Ulysses, sweeping rapidly through a cloud of randomly-inclined orbits, should see impacts strongly concentrated near the spacecraft apex direction. Over the time period shown in Fig. 2, this direction is visible to the dust detector between rotation angles of 150° and 180°. Such a concentration is not seen in the figure; instead, the impact directions concentrate strongly towards the interstellar direction (near 90° - see Fig. 2). The measured impact speeds are also compatible with the assumption that the dust particles are predominantly interstellar in origin. These results indicate that Divine (1993)'s model requires the addition of an interstellar dust population.

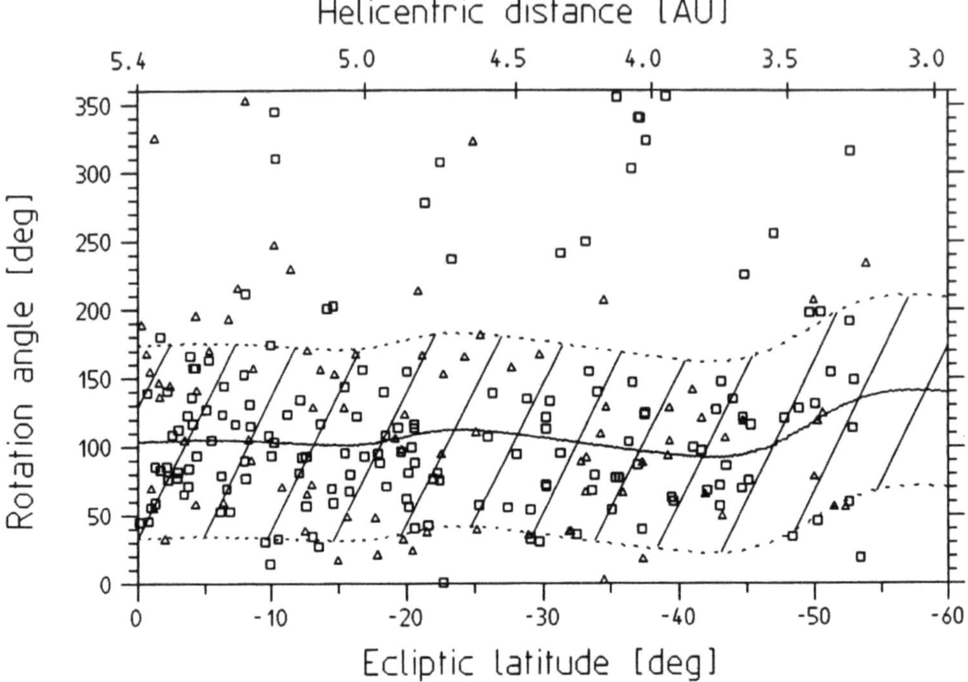

Fig. 2. Impact direction of dust particles versus the ecliptic latitude of Ulysses (For the definition of rotation angle see Grün et al. (1993)). The solid line gives the direction towards the upstream direction of the interstellar gas and the broken lines indicates the limits of the sensor's field of view for this direction. The squares denote particles with impact speeds compatible with an interstellar origin (v > 15 km/s, within the 1σ-error of the speed determination), triangles denote slower speeds.

3. Galileo dust measurements from 1 to 4.2 AU

We argued in the last section that Ulysses is currently seeing interstellar dust. If this is true, interstellar particles should also have been seen by Ulysses before the Jupiter encounter as well as by Galileo enroute to Jupiter. In the former case, Ulysses' highly eccentric orbit caused the impact directions of classical interplanetary and interstellar dust particles to largely overlap. In contrast, interstellar particles can be distinguished from dust on low-eccentricity low-inclination orbits over a large part of Galileo's orbit. This can be seen in Fig. 3, which gives the impact directions of all dust impacts registered after the second Earth encounter. The shaded region in Fig. 3 represents the arrival directions of interstellar particles. Nearly all impacts outside a distance of 2.6 AU have impact directions compatible with an interstellar origin. But between 1.0 and 2.8 AU, the direction is also consistent with that expected for low-eccentricity and low-inclination particles.

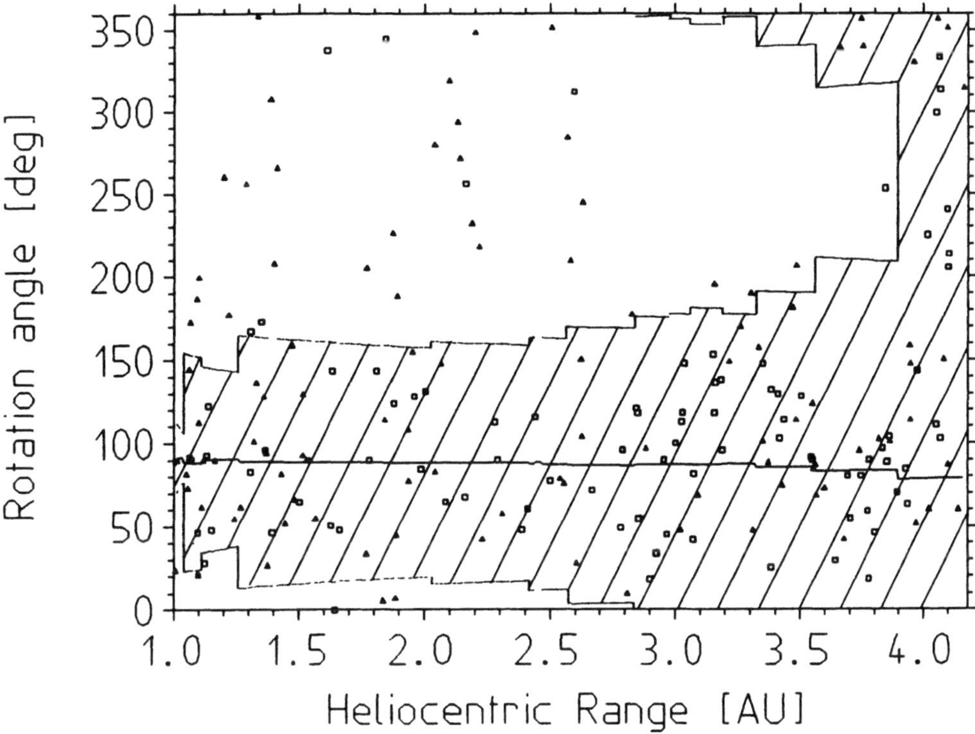

Fig. 3. Impact direction of dust particles versus the heliocentric range of Galileo. The solid line gives the direction towards the upstream direction of the interstellar gas and the dotted lines limit the sensor's field of view for this direction. Squares denote fast particles with impact speeds > 15 km/s, triangles represent slower particles.

Outside of 2.8 AU, however, particles on such orbits should come from near a rotation angle of 270° (see Fig. 3), clearly incompatible with the data. Although a distribution of bound orbits with large eccentricities and inclinations cannot be immediately ruled out, the distribution of impacts is more naturally explained with interstellar trajectories.

4. Discussion

From the data presented, it can be seen that dust of interstellar origin is the best explanation for the majority of the dust flux measured by Ulysses after the Jupiter flyby. Galileo too, appears to have sensed interstellar dust. Although speed measurements in both cases indicate hyperbolic velocities for the impacting particles, the best argument for an interstellar origin are the impact directions of the dust particles. For Ulysses, the combined effects of decreasing ecliptic latitude and radial distance do not significantly influence the flux rate and distribution of impact directions in agreement with the expected behavior of interstellar particles. Whereas during the in-ecliptic leg of the Ulysses orbit no distinction between interstellar and bound orbits can be derived from the impact directions, the out-of-ecliptic leg and the Galileo measurements show that interstellar dust is definitely present in the ecliptic plane outside about 2.8 AU. The situation within 2.8 AU is more complicated, but it is possible that interstellar dust may be present inward to 1 AU. Nevertheless, the contribution of interstellar dust to the flux at terrestrial distances should be lower than 3%, a value derived by McDonnell and Berg (1975) from flux isotropy arguments. Further analysis of recent data is necessary to improve the reliability of this estimate.

References

Divine N.: 1993, 'Five populations of interplanetary meteoroids', *JGR* **Vol.98 E9**, page 17029-17048.

Grün E., Zook H.A., Fechtig H. and Giese R. H.: 1985, 'Collisional Balance of the Meteoritic Complex', *Icarus* **Vol.62.**, page 244.

Grün E., Baguhl M., Fechtig H., Hanner M.S., Kissel J., Lindblad B.-A., Linkert D., Linkert G., Mann I., McDonnell J.A.M., Morfill G.E., Polanskey C., Riemann R., Schwehm G., Siddique N., Zook H.A.: 1992a, 'Galileo and Ulysses dust measurements: From Venus to Jupiter', *GRL* **Vol. 19.**, page 1311.

Grün E., Zook H.A., Baguhl M., Fechtig H., Hanner M.S., Kissel J., Lindblad B.-A., Linkert D., Linkert G., Mann I., McDonnell J.A.M., Morfill G.E., Polanskey C., Riemann R., Schwehm G., Siddique N.: 1992b, 'Ulysses Dust Measurements near Jupiter', *Science* **Vol. 257.**, page 1550-1552.

Grün E. et al.: 1993, 'Discovery of jovian dust streams and interstellar grains by the Ulysses spacecraft', *Nature* **Vol. 362.**, page 428-430.

Grün, E., Gustafson, B., Mann, I., Baguhl M., Balogh A., Morfill G.E., Staubach P., Taylor A., Zook H.A.: 1994, 'Interstellar Dust in the Heliosphere', *A&A* **in press**,

Humes D.H.: 1980, 'Results of Pioneer 10 and 11 Meteoroid Experiments: Interplanetary and Near-Saturn', *JGR* **Vol. 85.A11**, page 5841-5852.

McDonnell J.A.M., Berg O.E.: 1975, 'Bounds for the interstellar to solar system microparticle flux ratio over the mass range $10^{-11} - 10^{-13}$ g', in Space Research XV, Academie Verlag:Berlin, page 555-563.

Witte M., Rosenbauer H., Banaszkiewicz M., Fahr H.: 1993, 'The Ulysses Neutral Gas Experiment: Determination of the Velocity and Temperature of the Interstellar Neutral Helium', *Advances in Space Res.* **Vol. 13.**, page (6)121-(6)130.

SPATIAL DISTRIBUTION AND ORBITAL PROPERTIES OF INTERPLANETARY DUST AT HIGH LATITUDES

I. MANN
Max-Planck-Institut für Aeronomie, Katlenburg-Lindau, Germany

Abstract. Although the interplanetary dust cloud is assumed to be mainly concentrated in the ecliptic plane, there is a component of dust particles on highly inclined orbits that forms the out-of-ecliptic distribution. The ULYSSES mission for the first time makes this component accessible to in-situ detection. Evidence for this dust component is also provided from the analysis of the Zodiacal light brightness and especially from the spherical shape of the solar F - corona. An obvious source for an out-of-ecliptic dust population is the activity of comets on high eccentric, highly inclined orbits. We discuss the dynamical conditions of particles under the influence of the radiation pressure when released from the comet and discuss their input to the dust cloud based on brightness analysis and in-situ results.

Keywords: Ulysses – interplanetary dust.

1. Introduction

Our general picture of a planetary system is that of a number of small bodies surrounding its sun in one plane (i.e. the ecliptic plane) as indeed observed circumstellar systems display such a preferred plane of symmetry. But there is evidence for a small component with isotropic distribution. Since the lifetime of small dust particles in the solar system is limited due to the Poynting-Robertson effect (i.e. in the order of $10^6 - 10^7$ years under present conditions of solar irradiation), these - on the time scales of solar system history - must have been produced recently. So the out-of-ecliptic component is not a remnant from the origin of the solar system but rather a product of its evolution. Although the interaction with the solar magnetic field can change the inclination of orbits of small dust particles (Morfill and Grün 1979, Banaszkiewicz *et al.* 1994) it cannot produce an isotropic distribution of dust in circumsolar space. Single particles can be thrown from the initial orbit close to the ecliptic when encountering a huge planet (Dohnanyi 1978), a process similar to the Jupiter flyby of the ULYSSES spacecraft (see Wenzel *et al.* 1992), but this process will not have considerable effect on the global distribution of dust. Therefore the major sources are expected to have orbits similar to the dust particles. The concentration of the interplanetary dust cloud towards the ecliptic plane is evident from the reduced brightness of the zodiacal light at lower latitudes, which is already decreased by 58% at 0.3 au above the ecliptic (Levasseur-Regourd

Space Science Reviews **72**: 477–482.
© 1995 *Kluwer Academic Publishers.*

et al. 1991). Inversion of the visual and infrared Zodiacal light, mainly describing the distribution of particles in the 1 to 100 micron size range close to 1 au, results in a dust cloud with its main component concentrated to the ecliptic plane and most probable inclinations i_{max} of orbits $6.5° < i_{max} < 15°$. The contribution of the isotropic component also including particles on retrograde orbits amounts from 0 to 20% dependent on the applied model (see Kneißel and Mann 1991).

In contradiction to the Zodiacal light the F - corona appears spherical and most recent observations in the near infrared show an even steeper increase of brightness at the solar poles (MacQueen and Greeley 1994) than at the solar equator. Model calculations on the other hand have proven the high contribution of near solar dust to the F - corona brightness (Mann 1992), which therefore describes the most inner dust cloud. This means, that close to the sun the out-of-ecliptic component is of the same order of magnitude as the ecliptic component.

Further proof for the out-of-ecliptic dust production is the observation of long period comets and their activity. Among them, the exclusive spaceborne observations of several sungrazing comets detected from the SOLWIND and SMM coronagraphs (Marsden 1989) demonstrate the constraints and selection effects of usual ground-based observations. The in-situ detection of particles on inclined orbits from in-ecliptic measurements, although in principle possible, is limited by statistics due to the small position probability of these particles in the ecliptic plane (Kneißel and Mann 1991). The ULYSSES mission now provides the possibility for direct detection of such particles in the outer parts of the dust cloud.

2. ULYSSES Measurements

During the initial part of the ULYSSES mission the dust experiment detected an almost constant flux of particles (Grün *et al.* 1992a). The ongoing high flux rate at larger solar distance can be explained by detection of additional isotropic and high eccentric dust fluxes, whereas the ecliptic component (with dust particles on low inclined low eccentric orbits) is decreasing with solar distance. From the analysis of the ULYSSES data at larger solar distances, after separation of the dust streams originating from the Jupiter magnetosphere (see Grün *et al.* 1992b), at least 60% of detected particles were found, based on their high relative velocity and from their impact direction to be of interstellar origin (Grün *et al.* 1993). Again for the particles, which are not clearly interstellar the relative velocities are high and tend to indicate the majority of these particles also to be on unbound orbits. A preliminary analysis of new dust impacts in 1993 shows no indication for a strong change along the high latitude orbit (Baguhl 1994).

3. Conditions of Dust Emission from Comets

In a recent estimate (Thomas 1994), the average dust production for a short period comet amounts to $5 \cdot 10^{12}$ g per apparition. Dust trails observed in the infrared with the IRAS satellite give a dust production of 10^9 g yr^{-1} to 10^{12} g yr^{-1} (Sykes and Walker 1992). These fragments are in a size range from mm to cm. Kresak and Kresakova (1987) give values from 10^{11} g yr^{-1} to 10^{12} g yr^{-1} from visual observations. This can be even higher for long period comets, where Delsemme (1976) gives dust production rates up to 20 t/s. Also Fulle (1991) shows, that the dust production from long period comets can be significantly higher than from short period comets. Thus already the apparition of a small number of comets could replenish the out-of-ecliptic dust component, if the emitted particles stay in the solar system.

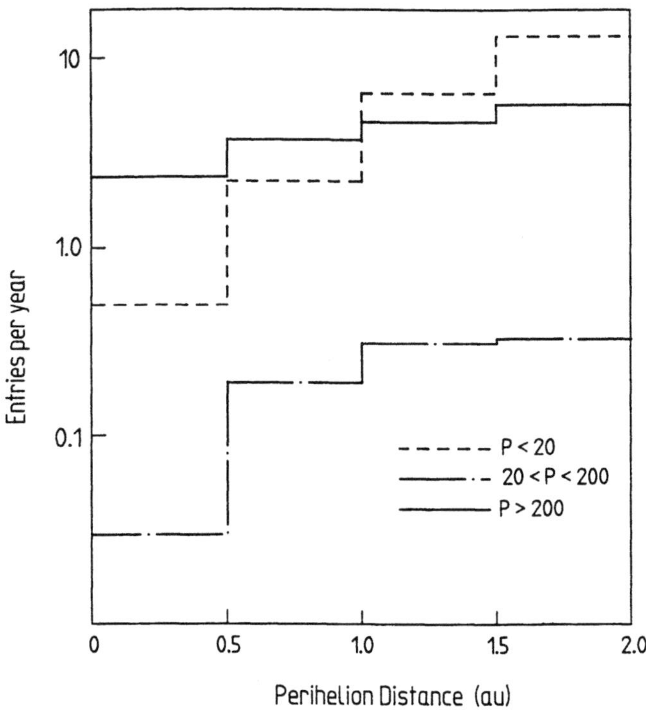

Fig. 1. The average number of entries per year for comets of the Jupiter family ($P < 20$), for Halley type comets ($20 < P < 200$) and for long period comets with perihelion distances $q < 0.5$ AE, 0.5 au $< q <$ 1 au, 1 au $< q <$ 1.5 au and 1.5 au $< q <$ 2 au according to the values given by Kresak (1992).

We regard the conditions for comets within 2 au around the sun as discussed by Kresak (1992). The average number of entries per year for comets of the Jupiter family ($P < 20$) is about 23 with a preference for the comets with large perihelion distances, for Halley type comets ($20 < P < 200$) the average number of entries is much smaller and close to 1 (see Fig. 1). In the case of the long period comets the average number of entries amounts to 16 per year and distances $q < 0.5$ au have still a significant contribution (see Fig. 2), the sungrazing comets of the Kreutz group (Marsden 1989) are not included (their mass contribution is only small).

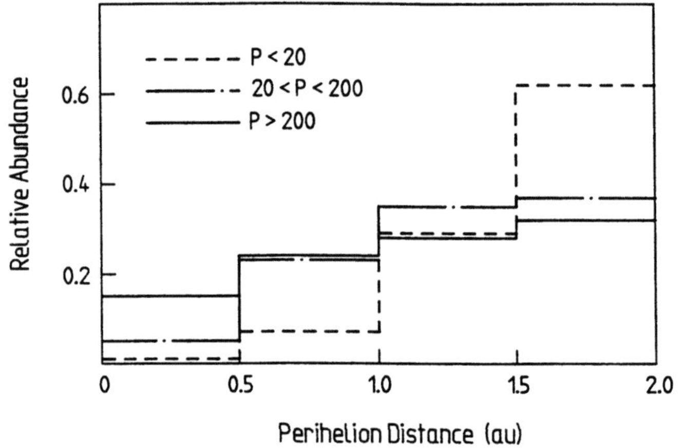

Fig. 2. The distribution of perihelion distances for comets with perihelion distance $q < 2$ au for comets of the Jupiter family ($P < 20$), for Halley type comets ($20 < P < 200$) and for long period comets given in the intervals $q < 0.5$ AE, 0.5 au $< q <$ 1 au, 1 au $< q <$ 1.5 au and 1.5 au $< q <$ 2 au (according to Kresak 1992).

When being released from the comet the particles suffer an increased force F_{rad} from radiation pressure that reduces the gravitational coupling F_{grav} to the sun. The radiation pressure forces acting on a particle are given in terms of $\beta = F_{rad}/F_{grav}$ (see Leinert and Grün 1991). Particles with $\beta \geq 0.005$ will be on unbound orbits as so-called β - meteorites when emitted from a comet at perihelion with orbital eccentricity 0.99. For e = 0.9 the critical value is $\beta \geq 0.95$. The conditions of expulsion are also effected by the emission velocity of fragments which, however is small compared to the orbital velocity and hence not regarded in an investigation of global behaviour.

From a common understanding of particle light scattering properties, which determine the radiation pressure (see Burns *et al.* 1978), we expect the particles with β-values in question to have masses $m < 10^{-6}$ g in the first case and $m < 5 \cdot 10^{-10}$ g in the second case. This implies, that especially in the range of particles detectable with the ULYSSES dust detector, the majority will be on unbound orbits and only contributing to the dust cloud for a short time period. Particles of larger sizes will stay coupled to the solar system and approach the sun on high eccentric orbits. In concern of the deceleration by the Poynting - Robertson effect (see Wyatt and Whipple 1950), particles on initially high eccentric orbits will keep their high eccentricity for a significant part of their lifetime during the Poynting-Robertson drift towards the sun. Especially particles emitted from long period comets keep their high eccentric orbits still in the inner solar system and therefore can cause a steep increase of the number density in the isotropic component. This could explain the significant contribution of out-of-ecliptic dust in brightness of the solar F -corona.

4. Summary

The measurements of the ULYSSES dust detector in the out-of-ecliptic leg of the orbit show a mainly constant flux of particles, under which about 60% are attributed to the interstellar flux. Besides this, there are mainly particles on unbound orbits which show no preferred orientation. This can be explained by dynamic conditions of particles emitted from long period comets: dust in the submicron range (i.e. the range of particles detected onboard ULYSSES) is on unbound orbits under the influence of radiation pressure wheras the majority of larger particles will stay in the solar system. These particles are a strong source of the out-of-ecliptic dust distribution in the inner solar system, which could be distinctly different from conditions for the ecliptic main component in its collisional evolution and therefore in its size distribution.

This work was supported by the Bundesminister für Forschung und Technologie BMFT.

References

Baguhl, M.: 1994, priv. communication.

Banaszkiewicz, M., Fahr, H.J. and Scherer, K.: 1994, *Icarus*, in press.

Burns, J.A., Lamy, P.L. and Soter, S.: 1979, *Icarus* **40**, 1.

Delsemme, A.H.: 1976, in H. Elsäßer and H. Fechtig (eds.), *Interplanetary Dust and Zodiacal Light*, Springer, Berlin, pp. 314-318.

Dohnanyi, J.S.: 1978, in J.A.M. McDonnell (ed.), *Cosmic Dust*, John Wiley Sons, New York, pp. 527-587.

Fechtig, H.: 1989, *Z. Naturforsch.* **44a**, 877.

Fulle, M. and Cremonese, G: 1991, in A.C. Levasseur-Regourd and H. Hasegawa (eds.), *Origin and Evolution of Interplanetary Dust*, Kluwer, Dordrecht, pp. 225-228.
Giese, R.H. and Kneißel, B.: 1989, *Icarus* **81**, 369.
Giese, R.H., Kneißel, B. and Rittich, U.: 1986, *Icarus* **68**, 395.
Grün, E., Zook, H.A., Fechtig, H. and Giese, R.H.: 1985, *Icarus* **62**, 244.
Grün, E. *et al.*: 1992a, *Geophys. Res. Lett.* **19**, 1311.
Grün, E. *et al.*: 1992b, *Science* **257**, 1550.
Grün, E. *et al.*: 1993, *Nature* **362**, 428.
Kneißel, B. and Mann, I.: 1991a, in A.C. Levasseur-Regourd and H. Hasegawa (eds.), *Origin and Evolution of Interplanetary Dust*, Kluwer, Dordrecht, pp. 139-146.
Kneißel, B. and Mann, I.: 1991b, *Adv. Space Res.* **11**, No 12, 123.
Kresak, L.: 1992, *Astron. Astrophys.* **279**, 646.
Kresak, L. and Kresakova, M.: 1987, in *Symposium on the Diversity and Similarity of Comets*, ESA SP 278, pp. 739-744.
Leinert, C. and Grün, E.: 1990, in R. Schwenn und E. Marsch (eds.), *Physics of the Inner Heliosphere, Physics and Chemistry in Space and Solar Physics*, Springer, pp. 207-275.
Levasseur-Regourd, A.C., Renard, J.B. and Dumont, R.: 1991, in A.C. Levasseur-Regourd and H. Hasegawa (eds.), *Origin and Evolution of Interplanetary Dust*, Kluwer, Dordrecht, 131-138.
MacQueen, R.M. and Greeley, B.W.: 1994, *Astrophys. J.*, submitted.
Mann, I.: 1992, *Astron. Astrophys.* **261**, 329.
Mann, I. and Grün, E.: 1992, in J.A.M. McDonnell (ed.), *Hypervelocity Impacts in Space*, The Unit for Space Science, The University of Kent at Canterbury, pp. 200-204.
Marsden, B.G.: 1989, *The Astron. J.* **98**, No 6, 2306.
Morfill, G.E. and Grün, E.: 1979, *Planet. Space Sci.* **27**, 1269.
Sykes, M.V. and Walker, R.G.: 1992, *Icarus* **95**, 180.
Thomas, N.: 1994, priv. communication.
Wenzel, K.P., Marsden, R.G., Page, D.E. and Smith, E.J.: 1992, *Astron. Astrophys. Suppl. Ser.* **92**, 411.
Whipple, F.L.: 1967, in J.L. Weinberg (ed.), *Zodiacal Light and the Interplanetary Medium*, NASA SP-150, pp. 409-426.
Wyatt, S.P. and Whipple, F.L.: 1950, *Astrophys. J.* **111**, 134.

SESSION 6
SUMMARY

ULYSSES' RACE TO THE POLE: SYMPOSIUM SUMMARY

MARTIN A. LEE
Institute for the Study of Earth, Oceans and Space
University of New Hampshire, Durham, NH 03824, USA

Abstract. This paper reproduces the Symposium Summary presentation at the 28th ESLAB Symposium on The High Latitude Heliosphere. The summary is not comprehensive but with considerable latitude covers the highlights of the conference. Featured are the observations of the high-speed wind from the growing southern polar coronal hole, new measurements of the interstellar gas and pickup ions, new measurements of cosmic ray modulation at high latitude, and our growing awareness of the three-dimensional heliosphere.

1. Introduction

During the last three days I have asked myself several times why I accepted this task. I am not a member of a Ulysses experiment team. I am not a Ulysses interdisciplinary scientist. I don't know much about the Sun. And I never lobbied for Solar Polar! Perhaps these credentials allow me to be objective. However, I must confess that I am involved in a controversy with J.R. Jokipii concerning the nature of the heliospheric magnetic field over the poles of the Sun (Hollweg and Lee, 1989; Jokipii and Kota, 1989), which may compromise my objectivity!

A symposium summary is necessarily personal. It is too short to give a balanced view. Hopefully it will convey the flavor of the Conference and provide a basis for discussion.

We have enjoyed three exciting days of science including expectations for the high-latitude heliosphere, tantalizing initial results from Ulysses, and a few new results from other missions and ground-based observations. The observations reported include many discoveries, but also phenomena which have become comforting old friends. As A. Balogh said in his study of 150 shocks observed by Ulysses: "They look as shocks have always looked!" However, many old friends can now be studied with new instrumentation, which has revealed some surprises reported at the Conference. Much of the reported work whets our appetite for the traversal of the last 20° of latitude to the southern polar region, Ulysses' traversal of the northern polar region, and with luck a chance in five years to observe a clumpy active Sun riddled with interspersed active regions and small coronal holes.

Even at this stage the mission has been a complete success in forcing us to view the heliosphere as a three-dimensional dynamic global entity. We are forced to place ourselves above (actually below) the Sun, for example, and contemplate the shape of a large coronal hole as shown in T. Hoeksema's plots of source-surface magnetic field strength. We must place ourselves at a great distance and ask whether the solar wind

Space Science Reviews **72**: 485–494.
© 1995 *Kluwer Academic Publishers.*

termination shock is a sphere or an ellipsoid, and imagine the density enhancement beyond the heliopause climbed by a shock wave as it excites the radio emission recently observed by Voyagers 1 and 2 and described by D. Gurnett. There are times when you feel the necessity of reviewing three-dimensional analytical geometry in order to understand the global structure of corotating interaction regions, the embedded shocks and flows, and why reverse shocks propagate to higher latitude than forward shocks as described by J. Gosling.

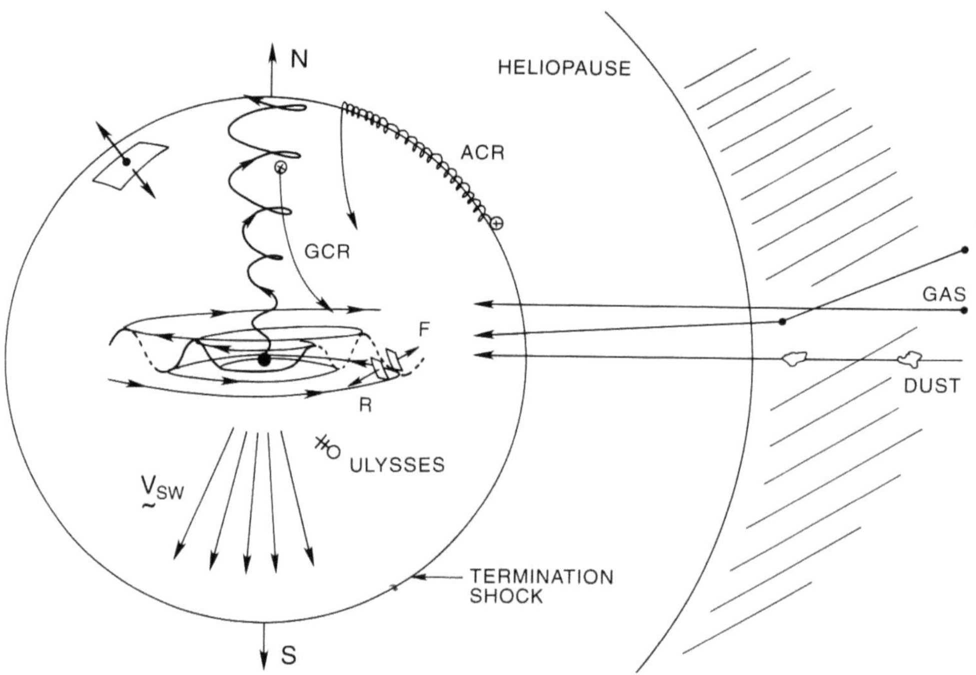

Fig. 1. Schematic view of the 3-D heliosphere showing Ulysses, high-speed wind from the southern polar coronal hole, sample Archimedes spiral magnetic field lines, the heliospheric current sheet, a corotating interaction region with forward and reverse shocks, the solar wind termination shock, the heliopause, interstellar gas and dust trajectories, and sample galactic and anomalous cosmic ray trajectories.

This global three-dimensional heliosphere is shown schematically in Figure 1, also depicting Ulysses as it travels toward the south pole of the Sun. The Sun is the solid dot near the center enveloped by the solar wind termination shock. Further out is the heliopause, dividing solar and interstellar plasma, with the pile-up of interstellar ionized gas beyond the heliopause denoted by hatched lines. The high-speed solar wind from the southern coronal hole now observed by Ulysses is shown. The warped current sheet separating magnetic fields of opposite polarity is shown along with three Archimedes spiral magnetic field lines. The interaction between fast and slow wind in the ecliptic is shown by the creation of a forward and a reverse shock on opposite sides of the current sheet. And, to be sure, the reverse shock moves to higher south latitude! The movement of the termination shock in response to varying wind ram pressure is denoted schematically in the upper left. Sample schematic trajectories of a galactic cosmic ray proton and an anomalous cosmic ray ion are shown indicating the possible easy access of galactic cosmic rays over the poles of the Sun, and the acceleration of the anomalous cosmic ray ion at the termination shock followed by its transport into the inner heliosphere. Interstellar neutral gas and dust are also shown as they enter the heliosphere due to the Sun's motion through the local interstellar cloud. The gas may charge-exchange with interstellar protons to alter the distribution function of the inflowing gas. Much of the gas is eventually ionized in the heliosphere to create interstellar pickup ions in the solar wind, which are the seed population of the anomalous cosmic rays.

2. The Solar Wind

Surprisingly the three-dimensional structure of the heliosphere appears to be fundamentally simpler than our earlier view from near the ecliptic , at least in this phase of the solar activity cycle. Once Ulysses was below the corotating interaction region in which fast wind from the southern polar coronal hole pushes against slow wind from adjacent longitudes, and overlooking a few high latitude coronal mass ejections, the simplicity and constancy of the high-speed wind from the coronal hole is remarkable. The constancy is in such contrast with the usual solar wind time profiles observed near the ecliptic which exhibit large variations in speed, density and temperature. When normalized to the same radial distance, the density, speed, mass flux and ram pressure observed by Ulysses were all nearly constant over the latitude range of 30°S to 60°S. Ulysses was clearly immersed in *pure* high-speed wind. The nearly constant ram pressure would not be expected to produce large variations in the location of the termination shock, in contrast to expectations based on ram pressure variations observed near the ecliptic (Suess, 1993; Belcher et al., 1993; Barnes, 1993).

An impressive continuous feature of the high-speed wind is the presence of large-amplitude Alfvén waves with periods up to 10-20 hours. These waves are clearly the agent required to accelerate the fast-wind up to the observed 750 km/s (Hollweg, 1978). E. Smith showed that the correlation between the magnetic field and velocity fluctuations implies that the waves propagate away from the Sun at a speed of about 0.7 times the Alfvén speed. The reduction in the wave phase speed appears to imply that the plasma pressure in the wind is larger parallel to the magnetic field than perpendicular to it. B. Tsurutani showed that the waveforms may be quite regular with a field fluctuation amplitude up to twice the ambient field strength, a hodogram indicating a rotation of the field with constant magnitude, and a tangential discontinuity at the leading edge of the wave.

The average magnetic field exhibits the expected Parker spiral behavior with the ratio of the azimuthal to radial components decreasing as sin θ, where θ is co-latitude. However, the spiral seems to be somewhat less tightly wrapped at high latitude than expected.

The SWICS instrument on Ulysses is able to identify over 20 ions in the solar wind and has an unprecedented ability to determine source temperatures based on the charge-state distribution of a given element, the composition of the solar wind, and the dynamics of the minor ions. In the fast wind the ions all have temperature proportional to mass and flow at the same bulk velocity, which was shown by B. Goldstein to be oriented outwards along the magnetic field at a fraction of the Alfvén speed relative to protons. Thus, the substantial intensity of Alfvén waves in fast solar wind appears to be responsible for the acceleration (or deceleration) of the ions to approximately the outward Alfvén velocity and for their heating, presumably by pitch-angle scattering of the ions on the waves. Solar wind composition in the high-speed solar wind also appears to reproduce solar abundances and does not show a marked dependence on First Ionization Potential (FIP), which would be indicative of complicated fractionation processes in the cool chromosphere.

The initial Ulysses observations leave the distinct impression that solar wind is fundamentally fast solar wind from coronal holes, extending over most of the Sun's solid angle. The slow wind, commonly observed from our restricted vantage point near the ecliptic for many years, is in fact a boundary condition on fast wind: slow wind is the result of the coronal hole's proximity to active region loops at low latitudes, in which the hot corona is trapped and unable to escape the confining magnetic field. Perhaps fast wind has a simple spherical expansion factor commensurate with the large uniform extent of the coronal hole, whereas slow wind has a larger expansion factor as it seeps out around the confined plasma of the loops and expands into the corona above. Perhaps the larger expansion factor in slow wind reduces the intensity of Alfvén waves, excited by turbulent photospheric convection below, to an intensity level incapable of providing the extended heating and acceleration present in fast wind above the coronal hole, and of controlling the behavior of the minor ions. Of course, during solar activity maximum, when the Sun is riddled with active loops, all wind may escape through more complex expanding routes, subject to fractionation in the chromosphere, and a less simple picture may emerge. We look forward to a Ulysses extended mission to observe our active Sun from high latitudes.

Although Ulysses' observations of the solar wind from the polar coronal holes, combined with low-Earth-orbit coronagraph and UV spectrometer measurements, and ground-based observations, will provide important constraints on theoretical modelling of fast solar wind, the ultimate origin of the solar wind will be elusive. The origin of the solar wind is intimately tied to the processes which heat the corona. Until we understand coronal heating (what role the Alfvén waves play, whether nanoflares are responsible, whether coronal heating arises from magnetic reconnection, why the temperature in coronal holes exhibits so little variation) our understanding of the solar wind will be incomplete. We hope that a Solar Probe may some day be able to explore directly the origins of the solar wind.

3. Interstellar Gas, Dust, Pickup Ions and the Anomalous Component

A vastly improved understanding of the coupled behavior of interstellar gas, interstellar pickup ions in the solar wind, and the cosmic ray anomalous component, has been a major success of Ulysses, even though these components are not restricted to high latitudes and some of the key results were not reported at this Conference. The success is due to new instrumentation on Ulysses which allows the interstellar neutral helium to be measured directly and the singly-charged interstellar pickup ions to be isolated from the same elements in the solar wind.

Interstellar helium has been detected indirectly by measuring the resonantly scattered UV radiation from the Sun (Chassefière et al., 1986). However, because the breadth of the solar emission line is comparable with the Doppler shift due to the flow of the helium through the heliosphere, it is difficult to infer the velocity, density and temperature of the interstellar helium. The neutral gas experiment on Ulysses (Witte et al., 1993) has partially resolved this ambiguity by direct detection of the helium atoms after traversing a baffle which determines their velocity. Although the density depends on instrument calibration and is difficult to measure, the bulk speed is measured to be 26 $\pm$ 1 km/s, the temperature is 6700 $\pm$ 1500 ^{O}K, and the direction of the flow agrees with previous measurements. The same velocity and temperature has been found for the local gas cloud by new sensitive measurements of absorption lines in the spectra from nearby stars (Lallement et al., 1993) which yield 25.7 $\pm$ 0.5 km/s and 7000 $\pm$ 1000 ^{O}K. The discrepancy between this speed and the value of 20 km/s inferred from the scatter of solar Lyman-α from interstellar hydrogen implies that a fraction of the inflowing hydrogen charge-exchanges with the ionized interstellar hydrogen beyond the heliopause, which decelerates and piles up as it flows around the heliopause as shown in Figure 1. The helium on the other hand is not affected by charge-exchange. These measurements should enable us to model the pile-up region and determine the interstellar hydrogen to helium ratio, the interstellar electron density, and indirectly perhaps the interstellar magnetic field. This proton pile-up beyond the heliopause is also thought to be the origin of the Voyager radio bursts (Gurnett et al., 1993) excited by electrons accelerated at a shock wave which propagates into the density ramp.

Interstellar pickup hydrogen has been observed for the first time by SWICS (Gloeckler et al., 1993; Geiss et al., 1994) along with low levels of interstellar pickup O^+, N^+, Ne^+ and He^{++}. Also pickup helium, first detected on AMPTE (Möbius et al., 1985), has been confirmed by SWICS. In general the observed densities of pickup hydrogen and helium confirm expectations based on solar UV backscatter and characteristic ionization rates. However, the pickup ions exhibit a mysterious variability and the hydrogen seems to correlate better with the pickup helium than with solar wind mass flux, another mystery in view of hydrogen's being primarily ionized by charge exchange and helium by photoionization.

Another interesting result reported at the Conference by N. Murphy and E. Smith is the presence of ULF fluctuations presumably excited by the isotropization of the interstellar pickup hydrogen. The waves have the expected frequency just above the proton gyrofrequency and left-hand polarization in the spacecraft frame (Lee and Ip, 1987). The waves are observed well away from Jupiter and are unlikely to be due to escaping ions or electrons from Jupiter. They are only observed when the field is quasi-radial. When the field is perpendicular to the radial direction the spacecraft frequency of the excited waves may be much lower, due to a negligible Doppler shift for waves propagating parallel to the magnetic field, and they may not be observable above the solar wind wave spectrum (Williams and Zank, 1994).

One major theme at this Conference was the origin of the energetic ions associated with the corotating interaction regions, presumably accelerated at the forward and reverse shocks bounding the regions. G. Gloeckler and E. Roelof presented SWICS measurements of suprathermal protons, α-particles and pickup He^+ in the interaction region. They showed that the protons, which appear to emerge smoothly from the pickup hydrogen distribution, and He^+ have extremely hard omnidirectional distribution functions proportional to v^{-4}, whereas the solar wind α-particles have a substantially softer spectrum and are dominated by the pickup ions. The extremely hard v^{-4} spectrum is completely unexplained; although a very strong shock could produce such a spectrum, the corotating shocks tend to be weak. The implication of the SWICS measurement is that these generally quasi-perpendicular shocks accelerate the small density of pickup hydrogen and helium far more efficiently than the large density of solar wind protons and α-particles. It is, however, puzzling that the α-particles have a softer spectrum than protons and He^+ which bracket the rigidity of the α-particles. At higher energies of a few MeV/nucleon both E. Keppler and L. Lanzerotti reported approximately solar abundances for the corotating ion events including an abundance of carbon comparable with oxygen. The presence of carbon rules out a pickup ion origin of these ions at higher energy. A possible resolution (Gloeckler et al., 1994) might be that the pickup ions more massive than He^+ have such high rigidities that the acceleration process is again inefficient; these elements are dominated by solar wind ions. The consequences of these Ulysses measurements and their interpretation are very important for the nature of ion injection and acceleration at the solar wind termination shock to form the cosmic ray anomalous component. The efficiencies of pickup and solar wind hydrogen acceleration determine the extent to which the termination shock is modified by the pressure of the anomalous component.

J. Simpson reported that the cosmic ray anomalous component at energies of 10-50 MeV/nucleon (helium, oxygen and nitrogen) is recovering, as expected during the onset of solar minimum with drift out along the heliospheric current sheet, to the same levels at which it was first discovered ~22 years ago (Garcia-Munoz et al.,1973; McDonald et al., 1974; Hovestadt et al., 1973).

Very similar to the interstellar gas is the behavior of interstellar dust, which was unambiguously detected for the first time by Ulysses. E. Grün described the detection of dust grains with speeds too great to be in orbit about the Sun and velocities consistent with the local cloud. Their detection should stimulate new studies of the role of dust in interstellar clouds.

4. Cosmic Rays

One of the earliest and strongest arguments in favor of the original Solar Polar, now Ulysses, Mission was the opportunity to observe galactic cosmic rays directly with a minimal amount of solar modulation. The argument was based on the fact that the nearly radial weak magnetic field over the poles of the Sun should result in a large radial spatial diffusion coefficient and easy access of cosmic rays. In contrast at low latitudes particles must either traverse the Archimedes spiral field, with stronger field magnitude and consequently smaller gyroradii and scattering mean free paths, or diffuse across field lines, generally a slower process. After drift transport was established to be important in the surprisingly regular heliospheric magnetic field (Jokipii and Levy, 1977), it was recognized that drift also provides easy access over the poles in the current phase of the solar magnetic cycle in which protons drift from the poles to the heliospheric current sheet. It was therefore expected that Ulysses would see substantial positive latitudinal gradients and a more nearly "interstellar spectrum".

Jokipii and Kota (1989) suggested that one effect had not been considered which could impede the otherwise easy access of cosmic rays. Transverse magnetic field frozen into a radial solar wind has a magnitude proportional to r^{-1} and soon dominates the radial field ($\propto r^{-2}$). They argued that the transverse field would be generated as a necessary consequence of photospheric motions. Hollweg and Lee (1989), on the other hand, argued that many transverse fields and motions are "combed out" by the accelerating wind which is deflected by the dominant pressure of the field. Whether very low frequency transverse convected structures dominate the radial average field or not, certainly Alfvén waves over the poles, which are observed to be intense in the fast wind observed by Ulysses, would rapidly saturate in the weak radial field and could scatter cosmic rays and partially inhibit their entry. Nevertheless a substantial latitudinal gradient was still expected.

In perhaps the biggest surprise of the Ulysses Mission, after the data are compared with IMP8 and corrected for a radial gradient, R.B. McKibben showed that the cosmic rays exhibit virtually no latitudinal gradient down to ~60° S! Another surprise, related by J. Simpson, was that when Ulysses was below the interaction region immersed in the high-speed solar wind with nearly constant speed, it still observed a strong variation in the cosmic ray intensity at the solar rotation period. The variation persisted even though drift should transport protons from the poles toward the interaction region with its corotating structure. The only obvious communication of the corotating structure to the cosmic rays would be if reverse shocks at larger heliocentric radial distance reached higher latitudes and intercepted the field lines passing through Ulysses.

M. Potgieter and J.R. Jokipii presented separate theoretical predictions of the cosmic ray intensity observed by Ulysses with increasing latitude. Both calculations model the solar activity and magnetic cycle dependence of drift and propagating structures; the model of Jokipii and J. Kota also includes the structure of the corotating interaction region in the wind velocity, drift and spatial diffusion coefficient. For both models the comparison with the Ulysses count rate down to 60° S latitude is not bad. However, both models predict a latitudinal gradient which becomes substantial beyond 60°, whereas the count rate shows no discernable latitudinal gradient. Clearly the crucial comparison will come this Summer and Fall. If there is still no latitudinal gradient observed, theory is "on the ropes". Jokipii and Kota's model did show a variation at the period of solar rotation at the latitude of Ulysses. However, it appeared to be smaller than that observed and it is not clear that the latitudinal extent of the corotating interaction region is comparable in the model and in reality.

What is the resolution!? In my view we are not ready to throw out modulation theory. The standard transport equation first written down by Parker (1965) and extended in important ways by Gleeson and Axford (1967) and Jokipii and Levy (1977) has worked admirably well for nearly-isotropic energetic particle populations throughout the heliosphere. There are terms left out including stochastic acceleration and cosmic-ray viscosity, but these are well-understood and very unlikely to resolve this issue. The fact is that we are putting theory to a stringent test to account for the Ulysses observations, or any particular series of observations. In order to calculate the modulation we need to know the solar wind velocity, drift velocity and the spatial diffusion tensor globally in both space and time. Even if we could infer these quantities locally from plasma observations, we cannot infer them globally. For example, if a reverse shock has propagated to high latitude from a corotating interaction region beyond Jupiter it will influence the wind velocity, drift and diffusion and subsequently can affect the modulated cosmic ray intensity at Ulysses. As another example, the high level of Alfvénic turbulence observed over the polar coronal holes could inhibit access of cosmic rays over the poles of the Sun. This effect is included heuristically in the calculations of Jokipii and Kota, but its magnitude is uncertain.

In addition the accurate specification of the spatial diffusion tensor based on the scattering turbulence in the solar wind has been a subject of active research for 20 years. Nevertheless fundamental uncertainties remain. The spatial diffusion coefficient parallel to the average field is very sensitive to the nature of the scattering turbulence. Throughout most of the heliosphere (heliocentric radial distances beyond 10 AU) scattering across the average magnetic field dominates the radial cosmic ray diffusive transport. This perpendicular diffusion has contributions from scattering by resonant fluctuations, random walk of individual field lines, and "stochastic" drift in the long wavelength fluctuations. Only the first of these contributions is relatively well understood but it is surely the least important. We need to better understand the microphysics of how particles interact with stochastic magnetic and electric fields at various length scales. Does the interaction divide conveniently into the above three "diffusive" components perpendicular to the "average" field, drift in the "average" field, and resonant scattering and "mirroring" parallel to the field, or does this division incorrectly simplify the interaction? Unfortunately the solar wind magnetic field does exhibit substantial fluctuations on all these scales. In the early 1970s a Workshop was held at NASA-Goddard Space Flight Center on this topic; perhaps it is appropriate to revisit these very difficult questions. R. Müller-Mellin expressed some of these same concerns at the Conference. The study of solar cosmic ray time intensity profiles provides information on the spatial diffusion coefficient parallel to the average field, but little information on perpendicular diffusion. However, we should not be pessimistic. As E. Roelof suggested, perhaps the Ulysses data can be used to pin down acceptable transport coefficients.

5. Conclusions

Perhaps I should conclude by revisiting Len Fisk's criterion for the success of Ulysses: How will the voyage of Ulysses be judged 20 years from now? While admitting that it is a mite early to judge the discoveries of Ulysses when the race to the southern pole has barely begun, I shall hazard an attempt. In the spirit of an Olympic year with memories of the mesmerizing Lillehammer figure-skating competition still vivid, I shall use the famous 6.0 scale in judging the discoveries of Ulysses.

1) Cosmic-ray modulation: 5.4. We do not have an unobstructed view of the interstellar galactic cosmic ray spectrum over the poles of the Sun. However, spherically-symmetric modulation has an aesthetic appeal.

2) Interstellar gas, dust and pickup ions, cosmic ray anomalous component: 5.8. The discovery of interstellar pickup H, O, N and Ne, interstellar dust, and the first direct detection of the neutral gas, opens a new field of study and enables us to determine interstellar abundances much more accurately. But we still must await Pioneer 10 or Voyager's discovery of the solar wind termination shock to open up studies of the interaction between the solar wind and the interstellar medium.

3) Origins of the solar wind, remote sensing the corona: 5.0. The high-speed polar coronal hole solar wind is impressive in its extensive uniformity and high levels of Alfvén waves. However, our ability to unravel the mysteries of coronal heating and the source regions of the solar wind may still be frustrated.

4) Morphology of the 3-D heliosphere: 6.0! (particularly on "artistic interpretation") This 3-D heliosphere which Ulysses has begun to reveal from below and above near solar minimum activity is a "real beauty"!

I refer you to J.L. Culhane's paper on recent results from Yohkoh for a picture of the emerging southern polar coronal hole, which is the origin of the fast wind in which Ulysses is immersed as it completes its race to the polar region.

Acknowledgements: The author is indebted to many colleagues at the Conference for fruitful discussions including L. Fisk, G. Gloeckler, B. Goldstein, J. Gosling, E. Grün, S. Habbal, J.R Jokipii, F. McDonald, I. Mann, E. Marsch, D. McComas, B. McKibben, G. Morfill, M. Potgieter, E. Roelof, J. Simpson, and E. Smith. He also wishes to express his gratitude to the organizers for an excellent conference, and to Richard Marsden for his patience in awaiting this paper. This work was supported, in part, by NASA Space Physics Theory Program Grant NAG5-1479, NASA Supporting Research and Technology Grant NAGW-2579, and NSF Grant ATM 9215279.

References

Barnes, A.: 1993, Motion of the heliospheric termination shock: A gas dynamic model, *J. Geophys. Res.* **98**, 15,137.

Belcher, J.W., A.J. Lazarus, R.L. McNutt, Jr. and G.S. Gordon, Jr.: 1993, Solar wind conditions in the outer heliosphere and the distance to the termination shock, *J. Geophys. Res.* **98**, 15,177.

Chassefière, E., J.L. Bertaux, R. Lallement and V.G. Kurt: 1986, Atomic hydrogen and helium densities of the interstellar medium measured in the vicinity of the Sun, *Astron. Astrophys.* **160**, 229-242.

Garcia-Munoz, M., G.M. Mason and J.A. Simpson: 1973, A new test for solar modulation theory: The 1972 May-July low-energy galactic cosmic-ray proton and helium spectra, *Astrophys. J.* **182**, L81-L84.

Geiss, J., G. Gloeckler, U. Mall, R. von Steiger, A.B. Galvin and K.W. Ogilvie: 1994, Interstellar oxygen, nitrogen and neon in the heliosphere, *Astron. Astrophys.* **282**, 924-933.

Gleeson, L.J. and W.I. Axford: 1967, Cosmic rays in the interplanetary medium. *Astrophys. J.* **149**, L115-L118.

Gloeckler, G., J. Geiss, H. Balsiger, L.A. Fisk, A.B. Galvin, F.M. Ipavich, K.W. Ogilvie, R. von Steiger and B. Wilken: 1993, Detection of interstellar pick-up hydrogen in the solar system, *Science* **261**, 70-73.

Gloeckler, G., J. Geiss, E.C. Roelof, L.A. Fisk, F.M. Ipavich, K.W. Ogilvie, L.J. Lanzerotti, R. von Steiger and B. Wilken: 1994, Acceleration of interstellar pickup ions in the disturbed solar wind observed on Ulysses, *J. Geophys. Res.*, in press.

Gurnett, D.A., W.S. Kurth, S.C. Allendorf and R.L. Poynter: 1993, Radio emission from the heliopause triggered by an interplanetary shock, *Science* **262**, 199.

Hollweg, J.V.: 1978, Some physical processes in the solar wind, *Rev. Geophys. Space Phys.* **16**, 689.

Hollweg, J.V. and M.A. Lee: 1989, Slow twists of solar magnetic flux tubes and the polar magnetic field of the sun, *Geophys. Res. Lett.* **16**, 919-922.

Hovestadt, D., O. Vollmer, G. Gloeckler and C.Y. Fan: 1973, Differential energy spectra of low-energy (<8.5 MeV per nucleon) heavy cosmic rays during solar quiet times, *Phys. Rev. Lett.* **31**, 650-653.

Jokipii, J.R. and E.H. Levy: 1977, Effects of particle drifts on the solar modulation of galactic cosmic rays, *Astrophys. J.* **213**, L85-L88.

Jokipii, J.R. and J. Kota: 1989, The polar heliospheric magnetic field, *Geophys. Res. Lett.* **16**, 1-4.

Lallement, R., J.-L. Bertaux and J.T. Clarke: 1993, Deceleration of interstellar hydrogen at the heliospheric interface, *Science* **260**, 1095-1098.

Lee, M.A. and W.-H. Ip: 1987, Hydromagnetic wave excitation by ionized interstellar hydrogen and helium in the solar wind, *J. Geophys. Res.* **92**, 11,041-11,052.

McDonald, F.B., B.J. Teegarden, J.H. Trainor and W.R. Webber: 1974, The anomalous abundance of cosmic-ray nitrogen and oxygen nuclei at low energies, *Astrophys. J.* **187**, L105-L108.

Möbius, E. et al.: 1985, Direct observation of He^+ pick-up ions of interstellar origin in the solar wind, *Nature* **318**, 426.

Parker, E.N.: 1965, The passage of energetic charged particles through interplanetary space, *Planet. Space Sci.* **13**, 9-49.

Suess, S.T.: 1993, Temporal variations in the termination shock distance, *J. Geophys. Res.* **98**, 15,147.

Williams, L.L. and G.P. Zank: 1994, The effect of magnetic field geometry on the wave signature of the pickup of interstellar neutrals, *J. Geophys. Res.*, in press.

Witte, M., H. Rosenbauer, M. Banaszkiewicz and H. Fahr: 1993, The Ulysses neutral gas experiment: Determination of the velocity and temperature of the interstellar neutral helium, *Adv. Space Res.* **13**, 6, 121-130.

List of Participants

Anglin, David	HIA, Ottawa, Canada
Baguhl, Michael	MPI für Kernphysik, Heidelberg, Germany
Balogh, André	Imperial College, London, UK
Bame, Sam	Los Alamos Nat. Lab., Los Alamos, USA
Banaszkiewics, M.	Polish Acad. of Sciences, Warsaw, Poland
Barnes, Aaron	NASA Ames Res. Center, Moffett Field, USA
Beech, Peter	ESOC / Jet Propulsion Laboratory, Pasadena, USA
Bird, Michael	University of Bonn, Germany
Bodmer, Roland	University of Bern, Switzerland
Bonnet, Roger	ESA Headquarters, Paris, France
Bothmer, Volker	MPI für Aeronomie, Lindau, Germany
Bougeret, Jean-Louis	Observatoire de Paris, Meudon, France
Brandt, Jack	University of Colorado, Boulder, USA
Burton, Marcia	Jet Propulsion Laboratory, Pasadena, USA
Bzowski, Maciej	Polish Acad. of Sciences, Warsaw, Poland
Canu, Patrick	CRPE, Issy les Moulineaux, France
Castro, Fernando	ESOC / Jet Propulsion Laboratory, Pasadena, USA
Coles, Bill	University of California, San Diego, USA
Cooper, John	NSSDC / GSFC, Greenbelt, USA
Crooker, Nancy	Boston University, USA
Culhane, Len	MSSL, Holmbury St. Mary, UK
Decker, Robert	Johns Hopkins Univ., APL, Laurel, USA
Ducros, Rene	Service d'Astrophysique, C.E. Saclay, France
Erdös, Geza	KFKI, Budapest, Hungary
Fahr, Hans-Jörg	University of Bonn, Germany
Ferrando, Philippe	Service d'Astrophysique, C.E. Saclay, France
Fichtner, Horst	University of Bonn, Germany
Fisk, Lennard	University of Michigan, USA
Fleck, Bernhard	ESA Space Science Department, ESTEC, NL
Forsyth, Robert	Imperial College, London, UK
Fränz, Marcus	MPI für Aeronomie, Lindau, Germany
Galvin, Antoinette	University of Maryland, College Park, USA
Garcia-Perez, Raul	ESOC / Jet Propulsion Laboratory, Pasadena, USA
Gazis, Paul	NASA Ames Res. Center, Moffett Field, USA
Geiss, Johannes	University of Bern, Switzerland
Giacalone, Joe	University of Arizona, Tucson, USA
Gloeckler, George	Univerity of Maryland, College Park, USA
Gold, Robert	Johns Hopkins Univ., APL, Laurel, USA
Goldstein, Bruce	Jet Propulsion Laboratory, Pasadena, USA
Gosling, Jack	Los Alamos Nat. Lab., Los Alamos, USA

Grün, Eberhard	MPI für Kernphysik, Heidelberg, Germany
Gurnett, Donald	University of Iowa, Iowa City, USA
Habbal, Shadia	Smithsonian Center for Astrophys., Cambridge, USA
Hamilton, D.	MPI für Kernphysik, Heidelberg, Germany
Hawkins III, Stephen	Johns Hopkins Univ., APL, Laurel, USA
Heber, Bernd	IFKKI, University of Kiel, Germany
Hoang, Sang	Observatoire de Paris, Meudon, France
Hoeksema, Todd	CSSA, Stanford University, USA
Hovestadt, Dieter	MPI für Extraterrestr. Physik, Garching, Germany
Huber, Martin	ESA Space Science Department, ESTEC, NL
Jokipii, Randy	University of Arizona, Tucson, USA
Jones, Vernon	NASA Headquarters, Washington D.C., USA
Kane, M.	Johns Hopkins Univ., APL, Laurel, USA
Karl, J.	University of Cologne, Cologne, Germany
Kecskemety, K.	KFKI, Budapest, Hungary
Kellogg, Paul	University of Minnesota, Minneapolis, USA
Keppler, Erhardt	MPI für Aeronomie, Lindau, Germany
Király, Peter	KFKI, Budapest, Hungary
Klecker , Berndt	MPI für Extraterrestr. Physik, Garching, Germany
Kohl, John	Smithsonian Center for Astrophys., Cambridge, USA
Krimigis, Stamatis	Johns Hopkins Univ., APL, Laurel, USA
Krupp, Norbert	MPI für Aeronomie, Lindau, Germany
Kunow, Horst	IFKKI, University of Kiel, Germany
Kyrölä, Erkki	FMI, Helsinki, Finland
Lantos, Pierre	Observatoire de Paris, Meudon, France
Lanzerotti, Louis	AT&T Bell Labs., Murray Hill, USA
Lee, Marty	University of New Hampshire, Durham, USA
Lin, Naiguo	University of Minnesota, USA
Löffler, Angela	University of Köln, Germany
Lorenz, Erwin	DARA, Bonn, Germany
MacDowall, Robert	NASA / GSFC, Greenbelt, USA
Maclennan, Carol	AT&T Bell Labs., Murray Hill, USA
Maksimouic, M.	Observatoire de Paris, Meudon, France
Mall, Urs	University of Bern, Switzerland
Mann, Ingrid	MPI für Aeronomie, Lindau, Germany
Marsch, Eckart	MPI für Aeronomie, Lindau, Germany
Marsden, Richard	ESA Space Science Department, ESTEC, NL
Massey, Ed	Jet Propulsion Laboratory, Pasadena, USA
McComas, Dave	Los Alamos Nat. Lab., Los Alamos, USA
McDonald, Frank	University of Maryland, College Park, USA
McKenzie, Jim	MPI für Aeronomie, Lindau, Germany
McKibben, Bruce	University of Chicago, USA

Meeks, Willis	Jet Propulsion Laboratory, Pasadena, USA
Meyer, Don	Jet Propulsion Laboratory, Pasadena, USA
Morfill, Gregor	MPI für Extraterrestr. Physik, Garching, Germany
Müller-Mellin, Reinhold	IFKKI, University of Kiel, Germany
Murphy, Neil	Jet Propulsion Laboratory, Pasadena, USA
Neubauer, Frits	University of Cologne, Germany
Pätzold, Martin	University of Cologne, Germany
Page, Edgar	ESA / Jet Propulsion Laboratory, Pasadena, USA
Paizis, Constantin	University of Milan, Italy
Phillips, John	Los Alamos Nat. Lab., Los Alamos, USA
Pick, Monique	Observatoire de Paris, Meudon, France
Ponyavin, Dmitri	St. Petersburg University, Russia
Potgieter, Marius	Potchefstroom Univ. for CHE, South Africa
Rastoin, Cecile	Service d'Astrophysique, C.E. Saclay, France
Raviart, Alain	Service d'Astrophysique, C.E. Saclay, France
Reiner, Michael	Hughes STX, Lanham, USA
Riley, Pete	University of Arizona, Tucson, USA
Roelof, Edmond	Johns Hopkins Univ., APL, Laurel, USA
Rucinski, Daniel	Polish Acad. of Sciences, Warsaw, Poland
Ruess, Matthias	MPI für Aeronomie, Lindau, Germany
Sanahuja, Blai	University of Barcelona, Spain
Sandback, O.	University of Oslo, Norway
Sanderson, Trevor	ESA Space Science Department, ESTEC, NL
Schulz, Michael	Lockheed Palo Alto Res. Labs., USA
Schwehm, Gerhard	ESA Space Science Department, ESTEC, NL
Schwenn, Rainer	MPI für Aeronomie, Lindau, Germany
Simnett, George	University of Birmingham, UK
Simpson, John	University of Chicago, USA
Siscoe, George	Boston University, USA
Smith, Edward	Jet Propulsion Laboratory, Pasadena, USA
Solomon, Jacques	University of Paris, Orsay, France
Sonett, Chuck	University of Arizona, Tucson, USA
Staines, Keith	Imperial College, London, UK
Steiger von, Rudolf	University of Bern, Switzerland
Summanen, T.	FMI, Helsinki, Finland
Szegö, Karoly	KFKI, Budapest, Hungary
Tappin, James	University of Birmingham, UK
Tranquille, Cecil	ESA Space Science Department, ESTEC, NL
Trattner, Karl-Heinz	ESA Space Science Department, ESTEC, NL
Tsurutani, Bruce	Jet Propulsion Laboratory, Pasadena, USA
Urbarz, H.	Astronomical Inst., Univ. of Tübingen, Germany
Usmanov, A.	St. Petersburg University, Russia

Velli, Marco	Università di Firenze, Italy
Veselovsky, Igor	Moscow State University, Russia
Wang, Yi-Ming	Naval Research Laboratory, Washington, USA
Wenzel, Klaus-Peter	ESA Space Science Department, ESTEC, NL
Wimmer, R.	University of Bern, Switzerland
Winterhalter, Daniel	Jet Propulsion Laboratory, Pasadena, USA
Witte, Manfred	MPI für Aeronomie, Lindau, Germany
Woch, Joachim	MPI für Aeronomie, Lindau, Germany
Woo, Richard	Jet Propulsion Laboratory, Pasadena, USA
Yamauchi, M .	Swedish Inst. of Space Physics, Kiruna, Sweden
Zlobec, P.	Trieste Astronomical Obs., Trieste, Italy

MIX
Papier aus verantwortungsvollen Quellen
Paper from responsible sources
FSC® C105338

If you have any concerns about our products, you can contact us on
ProductSafety@springernature.com

In case Publisher is established outside the EU, the EU authorized representative is:
Springer Nature Customer Service Center GmbH
Europaplatz 3, 69115 Heidelberg, Germany

Printed by Libri Plureos GmbH
in Hamburg, Germany